Laser Techniques in the Extreme Ultraviolet

(OSA, Boulder, Colorado, 1984)

AIP Conference Proceedings
Series Editor: Hugh C. Wolfe
Number 119
Subseries on Optical Science and Engineering
Number 5

Laser Techniques in the Extreme Ultraviolet

(OSA, Boulder, Colorado, 1984)

Edited by
S. E. Harris
Stanford University
and
T. B. Lucatorto
National Bureau of Standards

American Institute of Physics
New York 1984

L.C. Catalog Card No. 84-72128
ISBN 0-88318-318-8
DOE CONF- 840387

SECOND TOPICAL MEETING ON LASER TECHNIQUES IN THE EXTREME ULTRAVIOLET

BOULDER, COLORADO

MARCH 5-7, 1984

TECHNICAL PROGRAM COMMITTEE

S. E. Harris, Conference Cochairman
Stanford University, Stanford, CA

T. B. Lucatorto, Conference Cochairman
National Bureau of Standards, Washington, D.C.

D. T. Attwood, Jr.
Lawrence Berkeley National Laboratory
Berkeley, CA

J. Bokor
Bell Laboratories, Holmdel, NJ

R. N. Compton
Oak Ridge National Laboratory, Oak Ridge, TN

R. R. Freeman
Bell Laboratories, Holmdel, NJ

K. R. Manes
Lawrence Livermore National Laboratory
Livermore, CA

T. J. McIlrath
University of Maryland, College Park, MD

C. K. Rhodes
University of Illinois at Chicago
Chicago, IL

H. Takuma
University of Electro-Communications
Chofushi, Tokyo, Japan

C. R. Vidal
Max Planck Institute for Extraterrestrische Physik
Garching, West Germany

ADVISORY PANEL

T. J. McIlrath
University of Maryland, College Park, MD

R. R. Freeman
Bell Laboratories, Holmdel, NJ

Preface

This volume contains most of the papers presented at the Second Topical Meeting on Laser Techniques in the Extreme Ultraviolet. The objective of this conference, as that of its predecessor, was to bring together physicists, spectroscopists, chemists, and device engineers to review the current status of research and development in the rapidly advancing field of extreme ultraviolet physics and technology. The conference was held in Boulder, Colorado, on March 5-7, 1984 and was attended by approximately 125 participants representing an international group from the United States, Canada, England, France, West Germany, Italy, and Japan.

The conference program reflected the continued and dramatic development of new sources of coherent and incoherent extreme ultraviolet radiation and the application of these sources to new types of spectroscopic techniques. The manuscripts in this book appear in the same sequence as the talks presented during the conference.

The Program Chairmen wish to express their sincere appreciation for the invaluable help of the Program Committee. These meetings owe their existence to the enthusiastic and effective sponsorship of the Optical Society of America and, in particular, we wish to thank Barbara Hicks, M. Malzone, and Joan Carlisle for doing so much to organize and manage this conference. We also thank Thomas McIlrath and Richard Freeman for their continued help, support, and enthusiasm. This conference would not have been possible without the support and financial help of the Office of Naval Research and The Air Force Office of Scientific Research.

Stanford, California
Washington, DC
May, 1984

S.E. Harris
T.B. Lucatorto

TABLE OF CONTENTS

Broadly Tunable VUV Radiation Generated by Frequency Mixing in Gases

R. Hilbig, G. Hilber, A. Timmermann, and R. Wallenstein
Fakultät für Physik, Universität Bielefeld, 48 Bielefeld, FRG

Third order sum- and difference frequency conversion of laser radiation in rare gases or metal vapors is a well-known technique for generating coherent light in the spectral region of the vacuum ultraviolet (VUV) [1, 2]. In recent years it has been demonstrated that frequency tripling and -mixing of pulsed dye laser radiation produces intense tunable VUV light of narrow spectral width [2 - 4].

Phase-matching conditions between the generated VUV and the focused laser light restrict the tuning range of the sum frequency to spectral regions of negative phase-mismatch ΔK (defined as the difference between the wave vectors of the generated radiation and the driving polarization) [5]. Since the rare gases Ne, Ar, Kr and Xe provide the required negative dispersion in several portions of the wavelength range λ_{vuv} = 66 - 147 nm (Fig. 1) these gases have been used very successfully for phase-matched tripling and sum frequency mixing [3, 4, 6 - 12].

In Ar, for example, nonresonant frequency tripling of ultraviolet (UV) radiation of a frequency doubled pulsed dye laser produced coherent radiation [10, 11] in the region of negative mismatch ΔK λ_{vuv} = 97.4 - 104.75 nm located at the high energy side of the transition 3p-4s'[1/2,1]. In the same way VUV is obtained at shorter wavelengths (λ_{vuv} = 85.7 - 87 nm) in the negative dispersive regions at the transitions 3p-3d'[3/2,1] and 3p-5s'[1/2,1]. At pulse powers of the UV laser light P_{uv} = 1-2 MW the power of the generated VUV light pulses is typically 1-8 W [10].

Tunable radiation at even shorter wavelengths is obtained in Ne [12]. In Ne the mismatch ΔK is negative in the range λ_{vuv} = 66.2 - 73.6 nm (Fig. 1). Frequency tripling of ultraviolet radiation with λ_{uv} = 216 - 223 nm (generated by sum mixing of the outputs of a frequency doubled dye laser and of the Nd-YAG pump laser) produced coherent light at λ_{vuv} = 72.05 - 73.58 nm (Fig. 2). At UV input powers of

0094-243X/84/1190001-09 $3.00

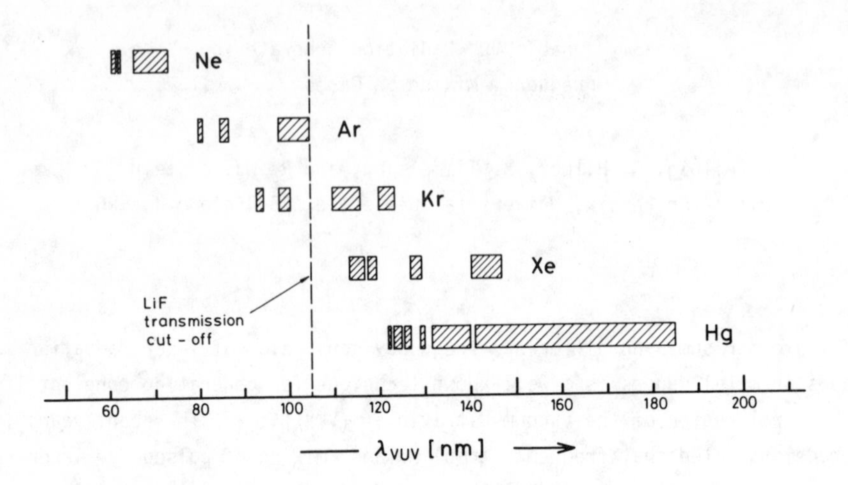

Fig. 1 Spectral regions with negative mismatch ΔK for frequency tripling ($\omega_{VUV}=3\omega_L$) in Ne, Ar, Kr, Xe and Hg.

0.1 - 0.3 MW the generated VUV power was typically P_{VUV} = 0.1 - 0.4 W ($1.5 - 6 \cdot 10^8$ photons/pulse). Since present UV dye laser systems provide at these UV wavelengths pulse powers of almost 1 MW the VUV output could easily be increased by more than one order of magnitude.

In contrast to the rare gases, Hg is negative dispersive in an exceptionally large spectral range which extends - at the high energy side of the $6^1S_0 - 6^1P_1$ resonance transition - from 143 to 185 nm (Fig. 1). Since ΔK changes little in the major part of this range third harmonic and sum-frequency generation produces - at constant vapor pressure - widely tunable radiation. Figure 3 displays the tuning of the sum frequency $\omega_{VUV} = \omega_{UV} + 2\omega_L$ where $\omega_{UV} = 2\omega_L$. Results obtained for different Rhodamine dyes - corrected for the wavelength dependence of the sensitivity of the detection system - are summarized in Fig. 4. This figure shows also the tuning of $\omega_{VUV} = \omega'_{UV} + 2\omega_L$ where $\omega'_{UV} = \omega_L + \omega_{IR}$; ω_{IR} is the fundamental of the Nd-YAG laser (Quanta Ray, Model DCR2) which excites the dye laser system (Quanta Ray, Model PDL1 with wavelength extender WEX-1). For these mixing processes the dye laser is tuned in the spectral range of DCM and the very efficient Rhodamine dyes. The doubling $\omega_{UV} = 2\omega_L$ and the mixing $\omega'_{UV} = \omega_L + \omega_{IR}$ is accomplished very conveniently by the wavelength extender (Quanta Ray, Model WEX-1). In addition to the results of Fig. 4 VUV has been generated in the same range by tripling the output of a Coumarin dye

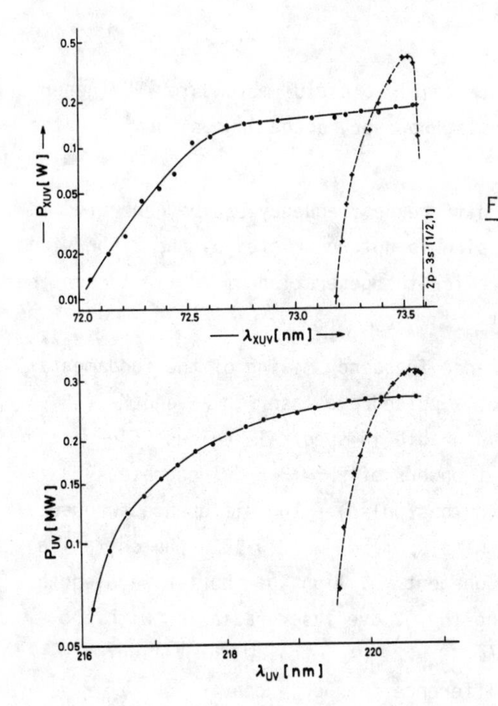

Fig. 2: Nonresonant frequency tripling in Ne in the negative dispersive region at the high energy side of the 2p-3s'[1/2,1] transition. Operating the dye laser system with Rh 6G (λ_L = 554.0-557.8 nm) and Fl. 27 (λ_L = 543.6-557.8 nm) the wavelength of the UV output at $\omega_{uv} = 2\omega_L + \omega_{IR}$ is in the range λ_{uv} = 219.8-221 nm (Rh 6G) and λ_{uv} = 216.5-221 nm (Fl. 27).

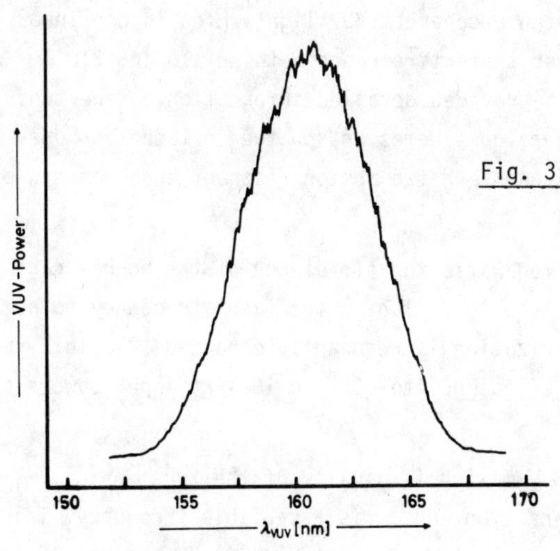

Fig. 3: Nonresonant frequency mixing $\omega_{VUV} = \omega_{uv} + 2\omega_L$ with $\omega_{uv} = 2\omega_L$ in Hg. The laser frequency ω_L is tuned in the range λ_L = 610-670 nm.

laser (λ_L = 430 - 550 nm).

Frequency tripling and sum mixing is certainly a useful method for VUV generation. The required negative mismatch prevents, however, a continuous tuning in the whole spectral region of the VUV.

In contrast to the sum-frequency the difference frequency can be generated in a medium with $\Delta K \gtrless 0$ [5]. Since this conversion is not restricted by the dispersion of the medium it should be of good advantage for the generation of VUV in the entire range between 105 and 200 nm.

This has been confirmed by the difference frequency mixing of the fundamental (ω_L) and the second harmonic (ω_{uv}) output of a pulsed dye laser in Xe and Kr [13]. While the sum-frequency $\omega_{vuv} = 2\omega_{uv} + \omega_L$ is tunable in spectral regions of negative dispersion between 110 nm and 130 nm - input powers of P_L = 3-5 MW generated VUV pulse powers of the order of 20 W ($5\cdot10^{10}$ photons/pulse) - the difference frequency $\omega_{vuv} = 2\omega_{uv} - \omega_L$ provides VUV light pulses with up to 60 W ($2.3\cdot10^{11}$ photons/pulse) at wavelengths between 185 nm and 207 nm. Coherent VUV light of shorter wavelength (159.5 nm to 186.6 nm) is obtained by mixing the UV dye laser radiation with the infrared output (ω_{IR}) of the Nd-YAG laser ($\omega_{vuv} = 2\omega_{uv} - \omega_{IR}$). With UV light of shorter wavelength ($\omega'_{uv} = \omega_{uv} + \omega_{IR}$) the difference frequency conversions $\omega_{vuv} = 2\omega'_{uv} - \omega_L$ and $\omega_{vuv} = 2\omega'_{uv} - \omega_{IR}$ allow the generation of VUV light with wavelengths between 122.6 nm and 160 nm. Thus with a single dye laser which is operated in the most efficient operating range of Nd-YAG laser pumped dye lasers (ω_L = 550 - 670 nm) these conversion schemes generated intense coherent VUV light which is continuously tunable in spectral regions which cover the entire range between 110 and 210 nm (Fig. 5). The performed investigations provided detailed information on the tuning characteristics of the different conversion schemes and on the influence of one- and two-photon-absorptions as well as one- and two-photon resonant enhancements of the conversion efficiency [14].

At laser pulse powers of a few megawatts the efficiency of the nonresonant frequency conversion is typically 10^{-5} to 10^{-6}. Tuning the laser frequency to a two-photon resonance the induced polarization is resonantly enhanced [2]. This enhancement increases the conversion efficiency to 10^{-3} to 10^{-4} at input powers of only 50 to 100 KW.

The two-photon resonant frequency mixing is usually of the type $\omega_{vuv} = 2\omega_1 \pm \omega_2$, where ω_1 is tuned to a two-photon transition and ω_2 is a variable frequency. This resonant frequency mixing has been investigated, for example, in Xe and Hg [15-17]. The experimental results provided tunable VUV of KW pulse power and detailed in-

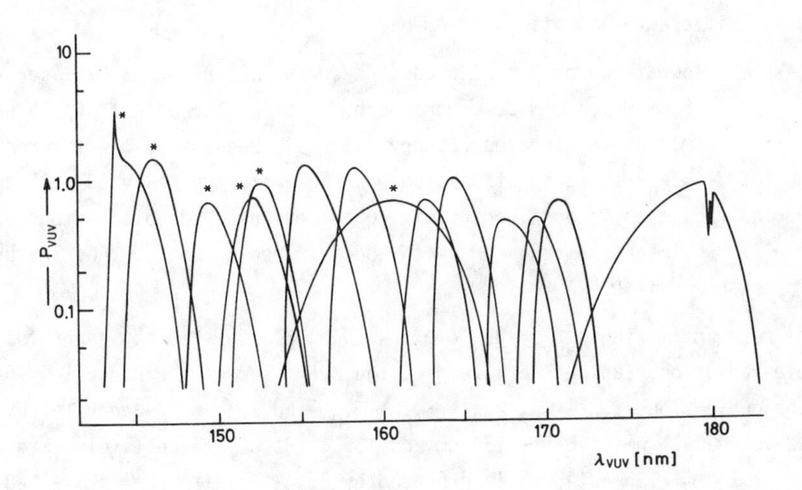

Fig. 4: Tuning of the sum frequency generated in Hg at $\omega_{VUV} = \omega_{UV} + 2\omega_L$ with $\omega_{UV} = 2\omega_L$ (indicated by stars) and $\omega_{VUV} = \omega'_{UV} + 2\omega_L$ with $\omega'_{UV} = \omega_L + \omega_{IR}$. The dye laser frequency ω_L is tuned in the range λ_L = 540 - 670 nm using Fluorescin 27, the Rhodamine dyes 6G, 610 (basic), 620, 640 (basic), 640, Sulforhodamine and DCM. Laser pulse powers $P_L \approx 1.2$ MW, $P_{UV} \approx 0.35$ MW and $P'_{UV} \approx 0.55$ MW generate VUV light pulses of 1 to 2 Watts.

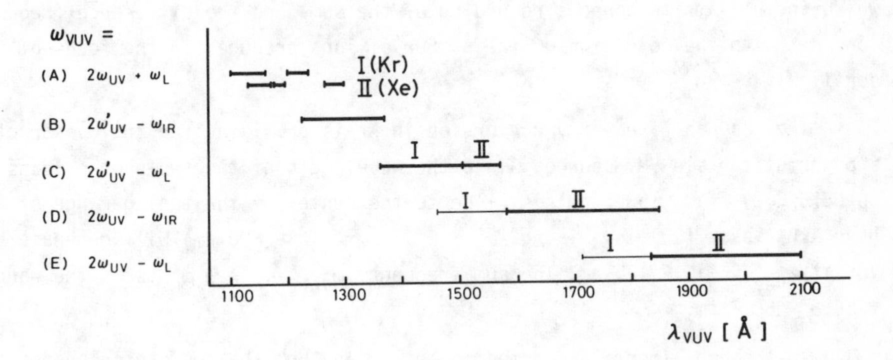

Fig. 5: Tuning range of the nonresonant sum- and difference-frequency conversion in Kr and Xe of UV dye laser radiation ($\omega_{UV} = 2\omega_L$ or $\omega'_{UV} + \omega_{IR}$), visible dye laser light ω_L and the IR radiation ω_{IR} of a Nd:YAG laser. The VUV generated by the mixing processes A-E require dye laser radiation of the following wavelength: A: λ_L = 5500-6500 Å (dye: Rh 6G, Rh 620, Rh 640, and DCM); B, CI, DII, and EII: λ_L = 5500-6300 Å (dye: Rh 6G, Rh 620, Rh 640); CII: λ_L = 6300-6700 Å (dye: DCM); DI and EI: λ_L = 5180-5500 Å (dye: coumarin 485 and 500).

formation on different saturation phenomena.

In Kr the lowest two-photon resonance 4p-5p [5/2,2] requires UV laser radiation at λ_R = 216.6 nm which can be generated by doubling the output of a blue dye laser (λ_L = 433 nm) in a deuterated KB5 crystal [18]. Because of the low conversion efficiency of $2\text{-}5\cdot 10^{-2}$ the generated pulse powers P_R are typically 60 - 150 KW. More powerful radiation is obtained by mixing the frequency doubled output of a Fluorescin 27 dye laser (λ_L = 544 nm) with the infrared of the Nd-YAG. The producable UV pulse power is close to 1 MW.

The resonant mixing $\omega_{vuv} = 2\omega_R - \omega_2$ (with λ_2 = 270 - 730 nm) generates widely tunable radiation. Tuning, for example, the dye laser in the range λ_L = 540 - 730 nm the conversions $2\omega_R - \omega_L$, $2\omega_R - (\omega_L + \omega_{IR})$ and $2\omega_R - 2\omega_L$ generate VUV at λ_{vuv} = 127.5 - 134.5 nm, 145.5 - 155 nm and 155 - 181 nm, respectively (Fig. 6). Radiation at λ_{vuv} = 135 - 145 nm is produced by $2\omega_R - \omega_L$ with λ_L = 428 - 548 nm which is in the tuning range of Coumarin dye lasers. The tuning curves displayed in Fig. 6 are measured at input powers P_R and P_L of about 70 KW. At optimum conditions an input of P_R = 200 KW and P_L = 1 MW generated VUV pulses close to 0.5 KW.

As observed previously [15] the VUV generated at $\omega_{vuv} = 2\omega_R - \omega_L$ is attenuated if the sum-frequency $\omega_{vuv} = 2\omega_R + \omega_L$ coincides with $4p^5ns$ and $4p^5nd$ Rydberg states (Fig. 6). At present these phenomena are subject to detailed measurements which include a simultaneous recording of the sum- and the difference frequency together with the detection of the number of ions produced in the focus of the laser light.

The resonant frequency conversion in Kr is promising for the construction of a powerful tunable VUV source. Since the wavelength of the two-photon transition 4p-6p[3/2,2] (λ_R = 193.5 nm) is close to the center of the tuning range of a narrowband ArF*-laser [19] the mixing $2\omega_R - \omega_L$ (λ_L = 216 - 800 nm) will generate tunable VUV at λ_{vuv} = 110 - 175 nm. The sum frequency $\omega_{vuv} = 2\omega_R + \omega_L$ is in the range λ_{vuv} = 66.8 - 86.3 nm.

In the experiments performed in Xe, Kr and Hg, the two-photon resonant enhancement of the induced polarisation provided conversion efficiencies of up to 0.2 percent at input intensities which can be produced easily with pulsed laser systems. With the additional enhancement of the third-order susceptibility by appropriate autoionizing states VUV radiation could even be generated in Sr-vapor by sum frequency mixing of multimode continuous-wave (cw) laser light [20].

Single frequency cw coherent VUV radiation is now generated for the first time by tripling the frequency of a stabilized dye ring laser (Spectra Physics Model

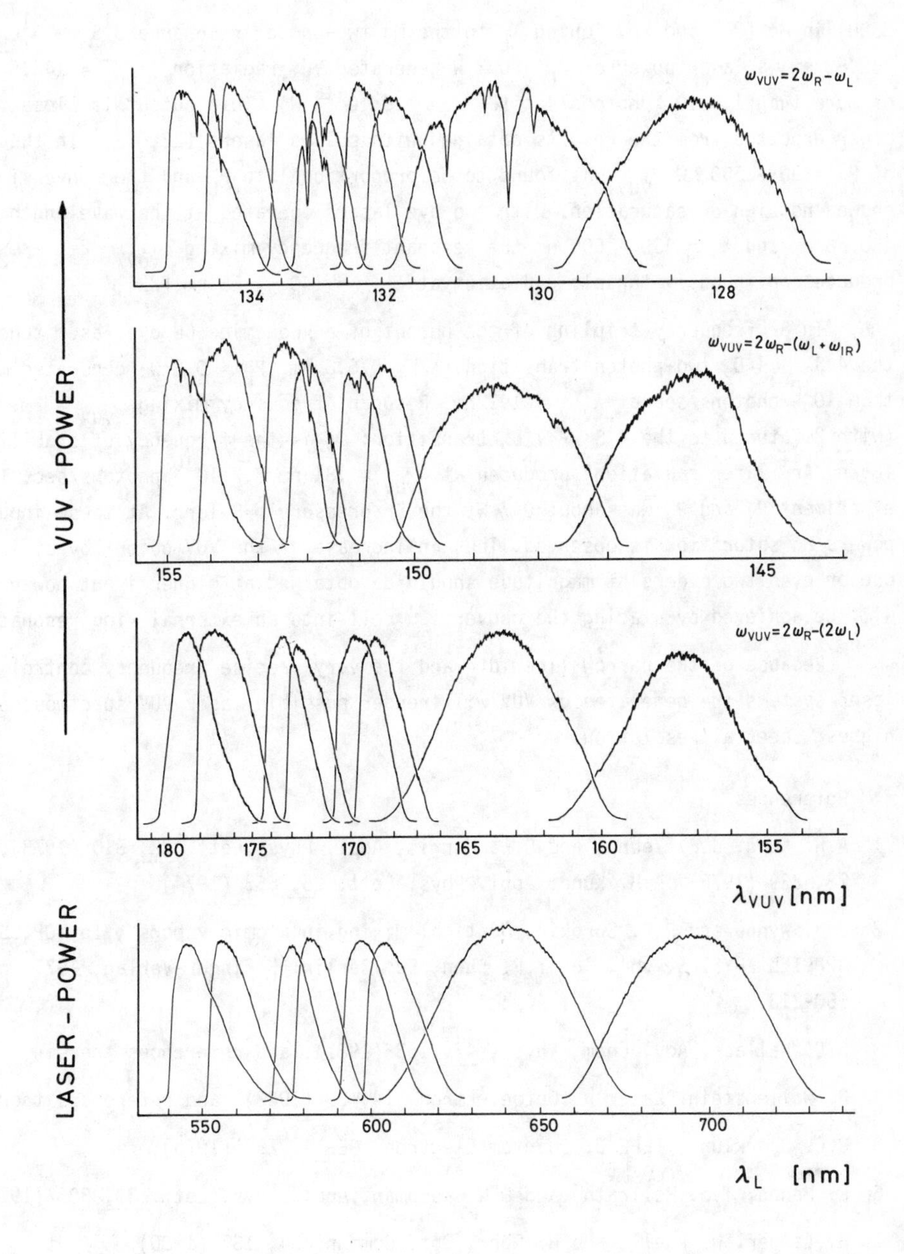

Fig. 6: Tuning of the VUV light generated by resonant frequency mixing in Kr.

380D) in Mg [21] and Sr. Tuning λ_L to the Mg two-photon resonance $3^1S_o - 3^1D_2$ (λ_L = 430.88 nm) a laser power of P_L = 0.2 W generated VUV radiation (λ_{vuv} = 143.6 nm) of more than $1.2 \cdot 10^5$ photons/sec ($P_{vuv} = 1.8 \cdot 10^{-13}$ W). This output is close to the power expected from the results obtained with pulsed lasers [22, 23]. In the range of P_L = 90 - 200 mW P_{vuv} was found to be proportional to P_L^3 and the conversion showed no sign of saturation. With two dye lasers operated at the wavelength λ_1 = 430.88 nm and λ_2 = 430 - 600 nm the resonant frequency mixing $\omega_{vuv} = 2\omega_1 + \omega_2$ will produce continuously tunable radiation at λ_{vuv} = 140 - 158 nm [22].

In Sr frequency tripling of the output of a Rhodamine 6G dye laser tuned to the $4^1S_0 - 4^1D_2$ two-photon transition (λ_L = 575.8 nm, P_L = 0.8 W) generated more than 10^{10} photons/sec at λ_{vuv} = 192 nm. Resonant frequency mixing $\omega_{vuv} = 2\omega_1 + \omega_2$ (with $2\omega_1$ tuned to the $4^1S_o - 4^1D_2$ transition; ω_2 is the frequency of stabilized 488 nm Ar^+ laser radiation) produced at λ_{vuv} = 181 nm $3.2 \cdot 10^{10}$ photons/sec. In this experiment P_1 and P_2 was about 0.7 W, the Sr pressure p=8 Torr. At these input powers no saturation is observed. Thus an increase of the VUV output by at least one or even two orders of magnitude should be obtained at higher input power which will be achieved by placing the conversion cell into an external ring resonator.

Because of the narrow linewidth and the very precise frequency control of cw laser systems the generated cw VUV will render possible laser VUV spectroscopy of highest spectral resolution.

References

1 A.H. Kung, J.F. Young, and S.E. Harris, Appl. Phys. Lett. 22, 310 (1973); 28, 239 (1976). A.H. Kung. Appl. Phys. Lett. 25, 653 (1974).

2 J.J. Wynne and P.P. Sorokin, "Optical mixing in atomic vapors", In TOPICS in APPLIED PHYSICS, vol. 16, Y.R. Shen, Ed. Berlin: Springer-Verlag, 1977, pp. 160-213.

3 S.C. Wallace, Adv. Chem. Phys., 47, 153 (1981), and references therein.

4 R. Wallenstein, Laser u. Optoelektron. 14, 29 (1982), and references therein.

5 G.C. Bjorklund, IEEE J. Quantum Electron. QE-11, 287 (1975).

6 R. Mahon, T.J. McIlrath, and D.W. Koopman, Appl. Phys. Lett. 33, 305 (1978).

7 H. Langer, H. Puell, and H. Röhr, Opt. Commun. 34, 137 (1980).

8 R. Hilbig and R. Wallenstein, IEEE J. Quantum Electron. QE-17, 1566 (1981).

9 W. Zapka, D. Cotter, and U. Brackmann, Opt. Commun. 36, 79 (1981).

10 R. Hilbig and R. Wallenstein, Opt. Commun. 44, 283 (1983).

11 E.E. Marinero, C.T. Rettner, R.N.Zare, and A.H. Kung, Chem. Lett. 95, 486 (1983).

12 R. Hilbig, A. Lago, and R. Wallenstein, Opt. Commun. (in press).

13 R. Hilbig and R. Wallenstein, Appl. Optics 21, 913 (1982).

14 R. Hilbig, G. Hilber, and R. Wallenstein (to be published).

15 R. Hilbig and R. Wallenstein, IEEE J. Quantum Electron. QE-19, 194 (1983).

16 R. Mahon and F.S. Tomkins, IEEE J. Quantum Electron. QE-18, 913 (1982).

17 R. Hilbig and R. Wallenstein, IEEE J. Quantum Electron. (in press).

18 J.A. Paisner, M.L. Spaeth, D.C. Gerstenberger, and I.W. Rudermann, Appl. Phys. Lett. 32, 476 (1978).

19 H. Schomburg, H.F. Döbele, and B. Rückle, Appl. Phys. B 28, 201 (1982).

20 R.R. Freeman, G.C. Bjorklund, N.P. Economou, P.F. Liao, and J.E. Bjorkholm, Appl. Phys. Lett. 33, 739 (1978).

21 A. Timmermann and R. Wallenstein, Opt. Lett. 8, 517 (1983).

22 S.C. Wallace and G. Zdasiuk, Appl. Phys. Lett. 28, 449 (1976).

23 H. Junginger, H.B. Puell, H. Scheingraber, and J.R. Vidal, IEEE J. Quantum Electron. QE-16, 1132 (1980).

XUV GENERATION IN PULSED FREE JETS: THEORY OF OPERATION AND APPLICATION TO H_2 DETECTION

A.H. Kung, N.A. Gershenfeld
San Francisco Laser Center, Chemistry Department
University of California, Berkeley, CA 94720

C. T. Rettner, D. S. Bethune, E. E. Marinero
I.B.M. Research Laboratories, San Jose, CA 95193

R.N. Zare
Chemistry Department, Stanford University
Stanford, CA 94305

ABSTRACT

The method for generating coherent XUV radiation by frequency tripling in a pulsed free jet is reviewed. A model is presented for this process which is able to predict successfully the output power as a function of different experimental parameters. Quantum-state-specific detection of H_2 is achieved at gas densities of $\sim 10^8$ molecules/cm^3 per vibration-rotation level, using time-of-flight detection of resonance-enhanced ionization.

1. INTRODUCTION

In recent years, the availability of commercial, reliable high-power laser sources has led to increased interest in accessing the XUV region using laser-based techniques. While a number of different techniques are being developed,[1] frequency upconverting a high-power dye laser has remained the workhorse in providing relatively intense XUV radiation. Here the combination of narrow bandwidth, broad tunability, and control of polarization of the resulting source are presently unrivaled. However, the high gas densities required for this nonlinear generation place stringent constraints on experimental design, making extensive differential pumping necessary for effective use of the radiation. We have recently demonstrated that pulsed free jets can greatly reduce these design requirements, by allowing the gaseous nonlinear medium to be highly localized in space and time.[2-4] These jets also provide for minimum self-absorption of the generated radiation and can permit the molecular samples to be concentrated in a few of their

0094-243X/84/1190010-13 $3.00

lowest-lying quantum states through expansion cooling. Gas densities up to a few hundred Torr can be attained in these jets equipped with modest pumping schemes. This has permitted the generation of useful amounts of XUV radiation for spectroscopic applications. Bokor et al have independently proposed the pulsed jet approach,[5] and have utilized Xe and He jets to generate the third, fifth and seventh harmonics of picosecond pulses at the KrF excimer laser wavelength to produce radiation deep into the XUV. Indeed, the seventh harmonic at 35.5 nm represents the shortest wavelength coherent radiation reported in the literature to date.

The experimental set-up for third harmonic generation in pulsed free jets is simple and straightforward. The collimated beam from a commercially available pulsed laser source is focused into the center of a vacuum chamber to intersect spatially and temporally with the gas pulse that emanates from a nozzle placed directly above the focal point of the laser (Fig. 1).

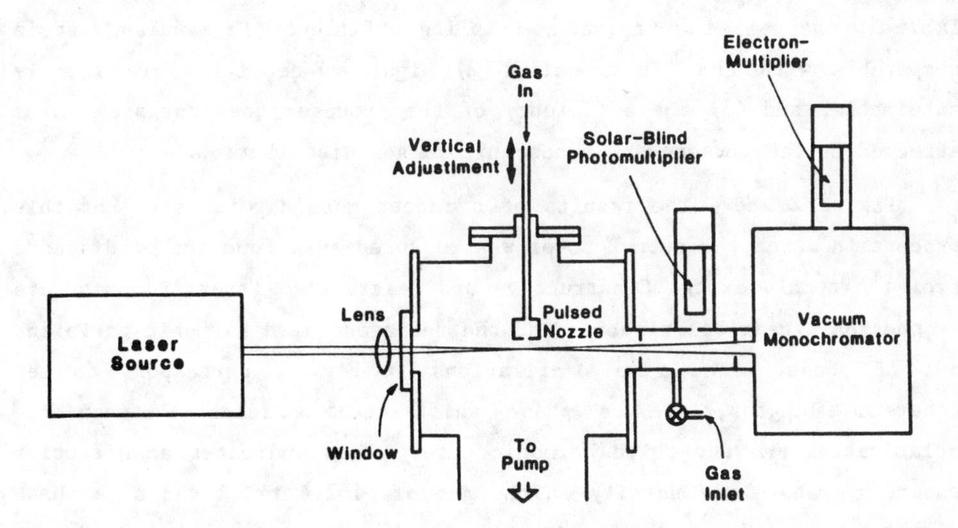

Figure 1 Schematic of apparatus for THG in a pulsed free jet and for quantum-state-specific detection of H_2. The laser source used was either the third harmonic output of a Nd:YAG laser when frequency tripling in Xe or a frequency-doubled Nd:YAG laser-pumped dye laser when frequency tripling in Ar or CO.

The generated radiation is detected through a vacuum monochromator attached to the main chamber. A photomultiplier or ion detector can also be placed downstream from the focal point of the laser beam to detect fluorescence signals or ion formation from samples excited by the XUV radiation.

In this paper we review our progress on the generation of XUV radiation using a pulsed free jet, present a brief description of a successful model for the method, and describe the quantum-state-specific detection of molecular hydrogen to exemplify the application of the XUV source.

2. THIRD HARMONIC GENERATION IN Xe, Ar AND CO.

The first experiment involving third harmonic generation in a pulsed supersonic jet was the tripling of the third harmonic of the Nd:YAG laser (354.7 nm) to 118.2 nm in xenon.[2] This process was chosen because it has been well-studied[6-9] and the results using the jet can be compared to those using a gas cell. It was demonstrated, as shown in Table 1, that (1) the frequency tripling efficiency in a pulsed jet is comparable with that in a cell, (2) high gas densities are readily attainable, and (3) the efficiency of the process does not seem to be affected by the presence of a non-uniform gas distribution.

Figure 2 shows the results of a recent more careful study of this process in which the output power was measured as a function of distance from the nozzle exit. Constructive and destructive interference occurs as the non-linear polarization and the radiated third harmonic go in and out of phase. The five oscillations in Fig. 2 correspond to ten coherence lengths, i.e. a phase shift of 5π radians between the polarization and the third harmonic wave. This indicates an effective room-temperature gas density in the jet of ~162 Torr, based on a phase mismatch of -6×10^{-17} radians per Xe atom.[9] We note that high gas density is essential for efficient conversion in processes where the phase mismatch per atom is small.

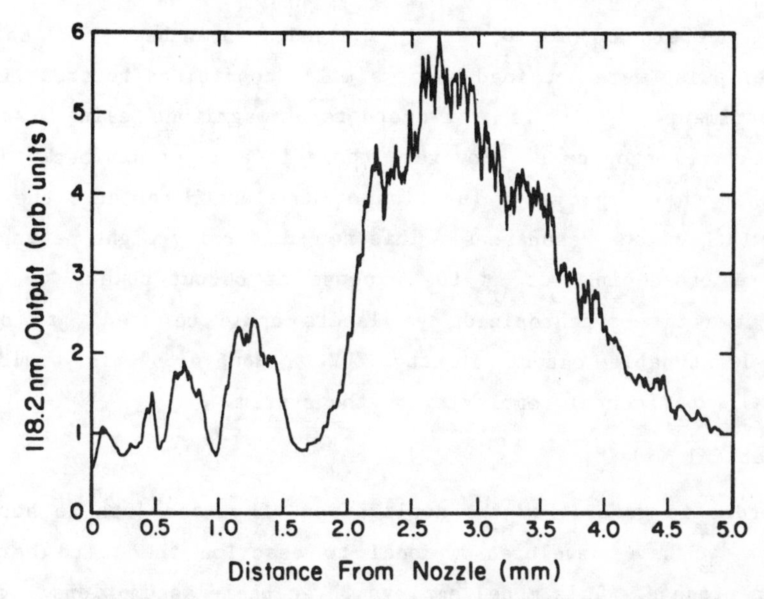

Figure 2 Measured 118.2 nm output vs. distance of focused laser from nozzle exit. The xenon backing pressure was 54.2 psia. The nozzle diameter was 1 mm and the confocal parameter of the 354.7 nm beam was 5 mm.

When a nanosecond, pulsed, frequency-doubled dye laser is used in place of the third harmonic of the Nd:YAG laser as the incident radiation, tunable XUV radiation can be achieved. Argon was found to be most effective of the rare gases for converting radiation from the dyes Kiton Red and Rhodamine 640 to the region around 100 nm. Uniform output tunable over a range of 5000 cm^{-1} was obtained. Approximately 10^{10} photons per pulse with a bandwidth of 1.7 cm^{-1} were measured over most of the range.[3] The uniformity of the output makes this source particularly attractive for spectroscopic applications.

It is well-known that the third harmonic output can be enhanced when the incident radiation is tuned onto a two-photon resonance of the generating medium.[10,13] This was demonstrated in a pulsed jet using CO[4]. In this case, the doubled dye laser wavelength was chosen so as to excite two-photon transitions of the CO $A^1\Pi - X^1\Sigma^+$ system. Dramatic enhancements in the third harmonic signal were observed when twice the laser frequency coincides with the line positions of the individual rotational members of the O, P, Q, R, and S branches. Up to 10^{12} photons per pulse were obtained under similar conditions to those in the argon experiment. This is a two-orders-of-magnitude gain in signal compared to the argon case. However, there is a major drawback in this case in that the output as a function of wavelength reflects the position and width of the resonances. This non-uniformity might be improved by pressure broadening but at the expense of output power.[11] Alternatively, two time-synchronized dye lasers could be used[12] to obtain continuously tunable output in the XUV. However, this would add substantial experimental complexity to the system.

3. Theoretical Model

In order to generalize the application of pulsed jets to harmonic generation, we have developed a model to describe the third harmonic generation process. This model employs three basic assumptions: first, the gas density along the direction of the laser beam has a rectangular profile; second, particle flux is conserved through all planes perpendicular to the direction of the expanding jet, and third, the laser field is propagating as a plane wave. The conversion efficiency is then given by [13,14]

$$\frac{P_3}{P_1} = \frac{0.08215}{\lambda_1^4} N^2 [\chi_3(\lambda_3)]^2 P_1^2 |\Phi|^2 \qquad (1)$$

where the power P is expressed in Watts, the wavelength λ in cm, the particle density N in cm^{-3}, the susceptibility $\chi(\lambda)$ in esu, and the subscripts 1 and 3 refer to the fundamental and third harmonic, respectively. In Eq. (1) $|\Phi|^2$ is a (dimensionless) geometric factor--the mismatch factor--which reduces in the plane-wave limit to

$$|\Phi|^2_{pw} = \left[(2L/b)\,\frac{\sin(\Delta kL/2)}{(\Delta kL/2)}\right]^2 \qquad (2)$$

where L is the sample length, b is the confocal parameter of the focused laser beam, and Δk is the wavevector-mismatch, defined in terms of the refractive indices at the generated and incident wavelengths as $\Delta k = k_3 - 3k_1 = 6\pi(n_3-n_1)/\lambda_1$. Miles[14] has shown that this relationship holds well up to $b \geq 3L$. Since b values tend to be of the order of several millimeters, this relationship should certainly apply to the case of focusing a laser beam close to the orifice of a typical free jet.

We have recently reconsidered this limit[15] and have shown that a more accurate result is given by

$$|\Phi|^2_{pw} = \left((2L/b)\ \frac{\sin[(\Delta k+4/b)L/2]}{(\Delta k+4/b)L/2}\right)^2 \qquad (3)$$

which has a maximum for $b\Delta k = -4$, as predicted by numerical evaluation[9] of the exact integral for $b \gg L$.

We apply Eq. (1) to calculate the P_3 generated by focusing the laser through a jet of Xe at a distance X downstream from the nozzle. For an idealized isentropic expansion, the Xe density N(X) at X is expressed in terms of the reservoir density N_o by

$$N(X) = N_o[1 + M(X)/3]^{-3/2} \qquad (4)$$

where

$$M(X) = 3.26(X/D - 0.075)^{2/3} - 0.61(X/D - 0.075)^{-2/3} \qquad (5)$$

is the Mach number expressed in terms of the nozzle diameter D. Based on assumption two we arrived at an effective jet width of

$$L(X) = 3/4\ D\ M(X)^{-1/2}\ [1+M(X)^2/3] \qquad (6)$$

Equations (4), (5), and (6) are valid for $X/D>3$.

Figure 3 shows the third harmonic power calculated using the above equations as a function of X for the experimental conditions of Fig. 2. Comparison of these two figures shows that the location and the intensity of the main peak as well as the subsequent peaks at $X/D > 3$ can be well represented by the above analysis.

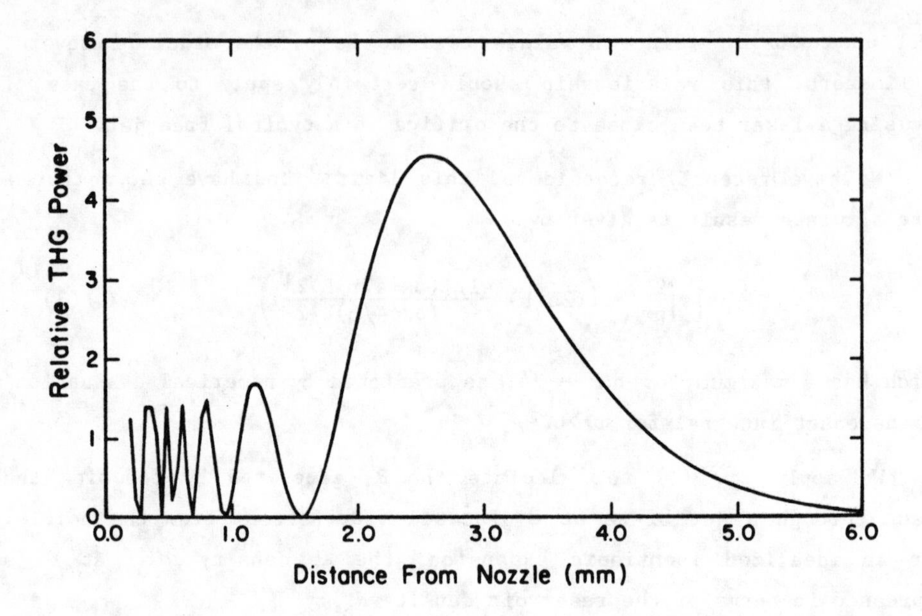

Figure 3 Calculated 118.2 nm output vs. distance of focused laser from nozzle exit, assuming rectangular gas jet profile. To obtain the fit with Fig. 2, we used a Xe backing pressure of 54.2 psia, effective nozzle diameter of 0.32 mm, and a laser confocal parameter of 5 mm.

This modelling can further be improved by using a more appropriate description of the gas density profile of the form

$$S(Z,X) = \left[1 + (2Z/L(X))^{2n}\right]^{-1} \tag{7}$$

where n is an integer, and Z is the distance along the laser beam measured from the jet axis. (Note that as n gets large this function rapidly approaches a square-well form, but is considerably "softer" for small n). A solution has been obtained for Φ using this profile as a "weighting factor" in the exact integral form of Φ. However, this treatment neglects the concurrent variation of Δk and thus is expected to apply best for large n. If Δk is allowed to vary across the jet profile, then it must be replaced by the integrated phase shift for any point in the profile. This complicates the analysis and will most likely require numerical evaluation and is not included in the present analysis. For <u>even</u> values of n and $\Delta k < 0$ which is appropriate for 118.2 nm generation in xenon, we find that

$$\Phi = \pi\ (\Phi_1 + \Phi_2) \tag{8}$$

where

$$\Phi_1 = \left(- b\Delta k + \frac{4n\ [b/L(X)]^{2n}}{1+\ [b/L(X)]^{2n}} \right) \times \frac{\exp(b\Delta k/2)}{1+[b/L(X)]^{2n}} \tag{9}$$

and

$$\Phi_2 = \frac{-i[b/L(X)]}{n} \sum_{m=1}^{n} \frac{r_m \exp[-i\Delta k L(X)\ r_m/2]}{[ir_m + b/L(X)]^2} \tag{10}$$

and

$$r_m = \exp\ [i\pi(2m-1)/2n] \tag{11}$$

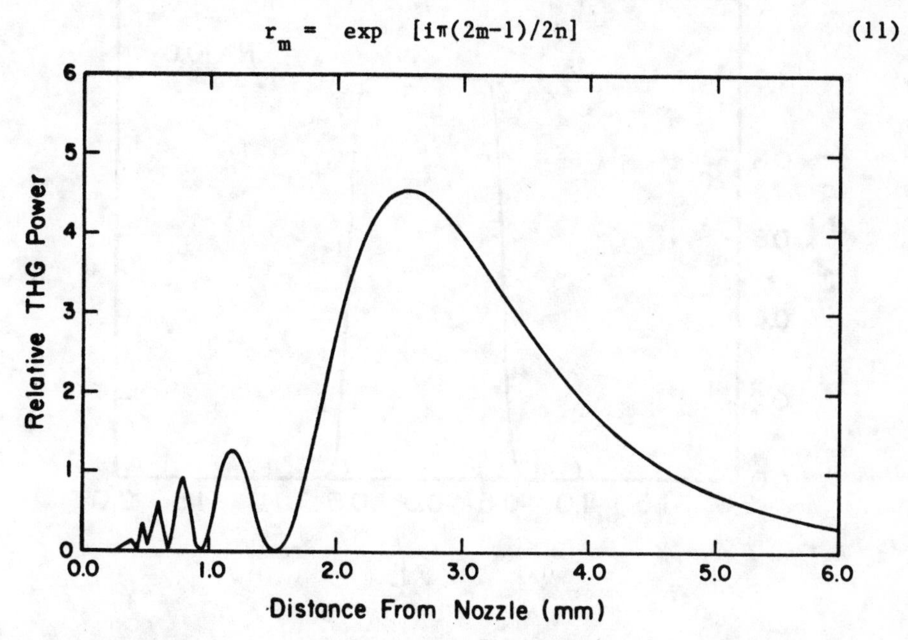

Figure 4 Calculated 118.2 nm output vs. distance of focused laser from nozzle exit, using gas jet profile given by Eq. (7) with n = 10. All other parameters remain the same as in Figure 3.

Substituting this result into Eq. (1) produces the curve shown in Fig. 4, where n has been optimized at a value of 10. Larger values of n overemphasize the early oscillations, converging on the plane-wave limit of Fig. 3. For small n, these features are under-represented. Figure 5 displays the function S(Z,X) for n=10. It suggests that the jet indeed has an approximately square-well profile, and explains the success of the plane-wave limit in fitting this process. Such a jet profile may be anticipated for fully hydrodynamic flow from the nozzle. For X = 1 mm we estimate that the mean-free-path for xenon atoms is more than a hundred times smaller than the jet width, so that hydrodynamic flow holds.

Based on the excellent agreement obtained between experiment and the predictions of this model, we believe that the above analysis permits the general design of pulsed-free-jet tunable XUV sources.

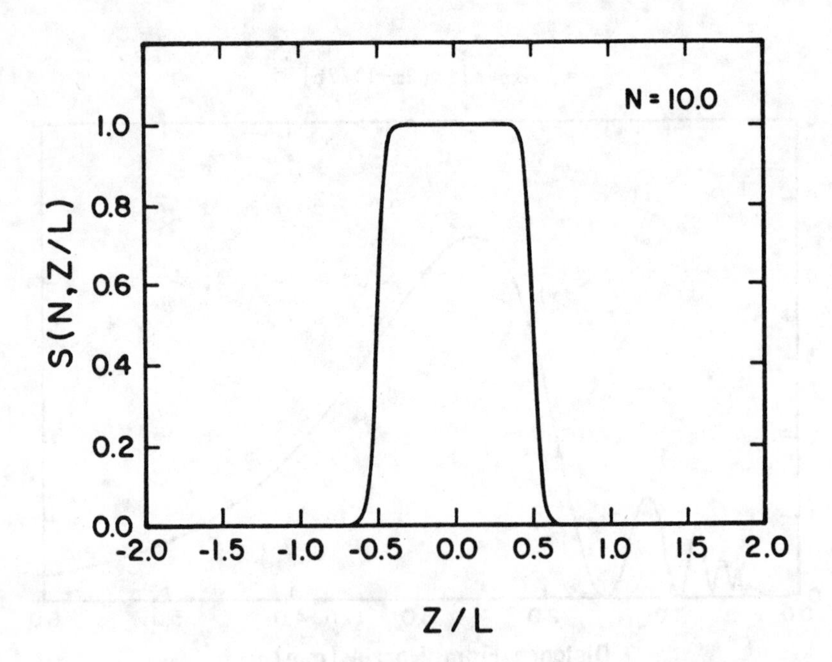

Figure 5 Gas density profile from the jet as given by Eq. (7) with n = 10.

4. MOLECULAR HYDROGEN DETECTION; FLUORESCENCE AND RESONANTLY EXCITED MULTIPHOTON IONIZATION

The quantum-state-specific detection of H_2 and its isotopic analogs under conditions of low concentration is of major importance in a diverse number of applications ranging from chemical physics to plasma diagnostics.[16-20] We have illustrated the use of our XUV source for this purpose by the technique of laser-induced fluorescence[3] (LIF). In this experiment, a small amount of molecular hydrogen was bled into the main chamber and the XUV radiation was tuned from 97.3 nm to 102.3 nm to excite four bands of the B-X Lyman system and two bands of the C-X Werner system of H_2. By monitoring the subsequent fluorescence, H_2 densities of ~2×10^8 molecules/cm^3 in a single vibration-rotation level were detected. With improvements in the fluorescence detection system a sensitivity of $\lesssim 10^7$ molecules/cm^3 per quantum state was predicted.

The LIF technique, while being very sensitive, suffers in general from the problems of scattered light and background molecular fluorescence in practical situations where large quantities of other molecules are present. Alternatively, one might detect H^+ or H_2^+ ions that are formed following absorption of additional photons by the excited H_2 molecules. This approach was first described by Rottke and Welge[21] where they resonantly excited the lowest vibrations of the Lyman system of H_2 followed by UV-photon ionization. They reported a detection sensitivity limit of ~4×10^{-7} Torr and projected that $\lesssim 10^{-10}$ Torr would be possible. We have explored this detection scheme using the pulsed nozzle set-up with the photomultiplier replaced by a time-of-flight (TOF) mass spectrometer[22]. In this spectrometer, a 5 mm diameter, 3 cm long cylinder of H_2 is ionized. The ions are sampled through a rectangular slit 1 mm wide and 8 mm long. The advantage of the pulsed-nozzle technique is that it can access the strongest ro-vibronic levels of H_2, thus enhancing the sensitivity of this approach.

Figure 6 shows the TOF spectrum obtained when the XUV source is tuned to excite the R(1) line of the (10,0) band of the Lyman system. The earliest ion peak corresponds to H^+, the next to H_2^+, while those at

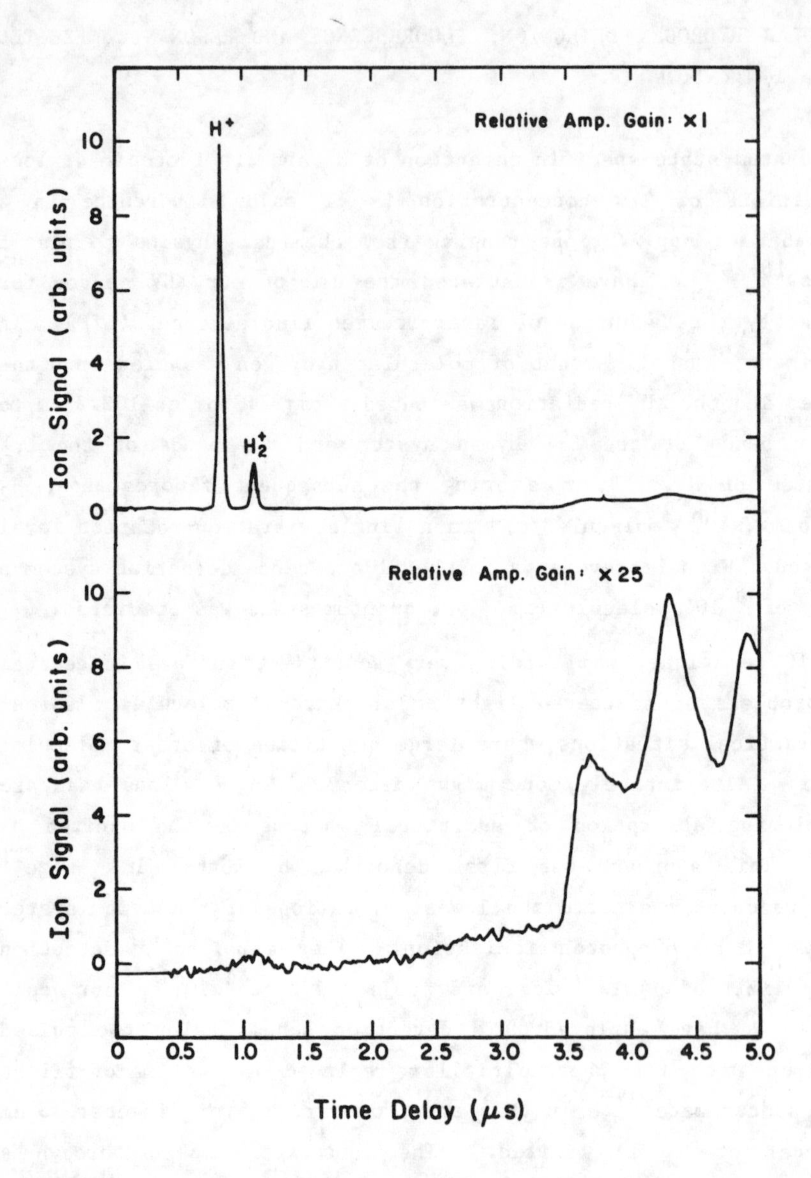

Figure 6 Time-of-flight spectra of resonantly-enhanced ionization of H_2. Upper trace: XUV radiation tuned onto resonance with the R(1) line of the B-X (11,0) band at 98.2 nm. Lower trace: XUV radiation detuned by 0.007 nm from the above line-center. The H_2 pressure in the vacuum chamber was 2 x 10^{-5} Torr. The broad peaks correspond to background gases.

later times result from ionization of background gases. It is estimated that a sensitivity of ~10^8 molecules/cm^3 per quantum state was attained in this first experiment. With additional efforts to optimize ion detection, it is anticipated that this detection limit could be substantially reduced. Because molecules ionize in the XUV but do not necessarily fluoresce, this method may prove to have a more universal application in the selective detection of molecular species.

ACKNOWLEDGMENTS

This work was supported in part by the Office of Naval Research under N00014-78-C-0403 and the National Science Foundation under NSF CHE 81-08823 and NSF CHE 79-16250. R.N.Z. also gratefully acknowledges support through the Shell Distinguished Chairs program, funded by the Shell Companies Foundation, Inc.

REFERENCES

1. See for example "Laser Techniques for Extreme Ultraviolet Spectroscopy" Editors T.J. McIlrath and R.R. Freeman, A.I.P. Conference Proceedings No. 90, American Institute of Physics, New York, 1982.

2. A.H. Kung, Optics Lett., 8, 24 (1983).

3. E.E. Marinero, C.T. Rettner, R.N. Zare, and A.H. Kung, Chem. Phys. Lett., 95, 486 (1983).

4. C.T. Rettner, E.E. Marinero, R.N. Zare, and A.H. Kung, J.Phys. Chem. (to appear in the Mar. 29, 1984 issue).

5. J. Bokor, P.H. Bucksbaum, and R.R. Freeman, Opt. Lett., 8, 217 (1983).

6. A.H. Kung, J.F. Young, and S.E. Harris, Appl. Phys. Lett. 22, 301 (1973); 28, 239 (E) (1976).

7. R. Mahon, T.J. McIlrath, V.P. Myerscough, and D.W. Koopman, I.E.E.E. J. Quantum Electron, QE-15, 444 (1979).

8. L.J. Zych and J.F. Young, I.E.E.E. J. Quantum Electron QE-14, 147 (1978).

9. G.C. Bjorklund, I.E.E.E. J. Quantum Electron. QE-11, 287 (1975).

10. R. T. Hodgson, P.P. Sorokin, and J.J. Wynne, Phys. Rev. Lett., 32, 343 (1974).

11. K.K. Innes, B.P. Stoicheff, and S.C. Wallace, Appl. Phys. Lett. 29, 715 (1976).

12. R. Wallenstein, preceding paper.

13. R.B. Miles and S.E. Harris, I.E.E.E. J. Quantum Electron. QE-9, 470 (1973).

14. R.B. Miles, "Optical Third Harmonic Generation in Metal Vapors", Ph.D. Thesis, Stanford University, (1972).

15. C.T. Rettner, D.S. Bethune, and A.H. Kung, to be published.

16. F.J. Northrup, J.C. Polanyi, S.C. Wallace, and J.M. Williamson, Chem. Phys. Lett. 105, 34 (1984).

17. M. Péalat, D. Debarre, J.-M. Marie, J.-P.E. Taran, A. Tramer, and C.B. Moore, Chem. Phys. Lett. 98, 299 (1983).

18. D.P. Gerrity, and J.J. Valentini, J. Chem. Phys. 79, 5202 (1983).

19. E.E. Marinero, C.T. Rettner, and R.N. Zare, J. Chem. Phys., in press; Phys. Rev. Lett. 48, 1323 (1982).

20. M. Péalat, J.-P. E. Taran, J. Taillet, M. Bacal, and A.M. Bruneteau, J. Appl. Phys. 52, 2687 (1981).

21. H. Rottke and K.H. Welge, Chem. Phys. Lett. 99, 456 (1983).

22. A.H. Kung and N.A. Gershenfeld, to be published.

FEASIBILITY OF COLLISIONAL XUV-LASER AND XUV RADIATION FROM AUTOIONIZING POTASSIUM LEVELS IN DISCHARGE.

V.O.Papanyan, A.E.Martirosyan, G.C.Nersisyan, Y.K.Gabrielyan
Institute for Physical Research of Armenian Academy of Sciences

Recently there were proposed 10-100 nm laser schemes with energy storage on a long-living autoionizing state of alkali metal atoms in discharge with fast excitation transfer on a doublet level in intense laser beam [1] . In Refs [2,3] was described selective pumping chemes of He-K and Ne-Rb vapor mixtures utilizing collisional energy transfer from metastable levels of the rare gases. In the present work theoretical and experimental investigations of the proposed schemes are carried out. It is shown thet in afterglow of impulse discharge (delay time 50-130 nsec) by 498 nm or 1.209 μm laser pumping it is possible to achive $\sim 10^{10}$ cm^{-3} consentration of upper laser level $3p^5 3d4s$ $^4P_{5/2}$ of K, and gain $\sim$2-5 cm^{-1} for 60 or 64nm wavelength. When using stimulated anti-Stokes Raman scattering process with 1.064 μm pump gain $\sim$1 cm^{-1} is possible.

Potassium emission spectrum in 50-80 nm wavelength region was investigated in hollow-cathode discharge. Seven lines arised from radiative decays of quartet autoionizing levels $^4F_{3/2}$, $^4P_{5/2}$, $^4S_{3/2}$ of $3p^5 3d4s$ and $3p^5 4s4p$ electron configurations of potassium were observed. Three lines 77.82 nm, 69.14 nm and 55.45 nm were detected and identified for the first time. This first observation of radiative emission from quartet high lieing levels in XUV spectrum of He-K mixture demonstrates their effective excitation in the hollow-cathode discharge. This indicates that radiation decay channel is important for K quartet series and that is of great interest for XUV-lasers development.

1. J.E.Rothenberg, S.E.Harris. IEEE J.QE-17, 415 (1981).
2. A.E.Martirosyan , V.O.Papanyan. Sov.J.Quant.Electr., 10,166 (1983) (Russian).
3. V.O.Papanyan et al., IEEE J.QE-19,n.12 (1983).

Soft X-ray Laser Experiments at the Novette Laser Facility*

D. Matthews, P. Hagelstein, M. Rosen, R. Kauffman,
R. Lee, C. Wang, H. Medecki, M. Campbell, N. Ceglio,
G. Leipelt, P. Lee+, P. Drake, L. Pleasance, L. Seppala,
W. Hatcher, G. Rambach, A. Hawryluk, G. Heaton, R. Price,+
R. Ozarski, R. Speck, K. Manes, J. Underwood, A. Toor, T. Weaver,
L. Coleman, B. DeMartini, J. Auerbach, R. Turner, W. Zagotta,
C. Hailey, M. Eckart, A. Hazi, and B. Whitten

Lawrence Livermore National Laboratory
Livermore, California 94550

G. Charatis, P. Rockett, G. Busch, D. Sullivan,
C. Shepard, and R. Johnson

KMS Fusion, Inc.
Ann Arbor, Michigan 48104

Brian MacGowan
Imperial College, United Kingdom

P. Burkhalter
Naval Research Laboratory
Washington, D.C.

We discuss the results of and future plans for experiments to study the possibility of producing an x-ray laser. The schemes we have investigated are all pumped by the Novette Laser, operated at short pulse ($\tau_L \sim 100$ psec) and an incident wavelength of $\lambda_L \sim 0.53$ µm. We have studied the possibility of lasing at 53.6, 68.0-72.0, 119.0, and 153.0 eV, using the inversion methods of resonant photo-excitation, collisional excitation, and three-body recombination.

The quest for producing an XUV wavelength laser has greatly escalated in recent years. Present researchers still strive for a high gain scheme since resonant cavities,[1,2] such as those used in optical lasers, are presently not demonstrated. Rapid progress is being made, though, on normal-incidence multilayer mirror assemblies[3,4] which should soon enable the possibility of multipassing through a gain medium.

There are numerous approaches to achieving a population inversion.[5-9] These include 1) photo- or collisional ionization of ions or atoms,[10-15] 2) photo- or collisional excitation,[16-25] 3) atom-ion resonance charge exchange,[26-29] and 4) collisional recombination of ions.[30-52] To date there have been few tests of these schemes.[25,43,49,50,53] As to the establishment of lasing, none of the

* This work performed under the auspices of the U.S. Department of Energy by the Lawrence Livermore National Laboratory under contract #W-7405-ENG-48.
+Currently at Los Alamos National Laboratory, Los Alamos, New Mexico.
0094-243X/84/1190025-20 $3.00

results have completely satisfied either the investigators or critics.

The purpose of this report is to discuss the results of a series of experiments that were attempted at the Novette Laser Facility. The schemes tested were of the type 2) and 4) as listed above. We chose them because they are best suited to the pump power characteristics supplied by a large optical laser. These schemes could have led to lasing on XUV wavelength atomic transitions in the interval from 230 to 80 Å.

The Resonant Photoexcitation Laser

The resonant photoexcitation scheme received the primary emphasis in these experiments. Gains predicted for the scheme are extremely high,[54] The design, however, is fairly complicated. The basic method for obtaining the population inversion is shown in Fig. 1. A

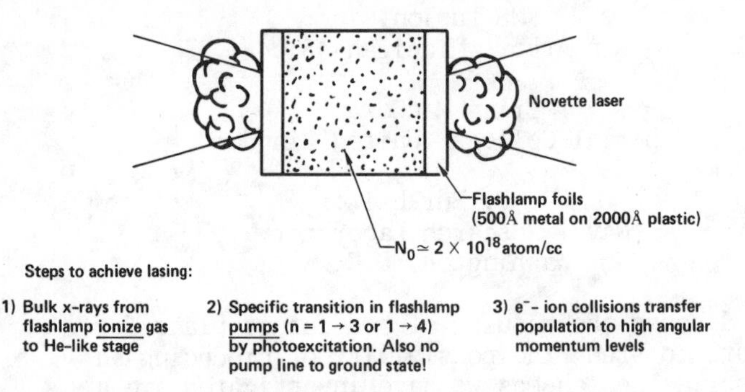

1. A simplified cross-sectional view of the resonant photoexcitation x-ray laser target.

rectangular cross section gas cell is bounded on two sides by x-ray flashlamp foils which are irradiated by the high power, 0.53μm wavelength Novette Laser. The plasma produced by the laser interaction (laser intensity ∿ 1×10^{14} W/cm^2) produces copious amounts of continuum and discrete x rays, which, in turn, ionize the lasant gas and resonantly-pump excited states of the ions. Electron-ion collisions then transfer the population to higher angular momentum states which cannot decay to the ground state from which the ion was photoexcited. A specific case for this scheme is shown in Fig. 2. Here we illustrate a level diagram for the He-like neon ion system. In one of the schemes attempted, we chose a flashlamp material which emitted 11.000 Å x rays which are necessary to resonantly photoexcite the 1s4p level of helium-like neon. We then depend on collisions in the plasma to populate the 1s4f level. This can lead to a population inversion of that level over the 1s3d, thus leading to a laser transition at 231 Å. Obviously this same idea can be used to pump the 1s3p level, with collisions again leading to a population inversion of the 1s3d over the 1s2p. This resonant pumping concept was also used to pump He-like fluorine to produce a possible laser transition at 104.2 Å. Finally, we also applied the technique to the H-like sequence of fluorine in order to pump the 3p level, again

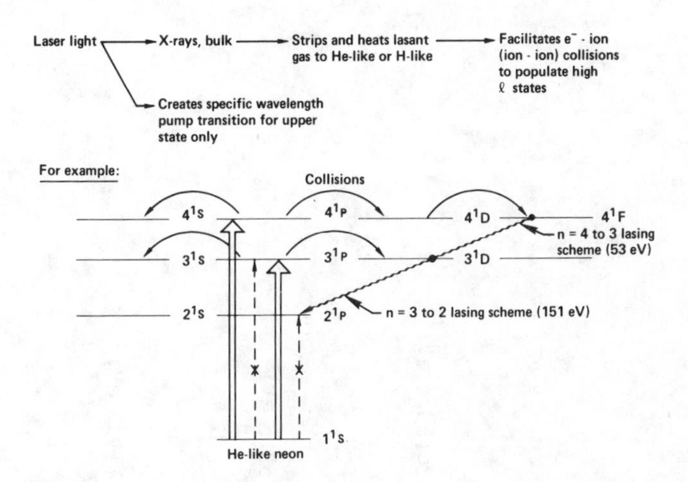

2. A simplified level scheme for the He-like neon laser, lines which should not be present in the pump, are shown dashed with an x.

with collisional transfer leading to an inversion of the 3d level over the 2p. Table I summarizes all the schemes attempted as well as the flashlamp material used to provide the resonant transition for pumping the lasant. This is a reduced list of the number of potential pump candidates that had been tabulated by Hagelstein.[54,55] This reduction comes as a result of some eliminations due to the high precision spectroscopy performed by P. Burkhalter et al.[56] in support of these experiments.

There are some serious physics and technical phenomena which can effect the outcome of a test of this laser scheme. Principally, the degree of resonance between the pump and lasant transition, the presence of lines that can pump the ground state (destroys the inversion), the brightness of the pump lines, integrity of the gas cavity during the course of the laser pulse, the lasant ion temperature, and the time of occurrence of the relevant ion charge state (ground state of photoexcitation). Out of this long list, only the degree of resonance of the Na-Ne scheme was known with sufficient accuracy prior to our experiments. In addition, we were only able to certify the brightness of the pump for one of the schemes, namely, the Cr-F at 14.458 Å. As for the plasma conditions inside the possible lasing channel, nothing was done to determine this before or during the measurements.

Target designs for this laser scheme placed some unusual constraints on both target fabrication and assembly. Most of the physics reasons for the critical dimensions have been recently reviewed by Hagelstein[55]. Some of the actual laser cavity dimensions are summarized in Figs. 3 and 4. Embodied in this design is the structural requirement that we support over a narrow rectangular opening a very thin flashlamp with an even thinner coating of metal vapor. The flashlamp had to seal the cavity to maintain a gas pressure of

Table I
Resonant Photoexcitation X-ray Laser Schemes

	PUMP		LASANT
(1)	$Na(1s2p\,^1P_1 \rightarrow 1s^2\,^1S_o)$	$\rightarrow$	$Ne(1s^2\,^1S_o \rightarrow 1s4p\,^1P_1)4d \rightarrow 3p(53.6\text{ eV})$
	$\lambda = 11.003$ Å		$\lambda = 11.000$ Å
	$Ni(N\text{-like}, 3d \rightarrow 2p)$	$\rightarrow$	$Ne(1s^2\,^1S_o \rightarrow 1s4p\,^1P_1)4d \rightarrow 3p(53.6\text{ eV})$
	$\lambda \overset{?}{=} 11.000$ Å		
(2)	$Cr(B\text{-like}, 3d \rightarrow 2p)$	$\rightarrow$	$F(1s^2\,^1S_o \rightarrow 1s3p\,^1P_1)3d \rightarrow 2p(119.0\text{ eV})$
	$\lambda \overset{?}{=} 14.456$ Å		$\lambda = 14.458$ Å
(3)	$Mn(Be\text{-like}, 3d \rightarrow 2p)$	$\rightarrow$	$F(1s\,^2S_{1/2} \rightarrow 3p\,^2P_{1/2,3/2})3d \rightarrow 2p(153.0\text{ eV})$
	$\lambda \overset{?}{=} 12.641$ Å		$\lambda = 12.643$ Å

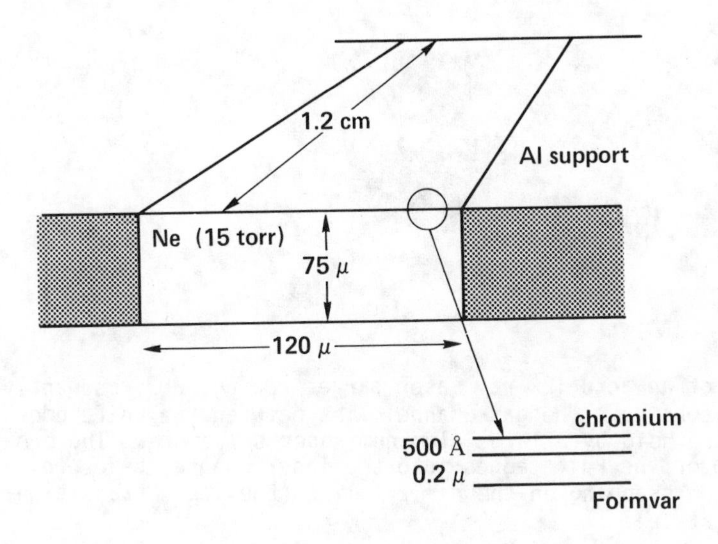

3. Actual laser cavity cross section for n = 4-3 schemes.

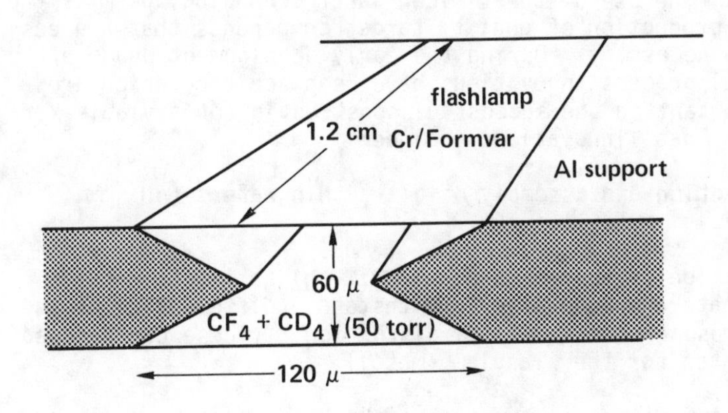

4. Actual laser cavity cross section for n = 3-2 schemes.

roughly 20-50 Torr of gases like neon or CF_4. The final target superstructure is shown in Fig. 5. This final design allowed us to accurately position the target, crudely verify the pressurization of the laser cavity, and discriminate against unwanted x-ray signals. A much improved version of the target is currently being tested. The mechanical design, fabrication, and assembly methods are radically changed. Basically, the micromachining methods previously employed are to be replaced with the microfabrication technologies of photolithography and anisotropic, chemical etching of single-crystal silicon wafers. This is the same precision technology that

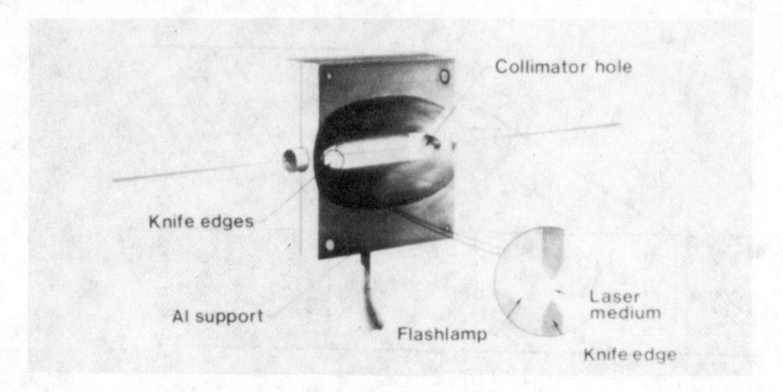

5. Detail of an actual x-ray laser target for the 3→2 resonantly-pumped schemes. The gas channel lies between two knife edges which are held by a large aluminum super structure. The protrusion of the knife edges into the laser channel helps relieve radiation trapping on the n = 2→1 line (the laser transition ground state).

is employed in the electronics industry for the production of miniature circuits. The use of the silicon wafer technology permits the reliable mass production of quality target components that are easy to modify when necessary. During the early development phase of the project several process innovations have been achieved which are extremely important to the successful construction of a viable x-ray laser target. These innovations include:

The fabrication and assembly of ultra thin wafers (40 μm thickness).

The generation of approximately 1500 Å-thick silicon nitride windows that have been able to withstand a differential pressure of one atmosphere with minimal distortion. (These can be used as substrates for the flashlamp metal).

The formation of a long, very straight knife edge by means of anisotropic etching.

The patterning of both sides of a thin silicon wafer with an alignment accuracy of better than 1 μm over the 1.2 cm length of the laser cavity.

This new target design is shown in Fig. 6.

The Collisional Excitation Laser

The collisional excitation laser design was initiated considerably later in the planning phase of this particular set of experiments. Considerable recent work has been done on the scheme,[57,58] particularly for the Kr atom, but we elected to avoid the difficulties of working with a gas target and chose to use a nearby solid element in the periodic table, selenium.

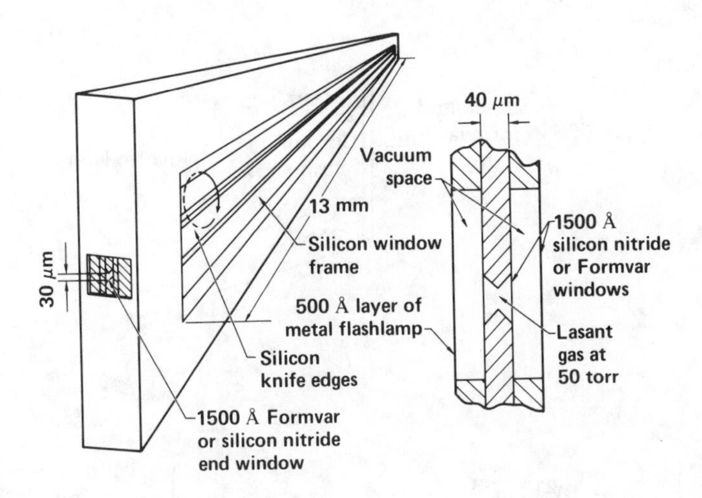

6. Schematic of the new target design which utilizes etched silicon wafer technology.

Figure 7 illustrates the principal physics mechanisms responsible for producing an inversion with this scheme. Laser light is used to produce a freely expanding coronal plasma having a specific density and temperature variation in the dimension parallel to the incident laser beam. Irradiation conditions are chosen to maximize the abundance of the neon-like ion species in the density regime around $1 \times 10^{21} cm^{-3}$. Collisional population of the 3s, 3d, and 3p levels leads ultimately to an inversion between the 3p and the 3s because the upper level (3p) cannot dipole decay back to the neon-like ground state configuration, whereas the 3s has a strong radiative decay to the 2p level. Actually the strongest candidates for lasing are those states having a j=0 initial and j=1 final state. A full level diagram showing some of the transitions emitted by the neon-like Se M-shell is shown in Fig. 8.

Unfortunately, the collisional population of the n=3 levels is non-trivial. Whereas producing neon-like ground state configurations is simple (it only requires the removal of weakly bound M-shell electrons), exciting one of the L-shell electrons to the n=3p state requires approximately 30% more plasma temperature (800 instead of 600 eV). This requires that a laser design incorporate plasma heating conditions which produce significant plasma temperature and density. Table II actually illustrates the sensitivity of the calculated gain to the plasma electron temperature for a Kr plasma (from the calculations of Feldman, Bhatia, and Suckewer).[58] Figure 9 shows the gain for our scheme as calculated by Hagelstein.[55] Note the region of plasma at a density of 1×10^{21} cm^{-3}, where the gain is of order 10 cm^{-1}.

The principal physics issues which can effect the collisional laser scheme are: a) the refraction of x rays due to the steep density gradient along the dimension parallel to the incident laser beam, b) the optimum plasma conditions necessary for producing the

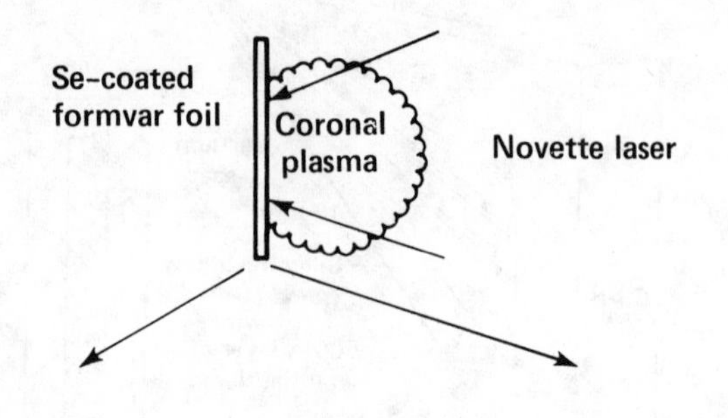

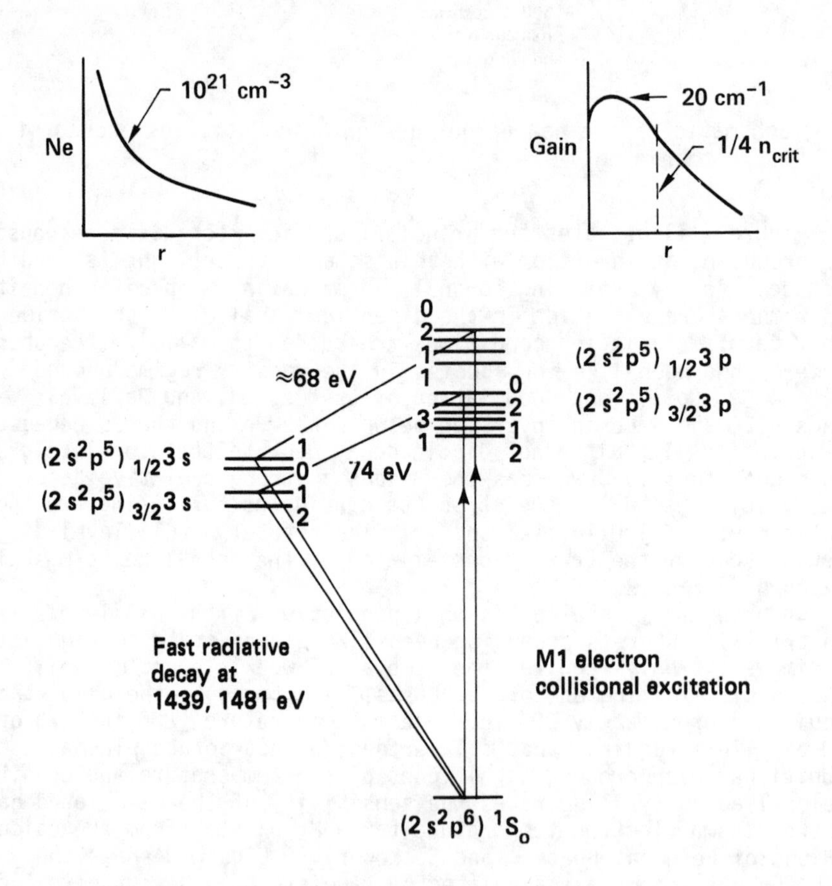

7. An illustration of the neon-like selenium collisional excitation laser scheme. Lasing can possibly occur between the 3p and 3s levels because of preferential radiative decay out of the 3s.

TABLE II

Calculated Gain Coefficients for the Neon-like Kr $2p^53p\ (^1S_0) \rightarrow 2p^53s(^3P_1)$ Transition as a Function of Two Electron Temperatures and Four Electron Densities.

	$\alpha(cm^{-1})$			
$Te(^0K)$	$N_e = 10^{19}cm^{-3}$	$N_e = 10^{20}cm^{-3}$	$N_e = 10^{21}cm^{-3}$	$N_e = 10^{22}cm^{-3}$
$1x10^7$	0.045	0.92	8.8	35
$3x10^7$	0.067	1.7	16.5	66

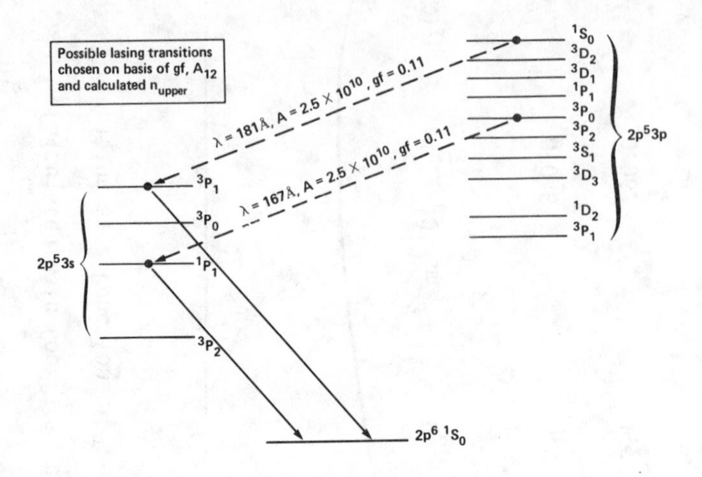

8. A simplified level scheme illustrating the wavelengths, radiative rates, and gf values for Se.

maximum population of j=0 neon-like states, and c) significance of absorption from the copious number of transitions present from other ionic species.

The Collisional Recombination Laser

The final scheme we studied, with only a few experiments, involves preferential recombination of electrons into high n-states of hydrogen-like oxygen ions. The original idea for this scheme comes from Gudzenko and Shelepin[27] with numerous later extrapolations to laser-pumped plasmas (Irons and Peacock[38], Pert[39], Korukhov et al.[53]). The basic recipe to produce the inversion is as follows: a) rapidly fully-ionize a plasma using a short pulselength high intensity optical laser, b) rapidly cool the plasma by adiabatic expansion, electron conduction, radiative power loss, or, preferably, some combination of the three.

Some of the physics questions that affect the performance of this laser scheme include: a) is population inversion maintained for this density and temperature plasma, i.e., can collisional recombination into the n=2 level dominate the collisional excitation into the n=2 level, and can the plasma be limited in size so that the trapping of the ground state for the laser (i.e., the n=2 to 1 Lyman alpha line) remains negligible, at least during the times of peak upper state population, b) can we truly model the atomic physics of this problem, i.e., the ionization kinetics, level populations, line widths, and radiation transport?

This scheme was not optimized before the experiments and our post-experiment modeling has definitely shown some flaws in our

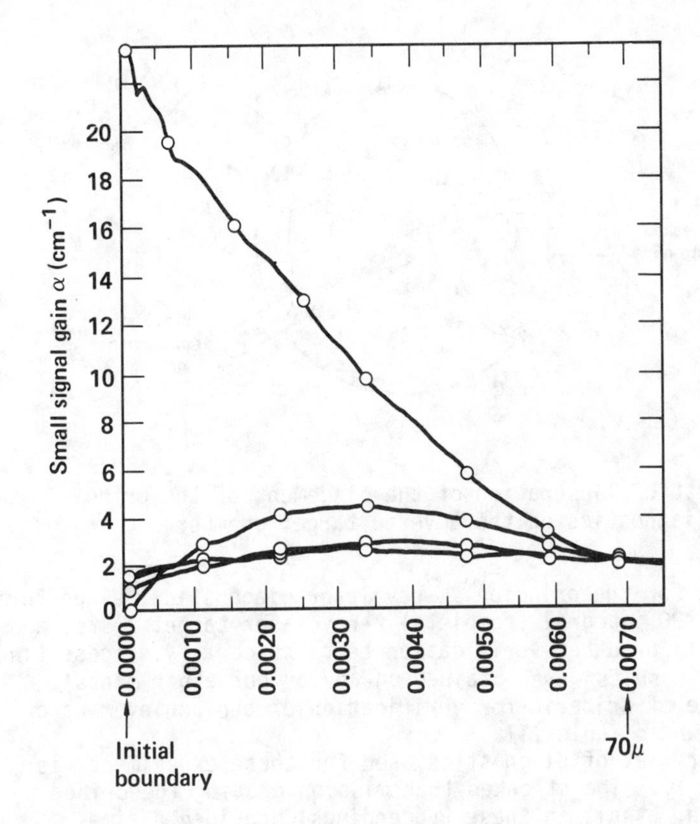

9. The small signal gain for the neon-like Se scheme as a function of distance off the surface of the Se-coated foil. The calculations were done by P. Hagelstein[5] using code XRASER. The curves illustrate the gains associated with different transitions. The highest gain is expected for the $1s^22s^22p^53p(^1S_0) \rightarrow 1s^22s^22p^53s(^3P_1)$ transition.

design. We overstripped the plasma, i.e., the electron temperature was too high for optimum production of H-like ion species during the lifetime of the plasma. Moreover, we did not attempt to cool the plasma by any exotic means other than that which comes naturally, i.e., due to adiabatic expansion (a cooling rate which scales as the expansion velocity of the plasma density distribution).

The Measurements

The attitude was adopted early in the planning for these experiments to perform integrated tests of the targets. Our strategy was to a) insure that the new Novette laser could produce 5 TW of green light per beam with a pulselength of 100 psec FWHM, b) verify that the new cylindrical lenses were functional, which means they produced a fairly uniform 1.2 by 0.01 cm focal spot, c) check flashlamp line brightness at green wavelengths (our previous studies[59] had been done at Shiva at 1.0 μm), d) validate proper operation and spectral

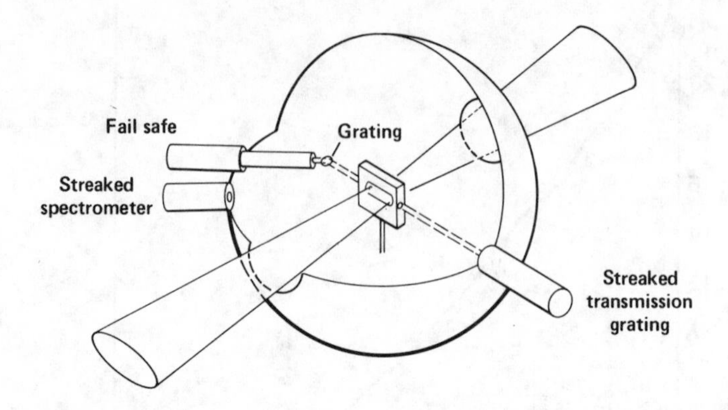

10. A schematic illustration of the placement of the principal x-ray laser diagnostics on the Novette target chamber.

line coverage for the principal x-ray laser diagnostics, e) perform approximately 20 integral (complete) x-ray laser target tests, and f) perform null or other verification tests to certify success (in case positive results were obtained on any of our experiments). A complete table of criteria for verification of the achievement of lasing is shown in Table III.

The primary set of diagnostics used for these experiments is shown in Fig. 10. The streaked transmission grating (described by R. L. Kauffman, et al, in these Proceedings) provided a time-resolved measurement of the brightness of a prospective laser line and also provided the alignment method via an autocollimation technique. The streaked crystal spectrograph provided the time history of the pump lines for the Resonant Photoexcitation scheme. The Fail-Safe was a reflection grating instrument utilizing a microchannel plate intensifier and so-named because of its tolerance to target misalignment (namely, a large acceptance angle).

These three main diagnostics were complemented by the following diagnostic array: x-ray microscopes and pinhole camera to measure the beam size from x-ray images; time-integrated x-ray crystal spectrographs to determine the absolute intensity of the flashlamp lines; optical diodes to determine laser absorption; and several other of our normal diagnostics for studying laser plasma conditions. The optical axes of the principal diagnostics (fail-safe and streaked transmission grating), together with the target chamber center, defined the line-of-sight along which we pointed the longitudinal axis of the laser focus, as well as the axis to which we aligned our x-ray laser target to high precision (+/- 0.5 mrad). The alignment scheme for the whole system is shown in Fig. 11.

Our experiments were divided into two parts: a) flashlamp shots to determine brightness and b) actual x-ray laser target shots. The results of a limited number of flashlamp shots are shown in Table III for Cr (14.458 Å He-like F scheme), Ni and Na (He-like neon scheme at 11.000 Å) and Mn (H-like F at 12.643 Å). The tables include the time-integrated intensity and x-ray pulselength data (when

TABLE III

Shot No. + Element	Pulsewidths, psec(FWHM) Laser	Line	Peak Line Intensity (keV/keV-Sphere) Front	Back	Line Width (eV)	Emission Area (cm^2)	Brightness (Photons/Mode) Front	Back
NaCl								
93012706	96	165	$7x10^{16}$	$3.3x10^{16}$	~2.5	$7.85x10^{-3}$	$8x10^{-4}$	$3.4x10^{-4}$
93012804	98	150	$8.3x10^{16}$	$4.3x10^{16}$	~2.5	$7.85x10^{-3}$	$1.04x10^{-3}$	$4.9x10^{-4}$
93020107	230	190	$9.5x10^{16}$	-------	~2.5	$7.85x10^{-3}$	$9.43x10^{-4}$	--------
93020204	102	60	$1.4x10^{17}$	$9.0x10^{16}$	~2.5	$7.85x10^{-3}$	$4.4x10^{-3}$	$2.6x10^{-3}$
93020206	100	170	$4.7x10^{16}$	--------	~2.5	$7.85x10^{-3}$	$5.2x10^{-4}$	--------
Ni								
93013110	116	190	$2.7x10^{16}$	--------	2.5	$7.85x10^{-3}$	$2.7x10^{-4}$	--------
Cr								
93011906	94	100	$4.0x10^{17}$	--------	2.0	$7.85x10^{-3}$	$1.1x10^{-2}$	--------
93012506	90	110	$4.2x10^{17}$	--------	2.0	$2.01x10^{-2}$	$4.1x10^{-3}$	--------
93012509	100	100	$2.3x10^{17}$	--------	2.0	$7.85x10^{-3}$	$6.3x10^{-3}$	--------
93012512	100	120	$1.1x10^{17}$	--------	2.0	$2.01x10^{-2}$	$9.8x10^{-4}$	--------
Mn								
93013119	105	140	$2.6x10^{17}$	--------	2.3	$7.85x10^{-3}$	$5x10^{-3}$	--------
93020104	101	160	$2.3x10^{17}$	--------	2.3	$7.85x10^{-3}$	$4.9x10^{-3}$	--------

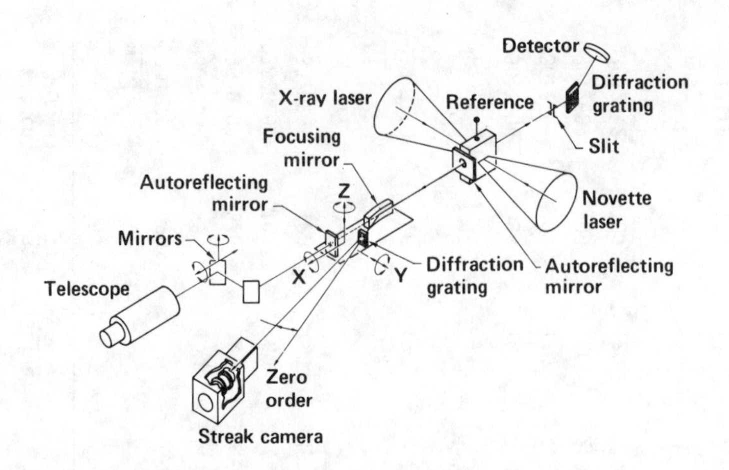

11. A schematic of the alignment scheme for the x-ray laser targets and the x-ray laser diagnostics: the transmission grating streak camera and the Fail-Safe.

available), in addition to an estimate of the brightness in photons/mode. In most cases, the brightness is less than optimum for achieving gain. For all cases, except Cr, the amount of data is too limited to conclusively state that sufficient brightness cannot be achieved. More experiments to determine the optimum laser irradiance and target thickness parameters for these flashlamps are planned in the near future.

Results

The results of the x-ray laser target experiments can best be discussed by the type of scheme. For the resonant photoexcitation scheme three potential laser scenarios were tested and are tabulated in Table I. Out of ∿ 22 target tests, the basic result: no lasing transitions were observed to within the sensitivity limits of our diagnostics (require achievement of gain lengths greater than 5-6). Although there was an enormous number of ways to fail in the proper execution of the experiments, we believe that a proper test of the schemes took place. Some possible reasons for the lack of positive results are 1) we did not properly certify a resonance between lasant and pump lines (a +/- 0.0015 Å overlap is required), 2) the brightness of our pump lines may have been too small to provide enough gain to observe with the sensitivity of our instruments, 3) the calculated laser kinetics may be slightly inaccurate or perhaps the lasant cavity ion temperature too high (neither was measured in the experiments), or 4) the laser cavity may have collapsed before the inversion was initiated.

The collisional excitation laser scheme with simple selenium coated on Formvar targets was also tested (See Fig. 12). No evidence of lasing lines was observed for this scheme either. Prior to finishing the experiments we already knew where one fault resided in

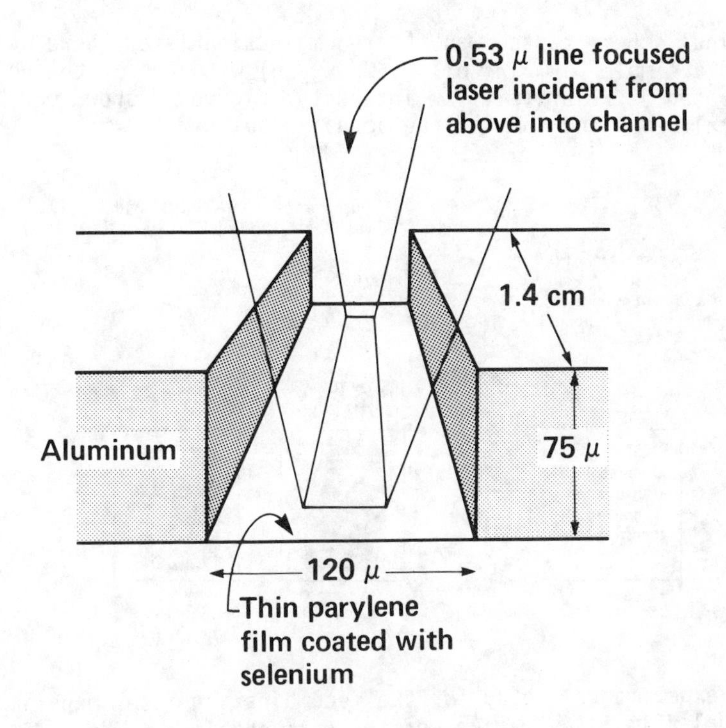

12. Schematic of Se x-ray laser target.

our design. Figure 13 illustrates the effects of a steep density gradient (which is produced with this type of target) on the trajectory of our potential x-ray laser beam. In short, our scheme would have been compromised by the effect of refraction bending the x-ray laser beam out of the maximum gain region, as well as out of the viewing area of our x-ray laser diagnostics. We have planned some major improvements to this scheme for our next series of experiments. These improvements should preclude the effects of x-ray refraction.

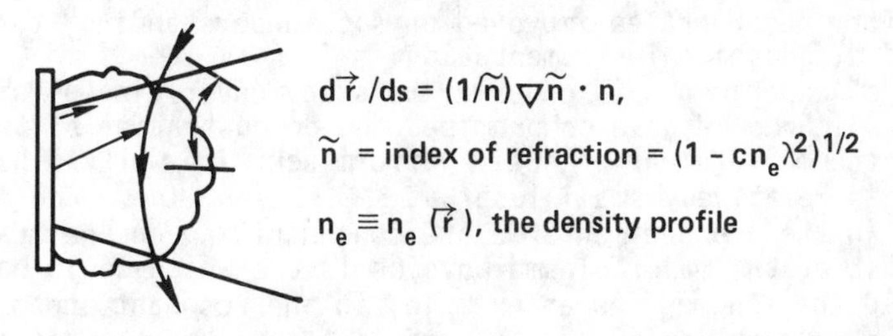

13. An illustration of the effects of a steep density gradient on the trajectory of a beam of x rays.

The final scheme tested was the recombination laser. Here we irradiated a simple Formvar foil (15% oxygen) with two of the Novette beams operated at relatively low intensity. Figure 14 shows our experimental setup and some of the actual results.

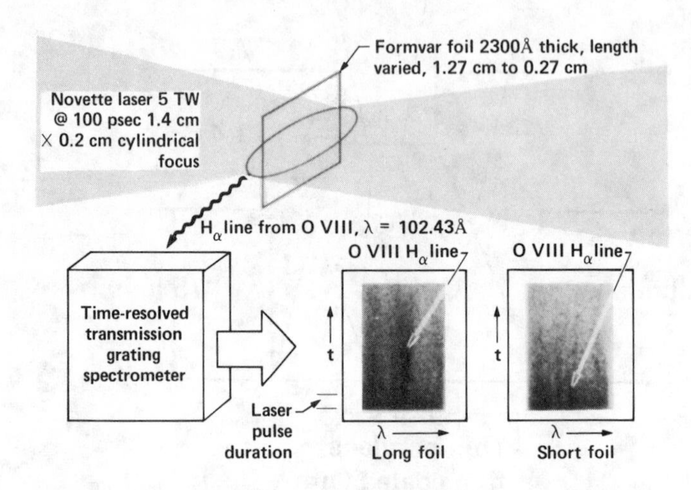

14. The experimental setup for the recombination laser experiment. Actual film illustrations of the data obtained for two different length plasmas are also shown.

We measured the time-dependent emission from the balmer α(n=3 to 2) transition of hydrogen-like O for both long and short foils. This determines if the signal is exponentiating in intensity with the length of the plasma, as would be the case if it were being amplified. An analysis of our data and comparison with theory indicates that a small amount of gain was achieved ($\alpha = 0.5\ cm^{-1}$). Since this scheme was not optimized prior to the experiments, we are encouraged by these results and hope to further test this scheme in a future experiment. Further details of these results are presented in a report.[60]

The results of all these experiments led us to conclude that another more rigorous series of measurements is warranted. Our latest effort incorporates improved physics understanding, target design and diagnostics instrumentation.

A set of experiments is underway to study underlying physical processes which could have compromised the demonstration of these lasers or the measurement of their performance. The work is being done using a relatively small laser at KMS Fusion, Inc., Ann Arbor, Michigan. In these experiments we are concentrating on the measurement of most of the major plasma physics issues which could have jeopardized the schemes. As an example, in the resonant photoexcitation scheme we must verify that a pump line overlaps the energy of an absorbing laser state to within a part in 10^4. This is currently being accomplished with some extremely accurate spectroscopic measurements.

Our strategy to achieve success includes more than a certification of the relevant plasma conditions. We also are constructing more sensitive diagnostics to permit detection of the laser lines, even if they are much weaker than predicted. In addition, we are improving our x-ray laser target construction techniques in order to make the critical components of a target more closely resemble the complex geometries that have been generated during redesign of the laser scheme. Improved target design also enables better execution of the experiment, since often we must certify alignment to our diagnostics to extremely close tolerances.

In the next Novette experiments we will test only one or two schemes. In these tests we will concentrate on iterating target design and optical laser pump parameters around the values predicted by our simulation codes. Hopefully, this method will reduce the possibility that a small uncertainty in the calculation might prevent a successful demonstration.

We envision our program expanding in the near future both due to an increased understanding of the relevant physics issues as well as the emergence in October of this year of an even bigger optical pump laser, Nova. This will enable us to study even more exotic laser design, as well as extend our measurements to shorter xray laser wavelengths.

References:

1. R. W. Waynant and R. C. Elton, Proceedings of IEEE 64, 1059(1976).

2. G. Chapline and L. Wood, Phys Today 28, 40(1975).

3. E. Spiller, AIP Conference Proceedings #75, Low-Energy X-ray Diagnostics-1981, edited by D. Attwood and B. Henke (Monterey, CA 1981) p.124.

4. T. Barbee, Jr., AIP Conference Proceedings #75, Low-Energy X-ray Diagnostics-1981, edited by D. Attwood and B. Henke (Monterey, CA 1981) p.131.

5. A. G. Molchanov, Sov Phys. - Usp. 15, 124(1972).

6. V. A. Bushuev and R. N. Kuzmin, Sov. Phys. - Usp. 17, 942(1975).

7. M. A. Duguay and P. M. Reutzepis, Appl. Phys. Lett. 10, 350(1967).

8. Y. L. Stankevich, Sov. Phys. - Dok l. 15, 356(1970).

9. R. McCorkle, Phys. Rev. Lett. 29, 982(1972).

10. W. W. Jones and A. W. Ali, Phys. Lett. 50A, 010(1974).

11. E. J. McGuire, Phys. Rev. Lett. 35, 844(1975).

12. R. C. Elton, Appl. Opt. 14, 2243(1975).

13. T. S. Axelrod, Phys. Rev. A13, 376(1976); A15, 1132(1977).

14. B. A. Norton and N. J. Peacock, J. Phys. B8, 989(1975).

15. A. V. Vinagradov, I. I. Sobelman, and E. A. Yukov, Sov. J. Quantum Electron. 5, 50(1975).

16. R. A. Andrews, Progress in Lasers and Laser Fusion, edited by A. Perlmutter and S. Widmayer (Plenum, New York, 1975), p.235.

17. R. W. Waynant, Appl. Phys. Lett. 22, 419(1973).

18. K. G. Whitney and J. Davis, J. Appl. Phys. 46, 4103(1975).

19. J. Davis and K. G. Whitney, Appl. Phys. Lett 29, 419(1976).

20. I. Knyazev and V. Letokhov, Opt. Comm. 3, 332(1971).

21. T. Bristow, M. Lubin, J. Forsyth, E. Goldman, and J. Soures, Opt. Comm. 5, 315(1972).

22. B. Lax and A. Guenther, Appl. Phys. Let. 21, 361(1972).

23. A. Vinagradov and I. Sobelman, Zh. Eksp. Teor. Fiz. 63, 2113(1973); Sov. Phys. JETP 36, 115(1973).

24. A. Zherikhin, K. Koshelev, and V. Letokhov, Sov. J. Quant. Mech. 6, 82(1976).

25. A. Illyukhin, G. Peregudov, E. Ragozin, I. Sobelman, and V. Chirkov, JETP Letters 25, 536(1977).

26. H. Zwally and D. Koopman, Phys. Rev. A2, 1850(1970).

27. R. M. Dixon and R. C. Elton, Phys. Rev. Lett. 38, 1072(1972).

28. R. Dixon, J. Seely, and R. Elton, Phys. Rev. Lett. 40, 122(1978).

29. L. I. Gudzenko and L. A. Shelepin, Sov. Phys. JETP 18, 998(1964); Sov. Phys. - DOKI 10, 147(1965).

30. L. I. Gudzenko, Y. K. Zentsov, and S. I. Yekovlenko, Sov. Phys. - JETP Lett. 12, 167(1970).

31. L. I. Gudzenko, L. A. Shelepin, and S. I. Yakovlenko, Sov. Phys. - Usp. 17, 848(1975).

32. B. F. Gordiets, L. I. Gudzenko, and L. A. Shelepin, Sov. Phys. - JETP 28, 489(1969).

33. E. Y. Konov and K. N. Koshelov, Sov. J. Quant. Electron. 4, 1340(1975).

34. J. Green and W. Silfvast, Appl. Phys. Lett. 28, 253(1976).

35. W. Jones and A. Ali, Appl. Phys. Lett. 26, 450(1975).

36. J. Peyraud and N. Pegraud, J. Appl. Phys. 43, 2993(1972).

37. W. Bohn, Appl. Phys. Lett. 24, 15(1974).

38. E. Y. Konov and K. Koshelov, Sov. J. Quantum Electron. 4, 1340(1975).

39. E. Y. Konov, K. Koshelov, Y. A. Levykin, Y. V. Sidelnikov, and E. S. Churilov, Sov. J. Quant. Electron. 6, 308(1976).

40. F. Irons and N. Peacock, J. Phys. B7, 1109(1974).

41. G. Pert, J. Phys. B9, 330(1976).

42. G. J. Tallents, J. Phys. B10, 1769(1977).

43. R. Dewhurst, D. Jacoby, G. Pert, and S. Ramsden, Phys. Rev. Lett. 37, 1265(1976).

44. K. Sato, M. Shiko, M. Hosokawa, H. Sugawara, T. Oda, and T. Sasaki, Phys. Rev. Lett 39, 1074(1977).

45. A. Skorupski and S. Suckewer, Phys. Lett. 46A, 473(1974).

46. S. Suckewer, R. Hawryluk, M. Okabayashi, and J. Schmidt, Appl. Phys. Lett. 29, 537(1976).

47. S. Suckewer and H. Fishman, J. Appl. Phys. 51, 1931(1980).

48. D. Jacoby, G. Pert, S. Ramsden, L. Shorrock, and G. Tallents 37, 193(1981).

49. J. G. Kepros, Appl. Opt. 13, 695(1974).

50. P. Jaegle, G. Jamelot, A. Carillon, A. Sureau, and D. Dhaz, Phys. Rev. Lett. 33, 1070(1974); P. Haegle, A. Carillon, D. Dhaz, G. Jamelot, A. Sureau, and M. Cukier, Phys. Lett 36A, 167(1971).

51. S. Slutz, G. Zimmerman, W. Lokke, G. Chapline, and L. Wood, Univ. of Calif., Lawrence Livermore National Laboratory Report UCID-16290 (1973).

52. W. Silfvast, J. Green, and O. R. Wood II, Phys. Rev. Lett. 35, 435(1975).

53. V. V. Korukhov, N. G. Nilulin, and B. I. Troshin, Sov. J. Quant. Elect. 12, 1099(1982).

54. P. Hagelstein, Ph.D. Dissertation, UCRL-53100, Jan. 1981, available from Lawrence Livermore National Laboratory Technical Information Department.

55. P. L. Hagelstein, Plasma Physics 25, 1345(1983).

56. P. Burkhalter, G. Charatis, and P. Rockett, Naval Research Laboratory Report #5022, June 31, 1983, and to be published.

57. R. Stewart (LLNL) private communication.

58. M. Feldman, A. K. Bhatia, and S. Suckewer, J. Appl. Phys., 54, 2188(1983).

59. D. L. Matthews, P. Hagelstein, E. M. Campbell, A. Toor, R. L. Kauffman, L. Koppel, W. Halsey, D. Phillion, and R. Price, IEEE J. Quant. Elec., Vol QE-19, 1786(1983).

60. D. L. Matthews, E. M. Campbell, K. Estabrook, W. Hatcher, R. L. Kauffman, R. W. Lee, and C. Wang, UCRL-90279 and submitted for publication (Appl. Phys. Lett.).

Tunable Sub-Angstrom Radiation Generated by Anti-Stokes Scattering from Nuclear Levels

C. B. Collins and B. D. DePaola
Center for Quantum Electronics
University of Texas at Dallas
P.O. Box 688, Richardson, TX 75080

ABSTRACT

It was the intent of this work to use states of nuclear excitation in order to demonstrate coherent multiphoton processes that were the analogs of well-known effects occurring at optical frequencies. Success in exciting radiofrequency sum and difference frequency sidebands on hyperfine components of the Mössbauer transition of ^{57}Fe near 0.086 nm is described. The transition energy of those sidebands was shown to be tunable over five linewidths by varying the radiofrequency component of the sum. A model is proposed that quantitatively explains these results in terms of coherently excited nuclear states and evidentally confirms the multiphoton nature of the process.

TEXT

This work concerns the tunable radiofrequency sidebands that can be produced on Mössbauer transitions [1]. The traditional explanations have attributed this effect to periodic Doppler shifts caused by drumhead vibrations of the absorber foil that were driven by magnetostriction and fortuitously amplified by mechanical resonances [2-7]. Quantitative agreement between theory and experiment was not obtained with that model.

Reported here are data obtained with a new experimental arrangement that was designed to eliminate drumhead vibrations. These results generally reproduce the literature phenomenology but also display significant differences. An evolution of multiphoton theories appears to confirm that these radiofrequency sidebands result from the generation of large amplitudes of *coherent* excitation of the nuclear states involved. This description accommodates all phenomena previously reported and suggests that sidebands developed in earlier experiments were driven by the same excitation of coherent states, perhaps benefiting from some enhancement contributed by physical motions that could have developed if acoustic resonances were present as postulated. Considerable significance can be attached to this demonstration that relatively long wavelength radiation can serve for the efficient production of coherent nuclear excitation. It would appear to support speculations [8] that the analogs of many of the interesting and useful processes of non-linear optics may be found at nuclear energies. Of particular interest in this work are the Raman processes that produce spectral lines of natural width at sub-Angstrom wavelengths.

By varying the frequency of the source producing the coherent excitation, these Raman lines can be tuned over a range of wavelength spanning many linewidths.

In 1978 West and Matthias [5] reported what seems to have been the only uncontested example of the prior excitation of a coherent nuclear state with long wavelength radiation. An extremely small effect at the level of the noise was obtained in ^{181}Ta at power inputs of 1 kW. As a consequence of a more general theoretical modeling of processes of coherent nuclear excitation [9-11], Olariu et al. [6] reported a multiphoton model for Mössbauer transitions comparable to the Matthias model that could explain both in magnitude and detail the much stronger development of sidebands in the earlier ^{57}Fe experiments.

The Olariu-Matthias representation had been constructed under the assumption that static magnetization was absent in the Mössbauer absorber and as a result that model was rigorously appropriate only for nuclei embedded in paramagnetic samples. Of concern here was the more usual circumstances for ^{57}Fe in which ferromagnetism of the sample aligns the hyperfine fields in which the nuclei are immersed. Thus, the upper, $J = 3/2$ and lower, $J = 1/2$ states of the 14.4 keV transition are split into Zeeman sublevels, each separated by $\nu_e = 25.87$ and $\nu_g = 45.49$ MHz, respectively, in pure iron.

The rather complicated response of a ferromagnetic sample to a time varying external field is a subject that has received considerable attention in studies of ferromagnetodynamics. The point of departure for calculations is usually the Landau-Lifshitz equation [12],

$$\partial\vec{M}/\partial t(\vec{M}) = -\gamma\mu_0(\vec{M}\times\vec{H}) - 4\pi\lambda_d(\vec{M}\times\vec{M}\times\vec{H})/|M|^2 \tag{1}$$

where $\vec{M}$ is the magnetization vector, $\vec{H}$ is the magnetic field vector, γ is the gyromagnetic ratio and λ_d is a phenomenological damping parameter. The first term on the right results in the usual precession of the magnetization around the $\vec{H}$ field and predominates in descriptions of NMR experiments [13]. The second term results in a further precession of $\vec{M}$ about $\vec{M}\times\vec{H}$, the direction of the torque, and usually contributes to a decay of the radius of the first precession, ultimately tending to bring $\vec{M}$ into alignment with $\vec{H}$, if $\vec{H}$ is static.

While Eq. (1) has been extremely successful in describing the dynamics of domain wall motion, the usual applications have assumed a geometry in which surfaces are remote and so $\vec{H}$ had to include only applied fields. However, that problem admits an equivalent description [14] in which the phenomenological term is replaced by the effect of demagnetizing fields arising from poles induced on the surfaces of the walls. Chen [15] has explicated that approach in a form that serves as the point of departure for our derivation.

It is important to emphasize that the model developed here is not the "domain wall passage" explanation that was refuted by earlier experiment. Rather, it was found that much of the computational machinery developed for the description of domain walls could be directly utilized in this work.

As shown in Fig. 1 the sample is assumed to be in the form of a thin lamina immersed only in a sinusoidally varying field, $\vec{H} = \vec{H}_0 \sin \omega_2 t$, lying in the plane of the foil. As shown by Chen [15] in such cases, the predominant effect is the precession of the saturation value of magnetization, $\vec{M}_s$ in each domain about the axis normal to the major surfaces of the foil. This precession is driven by the $\vec{M} \times \vec{H}_d$ torque in the direction of the applied field, $\vec{H}$. The $\vec{H}_d$ is the depolarizing field produced by the magnetic poles on the large surfaces that result from attempts of $\vec{M}_s$ to precess about $\vec{H}$. Precise computation of $\vec{H}_d$ is beyond the scope of this work, but Chen [15] gives an approximation sufficient for other applications,

$$H_d = M_s \omega t \sin \phi \sin \omega_2 t \quad , \tag{2}$$

where ω is the free space spin resonance frequency, $\omega = \mu_0 \gamma |H|$. The experiments to be done here concern applied frequencies, ω_2 of the order of tens of MHz so that $\omega >> \omega_2$ and thus for times of interest $|H_d| \sim |M_s|$. Since $|M_s|$ is of the order of a half of a megaGauss, the time-varying field of Eq. (2) that results from the surface poles completely replaces the applied $\vec{H}$ as the driving source of the mixing phenomena.

To be rigorously correct the subsequent precession of the atomic spins about $\vec{H}_d$ should be calculated and then the coupling of the nuclear spins to the resulting fields would determine the mixing of the nuclear states. It can be expected that the vector $\vec{M}_s$ would be found to swing back and forth within the plane of the foil in a complex manner and the coupling of the nuclear states to that vector would provide the preferable representation.

In the interest of obtaining a minimal estimate in order to determine whether the nuclear states would be affected appreciably, a much simpler approach was employed. It was assumed that the precession of $\vec{M}_s$ about $\vec{H}_d$ would be characterized by the slower Larmor precession of the nuclear spins rather than the more rapid precession of the atomic spins.

Then, the angular velocity of precession of $\vec{M}_s$ about $\vec{H}_d$ was approximated,

$$\dot{\vec{\phi}}_x = \gamma_x \vec{H}_d \mu_0 \tag{3}$$

where x denotes either e for an excited nucleus or g for a ground state. Substituting Eq. (3) into Eq. (2) and integrating gives

$$\tan[\phi_x(t)/2] = \tan(\phi_0/2) \exp[\beta_x(\omega\omega_g/\omega_2^2)(\sin \omega_2 t - \omega_2 t \cos \omega_2 t)], \tag{4}$$

where ϕ_0 is the angle describing the initial orientation of $\vec{M}_s$ in the domain, $\omega_g = \mu_0\gamma_g|M_s| = 2\pi \times 45.49$ MHz, $\beta_g = 1$ and $\beta_e = -3|\omega_e/\omega_g|$, the negative arising because the gyromagnetic ratios are opposed in ground and excited states of ^{57}Fe. Limits of validity to Eq. (4) are 0 and π and the expression for the return swing of $\vec{M}_s$ can be constructed from symmetry.

Following Slichter [13], we assumed that the spins are quantized along the axis of total magnetization, in this case $\vec{M}_s$; and that the effect of the rocking of the axis of quantization about $\vec{H}_d$ is to rotate the unperturbed nuclear states. The projection of the initial state onto the final state is given by the operator for relative rotation, R which can be expanded in a Fourier series of harmonics of the driving frequency, ω_2,

$$\hat{R} = \mathrm{EXP}[-\, i(\phi_e(t) - \phi_g(t))\, \hat{\sigma}_d/2] = \sum_{j=-\infty}^{\infty} c_j \mathrm{EXP}(ij\omega_2 t\hat{\sigma}_d/2) \quad (5)$$

where $\hat{\sigma}_d$ is the spin matrix for rotations about the axis of $\vec{H}_d$ and $\phi_e(t)$ and $\phi_g(t)$ are given by Eq. (4). The effect of rotating the initial state of the system with $\hat{R}$ can be seen to dress the state with $\pm j$ photons of the field, thus increasing its apparent energy relative to the excited state by $\mp j\hbar\omega_2$. The amplitude of the j-th dressed state is c_j and the relative strength of a transition between it and the reference, excited state is $c_j c_j^*$. In this model that is the relative intensity of the j-th sideband.

Notwithstanding, the obvious difficulty that the model treats the J = 3/2 state as a semiclassical entity with a spin of ±3/2 and neglects mixing of its other projections, it is surprisingly good in describing the results of this experiment.

As seen in Fig. 1, the apparatus was arranged to conform completely to the traditional pattern [2] for this type of experiment with one exception. This innovation lay with the foil carrier. One of the critical aspects had been the generally perceived need to provide a free surface for the absorber foil to execute the drumhead vibrations. The foil carrier used here was arranged to rigidly sandwich the absorber between graphite sheets to prevent mechanical movements while allowing a more rapid flow of air for cooling. The conductivity was sufficient to allow significant amounts of heat to be rejected into the face of the graphite sheet during the radiofrequency pulse, but not so great as to give a skin depth smaller than the physical thickness.

The radiofrequency power was applied to a coil containing the foil carrier. It was pulsed for a duration equal to the time required for the full sweep of the range of velocities being examined. The output from a proportional counter monitoring the transmitted γ intensity was gated into a multichannel scalar only during those times. Generally, duty cycles of 1/16 prevented excessive heating of the sample. The foil used in these experiments, was a 2 μ thick pure ^{57}Fe sample of about 1 cm^2 surface area.

Figure 2 shows typical data obtained at 10 Watts of input power at 61.7 MHz into a resonant circuit with a Q = 60 that contained the coil and foil carrier. In this case tuning was provided by the traditional Doppler shifting of the unsplit line source with respect to the absorber which was arranged to contain the structured transitions. To obtain the spectrum of Fig. 2, the relative velocities were arranged to increase linearly with channel number as plotted. Statistical scatter was reduced by averaging data from five successive channels. The marked asymmetry in the data of Fig. 2 resulted from a corresponding asymmetry in the velocity drive. Any resulting uncertainty in the identity of various features was resolved by varying the frequency.

Figure 3 shows a comparison between the experimental intensities observed at a fixed mixing frequency and the values for $c_j c_j^*$ obtained from Eq. (5). The calculated intensities depended upon experimental variables through $\omega \propto |\vec{H}| \propto \sqrt{P}$, where P is the applied radiofrequency power, and through ϕ_0 the static orientation of $\vec{M}_s$ and $\vec{H}$ in a particular domain. The dependence of $c_j c_j^*$ upon relative power can be seen in Fig. 3. The effect on the curves of varying ϕ_0 between domains closely approximates the translation of the scale of powers by a numerical factor. Thus, it can be expected that the effect of averaging the computed curves over a distribution of ϕ_0 for different domains will be about the same as averaging over some non-uniformity of P across the foil. The absolute scale for power should contain these spatial averages.

In making the comparison shown in Fig. 3 the single scale factor between the average input power and the abscissa was adjusted only once for optimal agreement and the relative intensities were used as measured, the parent lines being assigned a value of unity in the absence of applied power. Data from Ref. 3 was normalized at the lowest power to which it fit. Considering the rather rough, semiclassical nature of the model, the level of agreement shown in Fig. 3 is remarkable, somewhat exceeding that reported from magnetostrictive models [3].

The tunability of the sidebands that is predicted by the same theory was of paramount importance in this particular experiment. The two leftmost absorption lines in Fig. 2 are the sum frequency transitions involving the basic nuclear transitions plus one radiofrequency photon. To directly examine the potential utility of these Raman lines as tunable spectral features the relative velocity between the single-line source and the absorber whose spectrum is shown was set to provide a resonance in wavelength between the emission line from the source and the position of the second line from the left in Fig. 2. Maintaining the velocity as constant, the radiofrequency was varied causing the wavelength of that line sweep through the fixed wavelength of the source. The resulting data is shown in Fig. 4. For convenience, in this work the absorption line was tuned, but the system is symmetric so that a tunable source could be readily constructed by interchanging the environment of source and absorber nuclei.

It seems that the strong agreement between the experimental data obtained in this work and the model of coherently excited nuclear states confirms the multiphoton nature of the transitions, while the observed efficiency of the effect tends to support an optimistic perspective on many of the other exciting possibilities being proposed in nuclear quantum electronics. Even the preliminary theory described here suggests that the range of tuning should be readily extended, at least to the frequencies characteristic of ferromagnetic resonance phenomena. This would represent one to two orders of magnitude increase over the best range available from mechanically produced Doppler shifts.

The authors gratefully acknowledge the support of the Office of Naval Research under Contract N00014-81-K-0653.

REFERENCES

1. N. D. Heiman, L. Pfeiffer, and J. C. Walker, Phys. Rev. Lett. 21, 93 (1968).
2. L. Pfeiffer, N. D. Heiman, and J. C. Walker, Phys. Rev. B 6, 74 (1972).
3. C. L. Chien and J. C. Walker, Phys. Rev. B 13, 1876 (1976).
4. The most compelling experiments seem to have been Ref. 2 and 3.
5. P. J. West and E. Matthias, Z. Physik A288, 369 (1978).
6. S. Olariu, I. Popescu and C. B. Collins, Phys. Rev. C23, 1007 (1981).
7. Reviews of the extensive sideband literature are found in Refs. 5 and 6.
8. For example: Multiphoton nuclear transitions and induced neutron capture and fission - S. Olariu, I. Popescu and C. B. Collins, Phys. Rev. C23, 50 (1981); stimulated γ-ray emission - C. B. Collins, F. W. Lee, D. M. Shemwell, B. D. DePaola, S. Olariu and I. I. Popescu, J. Appl. Phys. 53, 4645 (1982); induced β-decay - N. Becker, W. H. Louisell, J. D. McCullen and M. O. Scully, Phys. Rev. Lett. 47, 1262 (1981); induced internal conversion - G. C. Baldwin and S. A. Wender, Phys. Rev. Lett. 48, 1461 (1982).
9. C. B. Collins, S. Olariu, M. Petrascu and I. Popescu, Phys. Rev. Lett. 42, 1397 (1979).
10. C. B. Collins, S. Olariu, M. Petrascu, and I. Popescu, Phys. Rev. C20, 1942 (1979).
11. S. Olariu, I. Popescu and C. B. Collins, Phys. Rev. C23, 50 (1981).
12. T. H. O'Dell, Ferromagnetodynamics (Wiley, New York 1981) Ch. I.
13. C. P. Slichter, Principles of Magnetic Resonance, 2nd ed. (Springer-Verlag, Berlin, 1978) p. 21.
14. T. H. O'Dell, *ibid*, pp. 46-47.
15. C. W. Chen, Magnetism and Metallurgy of Soft Magnetic Materials, (North Holland, Amsterdam, 1977) pp. 156-170.

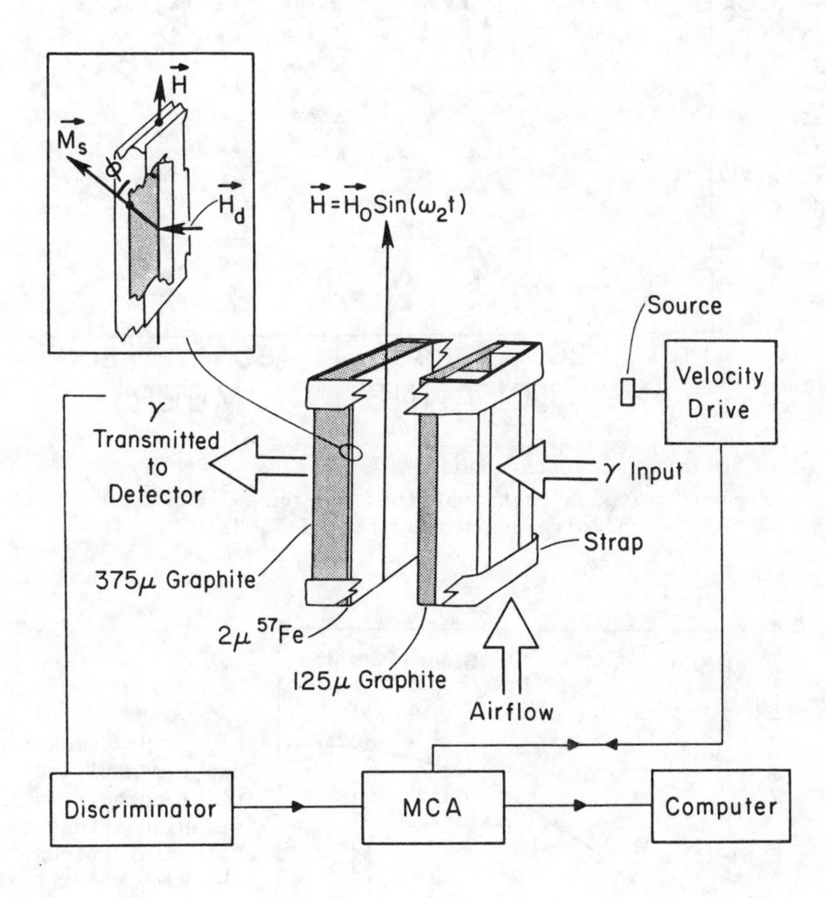

Figure 1: Schematic representation of the experimental apparatus, together with detailed views of the foil carrier and coordinate system used in the theoretical model.

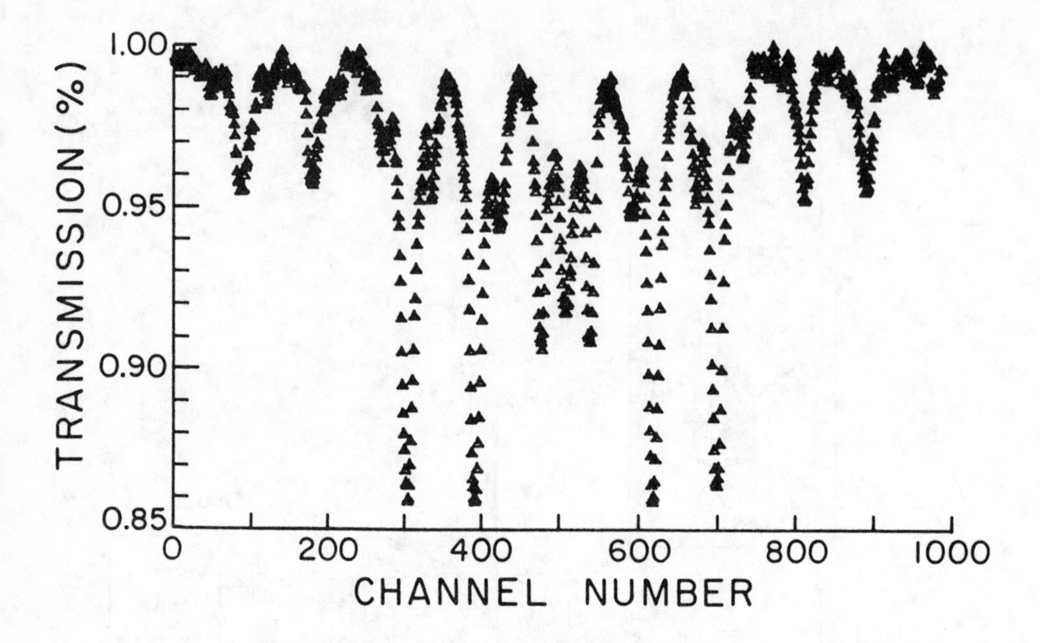

Figure 2: Typical data obtained for pure ^{57}Fe at 10.1 Watts of radiofrequency power at 61.7 MHz.

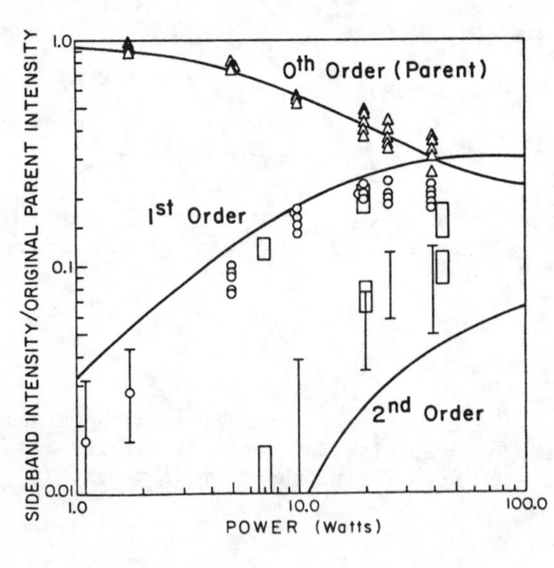

Figure 3: Comparison of experimental and theoretical computations of sideband intensities. Differing intensities at the same power reflect the measurement of sidebands from different parent transitions. Open square symbols from Ref. 3.

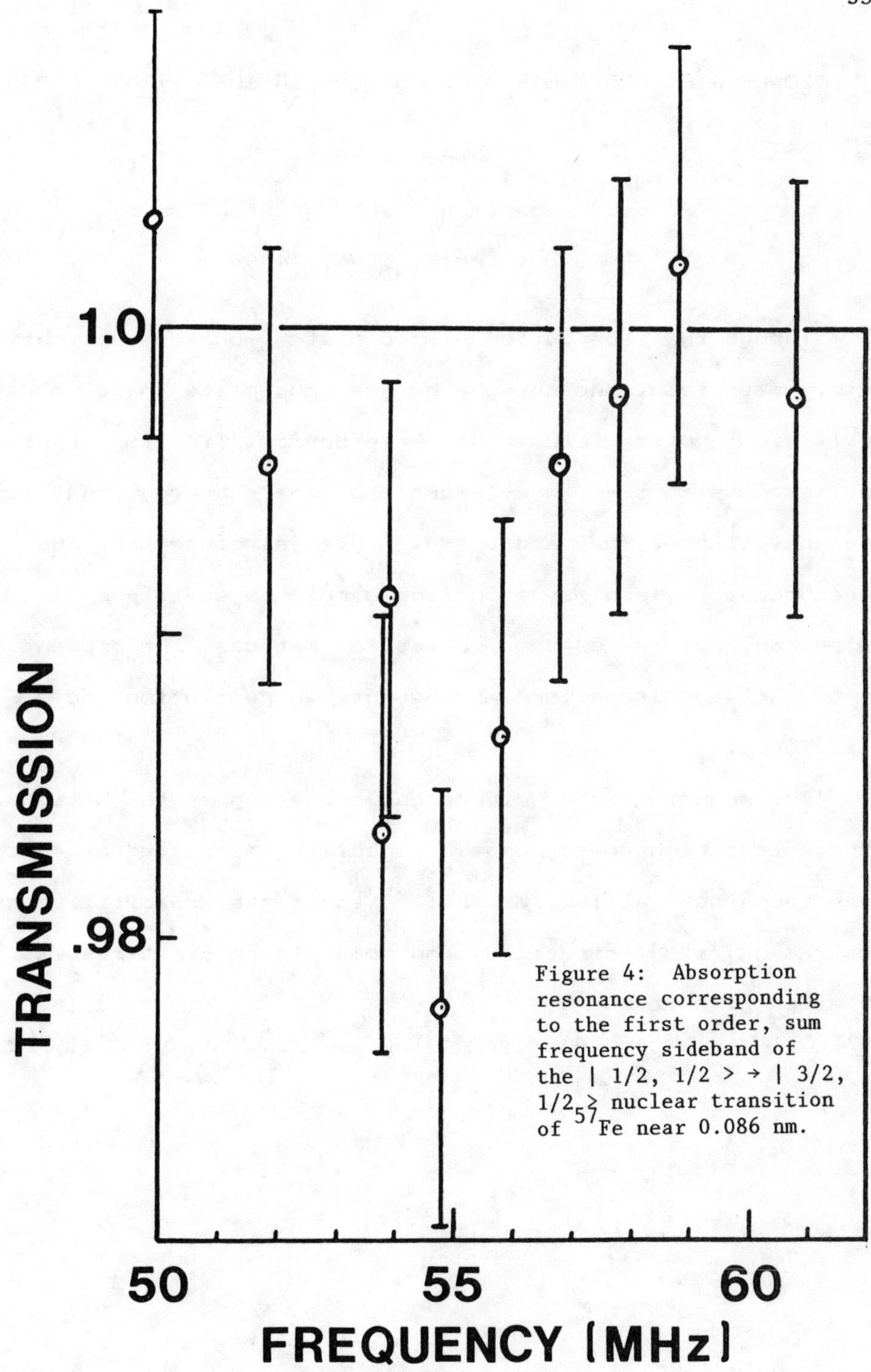

Figure 4: Absorption resonance corresponding to the first order, sum frequency sideband of the $| 1/2, 1/2 > \rightarrow | 3/2, 1/2 >$ nuclear transition of ^{57}Fe near 0.086 nm.

ZONEPLATES AND THEIR APPLICATIONS IN SOFT X-RAY IMAGING.

Janos Kirz

Physics Department, SUNY at Stony Brook

Stony Brook, N. Y. 11794

Though the fabrication of zoneplates for XUV applications is not an easy task, and though the best zoneplates made to date are inefficient optical elements, these devices find important applications in soft X-ray microscopy where conventional lenses are unavailable. They are are also used as condensers and monochromators, and may even find a role in soft X-ray holography. We review the fabrication methods, the present state of the art and ideas for improvements in resolution and efficiency.

The scanning X-ray microscope operating at the National Synchrotron Lightsource uses a zoneplate fabricated by electron beam techniques at IBM. We shall discuss the capabilities of this instrument, its limitations, and some of the first results obtained with it.

RECENT EXPERIMENTS ON SOFT X-RAY LASER DEVELOPMENT IN A CONFINED PLASMA COLUMN

S. Suckewer, C. Keane, H. Milchberg,
C.H. Skinner and D. Voorhees
Princeton University, Plasma Physics Laboratory,
Princeton, New Jersey 08544

ABSTRACT

We present studies of magnetically confined and expanding recombining plasma columns and measurements of gain for hydrogen-like (CVI) ions and population inversion for Li-like (NeVIII) ions.

I. INTRODUCTION

Extensive research on x-ray laser development has provided a number of interesting results for different schemes, e.g. recombination [1-4], line-coincidence[5-7], photo-pumping[8,9] and very recently multi-photon ionization schemes.[10] Each of these approaches has certain advantages and disadvantages. For example, recombination schemes which are one of the most promising approaches for soft X-ray laser development, require very fast plasma cooling and, at the same time, uniform conditions in the direction of the expected lasing action. To help solve this problem, we proposed to use a plasma column confined by a strong solenoidal magnetic field and cooled by radiation losses.[11]

We have shown experimentally[12] that radiation cooling of a magnetically confined plasma column can be more efficient than adiabatic cooling of a freely expanding plasma. The magnetic field prevents the radius of the column from increasing and hence prevents a rapidly decreasing electron density. The plasma was created by the interaction of a 1 kJoule (10 - 20 GW) CO_2 laser with a solid or gas target. Time integrated spectra in the EUV recorded simultaneously in the axial and transverse directions indicated an enhancement of CVI 182Å line intensity in the axial direction, and population inversion for Li-like CIV, OVI, FVII and NeVIII ions[12].

In this paper we present an overview of our recent time resolved measurements in the EUV of H-like CVI, and Li-like NeVIII line intensities spatial distributions, levels populations, and population inversions, as well as gain measurements for the CVI 182Å line. More detailed information about the time evolution and spatial distributions of EUV line intensities are presented in the next paper[13] of this proceedings, and on population inversions and gain measurements in the subsequent paper[14].

II. EXPERIMENTAL ARRANGEMENT

The experimental set-up is presented in Fig. 1. This differs from the one presented previously[12] in the focusing system of the

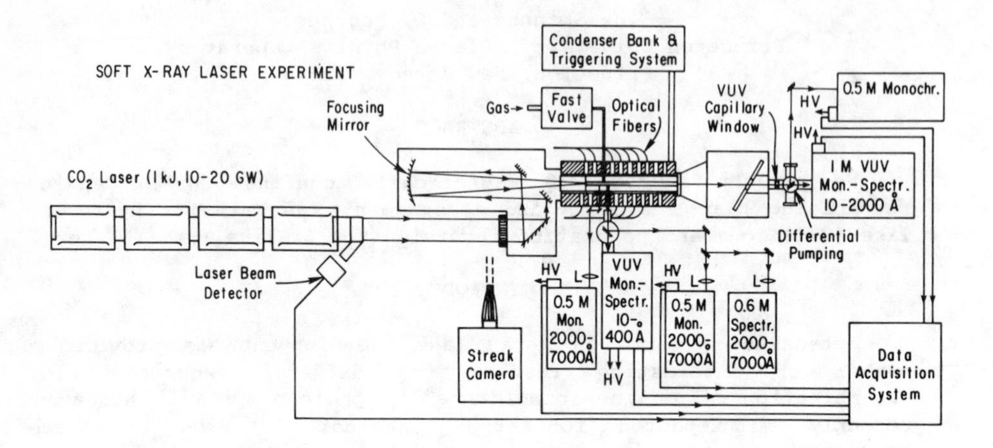

Fig. 1 Schematic of experimental arrangement.

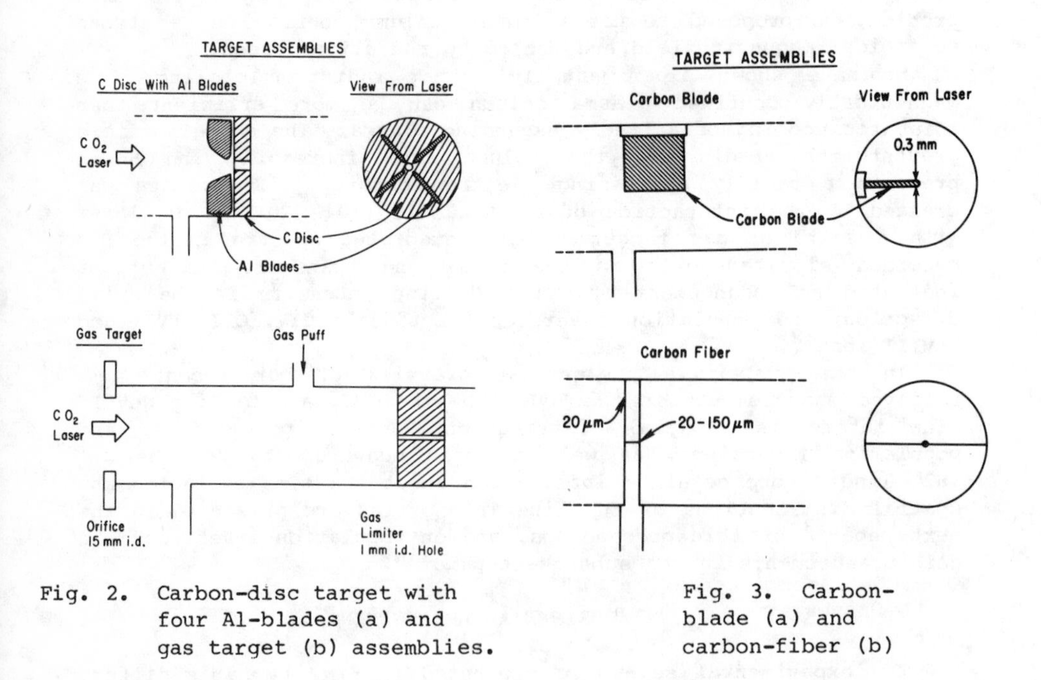

Fig. 2. Carbon-disc target with four Al-blades (a) and gas target (b) assemblies.

Fig. 3. Carbon-blade (a) and carbon-fiber (b)

CO_2 laser. Presently we are using a spherical mirror instead of a NaCl lens to focus the laser beam. This enables us to significantly decrease the focal spot size from 1 mm in diameter to an ellipse of 0.2 mm x 0.4 mm. In this way the intensity of laser beam in the focus is increased up to 10^{13} W/cm^2. The optical arrangement for the diagnostic instruments, except for 0.6m spectrometer, was the same as in the earlier work.[12] However, presently we are using a larger variety of targets. In addition to gas targets (CO_2, O_2, Ne) we have performed experiments with solid targets [C (carbon) - disc, C-disc with Al-blades, C-blade, C-fiber; teflon disc, quartz disc]. All the discs targets presented here had a 1mm or 1.5 mm hole in the center, with the laser beam focused on the edge of the hole. Our attention to solid targets was motivated by experimental conditions: with solid targets we expected to more easily obtain an electron density close to the critical density for the CO_2 laser wavelength of 10 μ ($n_e(cr) \approx 10^{19}$ cm^{-3}) than for gas targets. The gas targets provided us with a longer and more uniform plasma column, however with the present gas puffing system it was difficult to maintain a high gas pressure in a short windowless column.

In the upper part of Fig. 2 is shown a carbon disc (1.5 mm hole in center) with 4 aluminum blades. In the lower part is shown the schematic arrangement for the gas target. Carbon blade and carbon fiber assemblies (horizontal plane) are shown in Fig. 3. The CO_2 laser is located on the left hand side, axial monochromators on the right hand side and transverse instruments are on the bottom of Fig. 3. A EUV grazing incidence duochromator (transverse instrument) and monochromator (axial instrument), each one operating in the spectral range 10-350Å, were equipped with two channel electron multipliers and a 16 stage electron multiplier, respectively. These detectors were installed after the spectra produced by the carbon disc were obtained with the new focusing optics for CO_2 laser.

The rise time of the axial instrument was $\Delta t \approx 20$ nsec and approximately twice shorter for the transverse one. Both instruments were calibrated for absolute intensity at 150Å and 88Å by the branching ratio method.[15,16] This calibration is still in a preliminary stage and more measurements are required [13] (by using vacuum spark-gap). The instruments were also calibrated for relative intensity at 182Å and 28Å using the known line intensity ratio 182Å (3-2) / 28Å (3-1) in a relatively low density plasma (plasma optically thin for these transitions, no line intensity enhancement). The relative sensitivity of the axial and transverse instruments was obtained for the CVI 182Å and CVI 33Å lines by viewing simultaneously with both instruments the plasma created by the CO_2 laser interaction with a vertical fiber,[14] similarly to the Hull University experiment.[2] For a laser created plasma the last method we considered as more reliable then branching ratio method for the comparison of the sensitivity of the axial and transverse monochromators.

III. RESULTS

With the new focusing optics for the CO_2 laser we attempted to significantly increase the axial intensity of the CVI 182Å line in comparison to the CVI 33.7Å line using the same target (carbon disc with 1 mm hole in center), magnetic field (B = 50 kG), and spectrometers as in our earlier work.[12] We present in Fig. 4 a densitogram of a spectral plate in the vicinity of the 182Å and 33Å lines, obtained with the axial EUV spectrometer (the sensitivity of the spectrometer near 33Å was significantly lower than for 182Å, however the relative sensitivity for both wavelengths was the same as in Ref. 12). The intensity ratios of the CVI 182Å and CVI 33.7Å lines as well as the 182Å line and neighboring lines (particularly OVI 173 Å line) increased by a factor of 2-3 in comparison to our earlier data.[12]

Most of the measurements of line intensities presented here and in the next two papers[13,14] were performed with time resolution. Time resolved measurements provided us with much better information on level populations and intensity enhancements (gains) than the time integrated spectra. It is also important that from the time resolved measurements we have information from one discharge whereas for a spectral plate exposure we needed several discharges.

The time evolution of the CVI 182Å and CVI 135Å (4-2 transition) line intensities, measured in the axial and transverse directions are presented in Fig. 5 for two consecutive discharges. In the upper part of the figure the laser intensities

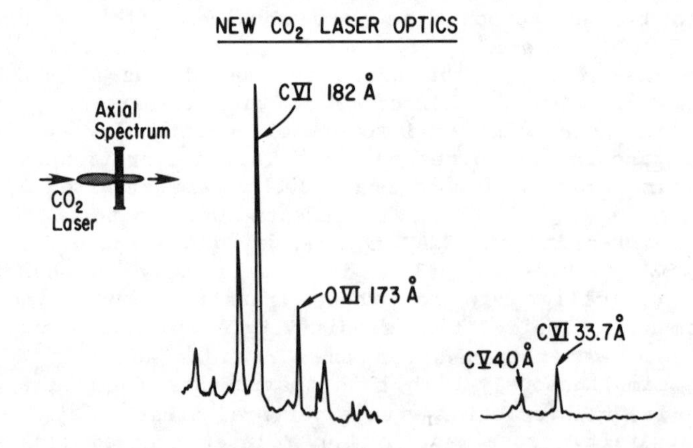

Fig. 4. Axial emission spectra in the vicinity of CVI 182Å and CVI 33Å lines.

obtained with a photon drag detector are shown as a function of time. The laser power was similar for both shots with the laser energy in the range of ~ 500J. The plasma column was created in a solenoidal magnetic field B = 90 kG by focusing the laser beam onto the edge of the 1.5 mm hole in the center of the carbon disc. One channel of the transverse EUV duochromator recorded the CVI 33.7Å line and showed very good reproducibility of CVI radiation.

The second channel of the EUV duochromator and axial EUV monochromator, both measured simultaneously the intensities of the CVI 182Å line for first shot (first row in Fig. 5) and CVI 135Å line for the second shot (second row in Fig. 5). The relative intensity of the CVI 135Å line in the transverse direction was obtained by subtracting the known contribution of the fourth order of the 33.7Å line, which was less than 30% of total intensity at 135Å (for the axial instrument this contribution is negligibly small due to the different angle of blaze of the grating). We had found that the ratio of axial to transverse intensity of the CVI 182Å line exceeded the same ratio of the CVI 135Å line by a factor larger than 10. Such enhancement of axial intensity of CVI 182Å line corresponds to a one pass gain $k.\ell \approx 3.5$.

The applied magnetic field plays a positive role in the generation of high 182Å line intensities. We observed that for magnetic field B = 90 kG the CVI emission in the transverse direction increased by up to factor of 5 and shows a faster decay[13] consistent with fast radiation cooling in the confined plasma column.

We have performed a number of experiments using aluminum blades attached to a carbon disc with a 1.5 mm hole in the center (Fig. 2). We expected to increase the CVI 182Å line enhancement in comparison to the enhancement obtained for the C-disc without Al-blades. Such an expectation follows from the computer calculation of effectiveness of medium Z elements (Al in this case) in cooling a transient high density plasma by radiation losses (Fig 6). A coupled set of rate equations for either carbon or carbon with aluminum is solved in time self consistently with a set of energy equations which describe the electron temperature. The peak value of the laser power is chosen to yield the same peak electron temperature for both cases; it is about 2.8 times greater for the carbon and aluminum system than for the case with carbon alone. In both instances the total number of ions is 4.5×10^{17} cm^{-3} and the electron density is approximately $2\text{-}3 \times 10^{18}$ cm^{-3}.

One may see from Fig. 6 that aluminum plays a significant role in the process of faster radiation cooling for electron temperature, T_e, below 100 eV. Aluminum is especially important for plasma cooling below 20-30 eV. Experimentally we observed higher CVI 182Å line intensity in the spectra and a faster decay time of the CVI line intensities. However, enhancement of the CVI 182Å line intensity did not differ substantially from the one with plain C-disc. One possible explanation may be an improper concentration of Al in the carbon plasma.

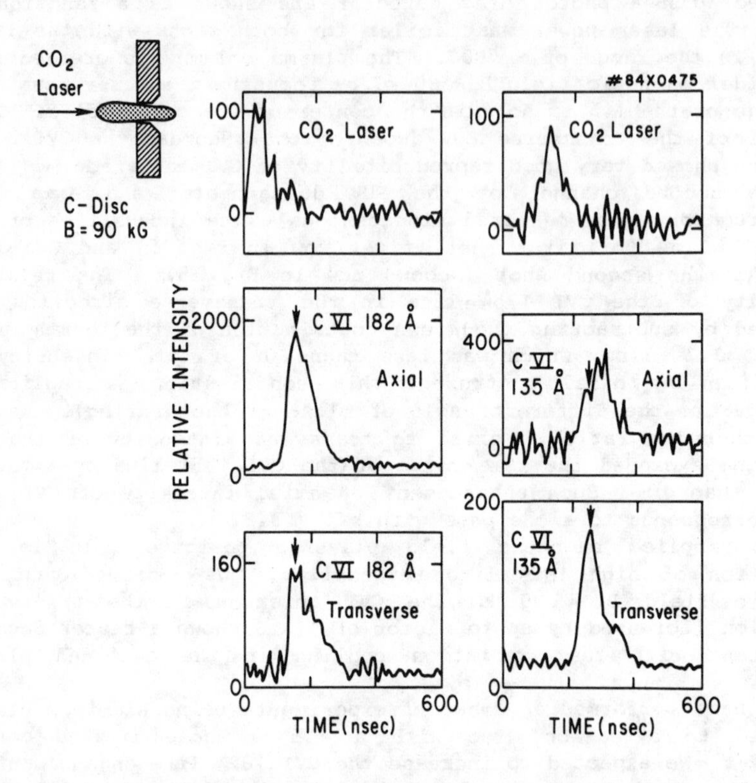

Fig. 5. Time evolution of CVI 182Å and CVI 135Å line intensities, measured in the axial and transverse directions, indicating gain-length product k.ℓ ≈ 3.5.

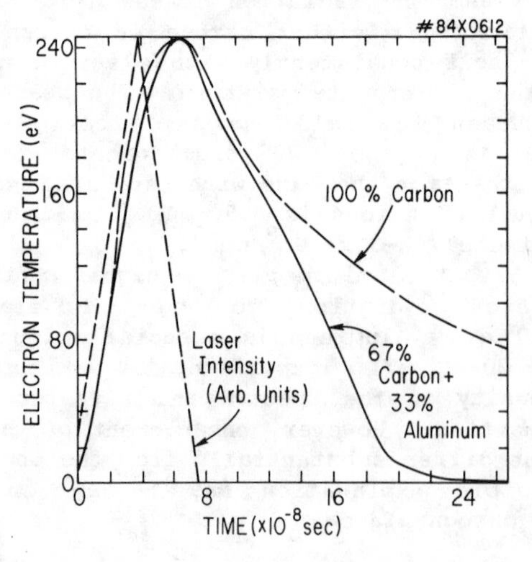

Fig. 6.
Computer calculations of effectiveness of radiation cooling for carbon and carbon-aluminum plasmas.

The radial distribution of the CVI line intensities shows much narrower profiles when the CO_2 laser beam interacts with a smaller target (e.g. carbon blade) than with a disc target, in agreement with the observations of Crawford and Hoffman.[17] With the CO_2 laser (energy ~ 120J) focused on the corner of the carbon blade, the measured radial profile of the CVI 182Å line emission was smaller than 0.3 mm, indicating that the total population of level n = 3 is $N_3 > 10^{14}$ cm^{-3} for this relatively low laser energy (for more details see following paper[13]).

A few preliminary experiments were conducted with carbon fibers (Fig. 3) located on the axis of the CO_2 laser beam. In contrast to the Hull University experiment,[2] we used fibers with a much larger diameter. The diameter of the fibers varied from 60 μ to 200 μ and the length from 2 mm to 10 mm. Up to now the most promising result was obtained with a 75 μ x 4 mm fiber, which was completely in view of both the axial and transverse EUV instruments. The measured enhancement of 5.3 of the 182Å line in the axial direction corresponds to a one pass gain $k.\ell \approx 2.8$ (Ref 14). In this preliminary experiment we had not yet used the magnetic field, which seems to be particularly suitable for thinner fibers (20-60 μ) for stabilization purposes. An estimation of the carbon ion density in the plasma column indicates that the core of the fiber remains cold and plays the role of a source of particles as well as a heat sink, providing extra cooling in addition to radiation cooling.

Our earlier[12] works on population inversions in Li-like CIV, OVI, FVIII and NeVIII ions were continued in more recent experiments with time resolution. In Fig. 7 are shown changes in the population inversion of the 4d and 3d levels of NeVIII as a function of initial neon (Ne) pressure.

Neon gas was puffed into the target chamber a few milliseconds before the laser pulse and was limited to a region 5 cm long by mechanical apertures. The initial Ne pressure was controlled by varying the delay of laser pulse with respect to the gas valve opening. The populations of the 4d and 3d levels were obtained from measurements of the intensities of the NeVIII 73.5Å (4d-2p transition) and NeVIII 98.2Å (3d-2p transition) lines in the transverse direction with the EUV duochromator. Column density populations were deduced from the absolute intensity.

Since these lines are close in wavelength corrections for the variation of instrument sensitivity as a function of wavelength is small and was neglected. Hence the relative population of the 4d and 3d levels can be much more accurately determined than the absolute populations, which depend on the reliability of the absolute intensity calibration. It can be seen that the 4d population exceeds the 3d population for higher pressure with the maximum population inversion at a gas pressure p = 12-15 torr. Since the 4d and 4f levels are expected to be closely coupled by collisions this leads to gain on the 4f-3d transition at 292Å. The effect of optical trapping on the observed 73Å / 98Å ratio was checked by measuring the ratio of the intensity of the fine

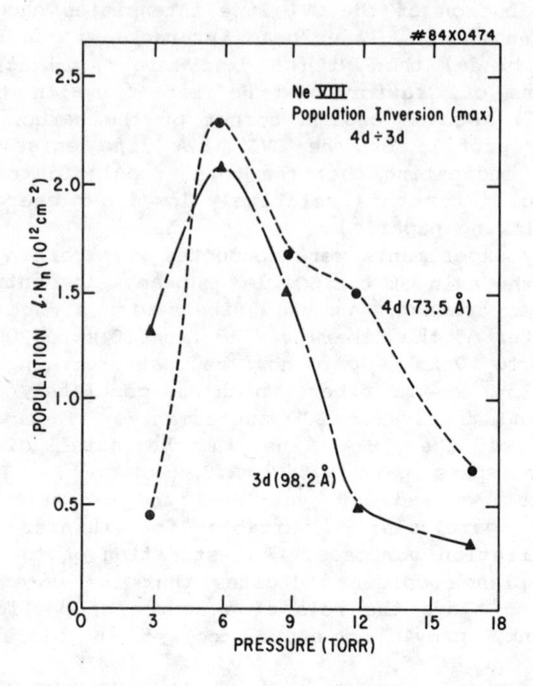

Fig. 7. Population inversion of the 4d and 3d levels of NeVIII as a function of initial neon pressure.

structure components of the 98Å line. This was close to the theoretical value for an optically thin plasma.[14] We also observed the NeVIII 292Å line emission in the axial direction but did not observe a significant increase at high neon pressures, possibly due to absorption in the neutral gas boundary. Further target development should help this problem.

IV. CONCLUSIONS

Enhancement of the CVI 182Å line intensity in the axial direction was measured for C-disc plasmas in a magnetic field B = 90kG, indicating a gain × length product $k.\ell \approx 3.5$ for CO_2 laser energy ~ 500-600 Joules.

A few preliminary experiments with carbon fibers showed promising results for CVI 182Å line: gain × length ≈ 3 for 75 μ × 4 mm fiber. The application of the magnetic field and a thin aluminum coating of the fiber for faster radiation cooling is expected to improve gain.

Population inversions of the 4d and 3d levels of Li-like NeVIII were reported as a function of Ne gas pressure. The line of the potential lasing transition NeVIII 292Å was monitored, however an enhancement in the axial direction was not yet observed. This may be due to absorption of 292Å line in Ne gas on the path between the plasma and the monochromator and/or the population density of the

4f level was too low. New gas target designs should help solve this problem.

ACKNOWLEDGMENTS

The authors are thankful to H. Furth for helpful discussions and encouragement. Productive discussions with J.L. Schwob and his help in operating EUV duochromator is highly appreciated. The authors would like also thank D. DiCicco and V. Vasilotas for technical assistance.

The earlier stage of this work was supported by DOE Basic Energy Sciences, and recently is supported by U.S. Air Force Office of Research, Contract No. AFOSR-84-0025.

REFERENCES

1. G. Jamelot, P. Jaeglé, A. Carillon, A. Bideau, C. Möller, H. Guennou, and A. Sureau, in Proc. Int. Conf. Lasers 81, New Orleans, L.A., Dec. 1981, p. 178.
2. D. Jacoby, G.J. Pert, L.D. Shorrock, and G.J. Tallents, J. Phys. B 15, 3557 (1982).
3. R.C. Elton, J.F. Seely, and R.H. Dixon, in Laser Techniques for Extreme Ultraviolet Spectroscopy, T.J. McIlrath and R.R. Freeman, Eds. (A.I.P., N.Y., 1982).
4. W.T. Silfvast and O.R. Wood, II, ibid, p. 128.
5. D. Matthews, P. Hagelstein, R. Kauffman, R. Lee, C. Wang, et al., Bull. Am. Phys. Soc. 28, 1193 (1983).
6. J. Trebes and M. Krishnan, Phys. Rev. Lett. 50, 629 (1983).
7. P.G. Burkhalter, G. Charatis, and P.D. Rockett, J. Appl. Phys. 54, 6138 (1983).
8. R.G. Caro, J.C. Wong, R.W. Falcone, J.F. Young, and S.E. Harris, Appl. Phys. Lett. 42, 9 (1983).
9. R.C. Elton and R.H. Dixon, Bull. Am. Phys. Soc. 28, 1069 (1983).
10. T.S. Luk, H. Pummer, K. Boyer, M. Shehidi, H. Egger, and C.K. Rhodes, Phys. Rev. Lett. 51, 110 (1983).
11. S. Suckewer and H. Fishman, J. Appl. Phys. 51, 1922 (1980).
12. S. Suckewer, C.H. Skinner, D.R. Voorhees, H.M. Milchberg, C. Keane, and A. Semet, IEEE J. Quant. Electr. QE-19, 1855 (1983).
13. C.H. Skinner, C. Keane, H. Milchberg, S. Suckewer, and D. Voorhees, in Proceedings of the Second Topical Meeting on Laser Techniques in EUV, Boulder, Co., March 1984.
14. H. Milchberg, J.L. Schwob, C.H. Skinner, S. Suckewer, and D. Voorhees, ibid.
15. E. Hinnov and F.W. Hofmann, J. Opt. Soc. Am. 53 1259 (1983).
16. F.E. Irons and N.J. Peacock, J. Phys. E. 6, 857 (1983).
17. E.A. Crawford and A.L. Hoffman, in Laser Interaction and Related Plasma Phenomena, Vol. 6, Eds. H. Hora and G. Miley (Plenum 1984).

COLLISION-FREE MULTIPLE IONIZATION OF ATOMS AND XUV STIMULATED EMISSION IN KRYPTON AT 193 nm

H. Egger, T. S. Luk, W. Müller, H. Pummer, and C. K. Rhodes
Department of Physics, University of Illinois at Chicago
P.O. Box 4348, Chicago, Illinois 60680

ABSTRACT

A description of studies involving collision-free multiple ionization of atoms and the observation of stimulated emission in krypton in the 90 - 100 nm region is given. The experiments, conducted at a wavelength of 193 nm, spanned the intensity range from $\sim 10^{12} - 10^{17}$ W/cm^2. On the basis of the coupling strength observed and the nature of the atomic states excited, a collective response of the atom involving the coherent motion of entire shells is strongly implicated.

I. INTRODUCTION

Studies of collision-free multiple ionization of atoms have shown the presence of very high order nonlinear amplitudes.[1-5] Indeed, the anomalously strong coupling observed[1] for certain elements with irradiation at 193 nm, mainly for heavier atoms, cannot adequately be accounted for by the standard theoretical treatments currently available. An important conclusion that emerged from the studies conducted at ultraviolet wavelengths[1,4] was that the atomic shell structure was the principal property governing the coupling to the radiation field.

The coupling strength and the character of the Z dependence observed both supported an interpretation involving a collective response of the atomic system. Indeed, collective motions of atomic shells,[6] and even groups of shells,[7] have been discussed widely in connection with single-photon photoionization.[8-11] A multiquantum analog of that mechanism is strongly suggested by the experimental results reported herein.

This work provides additional information on two different, albeit related, experimental studies. The first involves the atomic-number dependence of the general process

$$N\gamma + X \rightarrow X^{q+} + qe^- \tag{1}$$

at 193 nm in the intensity range spanning $10^{15} - 10^{17}$ W/cm^2. The earlier studies[1] had been confined to an intensity of $\sim 10^{14}$ W/cm^2.

The second concerns the properties of the stimulated emission observed in krypton in the spectral region spanning 91.6 - 100.3 nm.

II. EXPERIMENTAL APPARATUS

For the studies of multiple ionization, the 193 nm ArF* laser used for irradiation[12] ($\sim$ 5 psec, $\sim$ 3 GW) was focused by an appropriate lens to generate intensities in the range of $10^{15} - 10^{17}$ W/cm^2 in the experimental volume. In order to produce the highest intensities used, an f/2 aspheric focusing element was necessary. The ions are created in a vacuum vessel which is evacuated to a background pressure of $\sim 10^{-8}$ Torr. In contrast to the earlier studies,[1] the ion analyzer

0094-243X/84/1190064-15 $3.00

had a greatly extended time-of-flight drift region which permitted significantly superior mass and charge discrimination.[13] In this case, the isotopic signature of heavy atoms was readily distinguished. This aspect provided a clear identification of the signal and enabled unambiguous separation of the desired ion current from any spurious signals originating from the background gas. Figure 1 illustrates the isotopic characteristic pattern observed for Xe^{5+}. Note the close correspondence of the individual isotopic peaks to the strengths expected on the basis of the isotopic natural abundance. Under typical experimental conditions, the ions formed in the focal region were collected by the analyzer with an extraction field in the range of 50 - 500 V/cm. A microchannel plate located at the exit of the time-of-flight region served as the ion detector.

Studies of krypton were also performed in an experimental configuration designed to explore the possibility of stimulated emission in the extreme ultraviolet. In this case, the experimental arrangement used consisted of a windowless, differentially pumped chamber containing research grade krypton. The density of the krypton in the experimental volume in the vicinity of the pinhole, which separated the high pressure region from the monochromator, was estimated from pressure measurements on the gas plenum. A maximum krypton pressure of $\sim$ 1000 Torr was achievable while maintaining the pressure in the monochromator at a value less than 10^{-4} Torr. The experimental cell was attached directly to the entrance slit of a 1-m VUV monochromator (McPherson 225) with detection provided by an optical multichannel analyzer (OMA PAR) at the exit slit. The ArF laser radiation was focused into the pinhole with the use of a 50, 100, 200, or 300-cm focal length lens, producing intensities approximately in the range of 10^{12} - 10^{14} W/cm^2 in the focal region.

III. EXPERIMENTAL RESULTS AND DISCUSSION

A. Ion Production

The ion production in several elements has been studied for a range in atomic number spanning He (Z = 2) to U (Z = 92). A typical spectrum for xenon is illustrated in Figure 2. In this case, there is the clear presence of eight charge states, the first five of which are seen to have approximately comparable abundances. An overall summary of the species observed is presented in Figure 3. As represented in this figure, the maximum observed energy transfers are on a scale of several hundred electron volts for the heavy materials. As stated in an earlier publication,[1] conventional theoretical techniques are incapable of describing these results. Another clear characteristic of these data is the shell dependence manifested in the behavior of the heavier rare gases. For Ar, Kr, and Xe, the maximum charge states observed would correspond to the complete removal of atomic sub-shells. For these materials they are the 3p, the 4p and both the 5s and 5p shells, respectively. Similarly, if the I^{7+} signal is present under the H_2O^+ peak, then that also implies complete removal of the 5s and 5p shells.

The intensity dependence of these ion spectra has also been examined, and Figure 4 illustrates the nature of this response for xenon.

Fig. 1. ISOTOPIC SPECTRA OF Xe CHARGE STATES OBSERVED FOR Xe^{5+} and Xe^{6+}

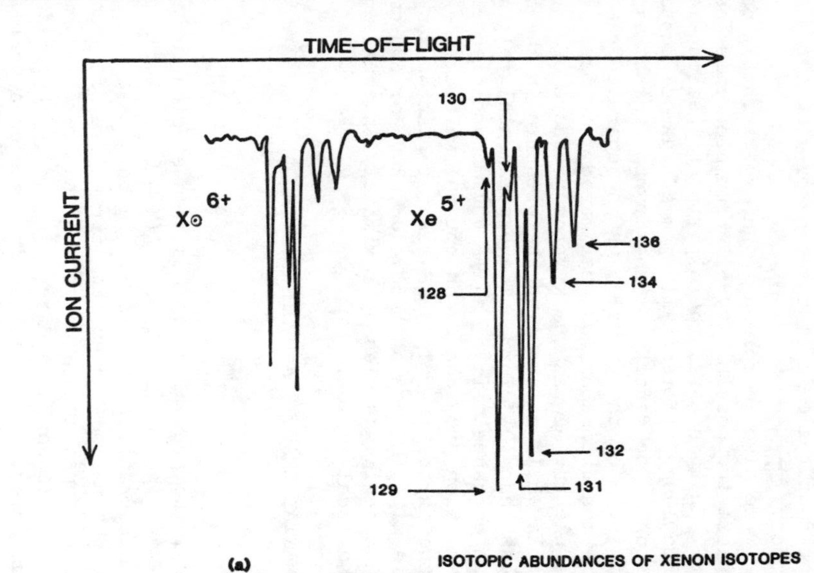

(a) ISOTOPIC ABUNDANCES OF XENON ISOTOPES

ISOTOPIC WEIGHT	128	129	130	131	132	134	136
NATURAL ABUNDANCE	0.07	0.98	0.15	0.8	1.0	0.39	0.33
OBSERVED Xe^{5+}	0.1	1.1	0.2	1.03	1.0	0.46	0.36

Insert (a) shows the comparison of the observed and natural abundances.

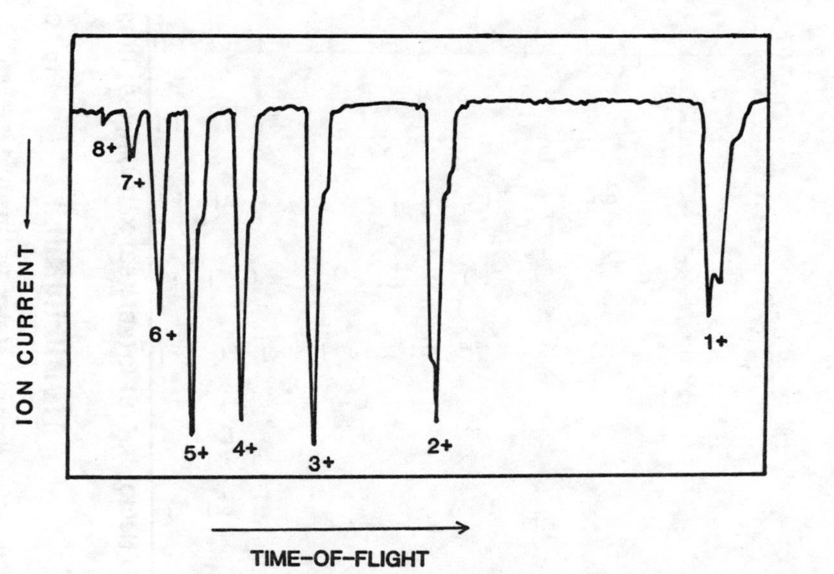

Fig. 2. Time-of-flight ion spectrum of xenon irradiated at $\sim 10^{16}$ W/cm^2 at 193 nm.

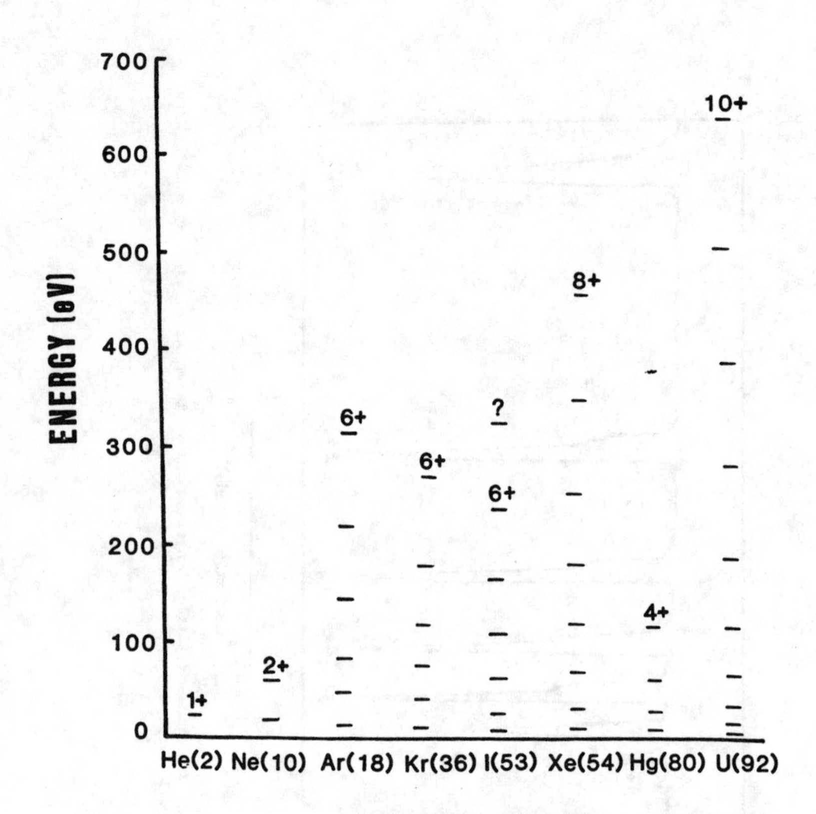

Fig. 3

Data concerning the multiple ionization of atoms for irradiation at 193 nm. Plot of total ionization energies of the observed charge states as a function of atomic number (Z). I^{7+} was not positively identified because it coincides with an H_2O^+ background signal.

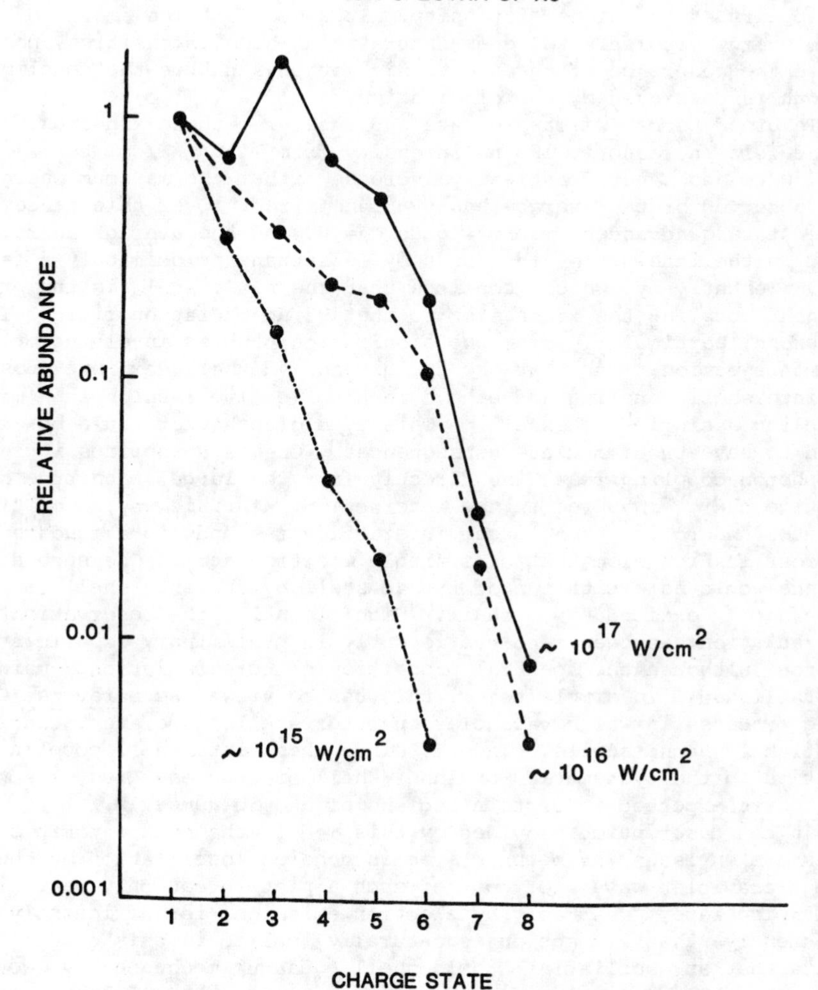

Fig. 4. Relative abundance of charge state distributions observed in the ion spectra of xenon in the intensity range 10^{15} - 10^{17} W/cm^2 at 193 nm.

Over the range of intensities studies ($\sim 10^{14} - 10^{17}$ W/cm^2), higher intensity translates generally into an increased yield of ions of a particular charge, although not necessarily an increase in the maximum charge state observed. For example, the ion Xe^{8+}, with ground state configuration[14] $4d^{10}$, is the greatest charge state detected at $\sim 10^{16}$ W/cm^2, and, although its abundance increases at $\sim 10^{17}$ W/cm^2, no Xe^{9+} appears at the higher intensity. It seems that the amount of 193-nm energy coupled into xenon saturates at high intensities, possibly because the coherence of the electronic motion and thus the coupling strength is destroyed by photoionization.

In broad terms, we can restate and interpret these findings. A one hundred-fold increase in 193 nm intensity from $\sim 10^{15}$ W/cm^2 to $\sim 10^{17}$ W/cm^2 does not drastically increase either the maximum charge state observed or the average energy transferred. Over this range, the charge state q advances from q = 6 to q = 8, and the average energy transferred in the interaction increases by less than approximately a factor of seven. Tentatively, we can conclude that the n = 5 shell is the principal agent coupling the xenon atom to the 193 nm radiation field. It is also known, particularly from photoionization studies involving multiple electron ejection,[15,16] that the 5p, 5s, and 4d shells exhibit substantial intershell coupling and behave in a collective fashion in a manner resembling a single supershell.[7] This type of behavior would be expected to have two immediate consequences. One is an obvious increase in multiphoton coupling resulting directly from the larger magnitude of the effective charge involved in the interaction. In this way, a multielectron atom undergoing a nonlinear interaction responds in a fundamentally different fashion from that of a single electron atom. The second consequence would be excitation of the 4d shell by the intershell coupling to the highly excited n = 5 shell. Significantly, the observation of such radiation has been reported recently in preliminary experiments[17] on xenon, although the spectral properties of this radiation remain to be established. In simple terms, this can be viewed as a <u>reverse</u> Coster-Kronig process[6] in which vacancies in outer shells generate vacancies in more tightly bound shells. Indeed, since these intershell couplings are sensitive to the systematics of inner-shell binding energies, resonance effects are expected in certain regions of atomic number Z.

In the description provided by this model, the outer atomic subshells are envisaged as being driven in coherent oscillation by the intense ultraviolet wave. Of course, such a picture can only be valid if the damping rate, presumably by electron emission, is sufficiently low. Consequently, that assumption is naturally implied in this description. We note that an oscillating atomic shell, quantum mechanically, would be represented by a <u>multiply</u> excited configuration. The simplest examples are doubly excited levels of the type commonly observed in the extreme ultraviolet spectra of the rare gases such as argon.[18] Naturally, higher stages of multiple excitation can be considered such as those discussed in the context of planetary atoms.[19] Therefore, if this type of description is a valid representation of the radiative coupling, then it would follow that multiply excited configurations would be prodigiously generated and, therefore, be prominent features in any excited state populations produced. We shall see below that there is strong evidence supporting this interpretation in the case of krypton.

B. Stimulated Emission in Krypton

In the experiment designed to observe amplification in Kr in the extreme ultraviolet range, intense stimulated emission was detected on five transitions spanning the range from 91.6 - 100.3 nm. An examination of the linewidths and tuning behavior of these transitions led to an identification[20-22] of the upper levels as autoionizing neutral levels involving both singly excited inner-shell excitations and doubly excited configurations. This is the first demonstration of stimulated emission arising from such electronically unstable states.

The particular classes of excited configurations indicated are $4s4p^6n\ell$ and $4s^24p^4n\ell n'\ell'$. Some of these are known experimentally, from both extreme ultraviolet spectroscopic measurements[23-25] and electron scattering studies,[26-28] to lie at $\sim$ 25.7 eV above the neutral krypton ground state $4s^24p^6$. This excitation energy falls exactly in the range equivalent to four 193 nm quanta.

The stimulated spectrum observed at an intensity of $\sim$ 2.8 x 10^{12} W/cm^2 at 193 nm, with a pressure of $\sim$ 300 Torr of krypton, is shown in Figure 5. Five distinct features are seen in the stimulated spectrum. Furthermore, it is apparent from the structure of the 93.1 nm component that it is composed of several overlapping transitions. A distinctive characteristic of these emissions is the lack of correspondence of the observed wavelengths with any transitions classified in the neutral or ionic spectra of krypton. Only the 93.1 nm line can be associated with any previously observed emission[29,30] in krypton which, in that case, is of unclassified origin. Basically, these transitions are making their initial appearance as stimulated emission. From this we conclude that the atomic configurations giving rise to these spectra are of an unusual character not normally produced in abundance with conventional techniques of excitation. States which satisfy this condition and which, moreover, would be expected to be inefficient spontaneous radiators are the autoionizing configurations $4s4p^6n\ell$ and $4s^24p^4n\ell n'\ell'$.

It should be stated that the general nature of the spectrum and the behavior manifested upon tuning the 193 nm radiation cannot be reconciled with the generation of the observed radiation with either parametric or Raman scattering processes. Conversely, the experimental observations can be attributed simply to the mechanism of conventional stimulated emission from an inverted population produced by multiquantum excitation, a process well established for two-quantum excitation in both the infrared[31] and ultraviolet[32] regions.

On the basis of all the data available, both from previous studies[23-30] as well as the current work,[20-22] the inversions are attributed to a four-quantum excitation at 193 nm leading directly to excited even parity states of the neutral atom at an energy $\sim$ 25.7 eV above the ground level by a reaction of the general type

$$4\gamma\ (193\ \text{nm}) + \text{Kr} \xrightarrow{\sigma_4} \text{Kr}^{**} \qquad (2)$$

in which Kr** represents the appropriate autoionizing level. On the basis of known information on states in that energy region and recent computations[33] of term values for selected configurations, the even parity configurations that appear to be involved are $4s4p^64d$, $4s^24p^44d5s$, and $4s^24p^45s6s$. Furthermore, it is expected for certain states that these

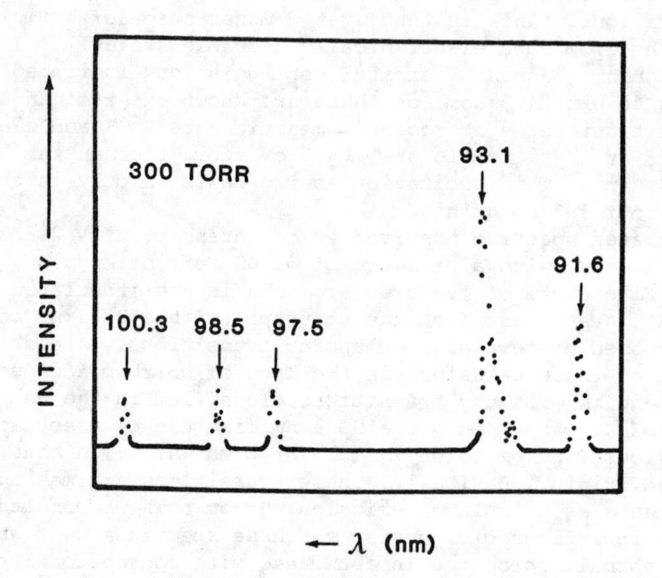

Fig. 5: The stimulated spectrum taken at a 193 nm intensity of $\sim 2.8 \times 10^{12}$ W/cm^2 at a krypton density of 300 Torr. Similar results were also obtained at $\sim$ 50 Torr.

configurations may be appreciably mixed, since it is known[6] that the energy of a single 4s vacancy is quite close to that of a double 4p vacancy.

The calculations[33] also indicate the presence of odd parity $4s^2 4p^4 4d5p$ and 5s5p states in the same region, some within $\sim 100\ cm^{-1}$ of the even parity configurations mentioned above. However, electron collisions could transfer population among the states in the excited manifold and provide a mechanism for population of certain odd parity levels. Electron collisions, particularly involving nearly degenerate states of opposite parity, can represent an efficient mechanism for population transfer. This is well established for the hydrogen atom[34] and has also been recently observed in the hydrogen molecule[35] for the $E,F^1\Sigma_g^+$ and $C^1\Pi_u$ states. If this process is occurring, the most likely case suggested by the configurations of the states tentatively believed to be involved in the stimulated emission is

$$Kr(4s^2 4p^4 4d5s) + e^- \xrightarrow{k_e} Kr(4s^2 4p^4 4d5p) + e^- . \qquad (3)$$

This simply involves a 5s → 5p transition in the configuration which, if viewed as an isolated electron in a hydrogenic orbit, has a cross section that can be estimated with the procedure of Purcell.[34] Since the 5s - 5p matrix element is extremely large,[36] the rate constant has a correspondingly large magnitude with a value $k_e \sim 10^{-4}\ cm^3/sec$. At an electron density $n_e \sim 10^{17}\ cm^{-3}$, the particle rate $n_e k_e \sim 10^{13}\ sec^{-1}$, a value comparable to the autoionization width[25] of these levels. The behavior of the 97.5 nm transition, which vanishes at pressures below ∼ 10 Torr, is consistent with a mechanism of this type.

In connection with the discussion given above, the density of doubly excited states deserves mention. For example, in the energy region of interest, the $4s^2 4p^4 4d5p$ levels have an average density of one state every 30 meV, an energy scale that is close to the level widths arising from autoionization.[25] Therefore, the level density is such that it is approaching the character of a continuum.

It is of interest to compare the spectral properties of the spontaneous emission recorded in separate experiments to that described above for the stimulated spectra. A conspicuous and unusual property of the spontaneous emission, observed under approximately comparable conditions of krypton density and 193 nm intensity, is the abundance of unknown lines. Figure 6 describes a region of the spontaneous spectrum recorded between 96 nm and 100 nm. Similar results are observed over other large spectral regions down to a wavelength of ∼ 47 nm. Approximately one half of the total radiated power appears in the unknown transitions. Given the density of states available, it seems quite probable that the principal origins of these emissions are the _multiply_ excited states. That class of levels apparently enjoys very favorable circumstances for excitation.

For emphasis, we restate these findings. In terms of radiated power, the entire stimulated spectrum and approximately one half of the spontaneous spectrum consist of unclassified emissions, and these are attributed principally to multiply excited or core excited configurations. The conventional transitions, either of the neutral or its ions, simply do not dominate the spectrum.

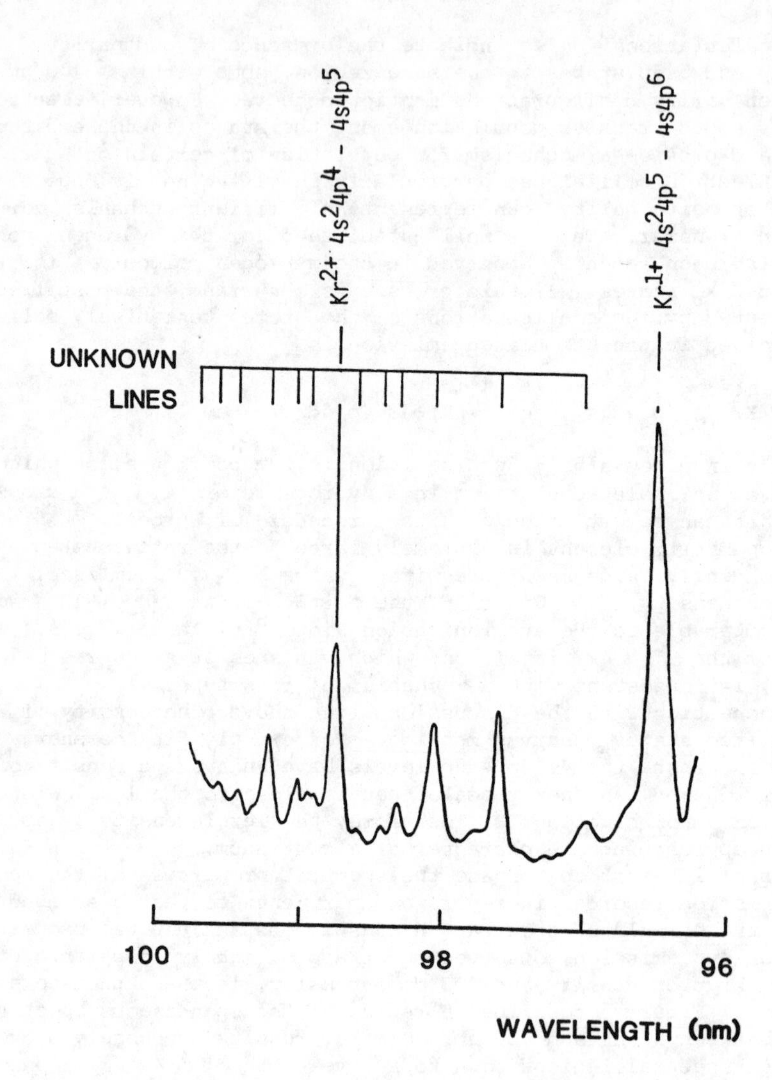

Fig. 6. Spontaneous emission spectrum of krypton observed in the 96 - 100 nm range. The emission was detected in the direction transverse to the irradiating 193 nm beam. An abundance of unknown transitions is seen in addition to the well established Kr^{+} and Kr^{2+} lines.

In connection with these facts concerning the observed emission properties and in light of the ion production studies described in Section III.A, it is significant to estimate the cross section σ_4 for process (2). An approximate value for the magnitude of the effective four-photon cross section σ_4 at the experimental intensity of $I \sim 2.8 \times 10^{12}$ W/cm^2 can be estimated from a previously studied two-quantum amplitude[37] by incorporating factors to account for the change in linewidth and the additional matrix elements involved in the fourth order process. If we take the state sequence $4s^24p^6 \rightarrow 4s^24p^55s \rightarrow 4s^24p^56p \rightarrow 4s4p^66p \rightarrow 4s4p^66d$ to represent the amplitude for the excitation process (2), we can write

$$\sigma_4 \sim (\alpha_2 I)\left(\frac{\Gamma_1}{\Gamma_2}\right)\left(\frac{\mu_\alpha E}{\hbar\Delta\omega_\alpha}\right)^2\left(\frac{\mu_\beta E}{\hbar\Delta\omega_\beta}\right)^2 \quad (4)$$

in which $\alpha_2 \sim 2.3 \times 10^{-31}$ cm^4/W is the previously established[37] two-quantum parameter for the $4p^6 \rightarrow 4p^56p$ process, $\Gamma_2 \sim 10^2$ cm^{-1} conforming to the linewidth arising from autoionization, and $\Gamma_1 \simeq 25$ cm^{-1} as used in the earlier[37] determination of α_2. The quantities μ_α, μ_β, $\Delta\omega_\alpha$, and $\Delta\omega_\beta$ denote the additional dipole moments and frequencies of detuning that appear in the fourth order amplitude. With $\mu_\alpha \simeq \mu_\beta \simeq 1$ debye, $\Delta\omega_\alpha \simeq 10^3$ cm^{-1}, and $\Delta\omega_\beta \simeq 6 \times 10^4$ cm^{-1} taken as approximately suitable values, it is found that $\sigma_4 \sim 7.7 \times 10^{-24}$ cm^2 with account taken for the broadening and shift of the near resonant $4s^24p^56p[\frac{3}{2}]$ $J = 2$ intermediate state.[37]

For given experimental conditions involving a krypton density of ~ 50 Torr and an autoionizing width of ~ 100 cm^{-1}, an excited state density of $\sim 3.3 \times 10^{12}$ cm^{-3} was obtained from this calculation. With a stimulated emission cross section of $\sim 5 \times 10^{-15}$ cm^2, estimated from a $\sim 10^{-10}$ sec radiative lifetime[38] and a 3.6 cm gain length, the total gain $G \simeq 6 \times 10^{-2}$ is obtained, a value more than one hundred fold too low to observe the stimulated output. This result suggests that the coupling estimated in this case is significantly underestimated by the conventional perturbation analysis. This, of course, is the same conclusion that was reached in the ion studies discussed above. Finally, if we reason that the anomalous increase in this coupling is connected with the presence of multiply excited configurations and if we recall that coherently excited shells, in quantum mechanical language, are described in terms of multiple excitations,[39,40] then the results of the ion production experiments and the observation of stimulated emission in krypton can be viewed in a unified manner.

IV. CONCLUSIONS

The study of multiply ionization of atoms at 193 nm reveals the presence of an anomalously strong coupling for heavy materials. From the nature of the measured behavior, a collective response of atomic shells is strongly implicated. Moreover, this type of motion is expected to lead to copious production of multiply excited configurations and, by intershell coupling, to the excitation of inner shell states. In very approximate terms, the latter can be considered as a reverse

Coster-Kronig process. Significantly, the spectrum of stimulated emission observed from krypton in the 90 - 100 nm region appears to involve exclusively doubly excited $4s^2 4p^4 n\ell n'\ell'$ and core excited $4s4p^6 n\ell$ autoionizing states. In addition, the strength of the coupling necessary to generate the stimulated signal is considerably above that estimated from a standard perturbation analysis. From these results, it appears that the basic characteristics of the radiative coupling leading to both the ion production and the stimulated emission, albeit observed at considerably different intensities, are essentially the same.

V. ACKNOWLEDGMENTS

This work was supported by the Air Force Office of Scientific Research under contract number F49630-83-K-0014, the Los Alamos National Laboratory under contract number 9-X53-C6096-1, the Department of Energy under grant number De AS08-81DP40142, the Office of Naval Research, the National Science Foundation under grant number PHY81-16626, the Defense Advanced Research Projects Agency, and the Avionics Laboratory, Air Force Wright Aeronautical Laboratories, Wright Patterson Air Force Base, Ohio.

REFERENCES

1. T. S. Luk, H. Pummer, K. Boyer, M. Shahidi, and C. K. Rhodes, Phys. Rev. Lett. 51, 110 (1983).

2. A. L'Huillier, L. A. Lompré, G. Mainfray, and C. Manus, Phys. Rev. Lett. 48, 1814 (1982).

3. A. L'Huillier, L. A. Lompré. G. Mainfray, and C. Manus, Phys. Rev. A27, 2503 (1983).

4. T. S. Luk, H. Pummer, K. Boyer, M. Shahidi, H. Egger, and C. K. Rhodes, in Excimer Lasers - 1983, edited by C. K. Rhodes, H. Egger, and H. Pummer, AIP Conference Proceedings No. 100 (AIP, New York, 1983) p. 341.

5. A. L'Huillier, L. A. Lompré, G. Mainfray, C. Manus, J. Phys. B16, 1363 (1983).

6. G. Wendin, Breakdown of the One-Electron Pictures in Photoelectron Spectra, Structure and Bonding Vol. 45 (Springer-Verlag, Berlin, 1981).

7. M. Ya. Amusia and N. A. Cherepkov in Case Studies in Atomic Physics 5, edited by E. W. McDaniel and M. R. McDowell (North-Holland, Amsterdam, 1975) p. 47.

8. A. Zangwill and P. Soven, Phys. Rev. A21, 1561 (1980).

9. A. Zangwill and P. Soven, Phys. Rev. Lett. 45, 204 (1980).

10. W. Ekardt and D. B. Tran Thoai, Phys. Scr. 26, 194 (1982).

11. M. Ya. Amusia in Vacuum Ultraviolet Radiation Physics, edited by E. E. Koch, R. Haensel, and C. Kunz (Pergamon Vieweg, Braunschweig, 1974) p. 205.

12. H. Egger, T. S. Luk, K. Boyer, D. R. Muller, H. Pummer, T. Srinivasan, and C. K. Rhodes, Appl. Phys. Lett. 41, 1032 (1982).

13. W. C. Wiley and I. H. McLaren, Rev. Sci. Instr. 26, 1150 (1955).

14. J. Blackburn, P. K. Carroll, J. Costello, and G. O'Sullivan, J. Opt. Soc. Amer. 73, 1325 (1983).

15. M. J. Van der Wiel and T. N. Chang, J. Phys. B11, L125 (1978).

16. M. Ya. Amusia in Advances in Atomic and Molecular Physics, Vol. 17, edited by D. R. Bates and B. Bederson (Academic Press, New York, 1981) p. 1.

17. T. S. Luk and K. Boyer, private communication.

18. R. P. Madden, D. L. Ederer, and K. Codling, Phys. Rev. 177, 136 (1969).

19. I. C. Percival, Proc. Roy. Soc. (London) A353, 289 (1977).

20. T. Srinivasan, H. Egger, T. S. Luk, H. Pummer, and C. K. Rhodes in Laser Spectroscopy VI, Proc. 6th Intl. Conf., Interlaken, Switzerland, edited by H. P. Weber and W. Lüthy (Springer-Verlag, Berlin, 1983) p. 385.

21. T. Srinivasan, H. Egger, T. S. Luk, H. Pummer, and C. K. Rhodes, "Observation of Stimulated Emission Below 100 nm in Krypton Arising from Selective Excitation of Autoionizing States," to be published.

22. K. Boyer, H. Egger, T. S. Luk, H. Pummer, and C. K. Rhodes, "Interaction of Atomic and Molecular Systems with High Intensity Ultraviolet Radiation," J. Opt. Soc. Amer., to be published.

23. K. Codling and R. P. Madden, J. Res. Natl. Bur. Std. 76A, 1 (1972).

24. K. Codling and R. P. Madden, Phys. Rev. A4, 2261 (1971).

25. D. L. Ederer, Phys. Rev. A4, 2263 (1971).

26. M. Boulay and P. Marchand, Can. J. Phys. 60, 855 (1982).

27. J. A. Baxter, P. Mitchell, and J. Comer, J. Phys. B15, 1105 (1982).

28. S. J. Buckman, P. Hammond, G. C. King, and F. H. Read, J. Phys. B16, 4219 (1983).

29. C. E. Moore, Atomic Energy Levels, Vol. II, NSRDS-NBS (U. S. Government Printing Office, Washington, 1971).

30. R. L. Kelly and L. J. Palumbo, Atomic and Ionic Emission Lines Below 2000 Angstroms, NRL Report 7599 (U. S. Government Printing Office, Washington, 1973).

31. H. Pummer, W. K. Bischel, and C. K. Rhodes, J. Appl. Phys. 49, 976 (1978).

32. H. Pummer, H. Egger, T. S. Luk, T. Srinivasan, and C. K. Rhodes, Phys. Rev. A28, 795 (1983).

33. R. L. Carman, private communication.

34. E. M. Purcell, Astrophys. J. 116, 457 (1952).

35. T. Srinivasan, H. Egger, T. S. Luk, H. Pummer, and C. K. Rhodes, J. Quantum Electron. QE-19, 1874 (1983).

36. H. A. Bethe and E. E. Salpeter, Quantum Mechanics of One- and Two-Electron Atoms (Springer-Verlag, Berlin, 1957).

37. J. Bokor, J. Zavelovich, and C. K. Rhodes, Phys. Rev. A21, 1453 (1980).

38. R. L. Carman, private communication.

39. M. E. Kellman and D. R. Herrick, Phys. Rev. A22, 1536 (1980); D. R. Herrick, Adv. Chem. Phys. 52, 1 (1982).

40. A. R. P. Rau, J. Phys. B16, L699 (1983).

MULTIPLY CHARGED IONS PRODUCED BY MULTIPHOTON ABSORPTION IN RARE GAS ATOMS

A. L'Huillier, L-A. Lompré, G. Mainfray and C. Manus

Centre d'Etudes Nucléaires de Saclay
Service de Physique des Atomes et des Surfaces
91191 Gif-sur-Yvette Cedex, France

ABSTRACT

This paper describes the production of multiply charged ions induced by collisionless multiphoton ionization of rare gases at 532 and 1064 nm. The production of multiply charged ions emphasizes atomic properties and laser characteristics. Multiphoton absorption is expected to selectively excite high ionic energy states, which could induce a stimulated emission in the extreme ultraviolet.

INTRODUCTION

Collisionless multiphoton ionization of atoms has been the subject of a large number of theoretical and experimental works in the past ten years [1-4]. Such processes have been described by the lowest order perturbation theory. In this picture, an atom is ionized by the absorption of the minimum number of photons to reach the first ionization threshold, and only one electron is assumed to be involved in the ionization process. Alkaline atoms and of course atomic hydrogen are the examples that best satisfy this condition. The agreement between theory and experimental results was excellent for resonance effects, coherence effects and multiphoton ionization cross sections in absolute values.

Multiphoton ionization of atoms which have several electrons in the outer shell is expected to occur although the basic interaction processes involved are foreseen to be considerably more complicated, and thus more difficult to understand and explain. Recent experiments emphasized the production of multiply charged ions in rare gas atoms [5-9]. The purpose of the present paper is to indicate broadly the different parameters which govern the production of multiply charged ions produced by multiphoton absorption. Finally, we will consider the developments that can be foreseen in the extreme ultraviolet radiation generated from excited multiply charged ions.

EXPERIMENTAL RESULTS

A mode-locked Nd-YAG laser is used to produce a 50-psec pulse which is amplified up to 5 GW at 1064 nm. The second harmonic can be generated at 532 nm up to 1.5 GW when needed. The laser pulse is focused into a vacuum chamber by an aspheric lens corrected for spherical aberrations. The vacuum chamber is pumped to 10^{-8} Torr and then filled with a spectroscopically pure rare gas at a static

0094-243X/84/1190079-10 $3.00

pressure of 5×10^{-5} Torr. At this pressure, no collisional ionization occurs, and no complications from charge exchange reactions are expected. Only collisionless multiphoton ionization occurs. The ions resulting from the laser interaction with the atoms in the focal volume are extracted with a tranverse electric field of 1 $kV.cm^{-1}$, separated by a 20 cm length time-of-flight spectrometer, and then detected in an electron multiplier. The laser intensity is adjusted in order to produce 1 to 10^5 ions. The experiment consists of the measurement of the number of ions corresponding to different charges formed as a function of the laser intensity.

Fig. 1 is a typical result of the multiphoton ionization of Xe at 532 nm. Up to Xe^{5+} ions are formed. Let us analyse the different processes which occur when the laser intensity I is increased. Fig. 1 can be divided into two parts. The first part ($I < 1.5 \times 10^{12}$ $W.cm^{-2}$) is characterized by a laser-neutral atoms interaction, while in the second part ($I > 1.5 \times 10^{12}$ $W.cm^{-2}$) a laser-ions interaction occurs. In the first part, the absorption of 6 photons by an atom leads to the removal of one electron and the formation of a Xe^+ ion. This process appears in Fig. 1 through experimental points joined by a straight line with a slope 6 because the 6-photon ionization rate varies as I^6. When the laser intensity is increased further, approaching the I_S value, the absorption of 15 photons by an atom induces the simultaneous removal of two electrons and the production of a Xe^{2+} ion. This process appears in Fig. 1 through experimental points joined by a straight line with a slope 15. The 6-photon and 15-photon ionization processes deplete the number of atoms contained in the interaction volume. A marked change appears in the slope of the curves for both Xe^+ and Xe^{2+} ions beyond the laser intensity I_S. This saturation is a typical effect which occurs in multiphoton ionization experiments when all the atoms in the interaction volume are ionized. The intensity dependence of both curves of Xe^+ and Xe^{2+} ions just beyond I_S arises from ions produced in the expanding interaction volume when the laser intensity is increased further.

The second part of Fig. 1, for $I > 1.5 \times 10^{12}$ $W.cm^{-2}$, describes the interaction of the laser radiation with ions, because the interaction volume is filled up with Xe^+ ions in the place of atoms. A sudden increase in the number of Xe^{2+} ions occurs when the laser intensity is increased further. This comes from the absorption of 10 photons by a Xe^+ ion. This removes one electron from the Xe^+ ion and produces a Xe^{2+} ion. This appears in Fig. 1 through experimental points joined by a straight line with a slope 10. When the laser intensity is increased further, the 10-photon ionization of Xe^+ ions also saturates and Xe^{3+}, Xe^{4+} and Xe^{5+} ions are formed most likely through stepwise processes. This means Xe^{3+} ions are produced from Xe^{2+} ions by absorbing 14-photons. In the same way Xe^{4+} ions are produced from Xe^{3+} ions by absorbing 20 photons, and likewise for Xe^{5+} ions. To sum up, Fig. 1 is a clear picture of the response of the electrons of Xe atoms to a high laser intensity. This response is highly non linear. Each step of increased intensity gives rise to the removal of an additional electron.

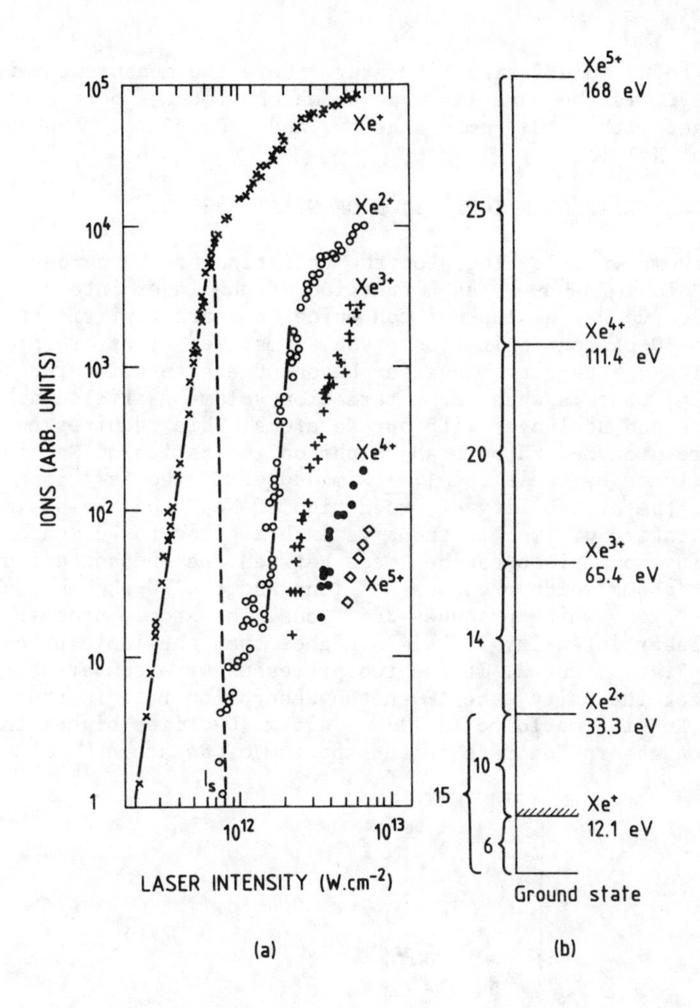

Fig. 1 - (a) A log-log plot of the variation in the number of Xenon ions formed at 532 nm as a function of the laser intensity. (b) Schematic representation of the number of photons involved in the production of multiply charged ions.

It should be noted that the 10-photon ionization of Xe^+ ions which produces Xe^{2+} ions can be investigated in a well defined laser intensity range. The corresponding 10-photon ionization cross section is measured to be $10^{-297 \pm 1}$. The uncertainty in this measurement comes mainly from the uncertainty of the laser intensity in absolute values. It should be recalled that a generalized N-photon absorption cross section is expressed in $cm^{2N}\ s^{N-1}$ units when the laser intensity is expressed in number of photons $cm^{-2}\ s^{-1}$. It is initially amazing to find that a multiphoton ionization cross section of a singly charged ion can be measured so easily while very few examples of measurements of photoionization cross section of ions are found in existing literature. Here the measurement is easy because it is the same laser pulse which produces both a clean Xe^+ ion target with a 10^{12} cm^{-3} density, and induces the 10-photon ionization of Xe^+ ions.

MULTIPHOTON IONIZATION OF Ne AT 532 NM

Fig. 2 shows in a log-log plot the variation of the number of Ne^+ and Ne^{2+} ions produced as a function of the laser intensity. The Ne^+ ion curve has a slope of ten which is characteristic of a non-resonant 10-photon ionization of Ne atoms. Ne^{2+} ions are produced in a laser intensity range far beyond the saturation intensity value I_S, that is when the interaction volume is filled up with Ne^+ ions and no longer with any Ne atoms. This requires that Ne^{2+} ions are produced through an 18-photon ionization of Ne^+ ions. This is confirmed by the slope 17 ± 2 measured on the Ne^{2+} ion curve. Here, the probability of production of Ne^{2+} ions by a simultaneous excitation of two electrons is much two low to be measured. An interesting comparison can be made between the 10-photon ionization of Ne atoms which produces Ne^+ ions and the 10-photon ionization of Xe^+ ions which produces Xe^{2+} ions. The atomic process requires a laser intensity 3.7 times higher than the ionic process, as shown in Figs. 1 and 2. If the two processes were compared at the same laser intensity, the 10-photon absorption rate in the spectrum of Xe^+ ion would be $(3.7)^{10} \simeq 4.5 \times 10^5$ times higher than the 10-photon absorption rate in the spectum of Ne atoms.

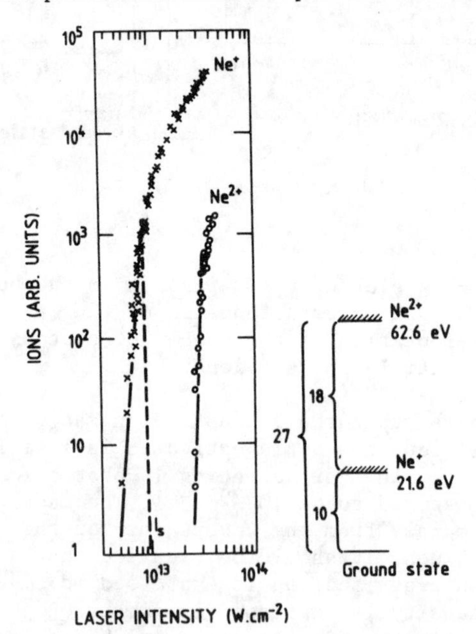

Fig. 2 - The laser intensity dependence of Ne^+ and Ne^{2+} ions formed at 532 nm.

MULTIPLY CHARGED IONS PRODUCED IN RARE GASES AT 1064 NM

The production of multiply charged ions has also been investigated in the five rare gases at 1064 nm. Let us consider here the two most different examples : Xe and He. Fig. 3 shows the variation of the number of Xe^+, Xe^{2+}, Xe^{3+}, Xe^{4+} and Xe^{5+} ions as a function of the laser intensity. The general behavior is similar to that observed at 532 nm, except for two points. First, the two different processes of production of Xe^{2+} ions, namely the simultaneous two-electron removal from Xe atoms, and the one-electron removal from Xe^+ ions, are not so well separated than at 532 nm. Second, the probability of creating Xe^{2+} ions through a simultaneous two-electron removal from Xe atoms is 30 times larger here than at 532 nm, at the reference intensity I_S. At saturation intensity I_S = 1.2×10^{13} W.cm^{-2} at 1064 nm the proportion of Xe^{2+} to Xe^+ ions is 1.5×10^{-2}, whereas it is only 5×10^{-4} at $I_S = 8 \times 10^{11}$ W.cm^{-2} at 532 nm.

Fig. 4 shows the variation of the number of He^+ and He^{2+} ions produced as a function of the laser intensity. 68 photons at least have to be absorbed by He to produce He^{2+} ions which most probably come from a simultaneous excitation of the two electrons. This conclusion is supported by the fact that saturation of both He^+ and He^{2+} ions occurs at the same laser intensity I.

DISCUSSION

A theoretical one-electron model has been used successfully to describe one electron removal in multiphoton ionization of atomic hydrogen and alkaline atoms in the past few years. However, such a model cannot be applied by merely extrapolating the lowest order perturbation theory to explain the production of multiply charged ions induced in many-electron atoms. In this respect, the following example of the production of doubly charged ions is very convincing. The one-electron removal in Xe atoms through 11-photon absorption at 1064 nm requires a laser intensity of 10^{13} W.cm^{-2}. The 29-photon absorption corresponding to the production of Xe^{2+} ions at 1064 nm would require a laser intensity of 10^{15} W.cm^{-2}, a value anticipated from the lowest order perturbation theory in the one-electron model. This is at variance with experimental results (Fig. 3) which show that a laser intensity of 1.5×10^{13} W.cm^{-2} is enough to produce Xe^{2+} ions. Fig. 3 also shows in other terms that the 29-photon absorption rate giving Xe^{2+} ions is only 100 times less than the 11-photon absorption rate giving Xe^+ ions at 1.5×10^{13} W.cm^{-2}.

The production of multiply charged ions through multiphoton absorption emphasizes both atomic properties and laser characteristics such as intensity, photon energy, pulse duration, etc... Consequently, the easy production of multiply charged ions could be explained either in terms of specific properties of many-electron atoms, or in terms of different mechanisms of absorption of photons in a continuum or in the spectrum of ions. Let us consider successively the two independent approaches.

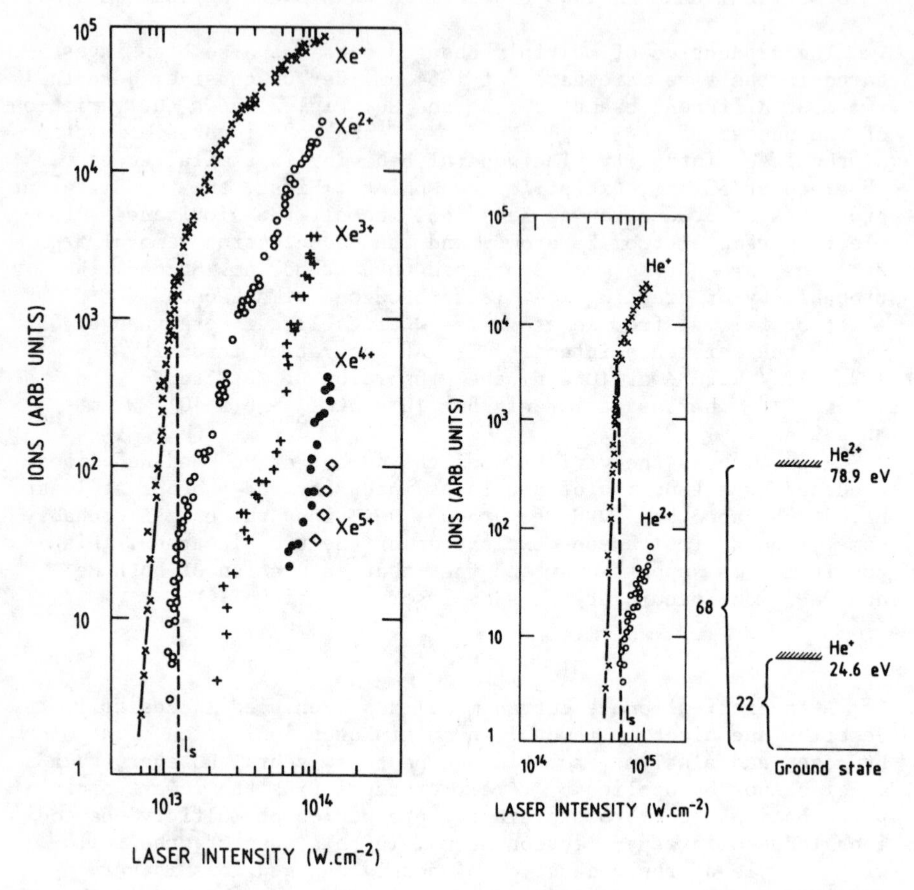

Fig. 3 - The laser intensity dependence of Xenon ions formed at 1064 nm.

Fig. 4 - The laser intensity dependence of He^{+} and He^{2+} ions formed at 1064 nm.

MULTIPHOTON ABSORPTION AND LASER CHARACTERISTICS

One of the most important questions that faces us is related to the mechanism of absorption of a very large number of photons, for example 68 photons at 1064 nm to produce He^{2+} ions. This problem could be somewhat similar to the multiphoton excitation or dissociation of large polyatomic molecules such as SF_6 by using CO_2 laser radiation. In such molecules, it is well known that 30 to 40 photons can be easily absorbed in the dense multitude of vibrational states. The quasicontinuum model postulates that due to coupling of the various normal modes of a large polyatomic molecule, there will always be quasi-resonant states available at each step of the excitation process. In the same way, the very large N-photon absorption rate giving rise to doubly charged ions at a high laser intensity could be explained in terms of a stepwise incoherent absorption process. At high intensity, one could describe the absorption in the continuum by the use of rate equations.

Some salient features can be drawn from experimental results. In a simplified way (Fig. 5), we have plotted the ionization threshold intensity required to produce singly and doubly charged ions in the five rare gases at 1064 nm, as a function of the number N of photons absorbed. By ionization threshold intensity, we mean the laser intensity which induces the production of a single ion and which is derived from original figures. Curve 5a is related to a one-electron removal from a neutral atom. Curve 5b is related to a one-electron removal from a singly charged ion, and curve 5c to a simultaneous two-electron removal from an atom. As a first observation, let us consider a fixed laser intensity, 1.8×10^{13} W.cm^{-2} for example, and let us move along a horizontal line. At this intensity, one electron can be removed from Ar through a 14-photon absorption, one electron can be removed from Xe^{+} ion through a 19-photon absorption, and two electrons can be simultaneously removed from a Kr atom through a 34-photon absorption. In the same way, considerations based on a vertical line would show that the absorption rate of a large number of photons is higher in the discrete spectrum of an ion than in the discrete spectrum of an atom, and is still higher when a simultaneous two-electron removal of an atom occurs.

Fig. 5 also illustrates laser intensity effects through the three lines a, b, and c which diverge from each other as N increases or as the laser intensity increases. Let us give an example drawn from Fig. 5 by a direct comparison of experimental data shown in Fig. 1 and 4. In a moderate intensity of 10^{12} W.cm^{-2} let us consider the one-electron removal from an Xe atom through a 6-photon absorption at 532 nm. The laser intensity has to be increased by a factor of 4 at the ion threshold detection to induce the simultaneous removal of two electrons from the Xe atom through a 15-photon absorption. In contrast, at a very high intensity of 6×10^{14} W.cm^{-2}, let us consider the one-electron removal from a He atom through a 22-photon absorption at 1064 nm. Increasing the laser intensity by 40 %, at the ion threshold detection, is enough to induce the simultaneous removal of the two electrons of He atom through a 68-photon absorption.

The photon energy seems to play an important role in the production of doubly charged ions through the simultaneous excitation of two electrons. For example, the probability of production of Xe^{2+} ions through a simultaneous two-electron removal from Xe atoms is 30 times less at 532 nm for 10^{12} W.cm^{-2} than at 1064 nm for 10^{13} W.cm^{-2}, as shown by a comparison of Figs. 1 and 3. This result can most likely be explained rather in terms of laser wavelength than in terms of laser intensity, as exemplified by Fig. 2. This figure shows that at 532 nm and at high laser intensity (10^{13} W.cm^{-2}) no Ne^{2+} ions are produced by a simultaneous excitation of two electrons.

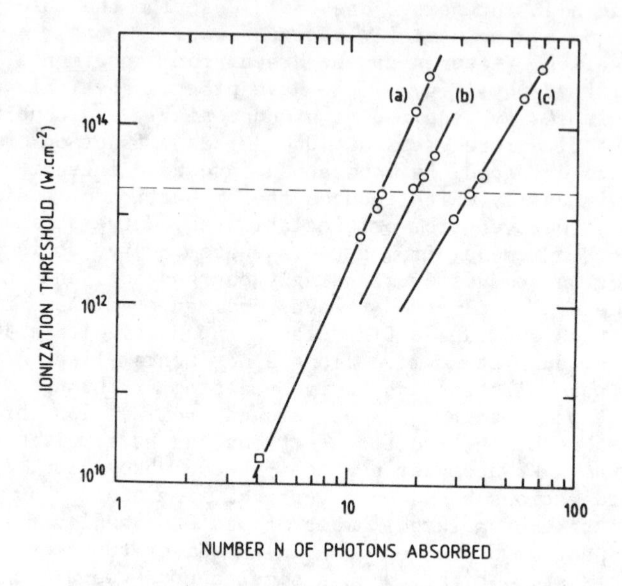

Fig. 5 - The ionization threshold intensity which induces the formation of one ion as a function of the number N of photons absorbed at 1064 nm. Curve (a) relates to a one-electron removal from an atom. (b) relates to a one-electron removal from a singly charged ion, and (c) to a simultaneous two-electron removal from an atom.

Such a wavelength dependence in the simultaneous excitation of two electrons looks like the well known wavelength dependence in the photoionization cross section of Rydberg atoms. In contrast, the sequential production of doubly charged ions through singly charged ions does not seem to depend so strongly on the photon energy.

ATOMIC PROPERTIES

In multiphoton ionization of a many-electron atom, one can no longer consider the interaction of the laser radiation with a single electron, as one could for a one-electron atom. Firstly two electrons can be simultaneously excited and removed. Experimental results have shown that such a process is very sensitive to the laser wavelength, and is especially important when a long wavelength laser radiation is used. Secondly, electron correlation effects can lead to a collective response of the outer shell irradiated by an intense laser pulse. In a closed shell, several electrons could be excited while the first or two first electrons are removed from the shell. This could explain why an additional electron can be removed by increasing further the laser intensity by a small amount.

Fig. 6 shows the variation of the maximum ion charge states produced at 1064 nm and 193 nm in the five rare gases, as a function of the atomic number Z. The 193 nm data used here are derived from a recent work by Luk et al.[9]. Fig. 6 shows that He^{2+} and Ne^{2+} ions are produced at 1064 nm and not at 193 nm. This result could be explained by considering that He^{2+} and Ne^{2+} ions are produced at 1064 nm through a simultaneous excitation of two electrons. It was shown previously that such a process is very sensitive to the laser photon energy and could have probability too low at 193 nm to be observed. Furthermore, a large amount of energy can be transmitted to a many-electron atom through multiphoton absorption processes. For example, 250 eV when Xe^{6+} ions are produced.

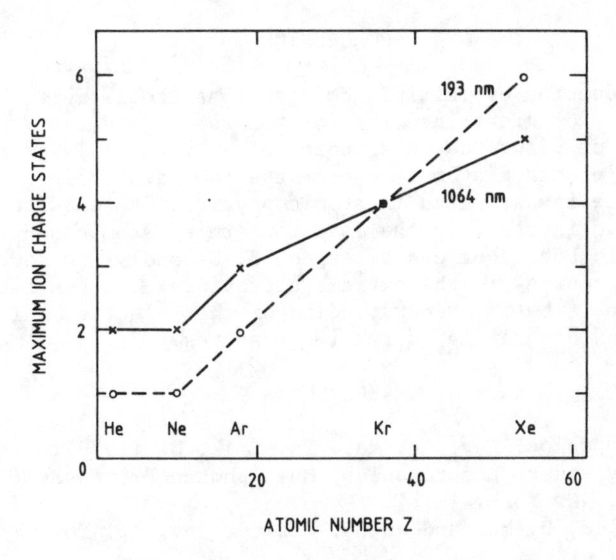

Fig. 6 - The maximum ion charge states produced in the five rare gases at 193 and 1064 nm as a function of the atomic number Z.

As is well known, inner shells have to be considered, especially in Kr and Xe, in multiple photoionization by using Synchrotron Radiation at about 100 eV. Here, it is not possible to draw any conclusion on the possible contributions of an inner shell in the production of multiply charged ions at 1064 nm, or 532 nm. It could be assumed that one goes gradually from a situation close to multiple photoionization when high energy photons (5-10 eV) are used to produce multiply charged ions with consequently few photons absorbed at moderate laser intensity, to a very different situation when low energy photons (1 eV) are used with a very large number of photons absorbed by atoms at very high laser intensity. Generally speaking, we have to consider how inner shell contribution varies in the response of an atom which absorbs one 100 eV photon or 100 photons with 1 eV each.

To conclude this discussion, it should be pointed out that resonance effects in multiphoton ionization rates of one-electron atoms have been investigated in detail over the past few years [10]. Their resonances are due to the selective multiphoton excitation of an atomic state. An obvious question arises here. Is it possible, in the same way, to observe selective multiphoton excitation of high lying energy ionic states ? If yes, is it possible to get a large population in a specific state inducing a population inversion and a stimulated emission in the extreme ultraviolet with a significant gain ? Measurements of the electron energy spectrum and of possible extreme ultraviolet radiation will be performed in the production of multiply charged ions in the near future to answer the afore-mentioned question.

CONCLUSION

The production of multiply charged ions though multiphoton absorption is of great interest for two reasons. Firstly, from a basic point of view, the interaction of an intense laser radiation with a many-electron atom is one of the few main topics which remains to be investigated in atomic physics. It is quite an open field, and an increase in the data concerning atomic and ionic spectroscopic behaviour can be expected. Secondly the development of coherent sources in the extreme ultraviolet can benefit significantly from a better understanding of the selective multiphoton excitation of high lying energy ionic states.

REFERENCES

1. P. Lambropoulos, Adv. At. Mol. Phys. 12, 87 (1976).
2. J. Eberly and P. Lambropoulos, Multiphoton Processes (John Wiley and Sons, New York, 1978).
3. J. Morellec, D. Normand and G. Petite, Adv. At. Mol. Phys. 18, 97 (1982).
4. G. Mainfray, J. Physique 43, C2-367 (1982).
5. A. L'Huillier, L-A. Lompré, G. Mainfray and C. Manus, Phys. Rev. Lett. 48, 1814 (1982).
6. A. L'Huillier, L-A. Lompré, G. Mainfray and C. Manus, J. Phys. B 16, 1363 (1983).
7. A. L'Huillier, L-A. Lompré, G. Mainfray and C. Manus, Phys. Rev. A27, 2503 (1983).
8. A. L'Huillier, L-A. Lompré, G. Mainfray and C. Manus, J. Physique 44, 1247 (1983).
9. T. Luk, H. Pummer, K. Boyer, M. Shahidi, H. Egger and C.K. Rhodes, Phys. Rev. Lett. 51, 110 (1983).
10. G. Mainfray and C. Manus, Appl. Optics 19, 3934 (1980).

Laser Enhanced Auto-Ionization and Dressed Resonances*

J.H. Eberly and D. Agassi

Department of Physics and Astronomy
University of Rochester
Rochester, New York 14627

Abstract

We describe the strong field dressing of atomic resonance states, and apply the dressed state description to the photo-excitation of an auto-ionizing resonance, including the effects of radiative and purely transverse line broadenings.

Summary

We discuss the photo-electron spectrum and the depletion of the initial state predicted by a simple model for laser-enhanced auto-ionization. Our model consists of an initially populated bound state, one auto-ionizing state, one electronic continuum and a continuous-wave laser field of arbitrary strength. Also included are phase jitter relaxation, as well as spontaneous decay from the continuum (and the auto-ionizing state) back to the ground state (coherent recycling). The model is exactly soluble, thus offering an opportunity to assess strong field effects and regimes of very rapid relaxation .

The analysis of the results is particularly simple and transparent in terms of "dressed resonances". These are a natural generalization of the usual dressed states appropriate for stimulated discrete-discrete transitions. A two-peak structure of the photo-electron spectrum reflects the two dressed resonances in the model, i.e. the "elastic" and the "inelastic" resonances. In addition, the spectrum exhibits a Fano zero. The oscillatory-decaying behavior in time of the initial state population can be simply related to the electron spectrum in the dressed-resonance representation (DRR).

The effects of several types of relaxation are also evaluated. Recycling, being a decay to the ground state, increases the fraction of time the atom spends in its ground state. Correspondingly the elastic peak is narrowed while the inelastic one is broadened as the recycling-frequency parameter increases. Phase-jitter relaxation, on the other hand, broadens both elastic and inelastic peaks. This latter feature can be precisely stated in terms of a substitution rule. Both relaxations have therefore a competing effect on the ground state. We derive in the DRR a simple analytical expression for the photo-electron spectrum. Qualitative understanding of the various effects can easily be supplemented by a systematic analysis of the exact solution of the model in the different regimes of the physical parameter space.

*Research partially supported by the U.S. AFOSR.

CORRELATIONS IN HIGHLY EXCITED TWO-ELECTRON ATOMS: 'PLANETARY' BEHAVIOR

W. E. Cooke
Physics Department
University of Southern California
Los Angeles, CA 90089

R. M. Jopson, L. A. Bloomfield, R. R. Freeman
AT&T Bell Laboratories
Murray Hill, NJ 07974

J. Bokor
AT&T Bell Laboratories
Holmdel, NJ 07733

ABSTRACT

We have experimentally observed the systematic onset of strong mixing of configurations in core-excited autoionization states of Ba as a function of excitation level of the 'core' electron relative to the Rydberg electron. This behavior was originally suggested by Percival in his discussion of 'Planetary Atoms'.

INTRODUCTION

In 1977, I. C. Percival coined the term "planetary atoms" to describe those atoms which had at least two highly excited electrons.[1] Percival used a semiclassical quantization theory to give some basic characteristics to these unsolved, three body quantum mechanical systems. Others were attempting to solve this problem from a purely quantum mechanical approach. Herrick and Sinanoglu[2] used group theory methods to classify energy levels for the He $3\ell 3\ell'$ manifold. Their approach was strongly suggestive that these atoms vibrated and rotated with the electrons localized on opposite sides of the nucleus - more like a "molecular atom". Fano and others[3] analyzed the problem using hyperspherical coordinates and found that, at least for the case of near threshold double ionization, both electrons were highly correlated in angle as they moved along a "Wannier ridge", a local maximum in the potential energy surface.

There are some characteristics of planetary atoms that all the theoreticians agree upon. (1) These states will typically have high energies, usually in the xuv, since two electrons are both excited. (2) In some of the planetary states the electrons will be correlated to produce long lifetimes, generating nearly bound states embedded in the ionization continuum. (3) Most of the planetary states will have little overlap with the ground state. (4) Single photon excitation will not efficiently excite these states (because of the low overlap with the ground state). (5) The structure of these states will depend primarily on the charge of the parent ion, and not its structure (again because of the lack of overlap with the ground state).

However, because of these characteristics, very little experimental observation of planetary atoms has been possible. Percival suggested that electron impact excitation might be the only efficient way to populate planetary atoms, although the immense phase space available would make it difficult to excite specific planetary states.[1] Recently, Buckman et al[4] have demonstrated the advantages and difficulties of this excitation method. They have used electron impact to excite doubly excited states of He^-, and have observed some interesting departures from normal Rydberg interval scaling. But it is very difficult to identify confidently the resonances they observe, and their energy resolution is limited to tens of meV, typical of electron monochromators.

Our approach has been to use multiphoton excitation, usually employing three different wavelength photons. In this way, we have maintained standard optical energy resolutions of fractions of cm^{-1} and we have retained the specificity of dipole selection rules. In fact, by using a multiphoton version of Isolated Core Excitation (ICE)[5] we have an even greater ability to identify specific configurations than is usually obtained in optical spectra.

EXPERIMENTAL METHOD

The idea behind multiphoton ICE is simply illustrated in figure 1. First, we use two lasers to excite barium atoms in an atomic beam from their ground state to a highly excited, or Rydberg, state. We use barium atoms because of their convenient visible transitions; the planetary barium atom should be insensitive to the xenon-like Ba++ core. These Rydberg states consist of primarily one single configuration which has one valence electron in the ground, 6s, state and one electron in a highly excited $n\ell$ state (where $\ell = 0$ or 2 by dipole selection rules). These bound states have been well characterized spectroscopically[6] and are easily resolvable with our pulsed lasers. This initial excitation isolates the core, insofar as further optical fields only affect the remaining 6s core electron, since the Rydberg electron is moving too slowly to respond to a rapidly oscillating field. Furthermore, since the Rydberg electron is localized far from the 6s electron, the 6s electron responds to frequencies very near the those that excite the 6s core electron to an ms excited state. This step can be accomplished stepwise,[7] with two different lasers, although it has been more convenient to use two photons from a single uv laser in most cases. Spectra are obtained by tuning the core laser frequency and monitoring the production of doubly excited states, while the first lasers remain fixed to a specific Rydberg state. This produces spectra for the $6sn\ell \rightarrow msn'\ell$ transitions, where n and n' need not be the same, although ℓ does not change.

The detection of a $6sn\ell \rightarrow msn'\ell$ transition is not trivial for any value of m since the atom will ionize if it absorbs one *or* both core transition photons. Consequently, we have used two different techniques to detect the two photon core transitions. For the lower core states (m = 7 or 8) we have analyzed the energy of the electrons produced by ionization.[7] Some of the doubly excited states will autoionized to produce Ba+(6s) ions, leaving the remainder of the energy with the free electrons. By only looking for electrons with enough energy to represent two photon absorption, we eliminated the single photon absorption background. For the electron energy analysis, we have used a simple drift tube with a retarding voltage to discriminate against the slower electrons. Our tube is mounted inside a solenoid ($\approx$5 Gauss) that guides the electrons to a microchannel plate detector.

For the higher core excitations, this technique is not efficient since the atom most often autoionizes to produce one of the many excited states of the Ba+ ion. The availability of excited ion states increases as m^3, with most of the states resulting in very slow electrons. Our second technique is to detect the excited ions by further ionizing them with the absorption of yet an additional core photon. We have easily separated the Ba++ ions from the Ba+ ions using a quadruple mass filter. This technique also provides an immediate laser wavelength calibration, since there are always some Ba+(6s) ions in our chamber which are three-photon ionized when the core laser is tuned exactly to the Ba+ 6s ms transition.

RESULTS AND DISCUSSIONS

The planetary atoms we have observed using these methods fall into two groups: states where n>>m; and states where $n \approx m$. In the first case, one expects that the independent electron model will still be a good approximation; in the second case it is not clear what to expect. Figure 2 shows a typical spectra for the $6s15s \rightarrow 10sn's$ transitions, which are examples of the first case. The very narrow peak at 2857Å corresponds to the laser calibrating ionic transition referred to above. The two broader features are transitions to doubly excited states with values of n' to 16 and 17. The shift of these lines from the ionic transition allows us to determine how much the binding energy of the Rydberg electrons changed by exciting the core. This binding energy, W, can then be easily related to the doubly excited state's effective quantum number, n^*, or quantum detect, δ

$$W = 1/n^{*2} = \approx 1/(n-\delta)^2 \quad (a.u.) \tag{1}$$

We do not observe large changes in the binding energy of the Rydberg electron because its binding energy is simply related to its most probable location - its classical turning point - and our core excitation cannot perturb the Rydberg electron much. The small change is, in fact, a direct result of the interaction between the core and Rydberg electrons. In addition to binding energies, the spectrum of figure 2 also shows the autoionization rates of the two states as the widths of their excitation lines. Since the lineshapes are symmetric, their linewidths can be easily measured.

When the Rydberg electron is inside the ms electron, its wavefunction oscillates faster so that there is a difference between the number of nodes it has (n) and the number of nodes a hydrogenic wavefunction of equivalent energy would have $(n)^*$. the quantum defect is a direct measure of this. Figure 3 shows how the measured quantum defect of $msn\ell$ states increases with increasing m. Since this effect just depends on the extent of the ms wavefunction, it is independent of ℓ (as long as m is large enough so that the $n\ell$ wavefunction penetrates it), and a simple evaluation of the expectation value of the potential due to the core electron predicts a linear dependence on m, as observed. The two cases of Rydberg electrons, ns or nd, do have different quantum defects for m=6, but this is primarily due to the xenon-like core of Ba++; note that the relative difference between quantum defects will become negligible as m becomes increasingly larger.

To analyze the autoionization widths of the $msn\ell$ states, it is convenient to consider scaled widths, i.e. autoionization widths multiplied by n^3. Since a Rydberg electron has a classical period proportional to $(n^*)^3$, this corresponds to the probability of autoionization per Rydberg electron orbit which should remain constant for an entire Rydberg series with a specific core configuration. A quantum mechanical argument also predicts this $(n^*)^3$ scaling, as a result of the normalization of the Rydberg wavefunction near the core. But there are few simple arguments for rules concerning the scaling of these autoionization widths as the core quantum number, m, is changed. One might expect that the scaled autoionization widths would increase dramatically for two reasons: (1) the core volume increases very rapidly with increasing m (as m^6), so that classically the core is a bigger target for the Rydberg electron; and (2) the number of final continuum states available increases as m^3. However, one could also argue that because the core size is increasing the magnitude of an $1/r_{12}$ interaction must decrease in inverse proportion to the distance scale increase.

Figure 4 summarizes our data on the scaling of autoionization widths times $(n^*)^3$ as a function of the core electron quantum number for msns and msnd doubly excited states. The msnd states show an increase of over an order of magnitude in the probability of autoionization per Rydberg orbit as m is increased from 7 to 11. However, the msns states show virtually no change in their scaled autoionization rates!

There is a classical argument for the insensitivity of the msns states to the core size. If one has a single classical solution to the two electron problem, one can generate an entire set of related solutions by scaling the radial coordinates of *both* the Rydberg and the core electronic by some factor γ^2. The geometry of this solution would be identical to that of the first solution, only its distance scale and time scale would be different. By Keppler's laws, the time scale would be increased by a factor γ^3. Consequently, the probability of autoionization per Rydberg electron orbit, which is a purely geometric factor with no time dependence, must be invariant under this distance scaling. This classical scaling also requires a scaling of the individual electrons' angular momenta, however, since both distance *and* time were scaled. Specifically, each electron's value of ℓ must be scaled by γ. For s states with ℓ=0, this scaling has no effect, and we can conclude that the classical autoionization rate per Rydberg electron orbit should be independent of m for msns series with $m << n$. Although the data support this prediction, we do not know of an equivalent quantum mechanical argument suggesting this behavior.

The second class of planetary atoms, those with $m \approx n$ have been much more difficult to observe and characterize. Since significant deviations from the independent electron model are expected, but the form of these deviations is not known, it is not clear where to search for such states. Moreover, it is expected that the interactions between the electrons will tend to mix a specific ms^2 configuration with a band of states that increases rapidly with m.[8] We have observed indirect evidence of this insofar as our signal became progressively weaker as we decreased n to values close to m, although the complications of our detection system do not allow a definitive measurement of oscillator strength to these states.

However, for states where n is within 3 of m, and m is large (10 or 11), we have observed striking, new structure. Figure 5 shows a typical $6s12s \rightarrow 10s13s$ excitation profile. Instead of a single line, the complex structure represents severe configuration mixing of the 10s13s, $^1S_0$ state with one or more Rydberg series. When an autoionizing state mixes with a Rydberg series, the result is a modification of the excitation profile such that additional resonances occur at the locations of the Rydberg states.[9] Near the center of the profile these resonances appear as dips while in the wings of the profile they appear as peaks. The relative strength of the configuration can be determined from the intensity of the structure compared to the unmodified background. Thus, the structure in figure 5 suggests that the 10s13s, $^1S_0$ states is mixed with the 6hnh Rydberg sequence as strongly as it is mixed with all other continuum configurations combined! The 6hnh Rydberg series is particularly significant because it represents the circular (ℓ=n≈1) state of the ion, i.e. the interaction has mixed in the highest values of ℓ possible, even though this requires a r^5 multipole interaction. This contrasts with the theoretical work on the "Wannier ridge" type states which should mix only with states of ℓ as high as $.5\sqrt{m}$.[10]

CONCLUSION

In conclusion, we have developed an efficient, selective method for populating "planetary" atoms. By exciting msns and msnd states with n>>m, we have observed simple behavior of the parameters associated with autoionizing Rydberg states, although much remains to be understood about the origins and the calculation of these parameters. We have also excited states with $n \approx m$ and find that, if m is large, new structure suggests a surprisingly strong mixing of states with high values of individual ℓ. Current theoretical models do not appear to describe the structure and mixings we are beginning to observe.

ACKNOWLEDGEMENTS

The portion of this work conducted at USC was supported in part by the National Science Foundation under grant PHY82-01688. One of us (WEC) wishes to acknowledge support from the Alfred P. Sloan Foundation.

REFERENCES

1. I. C. Percival, Proc. R. Soc. Lond. A *353*, 289 (1977).
2. D. R. Herrick and O. Sinanoglu, Phys. Rev. A *11*, 97 (1975).
3. U. Fano, J. Phys. B. *7*, L401 (1974); see also a review: U. Fano, Rep. Prog. Phys. *46*, 97 (1983).
4. S. J. Buckman, P. Hammond, F. H. Read, and G. C. King, J. Phys. B *16*, 4039 (1983).
5. S. A. Bhatti and W. E. Cooke, Phys. Rev. A *28*, 756 (1983).
6. J. R. Rubbmark, S. A. Borgstrom and K. Bockasten, J. Phys. B *10*, 421 (1977).
7. R. M. Jopson, R. R. Freeman, W. E. Cooke, and J. Bokor, Phys. Rev. Lett. *51*, 1640 (1983).
8. H. Taylor, Private Communication (March, 1984).
9. W. E. Cooke and S. A. Bhatti, Phys. Rev. A *26*, 391 (1982).
10. A. R. P. Rau, J. Phys. B *17*, L75 (1984).

FIGURE CAPTIONS

1. Schematic of Excitation Process. Two lasers are used to excite selectively a particular $6sn\ell$ Rydberg state. Subsequently, the transition $6s \rightarrow ms$ is driven in a two-photon process resulting in the population of states $msn'\ell$
2. Excitation Spectrum of the $6s\,15s \rightarrow 10sn's$ Transition. This is a "normal" shake-up spectrum, resulting in the population of $n' = 17$ and $n' = 18$.
3. Quantum Defects of the *ms* and *md* states as a function of core principal quantum number.
4. Core Scaling of Autoionization for *ms* and *md* states. The *md* states show increasing autoionization width with increasing core, while the *ms* states are independent of the core principal quantum number.
5. Excitation spectrum of $6s\,12s \rightarrow 10s\,13s$ Transition. This spectrum shows substantial mixing with the $6hnh$ Rydberg series. This mixing of the highest available angular momenta for each individual electron is indicative of the onset of correlated behavior.

'ISOLATED CORE' EXCITATION SCHEME

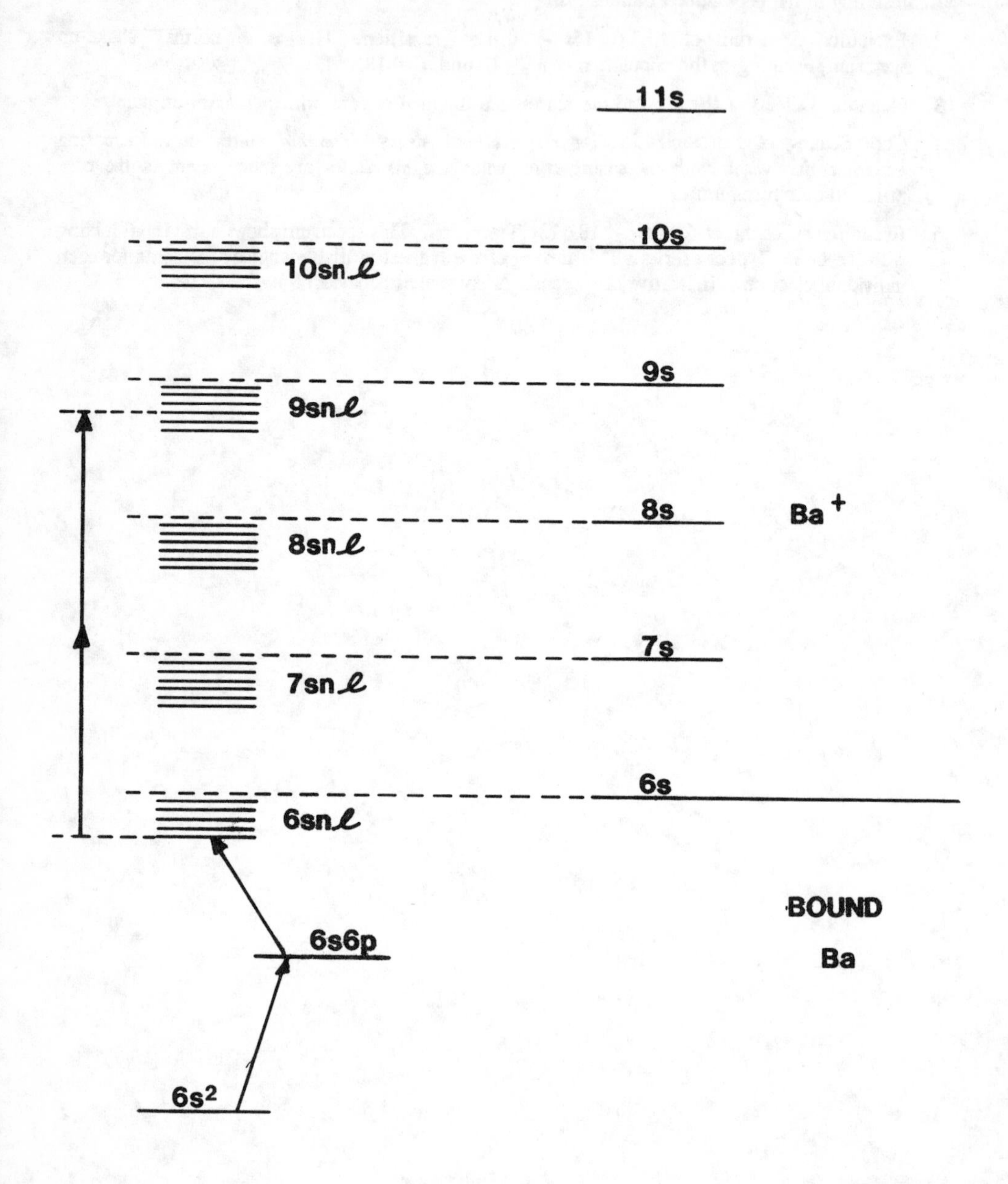

Fig 1

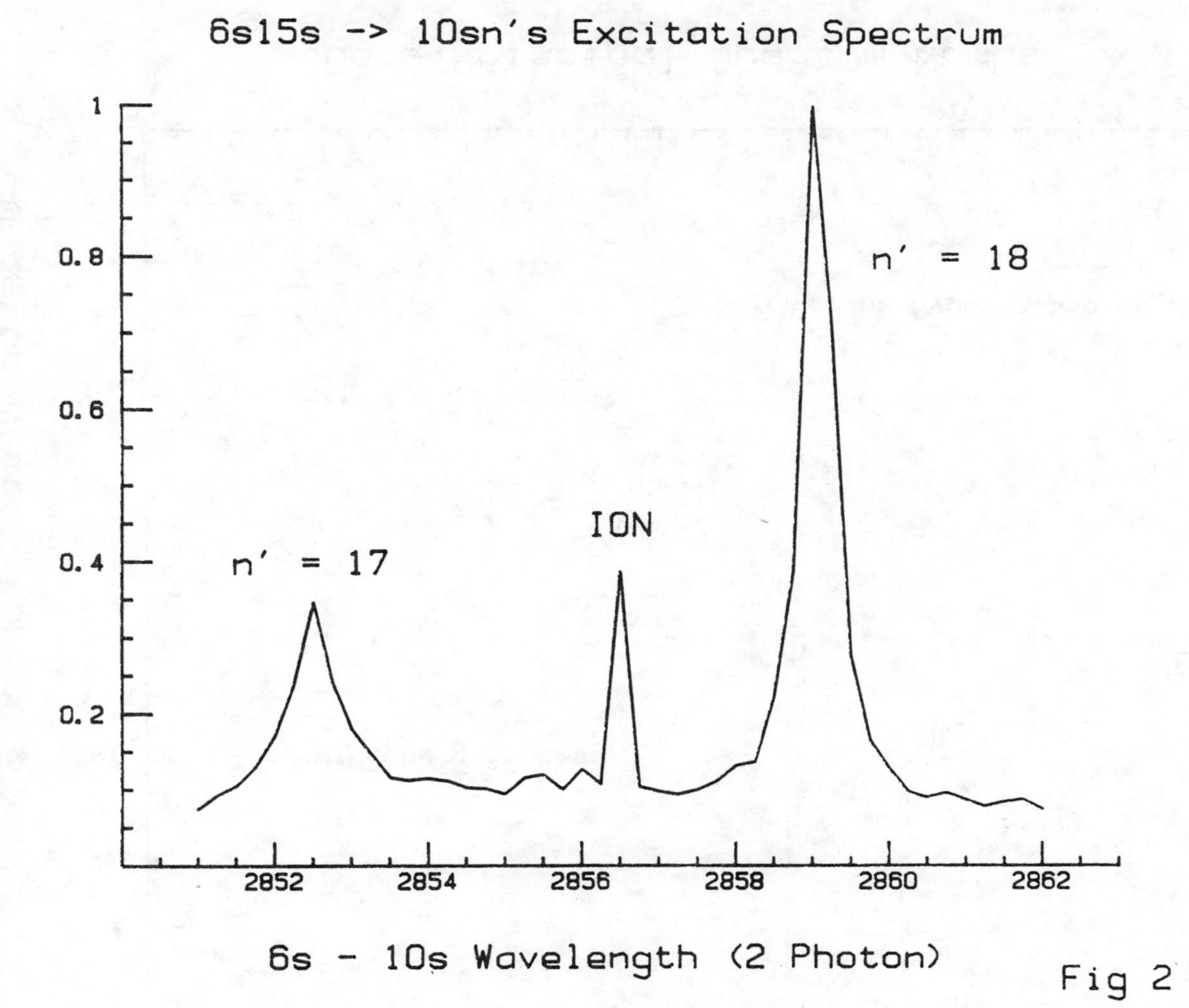
6s15s -> 10sn's Excitation Spectrum
Relative Autoionization Rate
1
0.8
0.6
0.4
0.2
n' = 17
ION
n' = 18
2852
2854
2856
2858
2860
2862
6s - 10s Wavelength (2 Photon)

Fig 2

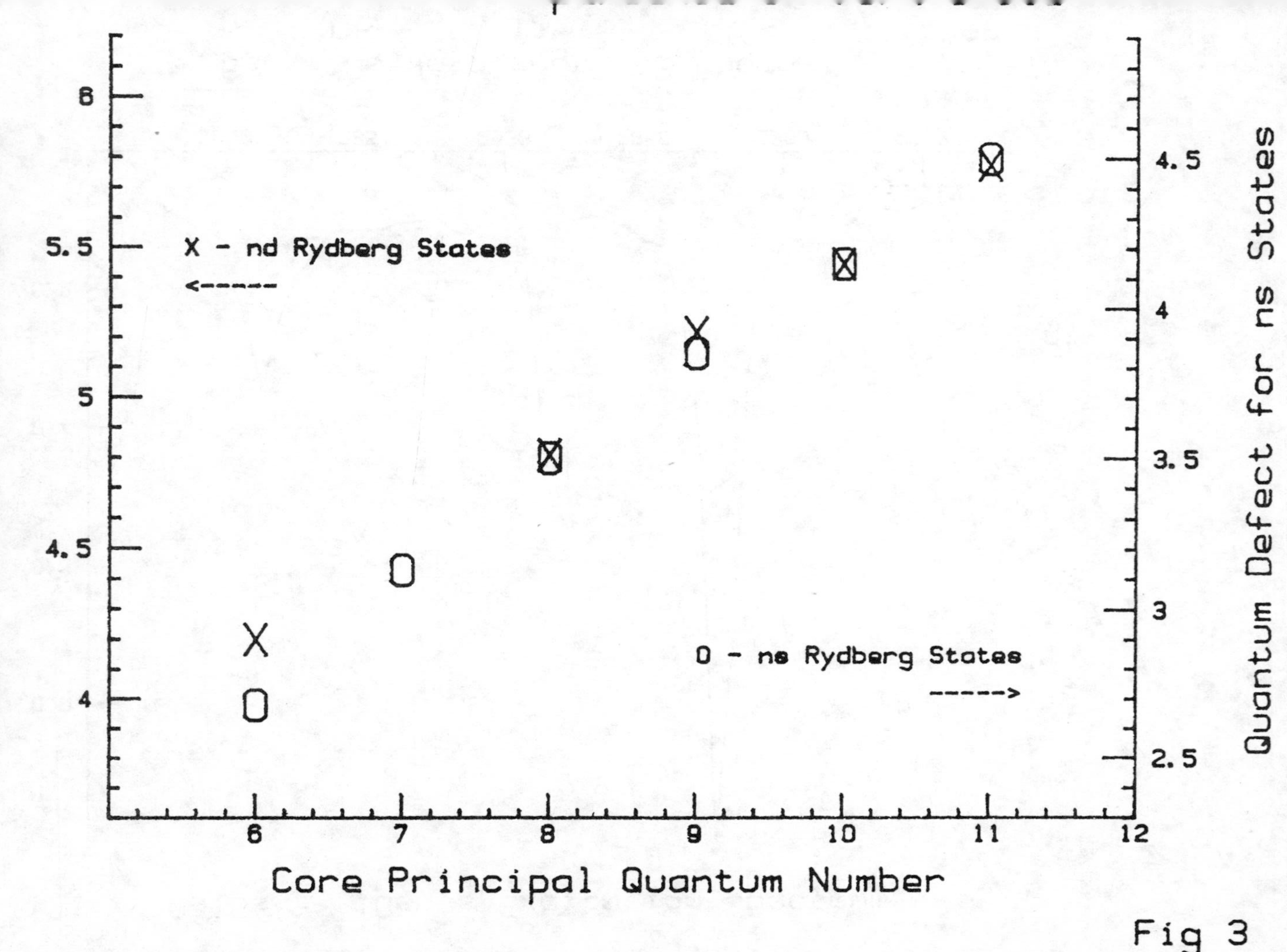

Quantum Defect for nd States
6
5.5
5
4.5
4
X - nd Rydberg States
<------
O - ns Rydberg States
------>
Quantum Defect for ns States
4.5
4
3.5
3
2.5
6
7
8
9
10
11
12
Core Principal Quantum Number

Fig 3

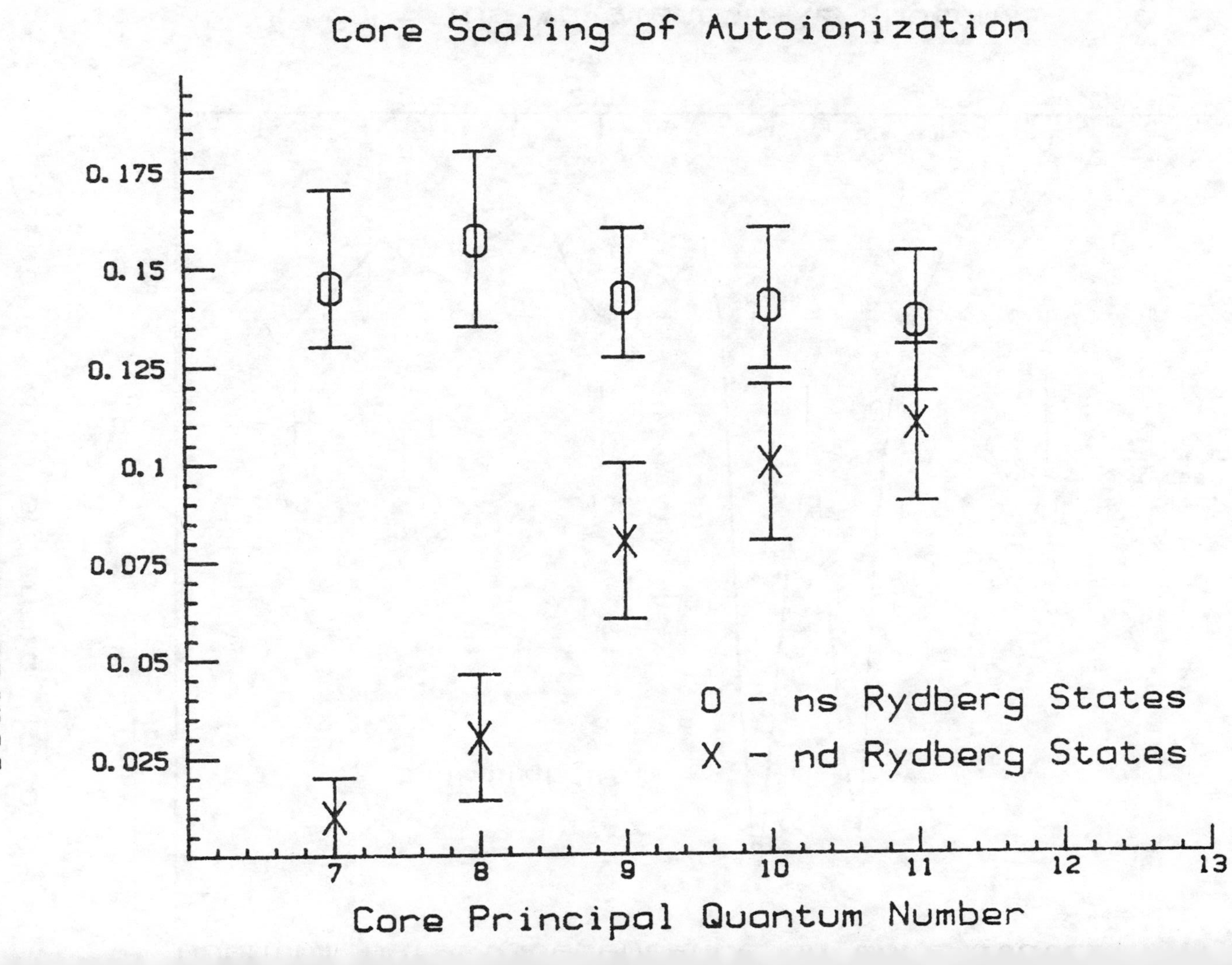

Core Scaling of Autoionization
Scaled Autoionization Rate
0.175
0.15
0.125
0.1
0.075
0.05
0.025
7
8
9
10
11
12
13
Core Principal Quantum Number
O - ns Rydberg States
X - nd Rydberg States

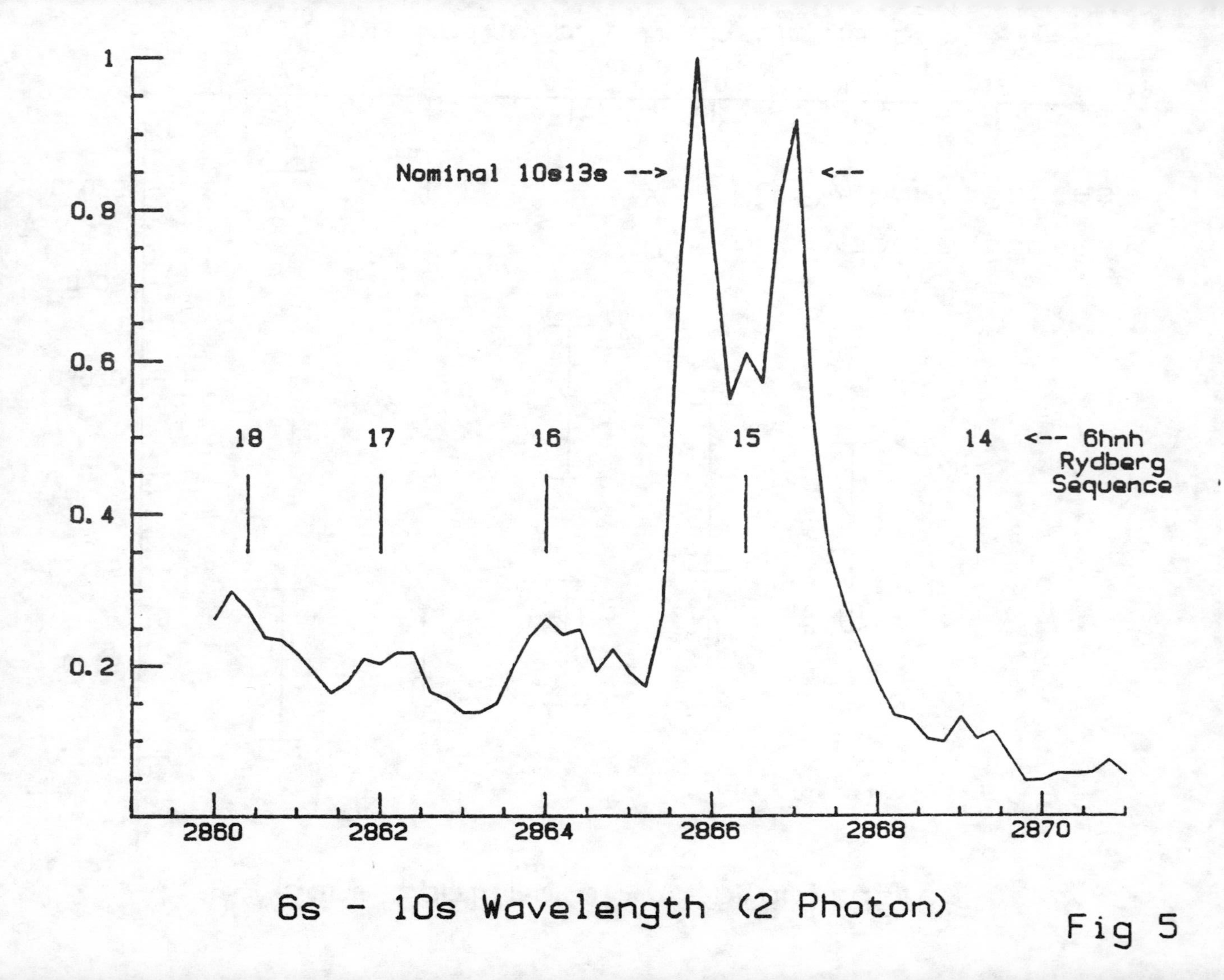

Relative Autoionization Rate
1
0.8
0.6
0.4
0.2
Nominal 10s13s -->
<--
18
17
16
15
14
<-- 6hnh
Rydberg
Sequence
2860
2862
2864
2866
2868
2870
6s - 10s Wavelength (2 Photon)

Fig 5

MULTIPHOTON IONIZATION OF ATOMIC CESIUM*

R. N. Compton, C. E. Klots, J.A.D. Stockdale, and C. D. Cooper**
Oak Ridge National Laboratory, Oak Ridge, Tennessee 37831

ABSTRACT

We describe experimental studies of resonantly enhanced multiphoton ionization (MPI) of cesium atoms in the presence and absence of an external electric field. In the zero-field studies, photoelectron angular distributions for one- and two-photon resonantly enhanced MPI are compared with the theory of Tang and Lambropoulos. Deviations of experiment from theory are attributed to hyperfine coupling effects in the resonant intermediate state. The agreement between theory and experiment is excellent. In the absence of an external electric field, signal due to two-photon resonant three-photon ionization of cesium via np states is undetectable. Application of an electric field mixes nearby nd and ns levels, thereby inducing excitation and subsequent ionization. Signal due to two-photon excitation of ns levels in field-free experiments is weak due to their small photoionization cross section. An electric field mixes nearby np levels which again allows detectable photoionization signal. For both ns and np states the "field induced" MPI signal increases as the square of the electric field for a given principal quantum number and increases rapidly with n for a given field strength.

Finally, we note that the classical two-photon field-ionization threshold is lower for the case in which the laser polarization and the electric field are parallel than it is when they are perpendicular.

INTRODUCTION

Multiphoton ionization (MPI) of alkali atoms has played a pivotal role in our understanding of the interaction of intense electromagnetic radiation with matter. The "hydrogen-like" energy levels and absence of low-lying autoionizing states make the alkali atom theoretically tractable. The low ionization potential and the ease with which one can produce atomic beams of alkali atoms facilitates the experimental studies. Many of the early MPI studies involved high-powered fixed-frequency lasers. Presently, the availability of tunable dye lasers with wavelength extension (i.e., frequency doubling, Raman shifting, etc.) makes it possible to study one- to six-photon ionization of the alkali atom with continuously tunable radiation. Also, harmonic generation in these systems which may be present during ionization can be recorded more easily than in other systems since the radiation generated is in the visible or

* Research sponsored by the Office of Health and Environmental Research, U.S. Department of Energy, under contract DE-AC05-84OR21400 with Martin Marietta Energy Systems, Inc.

** Department of Physics, University of Georgia, Athens, GA 30601.

near ultraviolet spectral region. The ability to tune the laser to high Rydberg states using either even or odd numbers of photons accesses electronic states of both parities. Stepwise excitation is capable of selectively producing states of high orbital angular momentum.

The angular distributions of photoejected electrons provide valuable information about the structure of atoms and molecules and the photoionization process itself. The photoelectron angular distributions depend upon the nature of the initial bound state and final continuum states. The continuum contribution involves the interference between the partial waves of the outgoing electron, and any interaction between this electron and the ion core. The electron angular distribution resulting from the electric-dipole interaction between a single photon and an isotropic distribution of atoms is given by

$$\frac{d\sigma(\lambda,\Theta)}{d\Omega} = \frac{\sigma_{TOT}(\lambda)}{4\pi} [1 + \beta P_2 \ (\cos \Theta)] \tag{1}$$

where σ_{TOT} is the generalized photoionization cross section for one-photon ionization at wavelength λ, $P_2(\cos \Theta)$ is the second Legendre polynomial, Θ is the angle between the polarization axis of the incident light and the direction of the photoelectron K and the value of β is the so-called asymmetry parameter. β can vary continuously from -1 to +2.

In MPI, the order of the process (i.e., number of photons involved) and the possible participation of real and virtual intermediate states are the two most important factors which determine the angular distribution. For the case of nonresonant MPI or for resonantly enhanced MPI and when the laser intensity is weak enough to validate the lowest order perturbation theory, the generalized cross section for N-photon ionization can be written as

$$\frac{d\sigma(\lambda,\Theta)}{d\Omega} = \frac{\sigma_{TOT}(\lambda)}{4\pi} \sum_{i=0}^{N} \beta_{2i} P_{2i} \ (\cos \Theta) \tag{2}$$

where $P_{2i}(\cos \Theta)$ is the 2ith order Legendre polynominal of order 2i and β_{2i} are functions of microscopic atomic parameters.

A number of theoretical studies of angular distributions of photoelectrons from MPI have been published.[1-5] Experimental measurements of angular distributions have been presented for two-photon ionization of sodium,[6,7] of titanium,[8] of cesium,[9] of strontium,[10] and five-photon (nonresonant) ionization of sodium.[11] The influence of nuclear spin on angular distributions has been studied for the case of sodium.[12] In addition, the effects of "quantum beats" due to the hyperfine levels on angular distributions have been observed.[13] Finally, angular distributions for so-called above threshold ionization of xenon at a fixed wavelength (0.53 μm) have been reported.[14] Both the experimental and theoretical studies have illustrated that measurements of angular distributions of photoelectrons from resonant MPI are complicated by the following:

1. The angular distributions may be laser power dependent due to saturation of some resonant level or due to a.c. Stark effects on the ground and excited states involved.[4,15]
2. So-called above threshold ionization[14,16] effects in which photoejected electrons gain energy from the radiation field may complicate the measurements.
3. If more than one hyperfine level is excited (which is most often the case), "quantum beat" interference effects produce angular distributions which are dependent upon the temporal characteristics of the laser beam. For example, in two-photon ionization, the photoelectron angular distribution for the two-photon ionization of sodium via the $3^2P_{3/2}$ state is found to depend upon the time delay between the exciting and ionizing laser pulses.[13]
4. Sometimes subtle background or surface ionization effects are observed which can interfere with (or even obscure) the real signal (see, e.g., Ref. 7).

The studies of photoabsorption in DC-electric fields is a well developed subject. First-order Stark shifts in hydrogen and second-order Stark shifts in other atoms have been exhaustively studied for the case of one-photon absorption. In this work we present the first experimental studies of the DC Stark effect upon MPI. As we shall see, the DC Stark effect can greatly enhance the MPI cross section both in the resonant excitation step and in the ionization step.

Experiments in which highly excited states are produced in external electric or magnetic fields allow one to test fundamental ideas about the electron-Coulomb system in the quasi-continuum.[17] In this work we describe studies of MPI in the presence of an electric field (E = 5 to 4000 V/cm) in which the plane of polarization of the laser can be rotated relative to the direction of the electric field. As we shall see, the Stark effect mixes nearby states of different parity, allowing for the detection of dipole forbidden states. Finally, we will show that the classical (two-photon) field ionization threshold is lower for the case when the laser polarization and the electric field are parallel than it is when they are perpendicular.

EXPERIMENTAL

The MPI photoelectron angular distributions were measured in an apparatus consisting of a collimated cesium beam which is directed into the entrance of a double focusing spherical sector electron energy analyzer. A laser beam is focused to a spot ~1 cm from the entrance hole. The acceptance angle is $\sim\pm2°$. The laser power density is $\sim10^8$ W/cm^2. Experiments involving the DC Stark effect on MPI were performed in a parallel field geometry. The laser beam was focused between two parallel plates separated by ~0.6 cm. The electric field could be varied from 0 to ~3 KV/cm. Ions were pushed through a grid in the negative plate and detected with a dual channel plate charged particle detector after drifting through a short time-of-flight mass spectrometer (~10 cm).

In both experiments, a Molectron model UV-24 nitrogen laser (~1 MW peak power) was used to pump the oscillator and amplifier of a Molectron model DL-14 dye laser. The bandwidth of the laser was ~0.2 A (FWHM). A Glan-Air prism (Carl Lambrecht) was used to purify the laser beam and the polarization was rotated through a full 2π angle with one of two different Fresnel rhomb polarization rotators.

In both experiments the laser ionization signal is processed with a Princeton Applied Research boxcar integrator model 165. Figure 1 shows a typical recorder trace of the electron energy spectrum for three of the resonant intermediate states reported herein. The absolute energy scale was not established. Excellent self-consistency among all of the peaks was observed. That is, by establishing an energy scale for one resonant intermediate level such as for the $7p^2P_{3/2}$ state (as was done in Fig. 1), all other peaks would correspond to the expected values [i.e., $2h\nu$-IP (Cs) or $3h\nu$ -IP (Cs)]. The best resolution obtained in any of the alkali studies was 0.07 eV with 0.1 to 0.2 eV resolution commonly used.

RESULTS AND DISCUSSION

A. Angular Distributions

Figure 2 shows the MPI photoelectron angular distribution for the case of two-photon ionization in which the first photon is in resonance with the $7p^2P_{1/2}$ state of cesium. Since no orientation of the $7p^2P_{1/2}$ state is expected from the transition $6s^2S_{1/2}$-$^27P_{1/2}$ the angular distribution will resemble a one-photon distribution, i.e., $I(\Theta) = 1 + \beta_2 \cos^2\Theta$ The experimental data shown in Fig. 2 are not corrected for the finite angular resolution of the electron spectrometer ($\Delta\Theta = \pm 2°$). The finite angular resolution was mathematically unfolded from all of the experimental data in this work and the points in the unfolded distribution were almost indistinguishable from the original distribution. The difference in the two distributions was well within the uncertainty in the measurements. Kaminski, Kessler, and Kollath[18] have published angular distributions for the same transition and these data are in good agreement with the data shown in Fig. 2. Tang and Lambropoulos[19] have calculated angular distributions for the $7p^2P_{1/2}$ state according to the method of Dixit and Lambropoulos[20] and their calculations are shown as a solid line in Fig. 2. There is excellent agreement between experiment and theory.

The angular distribution for photoionization of the $7p^2P_{3/2}$ state is shown in Fig. 3. In this case the agreement between theory and experiment is poor. Although the general features of the experimental data are predicted, the theory is always well below the experimental points when normalized at $\Theta = 0$ and π. The previous data of Kaminski, Kessler, and Kollath[18] in Fig. 4 shows even larger discrepancies. The rather large differences between the theory and the two conflicting experimental studies are believed to be due to the mixing of hyperfine levels in the $7p^2P_{3/2}$ state. The splitting between the two extreme hyperfine levels of the $7p^2P_{3/2}$ state is 198 MHz which corresponds to a hyperfine period of 5 ns. Our laser has

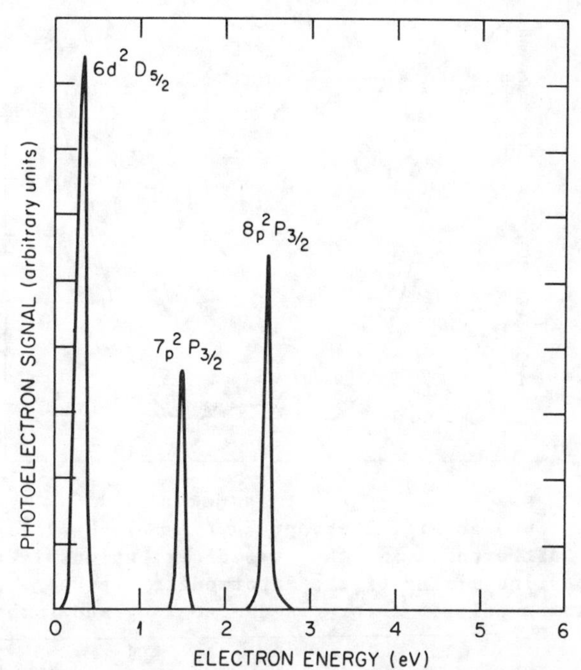

Fig. 1. Recorder traces of electron energy distributions for photoelectron produced by MPI of cesium. The energy scale was calibrated by fixing the $7p^2P_{3/2}$ peak at its calculated energy position. The laser was tuned to one- or two-photon resonance with the states shown.

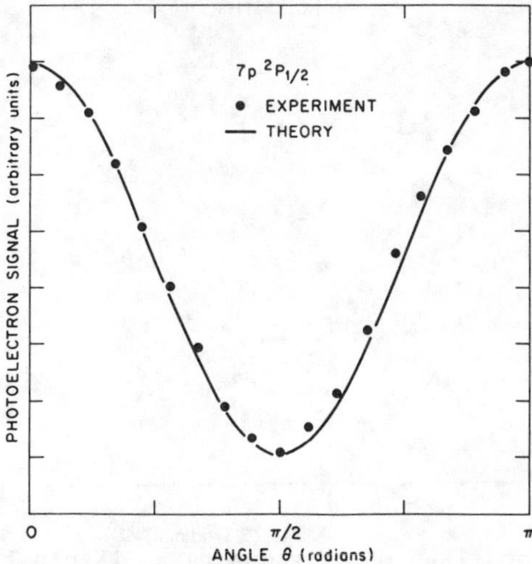

Fig. 2. Two-photon ionization photoelectron angular distributions in which the first photon is in resonance with the $7p^2P_{1/2}$ state. The theory is that of I. Tang and P. Lambropoulos.[19] The error bars in both intensity and angles are roughly twice the size of a data point.

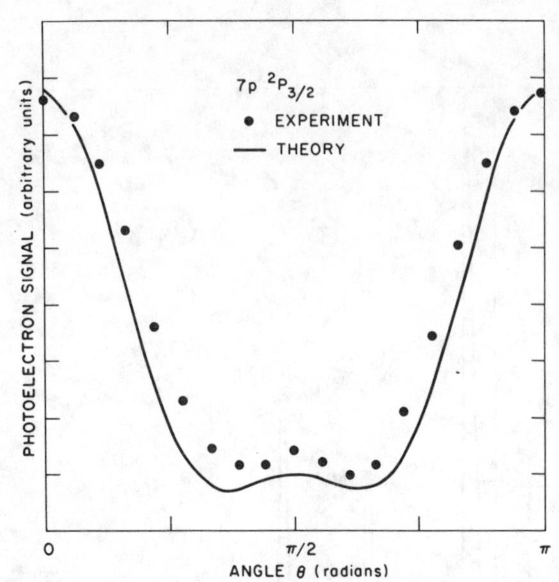

Fig. 3. Same as Fig. 2 except 7p1/2 → $7p^2P3/2$ is also added. The slight difference in the two distributions is attributed to partical hyperfine mixing of the intermedite $7p^2P3/2$ level during the ~10 μs laser pulse. Theory is due to Tang and Lambropoulos.[19]

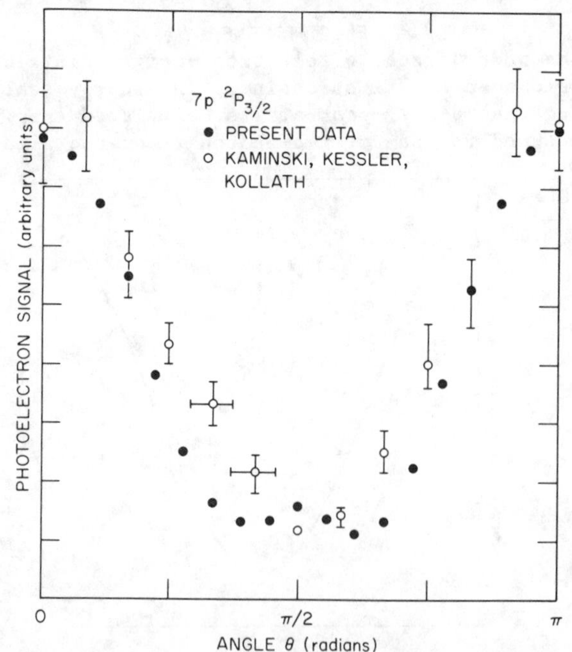

Fig. 4. Comparison of previous data [Kaminski, Kessler, and Kollath (KKK)][9] with the present data for the $7p^2P3/2$ state. The laser pulse width for the data of KKK is ~400 ns. The difference two sets of experimental in the data is attributed to complete hyperfine mixing during the laser pulse in the case of KKK. Theory is due to Tang and Lambropoulos.[19]

a pulse duration of ~10 ns which would allow for some hyperfine mixing during the excitation-ionization event. The flashlamp pumped laser of Kaminski, Kessler, and Kollath[18] had a pulse length of 400 ns allowing for complete mixing of the hyperfine levels thus producing the distribution shown in Fig. 3. The hyperfine period for the $8p^2P_{3/2}$ state is about twice as long as that for the $7p^2P_{3/2}$ and therefore we should expect better agreement between experiment and theory. Figures 5 and 6 show the experimental and theoretical results for the $8p^2P_{1/2}$ and $8p^2P_{3/2}$ states, respectively. The results further illustrate the effect of hyperfine coupling, i.e., the agreement for the $8p^2P_{1/2}$ state where no orientation occurs is excellent, whereas the theory for the $8p^2P_{3/2}$ again falls below the data.

Figure 7 shows the experimental angular distribution for the case of two-photon resonant three-photon ionization of cesium via the $8d^2D_{5/2}$ intermediate state. Here the hyperfine period is ~55 ns and the agreement between theory and experiment is good. In this case the laser pulse is over before hyperfine coupling can occur.

Tang and Lambropoulos[19] have developed a theoretical expression which takes into account the hyperfine effects on the angular distributions for the $7p^2P_{3/2}$ and $8p^2P_{3/2}$ resonant intermediate state. The theory satisfactorily accounts for the distributions presented in Figs. 3 and 4 and the data of Ref. 9.

B. Field Effects on Multiphoton Ionization

In these experiments an effusive atomic beam is crossed at right angles by tunable light from a nitrogen laser-pumped dye laser. Perpendicular to both beams, a variable, uniform electric field draws any resulting positive ions through a time-of-flight mass spectrometer into a channelplate charged particle detector. The angle θ between the applied electric field $\vec{F}$ and the electric vector $\vec{E}$ of the laser beam is also continuously variable using a double Fresnel rhomb rotator.

As Fig. 8 illustrates, the MPI signal at low electric fields is greatly enhanced when the second photon is resonant with a nd level. The MPI signal is seen to drop rapidly with increasing principal quantum number ($\sim 1/n^8$), mainly due to the fact that the photoabsorption ($\sim 1/n^3$) and photoionization ($\sim 1/n^5$) cross sections are decreasing rapidly with n. The signal reappears as an approximate "step function" where a third photon is no longer necessary to effect ionization. As illustrated, this step occurs below the zero-field two-photon ionization potential. The shift Δ from the true IP, as a function of the electric field, F follows closely the familiar relation for field ionization:

$$\Delta = \alpha|\vec{F}|^{1/2} \tag{3}$$

with Δ and $|\vec{F}|$ in atomic units, and $\vec{E}$ parallel to $\vec{F}$ we find $\alpha_{||}$ = 1.90 (±0.03), close to the semi-classical value of 2. Upon rotating $\vec{E}$ perpendicular to F, the ionization threshold shifts to lower energy, consistent with the classical model of Cook and Gallagher.[21]

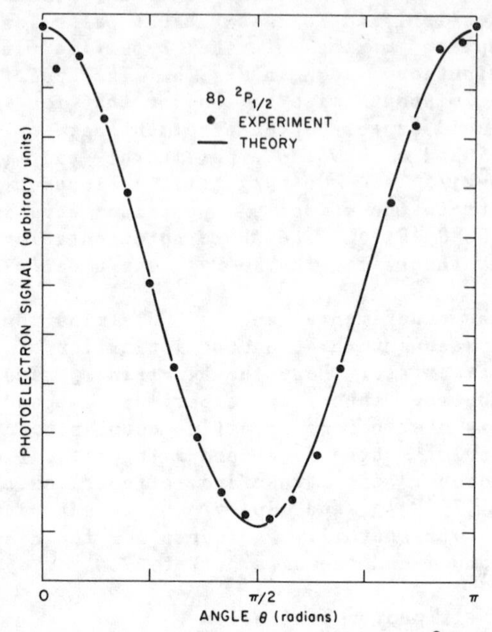

Fig. 5. Same as Fig. 2 except with the $8p^2P_{1/2}$ state.

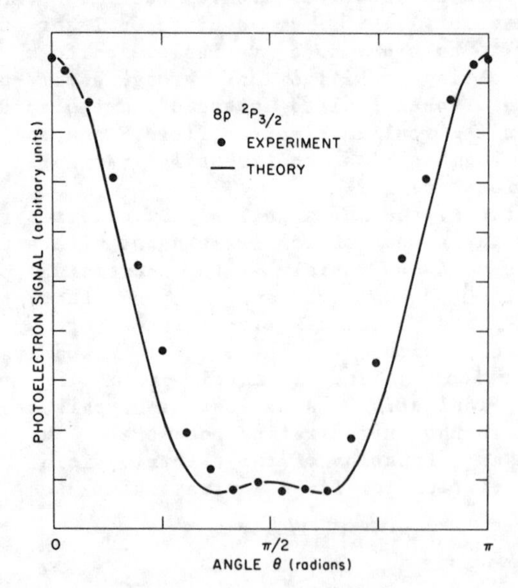

Fig. 6. Same as Fig. 3 except with $8p_{1/2} \rightarrow 8p^2P_{3/2}$.

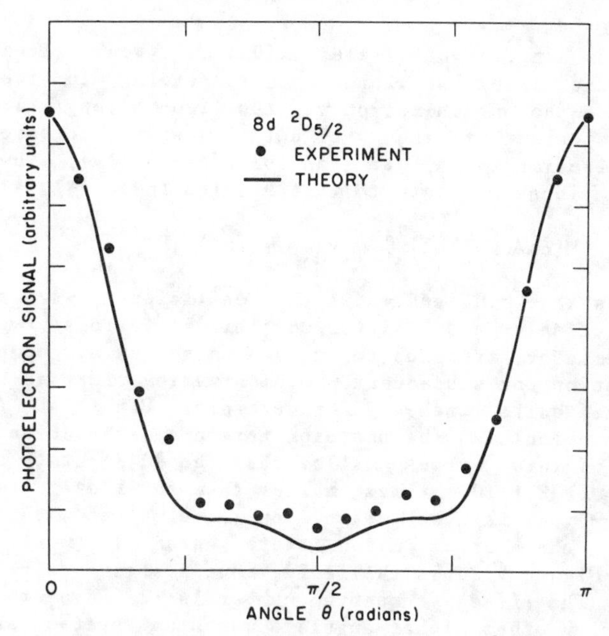

Fig. 7. Same as Fig. 2 except with the $8d^2D_{5/2}$ state.

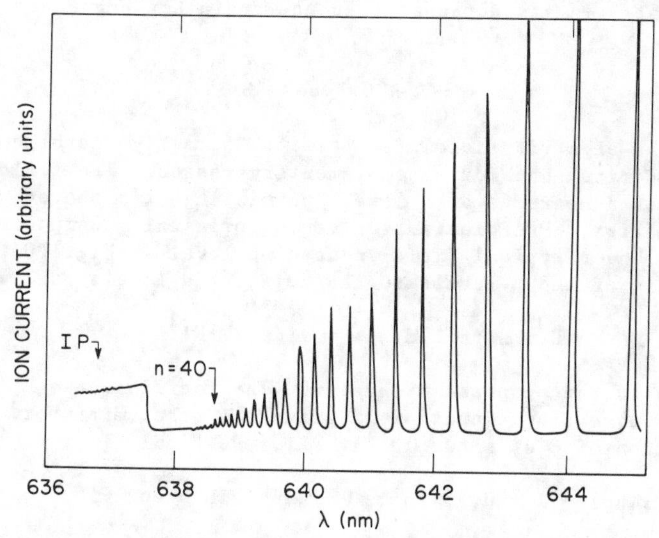

Fig. 8. Multiphoton ionization signal near the two-photon ionization threshold at 636.8 nm. The sharp peak represent two-photon resonant (nd level) three-photon ionization. The two-photon ionization "step" is observed below the zero field value at 636.8 due to field ionization.

For very low ion draw-out fields (<10 v/cm), weak signals are observed at the laser wavelength corresponding to two-photon excitation, three-photon ionization via the two-photon allowed ns levels. A component of the ns signal is seen to increase as the square of the electric field. We find for R[ns/(n-1)d], the ratio of MPI strength via an ns state to that via the (n-1) d5/2 state:

$$R[ns/(n-1)d_{5/2}] = \gamma_0 + \gamma\ (ns)\ |\vec{F}|^2 \tag{4}$$

where γ_0 is less than 0.01 and where the coefficient γ is a strongly increasing function of n. We interpret this as follows: two-photon excitation oscillator strengths for ns and nd series are comparable. The cross section for subsequent photoionization of an ns state is expected to be quite small. For example, Pindzola (private communication) calculates the photoionization cross section for the 8s state to be ~4 x 10^{-4} times smaller than the 8d3/2 state and the 12s state to be 8 x 10^{-3} times smaller than the 12d3/2 state. An electric field mixes in the nearby np levels and it is this component of the Stark mixed state which is more readily photoionized. Figure 9 shows the field-induced signal at the 16s energy level. The effective quantum number is 12 (quantum defect = 4) and therefore 9 other higher angular momentum states are seen (one is hidden under the 16s signal). The remaining 12d and 12p states are not shown.

Field-induced MPI is also readily distinguished from a field-free component via its dependence on the angle between $\vec{E}$ and $\vec{F}$. We find

$$\gamma(ns) \sim 1 + 0.8 \cos^2\theta \ . \tag{5}$$

Increasing the electric field also permits normally forbidden MPI via the np series, but for a complementary reason. The field mixes a p state with a nearby s or d level, permitting two-photon access to the p series. Photoionization of the original p component then ensues. We show a typical field-induced np level in Fig. 10 for n = 13. The ratio of the p levels to the adjacent d levels follows:

$$R(nP_J/(n-1)d_{5/2}) = \gamma(np_J)\ |\vec{F}|^2 \tag{6}$$

consistent with the proposed mechanism. The coefficient $\gamma(np_{3/2})$ is a rapidly increasing function of principal quantum number. It is also a function of θ as shown in Fig. 11. We find

$$\gamma(np) \sim 1 + 0.55 \cos^2\theta + 2.66(\sin\theta \cos\theta)^2 \tag{7}$$

and, for the ratio of spin-orbit components, $\gamma(np_{3/2})/\gamma(np_{1/2}) \simeq 4.5\ (\pm 0.5)$. This ratio, while not statistical is closer to the value 2 than the anomolously high values observed in one-photon absorption oscillator strengths.[22]

The relations reported here were obtained at sufficiently low electric fields that perturbation theory should still be valid. At higher fields we observe large-scale level shifts and the emergence

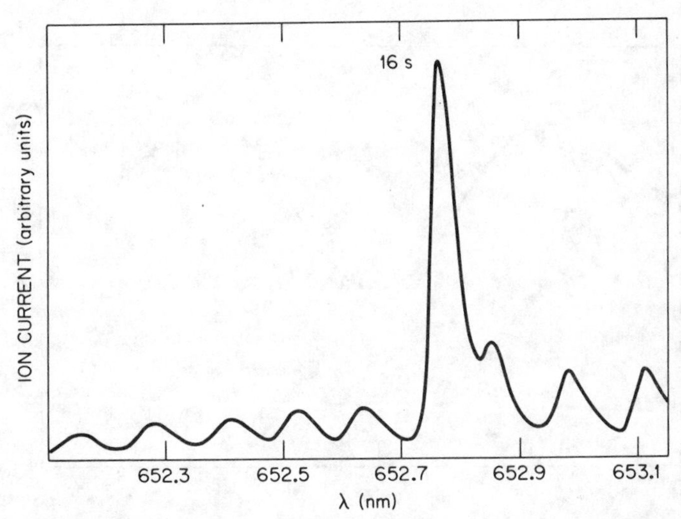

Fig. 9. Electric field enhanced three-photon ionization of nℓ states of cesium near the 16s level. Field strength 3500 V/cm.

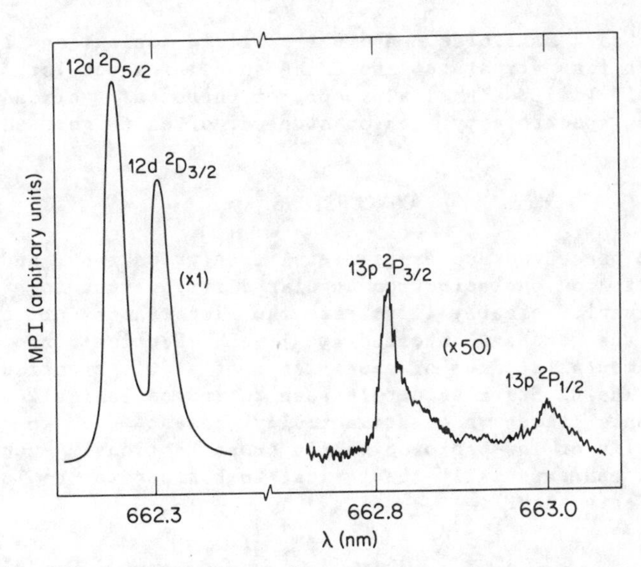

Fig. 10. Electric field induced three-photon ionization of atomic cesium in which the second photon is in resonance with the $13p^2P_{3/2,1/2}$ states. Electric field strength was 700 V/cm.

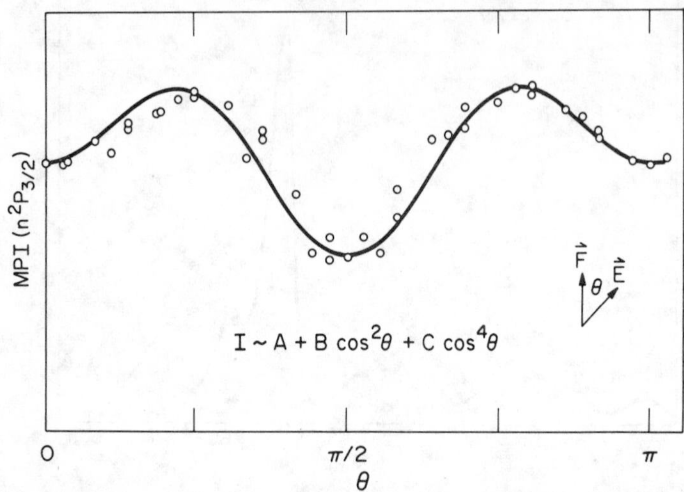

Fig. 11. Dependence of the field induced multiphoton ionization signal due to the $np^2P_{3/2}$ state (see Fig. 9) as a function of the angle between the laser polarization E and the electric field $\vec{F}$.

of complete Stark manifolds. Above the field-ionization limit we observe much fine structure, including some which persists to above the zero-field limit. Thus, a panoply of phenomena, already known in one-photon spectroscopy, can be studied, often to good advantage, via MPI.

CONCLUSIONS

We have presented new experimental results on two- and three-photon ionization photoelectron angular distributions for an alkali atom. The results clearly illustrate the importance of hyperfine mixing at the resonant intermediate level. Section B presents the first experimental studies of the effects of static electric fields upon MPI. The DC Stark effect is seen to induce ionization signals from electronic states which are normally absent due to symmetry selection rules or low photoionization cross sections. Much further experimental and especially theoretical work needs to be performed on Stark effects on MPI.

REFERENCES

[1] M. M. Lambropoulos and R. S. Berry, Phys. Rev. 8, 844 (1973).

[2] For reviews of various aspects of multiphoton ionization and photoelectron angular distribution see: P. Lambropoulos in ADVANCES IN ATOMIC AND MOLECULAR PHYSICS 12, 87-164 (Academic Press, New York, 1976); MULTIPHOTON PROCESSES (J. H. Eberly and P. Lambropoulos, Eds., Wiley, New York, 1978); and J. Morellec, D. Normands, G. Petite in ADVANCES IN ATOMIC AND MOLECULAR PHYSICS 18, 97-164 (Academic Press, New York, 1982).

[3] J. C. Tully, R. S. Berry, and R. J. Dalton, Phys. Rev. 176, 95 (1968).

[4] S. N. Dixit and P. Lambropoulos, Phys. Rev. Lett. 46, 1278 (1981).

[5] S. N. Dixit and P. Lambropoulos, Phys. Rev. A 27, 168 (1983).

[6] J. A. Duncanson, Jr., M. P. Strand, A. Lindgrard, and R. S. Berry, Phys. Rev. Lett. 37, 987 (1976).

[7] J. C. Hansen, J. A. Duncanson, Jr., R.-L. Chien, and R. S. Berry, Phys. Rev. A 21, 222 (1980).

[8] S. Edelstein, M. M. Lambropoulos, and R. S. Berry, Phys. Rev. A 9, 2459 (1974).

[9] H. Kaminski, J. Kessler, and K. J. Kollath, Phys. Rev. Lett. 45, 1161 (1980).

[10] D. Feldmann and K. H. Welge, J. Phys. B 15, 1651 (1982).

[11] G. Leuchs and S. J. Smith, J. Phys. B 15, 1051 (1982).

[12] M. P. Strand, J. Hansen, R.-L. Chien, and R. S. Berry, Chem. Phys. Lett. 59, 205 (1978).

[13] G. Leuchs, S. J. Smith, E. Khawaja, and H. Walther, Optic Comm. 31, 313 (1979).

[14] F. Fabre, P. Agostini, G. Petite, and M. Clement, J. Phys. B 14, L677 (1981).

[15] W. Ohmesorge, F. Diedrich, G. Leuchs, D. S. Elliott, H. Walther, Phys. Rev. 29, 1181 (1984).

[16] P. Agostini, F. Fabre, G. Mainfray, G. Petite, and N. K. Rahman, Phys. Rev. Lett. 42, 1127 (1979).

[17] See Richard R. Freeman, International Conference on Atomic Physics, 1980, in ATOMIC PHYSICS 7 (D. Kleppner and F. M. Pipkin, Eds.), Plenum Press, 1980.

[18] H. Kaminski, J. Kessler, and K. J. Kollath, Phys. Rev. Lett. 45, 1161 (1980).

[19] R. N. Compton, J.A.D. Stockdale, C. D. Cooper, X. Tang, and P. Lambropoulos, Phys. Rev. A (submitted).

[20] S. N. Dixit and P. Lambropoulos, Phys. Rev. A 27, 168 (1983).

[21] W. E. Cooke and T. F. Gallagher, Phys. Rev. A 17, 1226 (1978).

[22] E. Fermi, Z. Physik 59, 680 (1929).

PHOTOIONIZATION OF EXCITED MOLECULAR STATES USING MULTIPHOTON EXCITATION TECHNIQUES*

P. M. Dehmer, S. T. Pratt, and J. L. Dehmer
Argonne National Laboratory, Argonne, Illinois 60439

ABSTRACT

Photoelectron spectra are reported for three photon resonant, four photon ionization of H_2 via the B $^1\Sigma_u^+$, v=7 (J=2,4) and C $^1\Pi_u$, v=0-4 (J=1) levels and of N_2 via the o_3 $^1\Pi_u$, v=1,2, b $^1\Pi_u$, v=3-5, and c $^1\Pi_u$, v=0 levels. The results reflect both the spectroscopy and the dynamics of photoionization of excited molecular states and are discussed in terms of the selection rules for photoionization and the relative probabilities of photoionization from Rydberg and valence states. In some cases, in accordance with the Franck-Condon principle, the results demonstrate that resonant multiphoton ionization through Rydberg states may be a powerful technique for the production of electronic, vibrational, and rotational state selected ions. However, in other cases, systematic departures from Franck-Condon factors are observed, which reflect the more subtle dynamics of excited state photoionization.

INTRODUCTION

Resonantly enhanced multiphoton ionization (MPI) of molecules has been used to obtain detailed spectroscopic information on neutral intermediate states.[1] With the addition of photoelectron spectrometry (PES) to analyze the kinetic energy of the ejected electrons, it is possible to determine the branching ratios into different electronic, vibrational, and rotational levels of the product ion and to focus directly on both the dynamics of the multiphoton ionization process and the photoionization of excited state species.[2-6] In an (m+n) MPI/PES study, the excited state is prepared by m-photon excitation and subsequently is ionized by an additional n-photon excitation process. However, if n is greater than 1, significant complications may arise in the photoelectron spectrum as a result of accidental resonances at higher intermediate levels. In the present paper, we report several (3+1) MPI/PES studies of H_2 and N_2. These studies both illustrate the power of MPI/PES for the study of photoionization of excited molecular states and demonstrate the possibility of producing electronic, vibrational, and rotational state-selected ions using selective multiphoton ionization.

---

* Work supported by the U.S. Department of Energy and the Office of Naval Research.

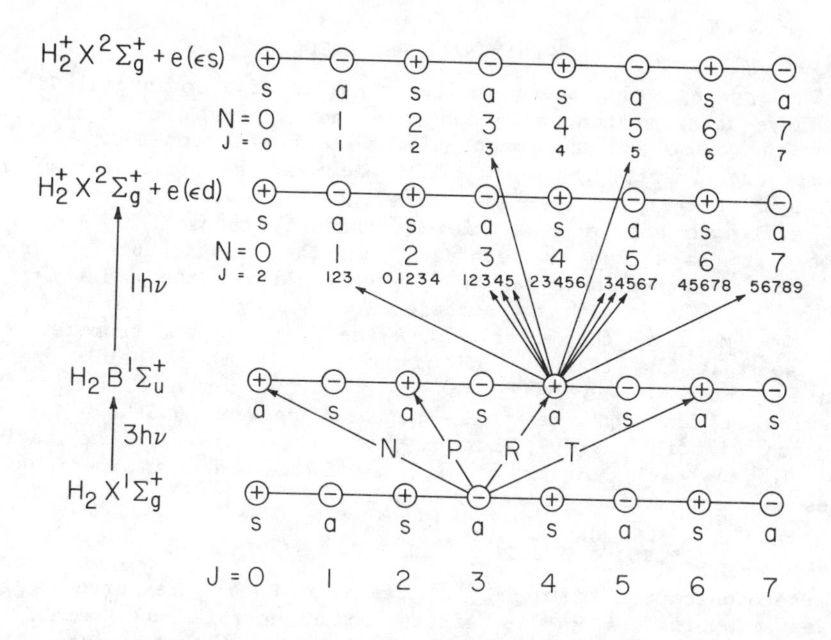

Figure 1. Schematic diagram of the allowed transitions for the (3+1) resonant multiphoton ionization of H_2 via the B $^1\Sigma_u^+$ state. Only those transitions from J=3 of the ground state and J=4 of the intermediate state are shown.

values differs for s and d waves, a rotationally resolved photoelectron spectrum can give detailed information on the relative importance of the different partial waves. We also note that in the angular momentum transfer (j_t) formulation,[9-11] the ejection of an s electron can lead only to a j_t value of 1, while the ejection of a d electron can yield j_t values of 1, 2, and 3. The j_t value of 2 is parity unfavored, and only values of 1 and 3 are important here.

Figure 2 shows the photoelectron spectra obtained by pumping the R(3) and P(3) transitions of the B $^1\Sigma_u^+$, v'=7 ← X $^1\Sigma_g^+$, v"=0 band. The most dramatic feature of the photoelectron spectra is that the observed rotational structure changes qualitatively with the intermediate rotational level of the B $^1\Sigma_u^+$ state, thus directly reflecting the selection rules for the ionizing transition.

In the photoelectron spectrum obtained by pumping the R(3) transition, the N=3 and 5 rotational levels of the ion clearly are observed. Although it is allowed by selection rules, the N=7 level is not observed in any of the vibrational bands, indicating that the $\Delta J=+1$, $\Delta N=+3$, $j_t=3$ ionizing transition, which corresponds to the ejection of a d-wave electron, is weak. Similarly, only the N=1 and 3 levels are observed in the photoelectron spectrum obtained by pumping the P(3) transition, whereas the N=5 level (which again requires $\Delta J=+1$, $\Delta N=+3$, $j_t=3$ and the ejection of a d-wave electron) is not observed. In the photoelectron spectrum obtained by pumping

EXPERIMENTAL APPARATUS

The apparatus for these studies consists of a commercial Nd:YAG pumped dye laser system, a 2-inch mean radius hemspherical electron energy analyzer, and a time-of-flight mass spectrometer.[3] In a typical experiment, the photoion spectrum is recorded as the wavelength of the dye laser is scanned. The laser is then tuned to the wavelength of a specific rovibronic transition and the photoelectron spectrum is recorded. In the measurements reported here, photoelectron spectra were recorded along the polarization axis of the laser with a photoelectron energy resolution of ~35 meV. For most of the experiments, the electron spectrometer was tuned so that the transmission function over the kinetic energy range of interest was nearly constant. The corrections for the transmission function are less than 20% between 0.5 and 1.6 eV (electron kinetic energy), but may be substantial in the region below 0.5 eV.[3] Under the conditions of the present experiments, nonresonant ionization was not observed.

PHOTOIONIZATION OF THE B $^1\Sigma_u^+$ STATE OF H_2

Photoionization of the B $^1\Sigma_u^+$ state of H_2 is presented as the initial example,[4] since it serves to illustrate the method of obtaining the selection rules for a (3+1) multiphoton ionization process. The selection rules appropriate for three photon transitions from the ground electronic state to the B $^1\Sigma_u^+$ state of H_2 are summarized schematically in Figure 1. The three photon selection rules based on symmetry properties are identical to those for single photon transitions ($g\leftrightarrow u$, $+\leftrightarrow -$, and $s\not\leftrightarrow a$).[7,8] In addition, for a three photon transition, $\Delta J \leq 3$. It is clear that the only transitions from the ground state to the intermediate B $^1\Sigma_u^+$ state that fulfill all of the selection rules have $\Delta J=\pm1,\pm3$.

The structure in the photoelectron spectrum obtained by ionizing a specific rotational level of the B $^1\Sigma_u^+$ state is determined by the single photon transition from the intermediate level to the ionization continua. The selection rules governing this transition can be deduced by considering the H_2^+ X $^2\Sigma_g^+$ ion and the ejected electron in Hund's case (d) coupling.[7] Here, the excited electron is completely uncoupled from the internuclear axis, and the rotational angular momentum of the ion core N and the orbital angular momentum of the ejected electron ℓ are added vectorially to give the total angular momentum J of the electron-ion complex. Owing to the $g\leftrightarrow u$ selection rule, the outgoing electron must have an even value of ℓ. In the present work, only s and d partial waves are considered; higher partial waves will have a large centrifugal barrier and may be ignored. By considering the various selection rules for the ionizing transition, the allowed final rotational levels of the ion can be derived. It is seen from Figure 1 that the ionizing transition must obey the selection rule $N(X\ ^2\Sigma_g^+)-J(B\ ^1\Sigma_u^+)=\pm1,\pm3$, and that for a given intermediate rotational level, the ion will be left in only odd or only even rotational states. Because the range of allowed H_2^+ rotational

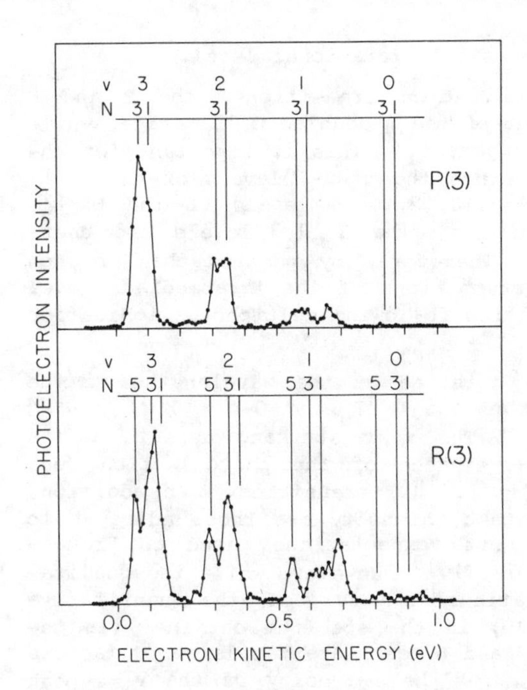

Figure 2. Photoelectron spectra of H_2 determined following (3+1) multiphoton ionization at the wavelengths of the resonant three photon $B\ ^1\Sigma_u^+$, v'=7, J'=2 [P(3)] and J'=4 [R(3)] $\leftarrow X\ ^1\Sigma_g^+$, v"=0, J"=3 transitions. The unassigned peaks at 0.69±0.01 eV in the R(3) spectrum and at 0.67±0.01 in the P(3) spectrum are believed to result from photoionization of H(2s or 2p) formed by photodissociation of $B\ ^1\Sigma_u^+$.

the R(3) transition, the N=1 rotational level of the ion appears to be missing. The N=1 level is allowed by selection rules, but requires $\Delta J=-1$, $\Delta N=-3$, $j_t=3$, and the ejection of a d-wave electron. Together with the other evidence, this indicates that either j_t values of 3 or the ejection of a d-wave electron are very weak in the photoionization processes studied here. Of these two possibilities, it is most likely that j_t values of 3 are weak in the photoionization process. Transitions of the type $2p\sigma\ B\ ^1\Sigma_u^+ \leftarrow nd\sigma$, $nd\pi$, $nd\delta$ have been observed in emission,[12,13] making it likely that the ejection of a d-wave electron will be observed in photoionization.

PHOTOIONIZATION OF THE $C\ ^1\Pi_u$ STATE OF H_2

In this experiment,[5] photoelectron spectra were determined following single photon ionization of the $C\ ^1\Pi_u^-$, J'=1 levels of H_2 accessed by Q(1) transitions from the ground state. The method of obtaining the selection rules for this process is identical to that for the case of ionization of the $B\ ^1\Sigma_u^+$ state described above. In this case, it is found that the $H_2^+\ X\ ^2\Sigma_g^+$ ion can be left with rotational angular momentum N=1,3. The N=1 and 3 levels are separated by ~30 meV (in $v^+=3$) and will not be resolved with the present spectrometer resolution. The production of $H_2^+\ X\ ^2\Sigma_g^+$, N=3 requires an outgoing d-wave and an angular momentum transfer $j_t=3$. As is discussed above, such a process was not observed in the study of photoionization from the $B\ ^1\Sigma_u^+$ state. Thus it is tempting to speculate that the H_2^+ ions resulting from photoionization of the $C\ ^1\Pi_u^-$, J'=1 levels are produced predominately in the N=1 level, i.e., the ions are produced in a single rotational level.

As in the case of single photon transitions, the P and R branches access the Π^+ component of the Λ-doubled C $^1\Pi_u$ state, while the Q branch accesses the Π^- component.[7] This is important for the present photoelectron studies, since the v'=0-4 levels of the C $^1\Pi_u$ state are perturbed by the v'=8,10,12,14,16 levels of the B $^1\Sigma_u^+$ state, respectively.[14] Fortunately, the C $^1\Pi_u^-$ levels are unaffected by this perturbation.[7] Therefore, by pumping the Q-branch transitions, the effects of perturbations of the intermediate level on the vibrational branching ratios following multiphoton ionization are avoided.

The photoelectron spectra obtained at the wavelengths of the three photon Q(1) transitions of the C $^1\Pi_u$, v'=0-4 $\leftarrow$ X $^1\Sigma_g^+$, v"=0 bands are shown in Figure 3. The most striking aspect of the photoelectron spectra is the dominance of the photoelectron peak corresponding to the $v^+(X\ ^2\Sigma_g^+)=v'(C\ ^1\Pi_u)$ transition. In addition, the weaker peaks with the greatest intensity are those adjacent to the $v^+=v'$ peak. This agrees with expectations based on Franck-Condon factor calculations (Table I).[15] However, while the qualitative agreement with the calculations is very good, the quantitative agreement is poor. For example, in the spectrum obtained via the C $^1\Pi_u$, v'=4 level, the v^+=3, 5, and 6 peaks are too large by factors of 3, 2, and 23, respectively, and the intensity of the v^+=4 peak accounts for only 43% of the total, rather than the predicted 90%.

Table I. Franck-Condon factors for H_2^+ X $^2\Sigma_g^+$, v^+ $\leftarrow$ H_2 C $^1\Pi_u$, v'

v^+	v' = 0	v' = 1	v' = 2	v' = 3	v' = 4
0	9.89×10^{-1}	1.10×10^{-2}	4.67×10^{-6}	1.45×10^{-8}	2.26×10^{-9}
1	1.06×10^{-2}	9.66×10^{-1}	2.29×10^{-2}	2.04×10^{-5}	2.98×10^{-8}
2	3.08×10^{-4}	2.16×10^{-2}	9.43×10^{-1}	3.55×10^{-2}	6.47×10^{-5}
3	1.46×10^{-5}	9.98×10^{-4}	3.22×10^{-2}	9.19×10^{-1}	4.80×10^{-2}
4	1.04×10^{-6}	6.53×10^{-5}	2.13×10^{-3}	4.16×10^{-2}	8.96×10^{-1}
5	1.00×10^{-7}	5.95×10^{-6}	1.82×10^{-4}	3.71×10^{-3}	4.92×10^{-2}
6	1.22×10^{-8}	7.24×10^{-7}	2.03×10^{-5}	3.98×10^{-4}	5.69×10^{-3}
7	1.87×10^{-9}	1.12×10^{-7}	2.92×10^{-6}	5.42×10^{-5}	7.43×10^{-4}
8	3.77×10^{-10}	2.04×10^{-8}	5.34×10^{-7}	8.99×10^{-6}	1.21×10^{-4}
9	1.04×10^{-10}	4.38×10^{-9}	1.19×10^{-7}	1.82×10^{-6}	2.43×10^{-5}
10	3.75×10^{-11}	1.12×10^{-9}	3.02×10^{-8}	4.65×10^{-7}	5.68×10^{-6}

The most likely causes of such deviations are: (1) a kinetic energy dependence of the electronic transition matrix element, which must be taken into consideration even within the Franck-Condon approximation; (2) an R-dependence of the same electronic transition matrix element, which, by definition, constitutes a breakdown of the Franck-Condon approximation; (3) a v^+-dependence of the photoelectron angular distribution; and (4) other effects such as perturbations or overlapping lines at the three photon level and accidental resonances with autoionizing states at the four photon level. Of these possibilities, accidental resonances and perturbations are not important in the present case.[5] Hence, the data in Figure 3 represent a well defined case for further experimental and theoretical investigations of the excited state dynamics outlined in (1)-(3) above.

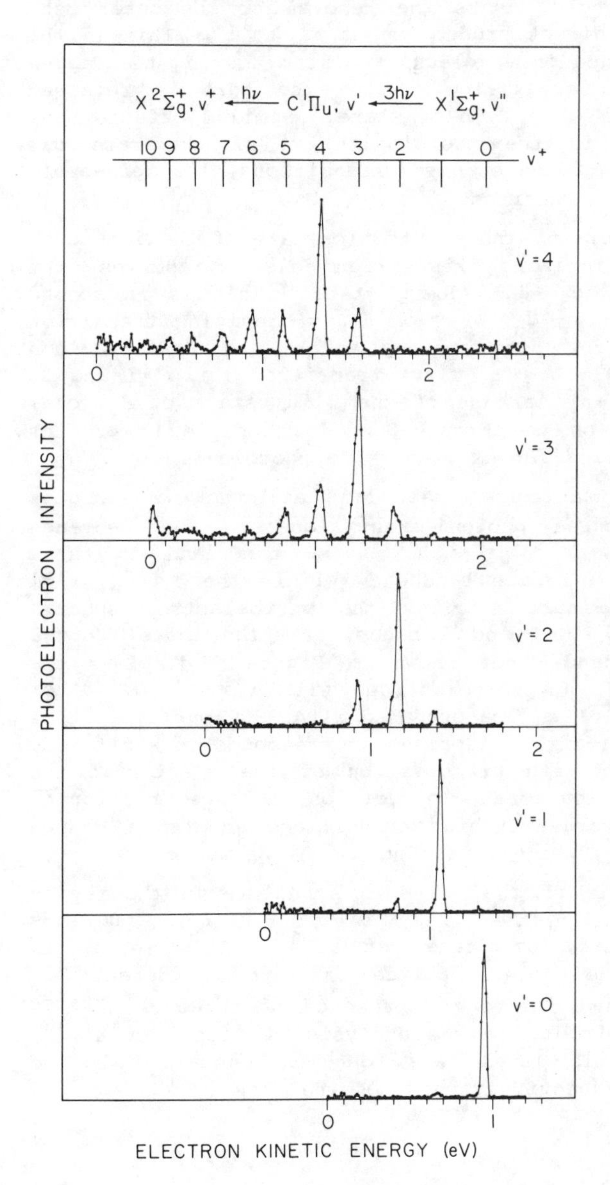

Figure 3. Photoelectron spectra of H_2 determined following (3+1) multiphoton ionization at the wavelengths of the resonant three photon $C\ ^1\Pi_u$, $v'=0-4 \leftarrow X\ ^1\Sigma_g^+$, $v''=0$, Q(1) transitions.

PHOTOIONIZATION OF THE $o_3\ ^1\Pi_u$ STATE OF N_2

This system serves to demonstrate the production of electronically excited, vibrational state-selected molecular ions by resonant multiphoton ionization via a Rydberg state with an electronically excited ion core. Specifically, it is shown that the dominant ionization pathway for the (3+1) multiphoton ionization of N_2 via the $o_3\ ^1\Pi_u$, v'=1,2 levels leads to the production of N_2^+ $A\ ^2\Pi_u$, v^+=1,2, respectively.[6] The $o_3\ ^1\Pi_u$ state of N_2 has the electron configuration $...(1\pi_u)^3(3\sigma_g)^2 3s\sigma_g$ and is the lowest member of Worley's third series,[16] which converges to N_2^+ $A\ ^2\Pi_{1/2u}$. The ionizing transition strongly favors the removal of the outer $3s\sigma_g$ electron leading to the direct production of N_2^+ $A\ ^2\Pi_u$. This is the first experimental evidence in a molecular system showing the degree to which the electronic excitation of the ion core is retained following photoionization of a Rydberg state. Similar effects have been observed recently in the photoionization of the rare gas atoms;[17] that is, in all of the observed transitions, the spin-orbit state of the ion core was preserved.

In addition to preserving the electronic state of the ion core, the $A\ ^2\Pi_u \leftarrow o_3\ ^1\Pi_u$ ionizing transition also preserves the vibrational level of the $o_3\ ^1\Pi_u$ Rydberg state. This is in accord with calculations for the $A\ ^2\Pi_u$, $v^+ \leftarrow o\ ^1\Pi_u$, v' ionizing transition using Morse potentials for both states, which give Franck-Condon factors greater than 0.99 for the v^+=v' transition for v'=1 and 2. However, as shown above in the case of the $C\ ^1\Pi_u$ state of H_2, one, in general, cannot rely on the Franck-Condon principle to predict vibrational branching ratios for excited state photoionization.

Photoelectron spectra were recorded at the wavelengths corresponding to the three photon transitions to the R-branch bandhead of the $o_3\ ^1\Pi_u$, v'=2 $\leftarrow X\ ^1\Sigma_g^+$, v"=0 band and to three positions, including the R-branch bandhead, within the $o_3\ ^1\Pi_u$, v'=1 $\leftarrow X\ ^1\Sigma_g^+$, v"=0 band. Figure 4 gives the photoelectron spectra obtained via the $o_3\ ^1\Pi_u$, v'=1 and 2 bands. Of the three spectra obtained via the v'=1 band, that shown in Figure 4 displays the greatest fraction (~20%) of photoelectrons with $v^+ \neq v'$. At other wavelengths, N_2^+ $A\ ^2\Pi_u$, v^+=1 can be produced with $\geq$90% purity. It is clear that the dominant ionization process corresponds to excitation of the Rydberg electron, with preservation of the electronic and vibrational state of the ion core. In addition, a large fraction of those ions produced by core-switching transitions undergo a change only in vibrational state, while retaining the $A\ ^2\Pi_u$ ion core.

An extensive analysis of the vibronic structure in the region of the N_2 spectrum containing the $o_3\ ^1\Pi_u$, v'=1,2 $\leftarrow X\ ^1\Sigma_g^+$, v"=0 bands has been reported recently by Stahel et al.[18] This region is extremely complicated due to the mixing of three different $^1\Pi_u$ states (see Figure 5), and due to the presence of three $^1\Sigma_u^+$ states that are also strongly mixed. In the analysis of Stahel et al.[18], the $o_3\ ^1\Pi_u$ v'=1,2 vibronic levels are found to contain 8.4% and 10.0% $b\ ^1\Pi_u$ character, respectively, and somewhat less $c\ ^1\Pi_u$

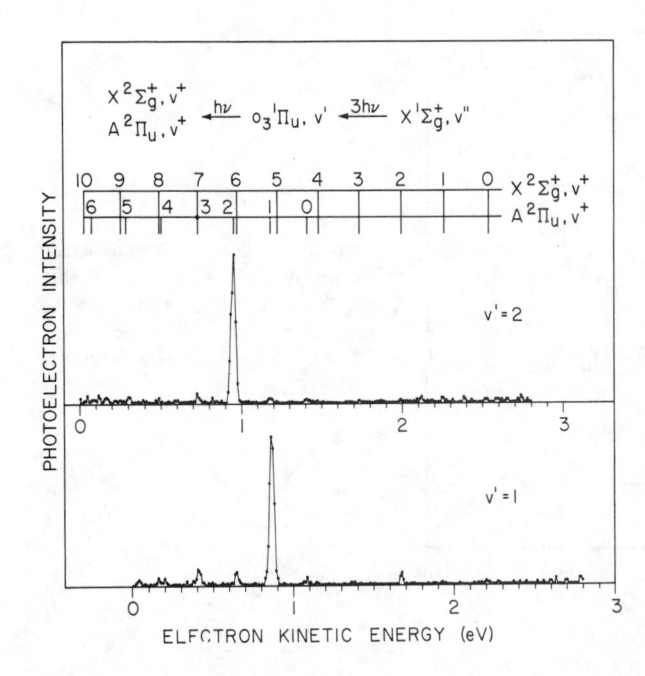

Figure 4. Photoelectron spectra of N_2 determined following (3+1) multiphoton ionization via the $o_3\,^1\Pi_u$, $v'=1,2 \leftarrow X\,^1\Sigma_g^+$, $v''=0$ bands.

character. In addition, a rotational analysis by Yoshino et al.[19] of the single photon absorption data for the $o_3\,^1\Pi_u \leftarrow X\,^1\Sigma_g^+$ transition indicates much stronger mixing at the R-branch bandhead of the v'=1 band due to a perturbation by the b $^1\Pi_u$, v'=12 level. One would expect these perturbations to strongly affect the vibrational branching ratios following multiphoton ionization. In particular, photoionization of high vibrational levels of the b $^1\Pi_u$ state, which is a valence state, would be expected to populate a wide distribution of N_2^+ vibrational levels in both the X $^2\Sigma_g^+$ and A $^2\Pi_u$ states. It is clear from Figure 4 that this is not the case. Calculations of Michels[20] indicate that two different electron configurations are important in describing the b $^1\Pi_u$ state of N_2. Both of these configurations, $...(2\sigma_u)^2(1\pi_u)^3(3\sigma_g)^1(1\pi_g)^2$ and $...(2\sigma_u)^1(1\pi_u)^4(3\sigma_g)^2(1\pi_g)^1$, differ from the X $^2\Sigma_g^+$ and A $^2\Pi_u$ states of N_2^+ by two orbitals. For this reason the photoionization cross section for the b $^1\Pi_u$ state may be relatively small, leading to a dominance of the $o_3\,^1\Pi_u$ character in the photoionization of the intermediate state. Similar behavior in the photoelectron spectra obtained following (3+1) multiphoton ionization of N_2 via the b $^1\Pi_u$, v'=3-5 and the c $^1\Pi_u$, v'=0 bands has been observed and is discussed below.

PHOTOIONIZATION OF THE b $^1\Pi_u$ AND c $^1\Pi_u$ STATES OF N_2

In preliminary studies of (3+1) multiphoton ionization of selected vibrational levels of the b $^1\Pi_u$ and c $^1\Pi_u$ states of N_2, we

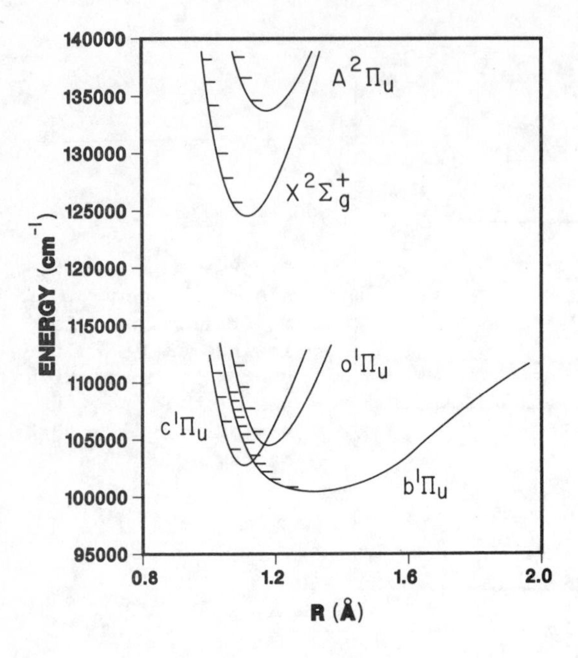

Figure 5. Potential energy curves for the $b\,^1\Pi_u$, $c\,^1\Pi_u$, and $o_3\,^1\Pi_u$ states of N_2 and the $X\,^2\Sigma_g^+$ and $A\,^2\Pi_u$ states of N_2^+.

see further evidence of the effect of Rydberg-valence mixing on the photoelectron spectra. Figure 6 shows photoelectron spectra recorded at the wavelengths of the three photon transitions to the R-branch bandheads of the $b\,^1\Pi_u$, $v'=3\text{-}5 \leftarrow X\,^1\Sigma_g^+$, $v''=0$ and $c\,^1\Pi_u$, $v'=0 \leftarrow X\,^1\Sigma_g^+$, $v''=0$ bands. The analysis of the vibronic structure of these states reported by Stahel et al.[18] shows that these levels are strongly mixed. The percentage vibronic character of each of these levels is given in Table II for the three main components of each wavefunction. (Note that the effects of local rotational perturbations, which are quite severe in some of these states, are not included in this analysis.[21]) The data of Table II show that each of these levels is a complex and unique mixture of $c\,^1\Pi_u$, $v=0$ and $b\,^1\Pi_u$, $v=2\text{-}5$, with smaller components (less than 10%) of other levels. In spite of this, all of the photoelectron spectra shown in Figure 6 look remarkably similar, displaying one dominant peak at the N_2^+, $v^+=0$ level. This is suggestive of ionization via a Rydberg state with a ground state core converging to $v^+=0$, i.e. the $c\,^1\Pi_u$, $v'=0$ level. It would therefore appear that the photoionization cross section from the $b\,^1\Pi_u$ electronic component of these levels is relatively small. However, while this interpretation appears to explain the results shown in Figure 6, it is less satisfactory for other bands in this region that are currently under investigation.[22] It is clear that more work is needed to fully understand these results. In particular, more complete and systematic studies are needed, which would include photoelectron angular distributions and photoelectron spectra determined as a function of the rotational level of the resonant intermediate state.

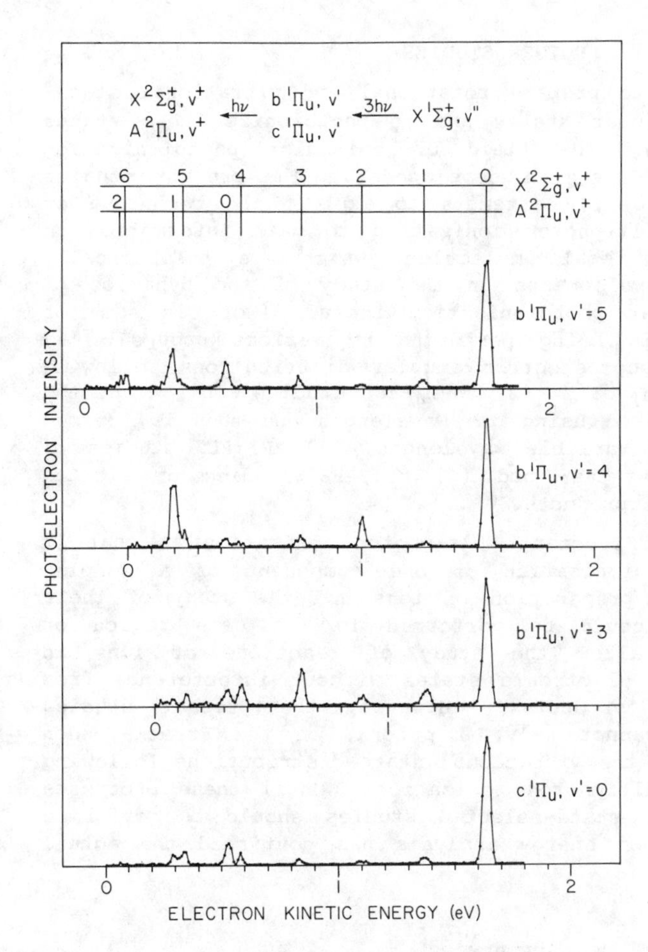

Figure 6. Photoelectron spectra of N_2 determined following (3+1) multiphoton ionization via the b $^1\Pi_u$, v'=3-5 ← X $^1\Sigma_g^+$, v"=0 and the c $^1\Pi_u$, v'=0 ← X $^1\Sigma_g^+$, v"=0 bands.

Table II. Percentage Vibronic Character for the c $^1\Pi_u$, v=0 and b $^1\Pi_u$, v=3-5 Levels of N_2 [(a)]

Level	c0	b2	b3	b4	b5	o0	Other
$c^1\Pi_u$, v=0	33.6			30.3	24.0		12.1
$b^1\Pi_u$, v=3	5.8	8.4	75.7				10.1
$b^1\Pi_u$, v=4	16.0		14.4	59.3			10.3
$b^1\Pi_u$, v=5	22.1				64.0	4.4	9.5

(a) Data from Reference 18. The column headings are appreviations of the appropriate term symbols, e.g. b2 is b $^1\Pi_u$, v=2.

FUTURE STUDIES

With the ability to prepare rotational and vibrational state-selected excited molecular states and to photoionize these states with a single photon, the field of molecular photoionization dynamics enters a new stage of development. The present studies represent some of the early attempts to exploit the technique of resonantly enhanced multiphoton ionization to gain information on the excited states of small molecules, which are theoretically tractable. Future developments in the study of the dynamics of excited state molecular photoionization include (work in some of these areas already is being performed by various groups): (1) determination of the photoelectron angular distributions following multiphoton ionization;[2e,23] (2) MPI/PES studies in which the excited state is prepared using one wavelength and then is ionized using an independently variable wavelength; (3) MPI/PES studies of autoionizing Rydberg states; and (4) MPI/PES studies of excited state photodissociation products.

Another aspect of resonant multiphoton ionization is that it allows the selective ionization of one component of a complex mixture. Thus, state preparation of ions and the study of their subsequent reactions could be performed in the same region of space. This would allow the study of reactions of ions in relatively short-lived electronic states without interference from extraneous ionization products. However, the present studies illustrate that one cannot rely, a priori, on the Franck-Condon principle to determine the vibrational state distributions following resonantly enhanced multiphoton ionization. Until these processes are better understood, state-selected studies should be performed only after photoelectron energy analysis has confirmed the actual state distribution.

REFERENCES

1. See, for example, P. M. Johnson and C. E. Otis, Ann. Rev. Phys. Chem. 32, 139 (1981).
2. For examples of MPI/PES studies of small molecules, see (a) J. C. Miller and R. N. Compton, J. Chem. Phys. 75, 22 (1981); (b) M. G. White, M. Seaver, W. A. Chupka, and S. D. Colson, Phys. Rev. Lett. 49, 28 (1982); (c) J. Kimman, P. Kruit, and M. J. van der Wiel, Chem. Phys. Lett. 88, 576 (1982); (d) J. H. Glownia, S. J. Riley, S. D. Colson, J. C. Miller, and R. N. Compton, J. Chem. Phys. 77, 68 (1982); (e) Y. Achiba, K. Sato, K. Shobatake, and K. Kimura, J. Chem. Phys. 78, 5474 (1983); (f) S. R. Long, J. T. Meek, and J. P. Reilly, J. Chem. Phys. 79, 3206 (1983); and S. T. Pratt, P. M. Dehmer, and J. L. Dehmer, J. Chem. Phys. 79, 3234 (1983). Also, for a review of MPI/PES processes in atoms and molecules to early 1983, see K. Kimura, Adv. Chem. Phys. (in press).
3. S. T. Pratt, E. D. Poliakoff, P. M. Dehmer, and J. L. Dehmer, J. Chem. Phys. 78, 65 (1983).

4. S. T. Pratt, P. M. Dehmer, and J. L. Dehmer, J. Chem. Phys. 78, 4315 (1983).
5. S. T. Pratt, P. M. Dehmer, and J. L. Dehmer, Chem. Phys. Lett. (in press).
6. S. T. Pratt, P. M. Dehmer, and J. L. Dehmer, J. Chem. Phys. (in press).
7. G. Herzberg, Molecular Spectra and Molecular Structure II. Spectra of Diatomic Molecules (Van Nostrand Reinhold, Princeton, 1950).
8. J. B. Halpern, H. Zacharias, and R. Wallenstein, J. Mol. Spectrosc. 79, 1 (1980).
9. U. Fano and D. Dill, Phys. Rev. A 6, 185 (1972).
10. D. Dill, Phys. Rev. A 6, 160 (1972).
11. D. Dill, in Photoionization and Other Probes of Many-Electron Interactions, edited by F. J. Wuilleumier (Plenum, New York, 1976), p. 387.
12. H. M. Crosswhite, The Hydrogen Molecule Wavelength Tables of Heinrich Dieke (Wiley-Interscience, New York, 1972).
13. K. P. Huber and G. Herzberg, Molecular Spectra and Molecular Structure IV. Constants of Diatomic Molecules (Van Nostrand Reinhold Company, New York, 1979).
14. I. Dabrowski and G. Herzberg, Can. J. Phys. 66, 5584 (1977).
15. Franck-Condon factors were calculated using the numerical potential of W. Kolos and J. Rychlewski, J. Mol. Spectrosc. 62, 109 (1976) for the C $^1\Pi_u$ state and that of H. Wind, J. Chem. Phys. 42, 2371 (1965) with the adiabatic corrections of D. M. Bishop and R. W. Wetmore, Mol. Phys. 26, 145 (1973) for the X $^2\Sigma_g^+$ ionic state.
16. A. Loftus and P. H. Krupenie, J. Phys. Chem. Ref. Data 6, 113 (1977).
17. (a) J. Ganz, B. Lewandowski, A. Siegel, W. Bussert, H. Waibel, M. W. Ruf, and H. Hotop, J. Phys. B. 15, L485 (1982); (b) A. Siegel, J. Ganz, W. Bussert, and H. Hotop, J. Phys. B 16, 2945 (983); and (c) K. Sato, Y. Achiba, and K. Kimura, J. Chem. Phys. 80, 57 (1984). See also R. F. Stebbings, F. B. Dunning, and R. D. Rundel, in Atomic Physics Volume 4, edited by G. zu Putlitz, E. W. Weber, and A. Winnacker (Plenum, New York, 1975), p. 713, and references therein.
18. D. Stahel, M. Leoni, and K. Dressler, J. Chem. Phys. 79, 2541 (1983).
19. K. Yoshino, Y. Tanaka, P. K. Carroll, and P. Mitchell, J. Mol. Spectrosc. 54, 87 (1975).
20. H. H. Michels, Adv. Chem. Phys. 45, 225 (1981).
21. P. K. Carroll and C. P. Collins, Can. J. Phys. 47, 563 (1969).
22. S. T. Pratt, P. M. Dehmer, and J. L. Dehmer, (to be published).
23. (a) M. G. White, W. A. Chupka, M. Seaver, A. Woodward, and S. D. Colson, J. Chem. Phys. 80, 678 (1983); and (b) S. L. Anderson, G. D. Kubiak, and R. N. Zare, Chem. Phys. Lett. (in press).

STATE-TO-STATE PHOTOCHEMISTRY OF GLYOXAL USING A TUNABLE VUV LASER FOR PRODUCT DETECTION

John W. Hepburn

Centre for Molecular Beams and Laser Chemistry
Department of Chemistry
University of Waterloo
Waterloo, Ontario, Canada N2L 3G1

and

N. Sivakumar and P. L. Houston
Department of Chemistry
Cornell University
Ithaca, New York, U. S. A. 14853

ABSTRACT

The predissociation of glyoxal has been studied by laser excitation of glyoxal molecules in a supersonic beam, followed by VUV laser-induced fluorescence (LIF) detection of the photofragments. Preliminary results on the time evolution of the photodissociation, as well as the internal energy distribution of the CO photoproduct will be discussed.

INTRODUCTION

In recent years there has been a surge of interest in learning how the initial state of a reactant affects the products in a chemical process. Knowledge of the state-to-state relationship between reactants and products not only allows practical control of reactions but often leads to a detailed understanding of the potential energy surfaces on which they occur. To date, however, there have been only a few systems on which state-to-state chemistry has been possible. One of the most convenient of these is a molecule undergoing predissociation. Since a predissociating molecule often possesses a structured absorption, state-selected excitation can easily be achieved. Furthermore, the unimolecular nature of the dissociation process makes it possible to determine the probability for formation of product states under isolated-molecule conditions, where collisions have not disturbed the nascent populations. Several examples of the power of state-to-state photochemistry have recently appeared, including studies of van der Waals compounds,[1] CF_3NO,[2,3] and HONO.[4] Most experiments, however, have been limited by the need to examine photodissociations in which the product fragment(s) absorb light in the visible or near ultraviolet regions accessible by commercial lasers.

Recently, vacuum ultraviolet sources have greatly expanded the scope of state-to-state experiments, particularly those involving the CO molecule. Shortly following the report by Wallace and Zdasiuk of four-wave mixing in Mg vapor to produce tunable VUV wavelengths in the 150-nm region,[5] several studies involving detection of products by VUV laser induced fluorescence near this wavelength appeared in the literature. Bromine atoms were detected as products of the F + HBr, H + Br_2 and H + HBr reactions by resonance atomic fluorescence,[6-8] while CO was detected by LIF on the $A^1\Pi \leftarrow X^1\Sigma$ transition as the product of both scattering from solid surfaces[9] and photodissociation of H_2CO.[10] In this paper we report preliminary results on the elucidation of the state-to-state photodissociation dynamics of glyoxal using VUV laser induced fluorescence to detect the CO produced following selective excitation of glyoxal to individual vibronic levels of its first excited singlet state.

While both the spectroscopy and photophysics of the S_1 state of glyoxal have been extensively studied,[11] its photochemistry is not fully understood. Along with formaldehyde, glyoxal stands as an important test case for theories of unimolecular photodissociation. Until recently it was doubted that photochemistry took place at all from low vibronic states of S_1 in the absense of collisions.[12] It is still not known whether the dissociation is a prompt unimolecular event or whether there might be a long-lived intermediate.[13-15] However, a recent molecular beam studycarried out by one of the authors[16] has demonstrated the presence of photochemistry in the absence of collisions and has provided detailed information on the branching ratios for different channels and on the microscopic fragmentation mechanism. Three possible photoproducts have been postulated:

$$(HCO)_2 \rightarrow HCOH + CO + \Delta E \simeq h\nu - 18{,}300\ cm^{-1} \quad (1)$$
$$\rightarrow H_2CO + CO + \Delta E \simeq h\nu + 1{,}600\ cm^{-1} \quad (2)$$
$$\rightarrow H_2 + 2CO + \Delta E \simeq h\nu + 2{,}600\ cm^{-1} \quad (3)$$

The yield for each of these three channels is thought to be 7%, 65%, and 28%, respectively.[16] From the large difference in available energy, particularly between channels (1) and (2), one might predict an interesting internal energy distribution in the CO product. One object of the current study was to determine that distribution. A second object was to measure the time dependent production of CO in order to confirm or disprove the presence of a long-lived intermediate in the collisionless dissociation.

EXPERIMENTAL

A schematic diagram of the experimental apparatus is shown in Fig. 1. The photolysis laser was an excimer-pumped dye laser (Lambda Physik 101, 2002) and was tuned to the 8^1 vibronic transition of glyoxal near 440.3 nm. The vacuum ultraviolet probe laser was similar to that described previously,[5] with the exception that the dye lasers (each a Lambda Physik 2002) were pumped by the two output beams of a dual cavity excimer laser (Lambda Physik 150ES). The two excimer pulses, derived from the same power supply, were locked in time to one another and separated by about 16 ns. By adjusting the optical delay caused by the two beam paths, the outputs of the dye lasers could be made to coincide temporally at the input of the magnesium heat pipe. The fixed frequency dye laser was made to operate at 430.9 nm, while the variable frequency dye laser scanned the range from 469.5-480 nm. The resulting VUV output in the range from 147.7-148.7nm was used to probe the (2,0) vibrational band of the CO A←X transition.

Glyoxal was seeded at 5% in helium and expanded through a pulsed valve of the design described by Adams et al.[17] The molecular beam chamber has been described in detail elsewhere.[18] The photolysis and probe laser beams were propagated collinearly in opposite directions and were perpendicular to the molecular beam. Fluorescence induced in glyoxal by the photolysis laser and fluorescence induced in CO by the VUV laser were collected in opposite directions, each mutually orthogonal to the laser and molecular beams. A Hamamatsu photomultiplier (R928) was used to detect the glyoxal fluorescence, while an EMR photomultiplier (S41G-09-17) was used to detect the CO fluorescence. A reflection from the exit window for the VUV laser was detected by a third photomultiplier (EMI; G-26-E314LF).

The signal from each photomultiplier was fed to a boxcar integrator (PAR 162 or Stanford Research Systems), while the outputs from each integrator was fed to one of three channels of an A/D converter. A computer (LSI 11/02) read the A/D converters and normalized the signal channel by dividing by voltages proportional to the VUV intensity and the glyoxal fluorescence. The normalized signal was plotted during each run and then stored for later analysis.

For some scans, one of the boxcars was operated in the active background subtraction mode. The photolysis laser was fired at half the repetition rate of the probe laser, and the boxcar was then made to average the difference between the signal with and without the photolysis laser. In this way, it was possible to correct the signal both for CO impurities in the glyoxal sample (which caused some interference in measuring low rotational levels of CO(v=0)) and for the possibility that the probe laser might itself dissociate some glyoxal and simultaneously excite the CO product. However, no evidence for the latter effect was observed.

The appearance rate of the CO photoproduct was measured by scanning the delay of the probe laser relative to the photolysis laser while keeping the boxcar gates for the CO fluorescence and VUV intensity locked to the probe laser.

Glyoxal was prepared from the trimer, obtained in 99% purity from sigma. The trimer was mixed thoroughly with P_2O_5 and heated to 160-180 °C. Water was trapped at -0°C, and the glyoxal monomer collected at liquid N_2 temperature. Immediately following preparation, the glyoxal monomer was distilled from a bath at 0°C to one at -73°C. At the latter temperature, formaldehyde and CO have high vapor pressures, whereas glyoxal is trapped as a solid.

RESULTS

Figure 2a displays a spectrum of the CO photoproduct of glyoxal dissociation. In the wavelength region of this diagram, the VUV laser probes primarily CO(v=0). For comparison, a spectrum of room temperature CO at a pressure of 2×10^{-6} torr is displayed in Fig. 2b. It is evident that, while the CO product of glyoxal dissociation is more rotationally excited than is a Boltzmann distribution at room temperature, it is not as highly exicited as the CO distribution from formaldehyde dissociation.[10,19] It is also clear that many more lines are present in the LIF spectrum of the CO photoproduct than in the spectrum of room temperature CO. Although a detailed analysis has not yet been performed, it is possible that these additional lines are due to absorption by vibrationally excited CO product molecules. For example, the (5,2) and (4,1) bands lie in the same spectral region.

The appearance time of the CO photoproduct is shown in Fig. 3. The risetime of the signal, about 1 μsec, is equal within the experimental uncertainty to the decay time of fluorescence from the 8^1 level of glyoxal[20]. The apparent decay of the signal in Fig. 3 is a result of the CO fragment leaving the viewing zone at the beam velocity ($\simeq 1.4 \times 10^5$ cm/sec), which will take on the order of 2-4 μsec, as observed.

DISCUSSION

The study of simple atom transfer reactions ($A+BC \rightarrow AB+C$) has shown that detailed knowledge of energy disposal in a chemical process can be translated into detailed knowledge of the potential energy hypersurface. Similarly, for a photofragmentation process, the detailed microscopic mechanism can be obtained from a thorough knowledge of the energy release in the fragmentation. As an example, the high level of rotational excitation of the CO product observed in H_2CO photodissociation indicates dissociation occurs through a very constrained, highly bent intermediate. The preliminary data for glyoxal seem to indicate the photodissociation process channels much less energy into rotation and instead energy is deposited into vibrational excitation. A detailed analysis of the CO photoproduct internal energy distribution, when combined with the data already available on translational energy in the products[16] will provide a fairly complete picture of the dissociation mechanism. Such an analysis will be available soon, when a more thorough study has been completed.

Even though two other groups have suggested the presence of a slowly decaying intermediate in the photochemistry of glyoxal,[12-15] the appearance time of CO displayed in Fig. 3 leaves little doubt that the collisionless photochemistry of glyoxal is prompt. The appearance time of the CO is in sensible agreement with the decay time of the 8^1 level of S_1 glyoxal (0.87 μsec[20]) The cause for the discrepancy between the previous results and those presented here is not yet clear.

Another result of this preliminary study is that it demonstrates the sensitivity of VUV laser-induced fluorescence for detection of CO photoproducts. Figure 2b shows that even with relatively little signal averaging (10 shots/point) the sensitivity for detection of CO is quite good. For these experiments, the CO detection limit was on the order of 10^7 molecules/cm^3 in a given quantum state, and improvements to the apparatus can increase this sensitivity by 10-100 times.

CONCLUSIONS

The data presented in this paper, though preliminary, provide new information about the detailed mechanism for glyoxal predissociation. The possibility of doing state to state dissociation experiments on glyoxal has been demonstrated, and it has been clearly shown that collisionless glyoxal predissociation does not proceed through a long lived intermediate.

Much work remains to be performed on the photodissociation of glyoxal. Detailed calculations of the vibrational and rotational distributions have yet to be made. While the present work has concentrated on vibronic excitation of the 8^1 transition in glyoxal, several other vibrational states may readily be examined in order to determine how the CO vibrational and rotational distributions depend on the internal energy of the glyoxal parent compound. Indeed, it would even be possible to examine the dependence of the final states on the initial rotational level excited in glyoxal. Finally, a recent publication has reported that the triplet state of glyoxal can be populated directly by laser excitation in a molecular beam.[21] It would be interesting to examine how the photochemistry from the individual levels of the triplet state differs from that of the singlet state.

It should be emphasized that the availability of VUV laser sources now opens up a wide range of experiments in molecular dynamics. In particular, the sensitive detection of CO demonstrated in this work should make it possible to examine in detail the dissociation mechanisms of several other aldehydes. Indeed, progress along these lines has been made concurrently with our work.[19] Detection of CO produced in the photodissociation of other compounds such as OCS, CO_2 and HCO can readily be envisioned. Finally, since CO is a prominent species in many interesting dynamical processes, it is likely that other state-to-state experiments, such as translation to vibration/rotation energy transfer in atomic collisions with CO and molecular beam scattering studies involving CO, will shortly appear. The possibilities for extending similar VUV detection techniques to other molecules (such as H_2) are endless.

ACKNOWLEDGEMENTS

This work has been supported by a grant and a University Research Fellowship to JWH from NSERC and by the National Science Foundation through grant CHE-8314146 to PLH. Travel between Ontario and New York were provided by a NATO grant for international collaboration. The experiments were performed in the Facility for Laser Spectroscopy in the Department of Chemistry at Cornell University with lasers supplied by a DoD Instrumentation grant (AFOSR-83-0279). We are grateful to D. Bamford and C. B. Moore for communicating their results to us prior to publication.

REFERENCES

1. W. Sharfin, K. E. Johnson, L. Wharton, and D. H. Levy, J. Chem. Phys. 71, 1292 (1978).
2. R. W. Jones, R. D. Bower, and P. L. Houston, J. Chem. Phys. 76, 3339 (1982).
3. R. D. Bower, R. W. Jones, and P. L. Houston, J. Chem. Phys. 79, 2799 (1983).
4. R. Vasudev, R. N. Zare, and R. N. Dixon, Chem. Phys Lett. 96, 399 (1983); J. Chem. Phys., to be published.
5. S. C. Wallace and G. Zdasiuk, Appl. Phys. Lett. 28, 449 (1976).
6. J. W. Hepburn, D. Klimek, K. Liu, J. C. Polanyi, and S. C. Wallace, J. Chem. Phys. 69, 4311 (1978).
7. J. W. Hepburn, D. Klimek, K. Liu, R. G. Macdonald, F. J. Northrup, and J. C. Polanyi, J. Chem. Phys. 74, 6226 (1981).
8. J. W. Hepburn, K. Liu, R. G. Macdonald, F. J. Northrup, and J. C. Polanyi, J. Chem. Phys. 75, 3353 (1981).
9. J. W. Hepburn, F. J. Northrup, G. L. Ogram, J. M. Williamson and J. C. Polanyi, Chem. Phys. Lett. 85, 127 (1982).
10. P. Ho and A. V. Smith, Chem. Phys. Lett.. 90, 407 (1982).
11. see Y. Osamura, H. F. Schaefer III, M. Dupuis, and W. A. Lester, Jr., J. Chem. Phys. 75, 5828 (1981) and references therein.
12. R. Naaman, D. M. Lubman, and R. N. Zare, J. Chem. Phys. 71, 4192 (1979) and references therein.
13. G. W. Loge, C. S. Parmenter, and B. F. Rordorf, Chem. Phys. Lett. 74, 309 (1980).
14. G. W. Loge and C. S. Parmenter, J. Phys. Chem. 85, 1653 (1981).
15. G. Loge and C. S. Parmenter, Abstract for the 14th Informal Conference on Photochemistry, March 1980.
16. J. W. Hepburn, R. J. Buss, L. J. Butler, and Y. T. Lee, J. Phys. Chem. 87, 3638 (1983).
17. T. E. Adams, B.H. Rockney, R.J.S. Morrison and E.R. Grant, Rev. Sci. Instrum., 52, 1469 (1981).
18. W.G. Hawkins and P. L. Houston, J. Chem. Phys. 86, 704 (1982).
19. D. Bamford and C. B. Moore, private communication.
20. B.G. MacDonald and E.K.C. Lee, J. Chem. Phys., 71, 5049 (1979).
21. L.H. Spangler, Y. Matsumoto, and D. W. Pratt, J. Phys Chem., 87, 4781 (1983).

Figure 1

Schematic diagram of the photodissociation apparatus described in the text. The beam chamber is on the right of the diagram, with the molecular beam perpendicular to the plane shown.

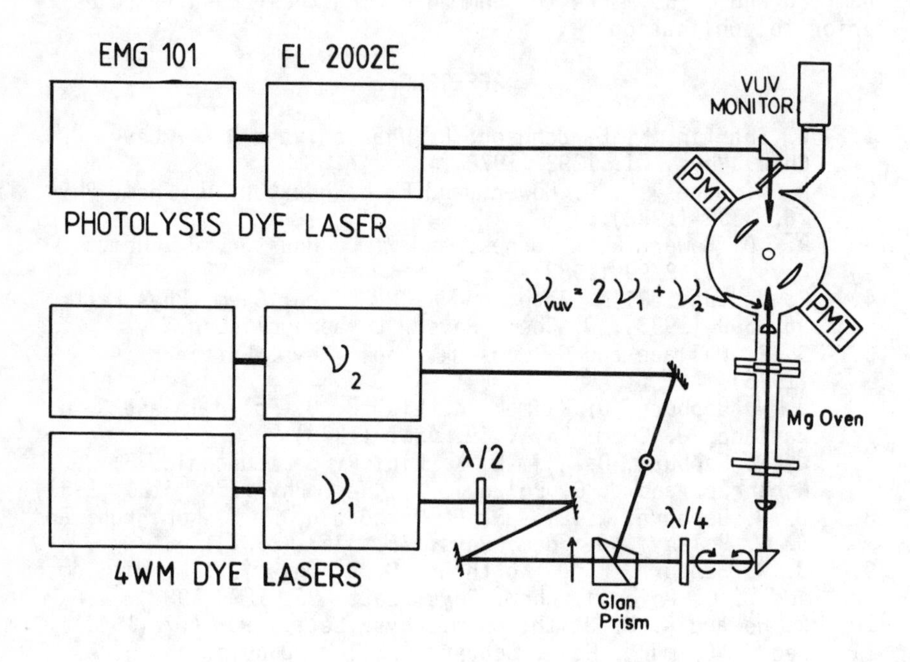

Figure 2

a) Spectrum of CO photoproduct from predissociation of the $^1_0 8$ band of glyoxal. The spectrum was recorded at 1 μsec delay after the photolysis laser and is uncorrected for background CO.

b) Spectrum of 300K CO in a gas cell at a measured pressure of 2 x 10^{-6} torr.

In both cases, the wavelengths scale is the wavelength of the tunable dye laser (ν_2).

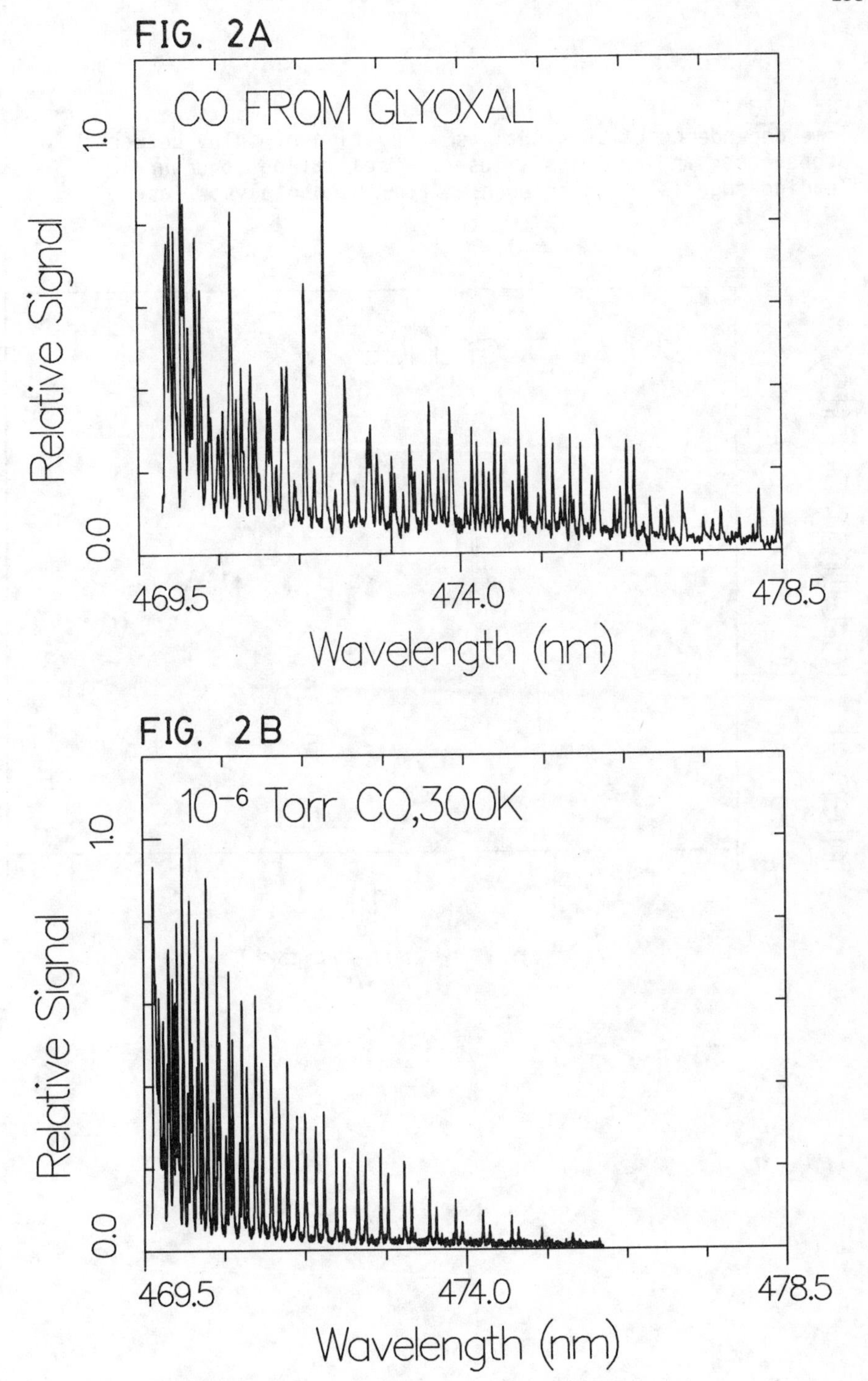
FIG. 2A
CO FROM GLYOXAL
Relative Signal
0.0
1.0
469.5
474.0
478.5
Wavelength (nm)
FIG. 2B
10⁻⁶ Torr CO,300K
Relative Signal
0.0
1.0
469.5
474.0
478.5
Wavelength (nm)

Figure 3
Time dependence of CO signal as a function of delay between VUV probe laser and photolysis laser. Oscillating structure on leading edge is rf interference from the photolysis laser.

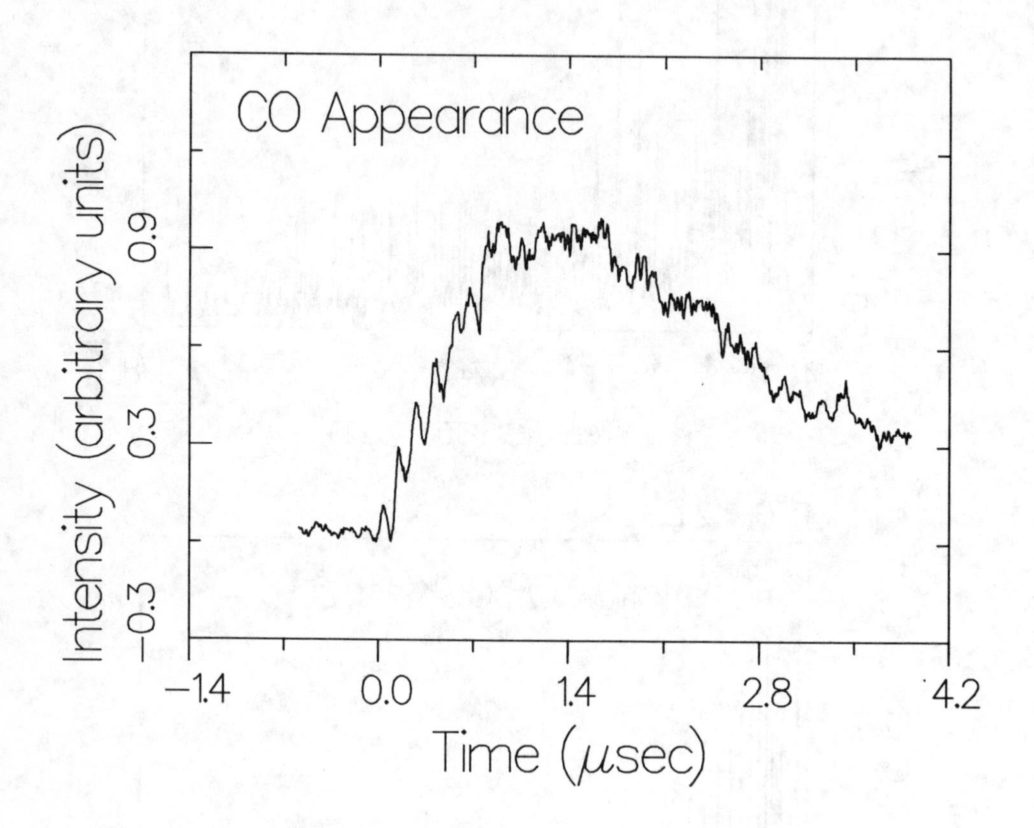

MULTIPHOTON IONIZATION OF BENZENE IN SOLUTION*

K. Siomos, H. Faidas,† and L. G. Christophorou†

Atomic, Molecular and High Voltage Physics Group
Health and Safety Research Division
Oak Ridge National Laboratory
Oak Ridge, Tennessee 37831

ABSTRACT

The multiphoton ionization spectrum of benzene in a dilute solution in n-pentane has been measured for laser excitation wavelengths from 360-560 nm. The observed spectrum, from 460-560 nm, has been measured using linearly and circularly polarized light and has been ascribed to multiphoton ionization via the two-photon resonant $^1B_{2u}-^1A_{1g}$ transitions of benzene. The order of the multiphoton ionization process has been determined, and the observed resonances in the spectral region from 360-445 nm are discussed. Power measurements at 337.1 nm indicate that two photons can ionize benzene in this energy region, giving an upper limit to the ionization threshold of benzene in n-pentane of ~7.36 eV

INTRODUCTION

Multiphoton absorption and excitation spectra of neat benzene or benzene in liquid and rigid solutions have been frequently reported (see Ref. 1 and references quoted therein). However, multiphoton ionization studies of neat benzene at room temperature are limited[2-4] and restricted to laser-excitation wavelengths, $\lambda \lesssim 440$ nm, which correspond to two-photon excitation energies $h\nu \gtrsim 45{,}500$ cm^{-1}. Multiphoton ionization studies of benzene in dilute solutions to our knowledge are not available. The latter studies are of significant importance in probing the effects of the liquid medium on the ionization process of atoms and molecules embedded in a dense medium.

In this paper we report our investigations of the multiphoton ionization of dilute solutions of benzene in liquid n-pentane at room temperature in the spectral region 360-560 nm which corresponds to two-photon excitation energies of 36,000-56,000 cm^{-1}. Linearly and circularly polarized light was used together with a laser conductivity technique we developed[5,6] for liquid-phase ionization investigations.

---

*Research sponsored by the Office of Health and Environmental Research, U.S. Department of Energy, under contract W-7405-eng-26 with the Union Carbide Corporation.

†Also Department of Physics, The University of Tennessee, Knoxville, Tennessee 37996.

EXPERIMENTAL

The experimental apparatus employed in the present study is shown schematically in Fig. 1. It consists of a frequency tunable dye laser pumped by a nitrogen laser, polarization optics, the photoconductivity cell, the electronic detection system, and the vacuum and liquid purification assembly.

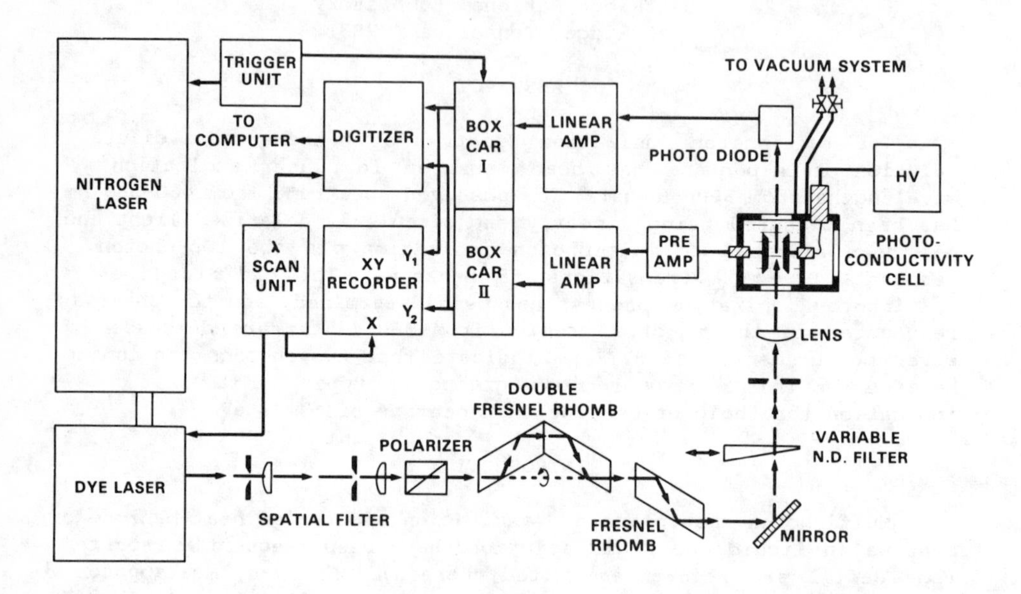

Fig. 1. Layout of the multiphoton ionization apparatus.

The laser beam, after spatial filtering, was linearly polarized by a Glan-Thomson prism polarizer and sent through a polarizing assembly comprised of a double- and a single-Fresnel rhomb. The plane of polarization of the laser radiation was oriented with respect to the fast axis of the single-Fresnel rhomb by rotating the double-Fresnel rhomb so that the light leaving the single-Fresnel rhomb was either linearly or circularly polarized. The laser beam, of well-defined spatial distribution[6] and a diameter of ~1 mm, was focused by a 130-mm focal length lens on the photoconductivity cell containing the charge-sensitive conductivity detector and the benzene solution. The laser beam leaving the conductivity cell is monitored by a photodiode of well-known spectral responsivity. The output signal of the diode was amplified and sampled by a boxcar integrator. The multiphoton ionization signal from the parallel-plate charge-sensitive conductivity detector after pre- and linear-amplification is averaged by a second boxcar integrator. The output signals of both integrators are digitized and fed into a PDP-10 computer for analysis.

The vacuum and liquid purification assembly is a combination of two separate high-vacuum pump systems with various molecular filters and distillation steps allowing for the simultaneous vacuum purification of two different liquid samples. The liquids (n-pentane and benzene) are distilled under vacuum ($P \sim 10^{-6}$ Torr) and sent through columns of silica gel and charcoal which have been previously activated under vacuum at 300 and 500°C, respectively. The liquids, upon exiting the columns, enter into a freeze-pump-thaw system which allows for further deoxygenation to a residual pressure of $P \lesssim 10^{-7}$ Torr. The liquid n-pentane is further stirred for 24 h under vacuum on liquid sodium-potassium alloy (NaK) to remove traces of oxygen and water. It was also treated for 72 h prior to vacuum distillation with sulfuric acid and barium carbonate in order to remove traces of olefinic components. Both liquids were obtained from Phillips Petroleum Company and were of a quoted purity of 99.99%.

RESULTS AND DISCUSSION

In the multiphoton ionization (MPI) spectroscopy of molecules in the liquid phase, the assignment of the observed resonances to specific energy levels is not trivial. In a liquid medium, sinks for rotational-vibrational energies alter the rates of the nonradiative molecular deactivation processes and the number of the nonradiative decay channels, thereby causing the well-known line broadening observed in the absorption spectra of molecules in the liquid phase. These spectra are, in addition, shifted to the red with respect to their gaseous-phase peaks due to dispersive forces in the liquid. Although these effects influence the transitions between valence states rather moderately, their influence on Rydberg transitions is dramatic.[7,8]

In general, resonances leading to electron-ion pair formation can occur at energies corresponding to any integer multiple of the laser-photon energy. If we consider the fact that the ionization threshold I_L of a molecule dissolved in a liquid is significantly lower (<3 eV)[9,10] than its corresponding ionization threshold in the gaseous phase, I_G, a number of excited states that lie below I_G in the gas phase lie above I_L in the liquid and can thus autoionize.[5,6,9]

In view of the aforementioned statements, it is necessary in liquid phase MPI studies to carefully consider the known spectroscopy of the solute under study, as well as to establish the order of the MPI process, by conducting laser-power dependence measurements.[6] It is also necessary to work in a laser-power regime in which the energy absorbed by the solute leads to excitation of the desired level, which is rather difficult to establish in liquid media. Rapid de-excitation mechanisms could lead to a situation in which, for example, the second photon in a two-photon excitation would be absorbed from an excited energy level different from the one initially prepared by the first photon.

In the one-photon spectrum of benzene in the ultraviolet region, the well-known 260 nm band ($\sim$37,000-45,000 cm^{-1}) is assigned to the $^1B_{2u}$-$^1A_{1g}$ transition whose upper state is the lowest excited singlet state of benzene. This transition is symmetry forbidden for one-photon excitation and parity forbidden for two-photon excitation. The spectrum in the 48,000-56,000 cm^{-1} energy region arises from two π-electron valence states of $^1B_{1u}$ (~48,000-53,000 cm^{-1}) and $^1E_{1u}$ (~54,300 cm^{-1}) symmetry.[11,12] In addition to these states a two-photon resonant band near 51,000 cm^{-1} has been reported by Johnson[13] in a two-photon resonant three-photon ionization experiment with gaseous benzene. This state has been variously assigned as a $^1E_{1g}$ Rydberg or a $^1E_{2g}$ valence state. The latter assignment, however, has apparently been ruled out by several recent studies in neat liquid benzene using thermal blooming and ionization techniques.[1,4]

The wavelength region in our investigation is 360-560 nm, corresponding to one-photon energy of ~18,000-28,000 cm^{-1}, a two-photon energy of ~36,000-56,000 cm^{-1}, and a four-photon energy of ~72,000-112,000 cm^{-1}. The last values lie well above the gaseous-phase ionization threshold of benzene (I_G = 9.247 eV[11]) and should be considerably above I_L.[14] In the one-photon region, no resonant transitions are expected, so that the first possible resonances are expected in the two-photon energy region (see Fig. 2). Figure 2 shows the linearly polarized multiphoton ionization spectrum of ~10^{-3} mol dm^{-3} solutions of benzene in n-pentane as a function of the laser-excitation wavelength from 360-560 nm. The spectrum is composed of nine individual spectra corrected by the second power (assumed second-order resonances) of the laser intensity and combined by normalization in the respective regions of overlap. The spectrum shows two interesting features: (1) distinct structure is observed over the entire spectral range investigated, and (2) the photocurrent versus λ declines sharply for $\lambda \geq 400$ nm and decreases slowly as λ increases, and it appears to set in at $\lambda \sim 550$ nm. The structure in the photocurrent versus λ spectrum for $\sim 440 < \lambda < 550$ nm (~36,000-45,000 cm^{-1}) reflects the linearly polarized two-photon absorption spectrum of the $^1B_{2u}$-$^1A_{1u}$ transitions of benzene. Power-dependence measurements of the observed photocurrent as a function of the laser-pulse energy (see Fig. 3) at different laser excitation wavelengths, λ = 445, 496, and 508 nm, indicate that the observed spectrum in the two-photon energy region (36,000-45,000 cm^{-1}) can be ascribed to a two-photon resonant four-photon ionization process. This interpretation is supported by the excellent agreement found between the spectrum in Fig. 2 and the two-photon excitation spectrum reported by Friedrich et al.,[15] as can be seen in Fig. 4. It is interesting to note the higher resolution obtained with our multiphoton ionization technique. In order to further scrutinize our interpretation that the observed resonances for $\lambda \geq 445$ in Fig. 2 are due to the $^1B_{2u}$-$^1A_{1g}$ transitions of benzene and free of other resonances in the additional two photons absorbed (bound-to-continuum transitions), we observed the $^1B_{2u}$-$^1A_{1g}$ transitions using circularly polarized light. The polarization ratios (Ω) defined as σ (circular)/ σ (linear) were calculated in the spectral region $480 \leq \lambda \leq 540$ nm and compared with the corresponding Ω ratios calculated from the data by Friedrich et al. The two Ω ratios are similar. An interpretation of the Ω ratios in the MPI spectrum is presently not possible, but the fact that they are similar to the Ω ratios obtained in two-photon excitation spectra indicate that the ionization from the resonant $^1B_{2u}$ state of benzene in solution is isotropic (bound-to-continuum transition) and dominated by the properties of the $^1B_{2u}$ state.

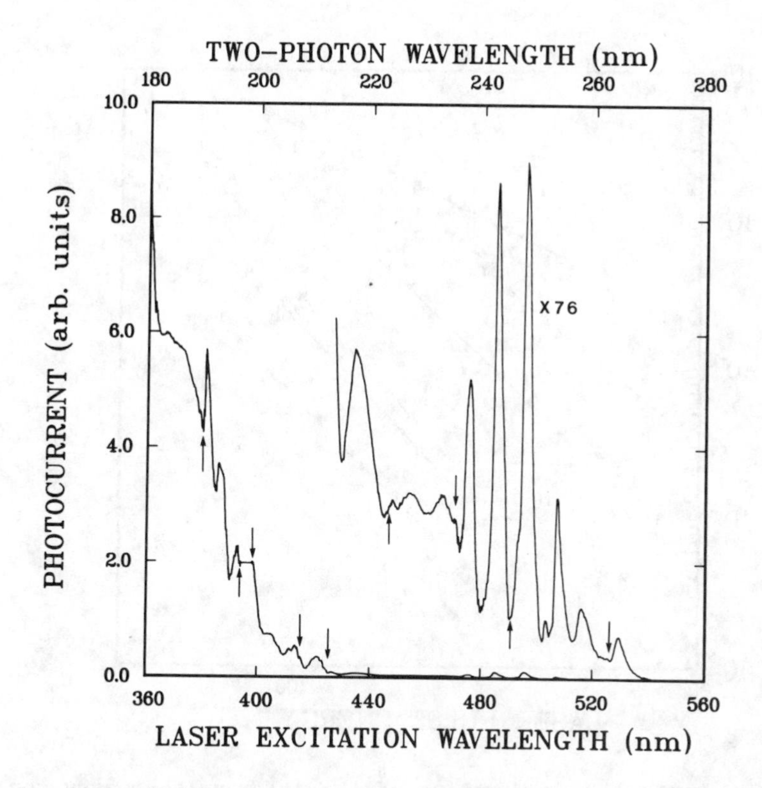

Fig. 2. The multiphoton ionization spectrum of $\sim 10^{-3}$ mol dm^{-3} solutions of benzene in n-pentane at room temperature. The spectrum above 420 nm has been magnified by a factor of 76 for the purpose of display. The spectrum is composed of nine individual corrected spectra combined by normalization in their respective region of overlap. The arrows indicate the points of normalization. The portion of the spectrum 394-397 nm could not be measured due to the low laser intensity.

Referring now to the spectral region $360 \leq \lambda \leq 445$ nm (45,000-56,000 cm^{-1}) of the MPI spectrum in Fig. 2, the following comments can be made: (1) The spectrum shows a number of relatively sharp peaks superimposed on a constantly increasing background as λ decreases. (2) Measurements of the photocurrent versus laser-pulse energy at $\lambda = 386$ and 365 nm (see Fig. 3) indicate that the ionization process in this spectral region involves a three-photon absorption. Whether the observed resonances can be assigned to two-photon intermediate resonant levels or to three-photon autoionizing states is a question that at the present time we cannot answer satisfactorily. It is, however, interesting to note that photocurrent versus laser-pulse energy measurements at $\lambda = 337.1$ nm show consistently that two photons are required to ionize the benzene molecule in this energy region ($2\,h\nu_{337.1}$ = 59,330 cm^{-1} or 7.36 eV). This could support an interpretation that the third photon absorbed by states in the energy

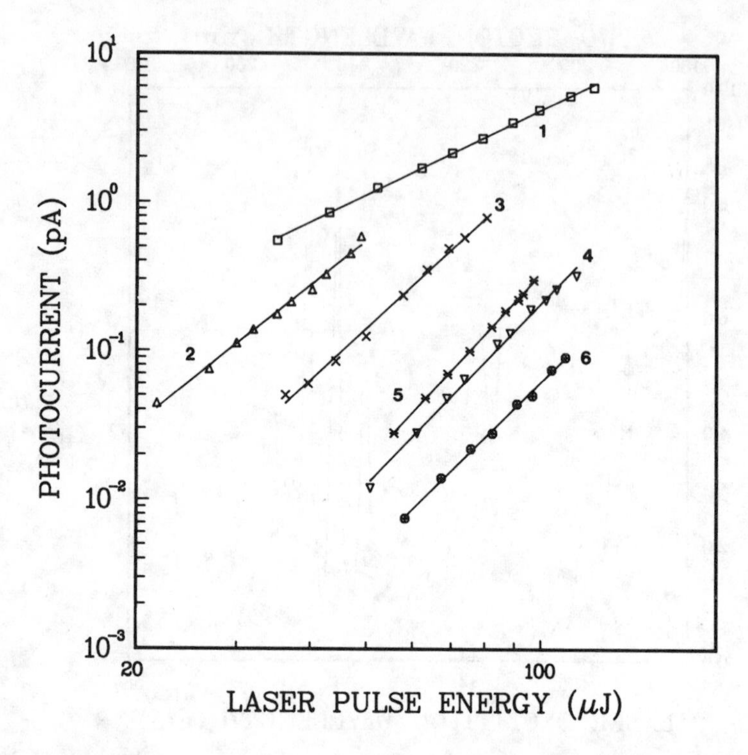

Fig. 3. Photocurrent as a function of the laser-pulse energy for benzene ($\sim 10^{-3}$ mol dm^{-3}, T = 300 K) in n-pentane at various excitation wavelengths: 337.1, slope = 1.91 (curve 1); 365, slope = 3.13 (curve 2); 386 nm, slope 3.62 (curve 3); 445, slope = 4.00 (curve 4); 496, slope = 4.22 (curve 5); and 508 nm, slope = 3.85 (curve 6). The solid lines were obtained by a linear least-squares-fitting procedure to the experimental points. The slope values are indicative of the MPI order.

region 45,000-56,000 cm^{-1} could possibly reach states in the energy region $\sim$67,000-84,000 cm^{-1} which are above I_L and are thus subjected to autoionization. But although the one-photon absorption spectrum of gaseous benzene in this energy region is rich in structure,[11] the one-photon ionization spectrum is practically structureless.[16]

The above arguments therefore cannot exclude the possibility that the observed resonances, or at least some of them, are due to the "elusive" $^1E_{2g}$ state; two-photon transitions from the ground state of benzene to the $^1E_{2g}$ state are allowed. A series of more systematic studies in this spectral region using linearly and circularly polarized light are thus needed. It is important in this connection to note that MPI and thermal-blooming studies conducted by other authors in neat liquid benzene[1-4] failed to observe any distinct structure in this spectral region. Although we did not carry out any measurements in neat liquid benzene, our experience with solutions

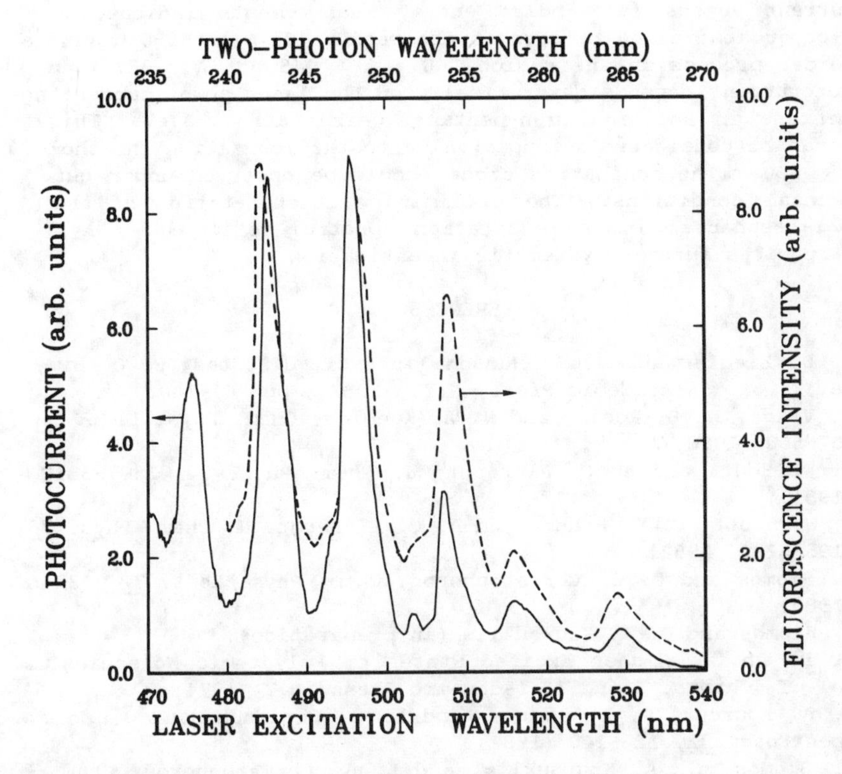

Fig. 4. The two-photon resonant multiphoton ionization spectrum (solid line) of ~10^{-3} mol dm^{-3} solution of benzene in n-pentane at room temperature. The dashed line represents the two-photon excitation spectrum of 10^{-2} mol dm^{-3} solution of benzene in cyclohexane (from Ref. 15). Both spectra are taken with linearly polarized light.

shows that changes in the concentration of benzene by a factor of ~4 in the 10^{-3} mol dm^{-3} range increases the photoionization background signal dramatically.

CONCLUSION

We have reported the MPI spectrum of benzene in dilute solutions of n-pentane using linearly and circularly polarized light in the spectral region $360 \le \lambda \le 560$ nm. We found that the structure in the MPI spectrum of benzene in the spectral region $445 \le \lambda \le 560$ nm reflects the ${}^1B_{2u}-{}^1A_{1g}$ transitions of benzene in a two-photon resonant four-photon ionization process. Calculated Ω ratios in this spectral region indicate that the two-photon ionization step from the resonant ${}^1B_{2u}$ state is isotropic allowing a tentative conclusion that we are dealing here with a bound-to-continuum transition. Although photocurrent versus laser-pulse energy measurements indicate a fourth-order ionization process in the region 445 ≤ λ ≤ 560 nm and a third-order process in the region 360 ≤ λ ≤ 445, at λ = 337.1 nm the photocurrent depends quadratically on the laser power suggesting that benzene in solution in n-pentane ionizes at ~7.36 eV. This energy can be considered an upper value to the ionization threshold. For λ > 560 nm no ionization signal could be observed under our experimental conditions. The preliminary interpretation of the observed resonances in the excitation spectral region 360 ≤ λ ≤ 445 nm requires further systematic investigation.

REFERENCES

1. L. D. Ziegler and B. S. Hudson, in Excited States (E. C. Lim, Ed.), Vol. 5 (Academic Press, N.Y., 1982), pp. 41-140.
2. V. Vaida, M. B. Robin, and N. A. Kuebler, Chem. Phys. Lett. 58, 557-560 (1978).
3. T. W. Scott and A. C. Albrecht, J. Chem. Phys. 74, 3807-3812 (1981).
4. T. W. Scott, C. L. Braun, and A. C. Albrecht, J. Chem. Phys. 76, 5195-5202 (1982).
5. K. Siomos and L. G. Christophorou, Chem. Phys. Lett. 72, 43-48 (1980).
6. K. Siomos and G. A. Kourouklis (in preparation).
7. M. B. Robin, Higher Excited States of Polyatomic Molecules, Vol. I, Chapts. I and II (Academic Press, N.Y. 1974).
8. G. A. Kourouklis, K. Siomos, and L. G. Chritophorou, J. Mol. Spectrosc. 92, 127-140 (1982).
9. K. Siomos, G. A. Kourouklis, and L. G. Christophorou, Chem. Phys. Lett. 80, 504-511 (1981).
10. K. Siomos and L. G. Christophorou, J. Electrost. 12, 147-152 (1982).
11. E. E. Koch and A. Otto, Chem. Phys. Lett. 12, 476-480 (1972).
12. R. L. Whetten, Ke-Jian Fu, and E. R. Grant, J. Chem. Phys. 79, 2626-2640 (1983).
13. P. M. Johnson, J. Chem. Phy. 62, 4562-4563 (1975).
14. The ionization threshold of benzene in solution has not yet been established; see, however, later this section.
15. D. M. Friedrich, J. Van Alsten, and M. A. Walters, Chem. Phys. Lett. 76, 504-509 (1980).
16. V. H. Dibeler and R. M. Reese, J. Research NBS 68A, 409-417 (1964).

MULTIPHOTON IONIZATION OF CESIUM ATOMS ABOVE AND BELOW THE TWO-PHOTON IONIZATION THRESHOLD*

A. Dodhy,** J.A.D. Stockdale, and R. N. Compton
Oak Ridge National Laboratory, Oak Ridge, Tennessee 37831

ABSTRACT

Two- and three-photon ionization processes in cesium atoms using a single pulsed-dye laser and a cesium atomic beam have been investigated. Photoelectron angular distributions have been measured for two-photon resonant, three-photon ionization via the nd states (n = 12, 15, and 21), for two-photon nonresonant transitions over the photoelectron energy range from ~26-100 meV, and for "quasi-free-free" transitions.

INTRODUCTION

In the last several years research in the field of multiphoton ionization (MPI) has rapidly advanced. A number of review articles have been published concerning theory and experiment for both resonant and nonresonant MPI processes (see Refs. 1-4 and references cited therein). In particular, the study of photoelectron angular distributions utilizing MPI processes in atoms has received considerable attention.[5-8] Photoelectron angular distributions depend on the nature of the quantum states involved and hence prove important in obtaining information about both the initial bound state and final continuum states of the photoejected electron and the microscopic parameters of the particular system under study. To our knowledge, photoelectron angular distribution measurements via MPI of cesium in an atomic beam are limited to MPI via 7^2P states.[9]

In this paper we report measurements of photoelectron angular distributions from cesium atoms for (1) two-photon resonant three-photon ionization via the n = 12, 15, and 21d states, (2) two-photon nonresonant ionization, and (3) "quasi-free-free" transitions, using a single laser and a cesium atomic beam.

EXPERIMENTAL

The experimental apparatus employed in the present study is shown schematically in Fig. 1. It consists of a frequency tunable dye laser, the polarization optics, and the vacuum chamber. The latter includes the atomic beam assembly and the electron energy analyzer and detector. The dye laser was pumped by a Nd:YAG laser

---

*Research sponsored by the Office of Health and Environmental Research, U.S. Department of Energy, under contract DE-AC05-84OR2100 with Martin Marietta Energy Systems, Inc.

**Physics Department, Auburn University, Auburn, AL 36849.

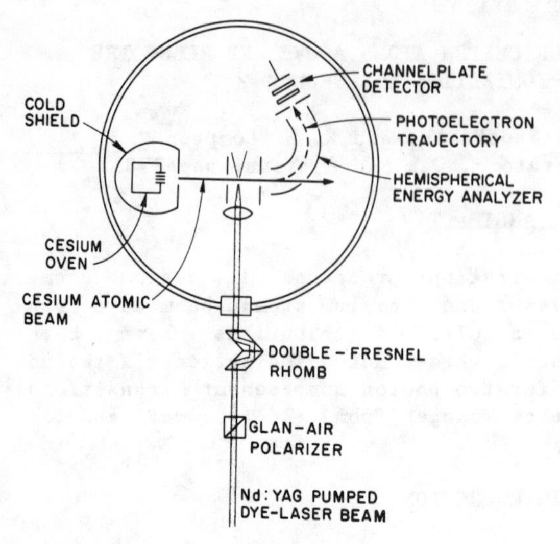

Fig. 1. Experimental apparatus.

(Quanta Ray) and had a pulse duration of 8 ns and a bandwidth of ~0.02 nm in the spectral region of interest ($610 \leq \lambda \leq 685$ nm). The laser beam was linearly polarized using a Glan-air polarizer and was focused by a 35 mm focal length lens into the cesium atomic beam ~1 cm from the entrance aperture of the energy analyzer. The laser beam crossed the cesium atomic beam at 90° and the interaction volume defined solely by the laser beam focus was estimated to be $\sim 10^{-4}$ cm^3. The laser power density at the focal point was $\sim 10^8$ W/cm^2. The polarization direction of the laser light was rotated using a double-Fresnel rhomb.

Metallic cesium was evaporated in a resistively heated stainless steel oven. The cesium atoms exited the oven through a multichannel array and were directed into the entrance of the energy analyzer after passing through a water-cooled baffle and a second 3-mm-dia defining aperture. The cesium beam finally passed through a small hole in the back of the analyzer. The cesium atom number density in the interaction volume was estimated to be $\sim 10^{13}$ $atoms/cm^3$ at an oven temperature of 120°C.

The electrons produced by MPI of the cesium and ejected perpendicular to the propagation vector of the laser beam were energy analyzed by a spherical sector electrostatic energy analyzer and were detected by a dual channelplate charged particle detector. The amplified output of the detector was sampled by a boxcar integrator (Princeton Applied Research, model 162) and plotted on a x-y recorder as a function of the laser excitation wavelength or the angle θ between the polarization direction of the laser light and the direction of the detected photoelectrons. The analyzer had an acceptance angle of ~3° and a resolution of ~0.4 eV at a transmission energy of ~38 eV. This resolution was sufficient for the present work where either single or well separated (~2 eV) atomic peaks were being studied. Space charge broadening was minimized by reducing both the laser power density ($\sim 10^8$ W/cm^2) and the cesium atom number density ($\sim 10^{13}$ $atoms/cm^3$).

Normal operation was under electric field-free conditions and the earth's magnetic field was compensated for by using 40-cm radius Helmholtz coils. An electrode located ~1 cm from the interaction volume and in front of the energy analyzer allowed the application of an external electric field.

Photoelectron angular distributions were obtained by monitoring the photoelectron intensity as a function of angle Θ. Spectra of photoelectron signal intensity as a function of the laser excitation wavelength at a fixed transmission energy of the energy analyzer were also measured.

RESULTS AND DISCUSSION

Referring to the MPI theory of photoelectron angular distributions, we note that the differential cross section for photoelectron emission can be represented by

$$\frac{d\sigma^{(N)}(\lambda,\Theta)}{d\Omega} = \frac{\sigma^{(N)}_{TOT}(\lambda)}{4\pi} \sum_{i=0}^{N} \beta_{2i} P_{2i} (\cos\Theta) \tag{1}$$

where N is the number of absorbed photons, $\sigma^{(N)}_{TOT}$ is the generalized total cross section for ionization at wavelength λ, P_{2i} (cos Θ) are the Legendre polynominals of the order 2i, and β_{2i} are the asymmetry parameters. The β_{2i} are functions of the microscopic atomic properties of the system under study and in general depend on the laser light intensity. However for the power densities ($\sim 10^8$ W/cm^2) used in the present experiment, the β_{2i} parameters were independent of the laser intensity. For mathematical convenience and analysis of the experimental data, Eq. (1) can be written as a summation of even powers of cos Θ and expressed as

$$\frac{d\sigma^{(N)}(\lambda,\Theta)}{d\Omega} = \frac{\sigma^{(N)}_{TOT}(\lambda)}{4\pi} A_N \sum_{i=0}^{N} \beta_{2i} \cos^{2i}(\Theta) \tag{2}$$

where A_N is a normalizing constant corresponding to N photon ionization.

Two-Photon Resonant Three-Photon Ionization via the nd States

Figure 2(a) shows the two-photon resonant three-photon ionization scheme used to study the nd states of cesium. Using this excitation scheme, photoelectron angular distributions were measured for the n = 12d, 15d, and 21d states and the result for n = 12d is shown in Fig. 3. The solid line through the experimental points was obtained by a least squares fitting procedure performed to Eq. (2). The minimum value of χ^2 was used as a criterion for the best fit. This gave an intensity distribution containing powers of cos Θ up to the sixth order and the β coefficients calculated using this method were $\beta_0 = 1$, $\beta_2 = 15.09$, $\beta_4 = -39.57$, and $\beta_6 = 38.77$.

Nonresonant Two-Photon Ionization

The excitation scheme used for two-photon nonresonant ionization of cesium atoms is shown in Fig. 2(b). Figure 4 shows the photoelectron signal as a function of laser excitation wavelength ($630 \leq \lambda \leq 646$ nm) for: E = 0 (spectrum I), E = 0.1 V/cm

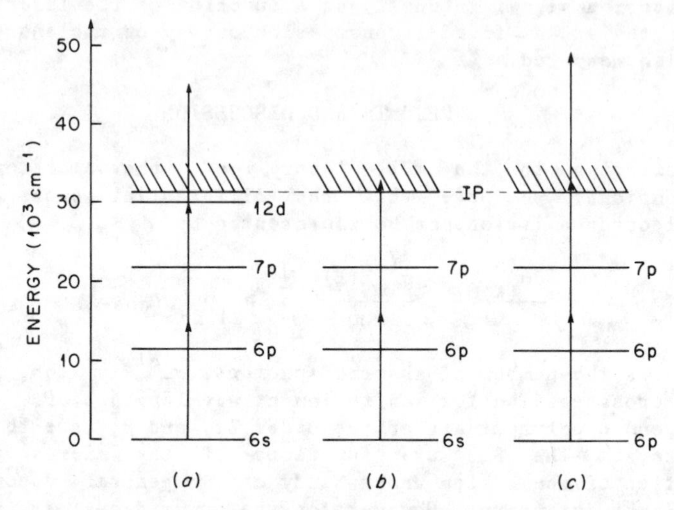

Fig. 2. Energy level diagrams for multiphoton ionization of cesium atoms (a) two-photon resonant three-photon ionization via the nd state, (b) two-photon nonresonant ionization, and (c) quasi-free-free transitions.

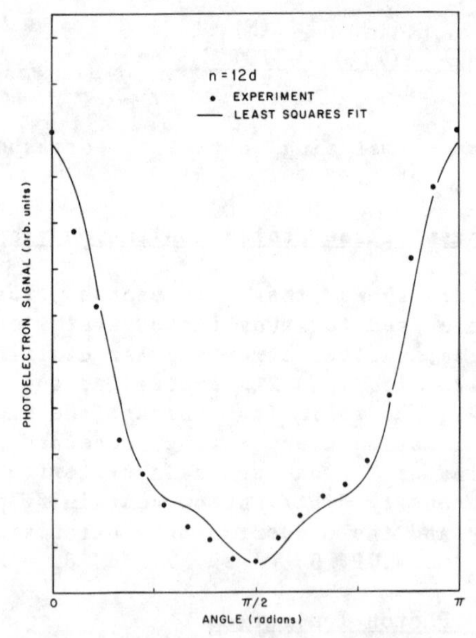

Fig. 3. Photoelectron angular distributions for two-photon nonresonant three-photon ionization via the n = 12d state in cesium atoms.

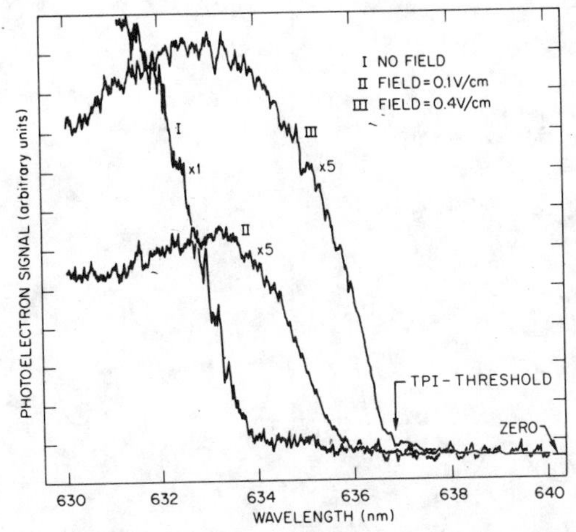

Fig. 4. Spectrum of photoelectron signal as a function of laser excitation wavelength for two-photon nonresonant ionization in cesium atoms. The arrow indicates the position of the two-photon ionization (TPI) threshold which corresponds to one-photon energy of 1.946 eV.

(spectrum II), and E = 0.4 V/cm (spectrum III) where the electric field was used to repel the photoelectrons with the energy analyzer. In spectrum I of Fig. 4 it was observed that the ionization threshold was blue shifted by ~140 cm^{-1}, but the application of a small dc field (spectra II and III) clearly moved this threshold towards the two-photon ionization (TPI) threshold which corresponds to a one-photon energy of 1.940 eV. This leads to the conclusion that this apparent shift in the photoelectron threshold can be attributed to the inability of our detection system to detect electrons with energy less than ~20 meV. Note that the 0.4 eV resolution of the energy analyzer prevents the simultaneous transmission of photoelectrons from both two- and three-photon ionization processes since the difference in photoelectron energies is ~2 eV. This explains the absence of the three-photon ionization signal below the two-photon ionization threshold (λ = 636.8) in the spectra of Fig. 4.

Photoelectron angular distributions were measured at seven excitation wavelengths, λ = 632.5, 630.5, 628.5, 626.5, 624.5, 622.5, 620.0 nm corresponding to a photoelectron energy range from ~26-100 meV. A typical distribution is shown in Fig. 5 where λ = 622.5 nm and the photoelectron energy is ~89 meV. The solid line through the experimental data points is a least squares fit to Eq. (2). This gives an intensity distribution containing $\cos^2\theta$ and $\cos^4\theta$ terms and the β parameters calculated using this method are $\beta_0 = 1$, $\beta_2 = -3.72$ and $\beta_4 = 5.47$. It was observed that if all the photoelectron angular distributions were normalized to the same maximum value at $\theta = 0°$, then the peak at $\theta = \pi/2$ decreased as the exciting photon energy is increased. Preliminary theoretical analysis[10,11] of these measurements shows a good qualitative agreement, and a detailed quantitative comparison between theory and experiment will be published elsewhere.[12]

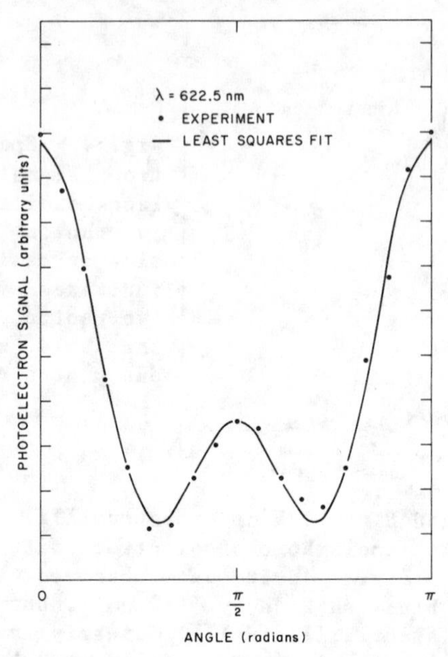

Fig. 5. Photoelectron angular distribution for two-photon nonresonant ionization in cesium atoms at a laser excitation wavelength λ = 622.5.

"Quasi-Free-Free" Transitions

The energy level scheme for the "quasi-free-free" transition studied in this experiment is shown in Fig. 2(c). Two photons (λ = 633.66 nm) were used to nonresonantly ionize a cesium atom and a third photon was absorbed in the continuum. The corresponding photoelectrons were observed at an energy of 3hν -IP where hν is laser photon (λ = 633.66 nm) energy and IP is the single photon ionization threshold (3.893 eV) of cesium. An angular distribution measured for this process is shown in Fig. 6 where the solid line through the experimental data is a least squares fit to Eq. (2). This procedure yields a sixth order polynominal of cos θ with the β coefficients given by $\beta_0 = 1$, $\beta_2 = 3.07$, $\beta_4 = -6.36$, and $\beta_6 = 8.47$. According to the selection rules in the electric dipole approximation, the final continuum state is composed of p and f partial waves, and the clear deviation from a $\cos^2\theta$ type function in the present photoelectron angular distribution indicates a substantial contribution from the $\cos^4\theta$ and $\cos^6\theta$ terms. Further experimental investigation is proceeding and efforts are being made to achieve a theoretical analysis for these measurements.

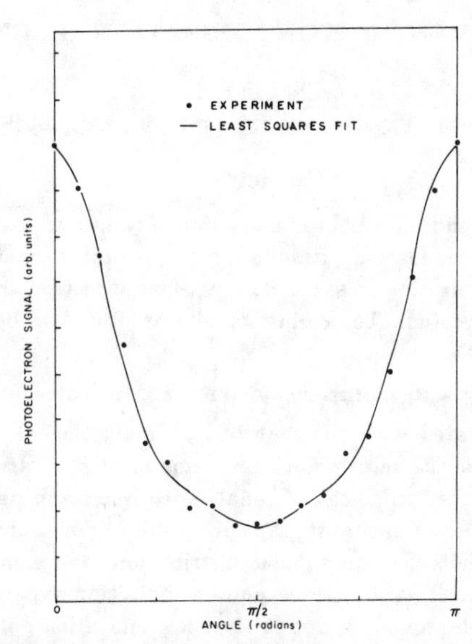

Fig. 6. Photoelectron angular distribution for two-photon nonresonant ionization with the absorption of an additional photon in the continuum (quasi-free-free transition) at a laser excitation wavelength λ = 633.66 nm.

REFERENCES

[1] A. T. Georges and P. Lambropoulos, "Aspects of Resonant Multiphoton Processes," in ADVANCES IN ELECTRONICS AND ELECTRON PHYSICS (L. Marton and C. Marton, Eds.), 54, 191 (1980).

[2] N. B. Delone, Sov. Phys. Usp, 18(3), 169 (1975).

[3] J. S. Bakos "Multiphoton Ionization of Atoms," in ADVANCES IN ELECTRONICS AND ELECTRON PHYSICS (L. Marton, Ed.), 36, 57 (1974).

[4] J. Morellec, D. Normand, and G. Petite, ADVANCES IN ELECTRONICS AND MOLECULAR PHYSICS, 18, 97 (1982).

[5] M. Lambropoulos and R. S. Berry, Phys. Rev. A 8, 855 (1973).

[6] G. Leuchs and S. J. Smith, J. Phys. B 15, 1051 (1982).

[7] S. N. Dixit and P. Lambropoulos, Phys. Rev. A 27, 861 (1983).

[8] G. Leuchs, S. J. Smith, E. Khawaja, and H. Walther, Optic Comm. 31, 313 (1979).

[9] H. Kamiski, J. Kessler, and K. J. Kollath, Phys. Rev. Lett. 45, 1161 (1980).

[10] P. Lambropoulos and X. Tang (private communication).

[11] M. Pindzola (private communication and these proceedings).

[12] A. Dodhy, J.A.D. Stockdale, R. N. Compton, X. Tang, P. Lambropoulos and M. Pindzola (to be published).

TWO AND THREE PHOTON IONIZATION OF CESIUM

Michael S. Pindzola
Department of Physics, Auburn University, AL 36849

Abstract

Probabilities for the two and three photon ionization of cesium are calculated in the non-relativistic Hartree-Fock approximation. Results for two photon excitation cross sections, excited-state single photon ionization cross sections, and photoelectron angular distributions to multiple continua are presented. The cesium results are found to be in fair agreement with recent experimental measurements.

Theory and Comparison with Experiment

The spectroscopy associated with the multiphoton ionization of atoms shows great promise in providing new insights into various problems of atomic structure and dynamics. Recent developments in the technology of tunable dye lasers and photoelectron energy spectrometers have provided experimentalists the opportunity to measure not only total ionization cross sections, but photoelectron angular distributions and spin polarizations over a wide wavelength spectrum. Two and three photon ionization experiments have generally concentrated on the alkali atoms[1-4], due to their low ionization potential and the ease with which one can produce intense atomic beams. Six photon ionization experiments on xenon atoms have also investigated photoelectron angular distributions for above threshold absorption[5].

For non-resonant two photon ionization of an atom, the application of time dependent perturbation theory results in a cross section, σ_2, given by

$$\sigma_2 = \frac{16\pi^2\omega^2}{c^2k}\left|\sum_n \frac{\langle\psi_k|\sum_{i=1}^N \mathbf{e}\cdot\mathbf{r}_i|\psi_n\rangle\langle\psi_n|\sum_{i=1}^N \mathbf{e}\cdot\mathbf{r}_i|\psi_g\rangle}{E_g - E_n + \omega}\right|^2, \tag{1}$$

where ω is the frequency of the radiation, c is the speed of light, k is the photoelectron wavenumber, N is the number of electrons in the atom, $\mathbf{e}$ is the direction of radiation field polarization, and atomic units are used. If one makes a partial wave expansion of $|\psi_k\rangle$, the differential cross section in photoelectron emission angle, $\sigma_2(\theta)$, is given by

$$\sigma_2(\theta) = \frac{4\pi\omega^2}{c^2k}(A_0 + A_2\cos^2\theta + A_4\cos^4\theta), \tag{2}$$

where we have restricted ourselves to LS-coupled alkali atom wavefunctions and linearly polarized radiation[6,7]. The coefficients, A_i, are given by

$$\begin{aligned} A_0 &= a_s^2 + 5a_d^2/4 + \sqrt{5}a_s a_d\cos(\delta_s - \delta_d), \\ A_2 &= -30a_d^2/4 - 3\sqrt{5}a_s a_d\cos(\delta_s - \delta_d), \\ A_4 &= 45a_d^2/4, \end{aligned} \tag{3}$$

where δ_s and δ_d are the s and d wave phase shifts of the photoelectron. The matrix elements, a_l, are given by

$$a_l = c_l \sum_{np} \frac{\langle kl|r|np\rangle\langle np|r|gs\rangle}{\epsilon_{gs} - \epsilon_{np} + \omega}, \tag{4}$$

where $c_s = \sqrt{1/9}$, $c_d = \sqrt{4/45}$, and $\langle np|r|gs\rangle$ are radial dipole integrals. The continuum normalization is chosen as one times a sine function. Differential cross section expressions for other atoms and radiation field polarizations may be derived in a straightforward manner[8].

We generated radial wavefunctions for the ground state of cesium in the Hartree-Fock approximation[9]. The sum over intermediate states found in Eq.(4) was determined using the inhomogeneous differential equation technique[10]. An iterative Hartree-Fock method generated perturbed orbitals, $|Np\rangle$, by solving the equation

$$(\epsilon_{6s} + \omega - H)|Np\rangle = r|6s\rangle - \sum_{n'}^{occ} |n'p\rangle\langle n'p|r|6s\rangle, \tag{5}$$

where H is the Hartree-Fock operator for excited $|np\rangle$ orbitals and the $|n'p\rangle$ orbitals are occupied in the ground state configuration. The $|Np\rangle$ orbital is approximately equal to a $|6p\rangle$ orbital divided by the small denominator of Eq.(4). The coefficients, a_l, are then found by calculation of $\langle kl|r|Np\rangle$. A non-iterative Hartree-Fock method is used to solve the continuum equation for the $|kl\rangle$ orbital.

Our theoretical results for the differential cross section for two photon ionization of cesium are shown in Fig.1. The experimental measurements of A.Dodhy et. al.[11] are shown as open circles in the figures. Since no absolute cross section scale was established, the experimental points are normalized to theory at zero degrees. The multiple minima predicted by theory is borne out in the experimental results. The ratio of the cross section at ninety degrees to that at zero degrees decreases as the photon wavelength decreases. Better overall agreement between theory and experiment is obtained at the shorter wavelengths.

At a wavelength of 624.5 nm our total cross section for the two photon ionization of cesium is $3.37 \times 10^{-48} cm^4 sec$ in the dipole-length gauge and $1.07 \times 10^{-48} cm^4 sec$ in the dipole-velocity gauge. The theoretical calculations of M.R. Teague et. al.[12] yielded $8.10 \times 10^{-49} cm^4 sec$ at the same wavelength. Relativistic corrections to our Hartree-Fock orbitals may change our total cross sections by a factor of two.

For two photon resonant, three photon ionization of an atom, the probabilities for ionization may be calculated using the density matrix formalism[13]. Perturbation theory breaks down since population probability may cycle between the gound and resonant bound state. The shape of the photoelectron angular distribution is governed by the excited state photoionization cross section, while the magnitude may be determined by solving coupled first order differential equations for the elements of the density matrix. Complications arise when one tries to describe the temporal and spatial behavior of most multimode laser fields.

In recent experiments on cesium[4,11], it is interesting to note that the two photon resonant excitation of ns states is very weak when compared to the nd states in the three photon ionization spectra. This can be explained in part by examining two important atomic parameters in the density matrix equations. The first parameter is the two photon excitation cross section, Ω, given by

$$\Omega = \frac{4\pi\omega}{c} \left| \sum_n \frac{\langle\psi_f| \sum_{i=1}^N \mathbf{e}\cdot\mathbf{r}_i |\psi_n\rangle\langle\psi_n| \sum_{i=1}^N \mathbf{e}\cdot\mathbf{r}_i |\psi_g\rangle}{E_g - E_n + \omega} \right|, \tag{6}$$

which for an isolated two level system under a monochromatic radiation field implies a time dependent population of the upper level, $P_f(t)$, given by

$$P_f(t) = \sin^2(\Omega F t/2), \tag{7}$$

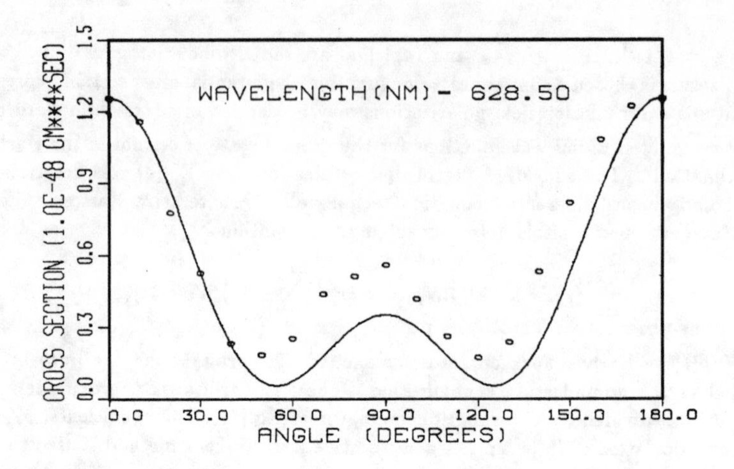

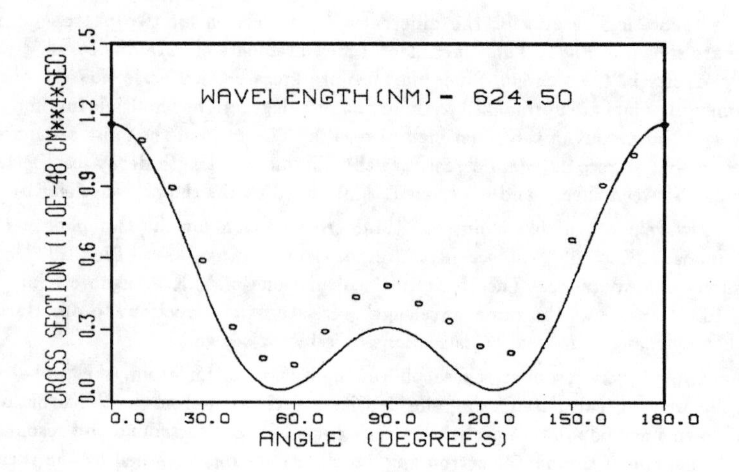

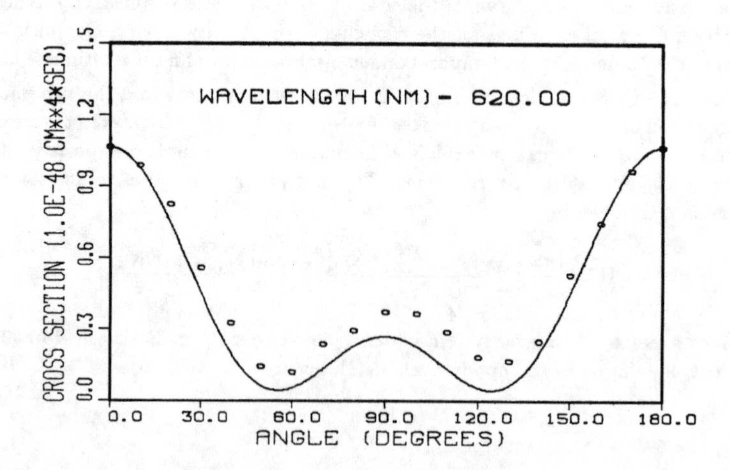

Figure 1: Two Photon Differential Cross Sections for Cesium

Table 1: Cesium Excitation and Ionization Cross Sections

Level designation (nlj)	Excitation Cross Section ($10^{-18} cm^2$)	Ionization cross section ($10^{-18} cm^2$)
8s1/2	121.0	0.0024
8d3/2	51.5	4.98
8d5/2	63.1	5.10

where F is the photon flux. The second parameter is the excited-state photoionization cross section, σ, given by

$$\sigma = \frac{8\pi\omega}{ck}\left|\langle\psi_k|\sum_{i=1}^{N}\mathbf{e}\cdot\mathbf{r}_i|\psi_n\rangle\right|^2 . \tag{8}$$

Since the fine structure levels of cesium are observed in the experiment, we reduce the matrix elements in Eqs.(6) and (8) to radial form using LSJ-coupled wavefunctions. We also assume linearly polarized radiation. The sum over intermediate states found in Eq.(6) is again determined by the inhomogeneous differential equation technique. In Table 1 we present our results for the n=8 levels of cesium. The ionization cross section for the 8s1/2 level is seen to be over three orders of magnitude smaller than the 8d levels. For three photon ionization of cesium through a two photon resonant 8s1/2 level the wavelength of the radiation is 821.2 nm. Furthur photoionization calculations reveal that this three photon ionization wavelength is very close to a minimum in the cross section versus wavelength curve for the 8s1/2 level. Additional ionization cross section calculations for the n=12 levels yield a 12s/12d ratio of 0.008. We conclude that the very small peak heights of ns states in the experimental ionization spectra are due in part to their small photoionization cross sections.

For non-resonant three photon ionization of an atom, the application of time dependent perturbation theory results in a cross section, σ_3, given by

$$\sigma_3 = \frac{32\pi^3\omega^3}{c^3k}\left|\sum_m\sum_n\frac{\langle\psi_k|\sum_{i=1}^{N}\mathbf{e}\cdot\mathbf{r}_i|\psi_m\rangle\langle\psi_m|\sum_{i=1}^{N}\mathbf{e}\cdot\mathbf{r}_i|\psi_n\rangle\langle\psi_n|\sum_{i=1}^{N}\mathbf{e}\cdot\mathbf{r}_i|\psi_g\rangle}{(E_g-E_m+2\omega)(E_g-E_n+\omega)}\right|^2 . \tag{9}$$

If one makes a partial wave expansion of $|\psi_k\rangle$, the differential cross section, $\sigma_3(\theta)$, is given by

$$\sigma_3(\theta) = \frac{8\pi^2\omega^3}{c^3k}(A_2\cos^2\theta + A_4\cos^4\theta + A_6\cos^6\theta), \tag{10}$$

where we have restricted ourselves to LS-coupled alkali atom wavefunctions and linearly polarized radiation. The coefficients, A_i, are given by

$$\begin{aligned} A_2 &= 3a_p^2 + 63a_f^2/4 + 3\sqrt{21}a_pa_f\cos(\delta_p-\delta_f), \\ A_4 &= -210a_f^2/4 - 5\sqrt{21}a_pa_f\cos(\delta_p-\delta_f), \\ A_6 &= 175a_f^2/4, \end{aligned} \tag{11}$$

where δ_p and δ_f are the p and f wave phase shifts. The matrix elements, a_l, are given by

$$a_l = \sum_{ml'}c_{ll'}\sum_{np}\frac{\langle kl|r|ml'\rangle\langle ml'|r|np\rangle\langle np|r|gs\rangle}{(\epsilon_{gs}-\epsilon_{ml'}+2\omega)(\epsilon_{gs}-\epsilon_{np}+\omega)}, \tag{12}$$

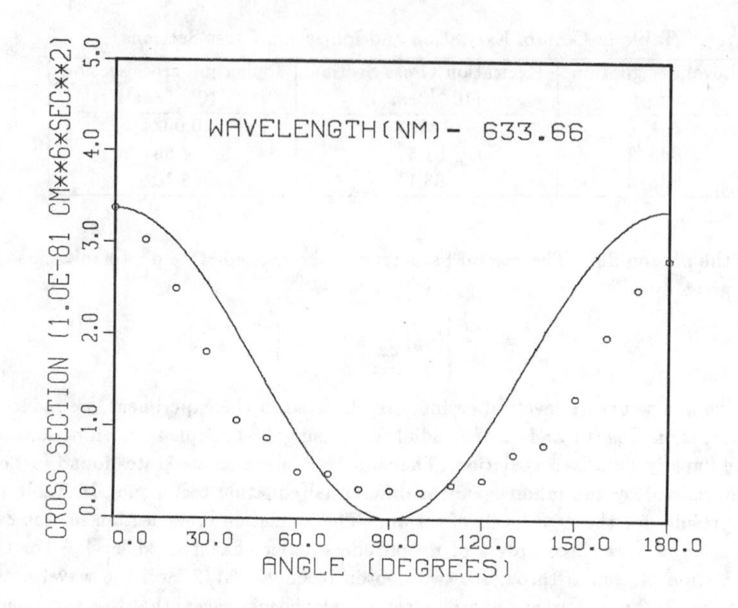

Figure 2: Three Photon Differential Cross Section for Cesium

where $c_{ps} = \sqrt{1/27}$, $c_{pd} = \sqrt{16/675}$ and $c_{fd} = \sqrt{12/525}$.

If the quantity $(\epsilon_{gs} + 2\omega)$ in the denominator of Eq.(12) exceeds the ionization potential, I_p, of the alkali atom, we have above threshold three photon ionization. What is observed experimentally is a slow photoelectron at $(2\omega - I_p)$, and a fast photoelectron at $(3\omega - I_p)$. The angular distribution of the fast electron is given by Eq.(10), while Eq.(2) gives the distribution for the slow electron. The multiple sums over intermediate states found in Eq.(12) are determined using a step by step inhomogeneous differential equation technique[14]. For cesium the first step is the solution of Eq.(5) for the $|Np\rangle$ orbital at the appropriate frequency ω. The next step is the solution of the equation

$$(\epsilon_{6s} + 2\omega - H)|Ml'\rangle = r|Np\rangle - \sum_{m'}^{occ} |m'l'\rangle\langle m'l'|r|Np\rangle, \tag{13}$$

where H is the Hartree-Fock operator for excited $|ml'\rangle$ orbitals. The $|Ml'\rangle$ orbitals are oscillatory at infinity and their normalization is determined by making them ninety degrees out of phase with the homogeneous solution of Eq.(13). Our non-iterative Hartree-Fock continuum program was modified to solve Eq.(13). The coefficients, a_l, are then found by calculation of $\langle kl|r|Ml'\rangle$, which can be made absolutely convergent [14].

Our theoretical results for the differential cross section of the fast photoelectron in above threshold three photon ionization of cesium are shown in Fig.2. The experimental measurements of A. Dodhy et. al.[11] are shown as open circles in the figure. Again the experimental points are normalized to theory at zero degrees. The agreement between theory and experiment is not as good as that found in two photon ionization. Substantial wavefunction cancellation forces the coefficient, a_f, to be quite small in Eq.(11), effectively eliminating the $\cos^4\theta$ and $\cos^6\theta$ terms in Eq.(10). The experiment, however, seems to indicate more substantial contributions from the higher power $\cos\theta$ terms. We note that convergence of the free-free dipole matrix element, $\langle kl|r|Ml'\rangle$, is extremely poor for the near threshold wavelength of 633.66 nm chosen by the experimentalists. Above threshold ionization at shorter wavelengths may bring better agreement between theory and experiment.

In conclusion we find that various multiphoton atomic parameters calculated in the non-relativistic Hartree-Fock approximation are in fair agreement with recent experimental measurements in cesium. Furthur experimental work in multiphoton ionization spectroscopy promises to force a more careful consideration of many-electron atomic theory.

REFERENCES

[1] J.C.Hansen, J.A.Duncanson,Jr., R.L.Chien and R.S.Berry, Phys.Rev.A21,222(1980).

[2] H.Kaminski, J.Kessler, and K.J.Kollath, Phys.Rev.Lett.45,1165(1980).

[3] W.Ohensorge, F.Diedrich, D.S.Elliott, G.Leuchs and H.Walther, Abstracts of Contributed Papers, XIII ICPEAC, (Berlin,1983),p62.

[4] J.A.D.Stockdale, R.N.Compton and C.D.Cooper, Abstracts of Contributed Papers, XIII ICPEAC, (Berlin,1983),p65.

[5] F.Fabre, P.Agostini, G.Petite and M.Clement, J.Phys.B14,L677(1981).

[6] P.Lambropoulos, Phys.Rev.Lett.28,585(1972).

[7] J.Mizuno, J.Phys.B6,314(1973).

[8] M.M.Lambropoulos and R.S.Berry, Phys.Rev.A8,855(1973).

[9] C.F.Fischer, Comp.Phys.Commun.14,145(1978).

[10] A.Dalgarno and J.T.Lewis, Proc.R.Soc.(London)A233,70(1955).

[11] A.Dodhy, J.A.D.Stockdale and R.N.Compton, (these proceedings).

[12] M.R.Teague, P.Lampropoulos, D.Goodmanson and D.W.Norcross, Phys.Rev.A14,1057(1976).

[13] S.N.Dixit and P.Lambropoulos, Phys.Rev.A27,861(1983).

[14] M.Aymar and M.Crance, J.Phys.B14,3585(1981).

MULTIPHOTON IONIZATION SPECTROSCOPY OF THE FLUOROMETHYL RADICAL

by

C. S. Dulcey, Denis J. Bogan, and Jeffrey W. Hudgens
Chemistry Division, Code 6185, Naval Research Laboratory
Washington, D.C. 20375

Using mass resolved multiphoton ionization spectroscopy, we have detected the first electronic spectrum of the CH_2F free radical. The CH_2F radicals were produced in a discharge fast flow reactor[1] and introduced into a mass spectrometer's ion optics where they were irradiated with the tightly focussed output of a Nd:YAG pumped dye laser. CH_2F radicals were first produced via the reaction of F + ketene. At laser wavelengths between 345-363 nm a spectrum was observed from m/z 33 ions. To identify the spectrum's carrier these experiments were repeated with the reaction, F + CH_3F. Identical spectra were observed.

Major features of the spectrum are two triplet bands centered at 358 nm and 349.5 nm which represent (in two photons) a separation of 1560 cm^{-1}. The spacing of the sub-bands within each of these triplets is ~155 cm^{-1}. Several broad, weaker bands are also present. The most probable origin of these bands is a two photon resonance with a 3p Rydberg state (near 55700 cm^{-1}) with a quantum defect of 0.56. After preparation of the Rydberg radical, absorption of an additional photon generates the ion--but with insufficient energy to cause ion fragmentation. No fragmentation was observed. Studies of the multiphoton ionization of CD_2F radicals are underway to resolve the nature of the spectrum.

[1]C. S. Dulcey and J. W. Hudgens, J. Phys. Chem. 87 , 2296(1983).

GROTRIAN DIAGRAM OF THE QUARTET SYSTEM OF Na I

D. E. Holmgren, D. J. Walker, D. A. King, and S. E. Harris
Edward L. Ginzton Laboratory
Stanford University, Stanford, California 94305

ABSTRACT

Laser enhanced fluorescence experiments have been used in conjunction with Hartree-Fock calculations and emission studies to identify and locate Na I core-excited quartet levels. Twenty-eight Na emission lines are identified as transitions between these levels.

INTRODUCTION

Because of their metastability against autoionization, the core-excited quartet levels of alkali atoms and alkali-like ions are of interest for the construction of XUV lasers[1,2] and as base levels for studies of atomic and molecular collisional phenomena.

In this paper we develop a partial Grotrian diagram for the quartet manifold of neutral Na. By using a pulsed hollow-cathode discharge, together with laser enhanced fluorescence experiments, and the atomic physics code RCN/RCG,[3] we have identified 28 new visible and near-UV transitions between quartet levels of the $2p^5 3s3p$, $2p^5 3s4s$, and $2p^5 3s3d$ configurations. Prior to this work, Berry, et al.[4] and Frohling and Andra[5] identified the emission lines at 3883 Å and 5071 Å as arising from the transitions $2p^5 3s3p$ $^4D_{7/2}$ – $2p^5 3s3d$ $^4F_{9/2}$ and $2p^5 3s3p$ $^4D_{7/2}$ – $2p^5 3s4s$ $^4P_{5/2}$, respectively.

EMISSION SPECTRA AND THEORETICAL CALCULATIONS

Our study is based upon emission spectra of Na, recorded over the spectral region of 2500 Å to 6500 Å. The emission source was a pulsed hollow-cathode discharge, described elsewhere,[6] operating at a Na vapor pressure of several torr. A 1-meter Spex monochromator, with a 1200 line/mm grating and a slit width of 50 μm, provided a wavelength accuracy of about 0.2 Å. Known Na I and Na II lines were used for calibration. Over this spectral region, we observed approximately 120 emission lines that could not be attributed to known lines of Na I or Na II. The weakest of these lines was approximately 200 times less intense than the bright quartet line at 3882 Å (reported in Ref. 4 at 3882.8 ± 2.0 Å).

Using the RCN/RCG code, autoionization and dipole radiative rates were calculated for all possible transitions between levels of the core-excited configurations $2p^5 3s3p$, 3s4p, 3p4s, 3p3d, 3s3d, 3s4d, 3s5d, $3s^2$, $3p^2$, and 3s4s. The average configuration energies and Coulomb integrals were adjusted to give reasonable agreement with the level energies of Sugar, et al.,[7] and of Ref. 4. In particular, the $2p^5 3s3p$ $^4D_{7/2}$ – $2p^5 3s3d$ $^4F_{9/2}$ transition wavelength was placed at 3882 Å. From these calculations, the branching ratios for photoemission of the $2p^5 3s3d$ and $2p^5 3s4s$ quartet states were tabulated. About 30 transitions from these levels to $2p^5 3s3p$ quartet levels were predicted to be observable in our emission spectra.

0094-243X/84/1190157-05 $3.00

We believe that the absolute energies we calculated are accurate to several hundred to a few thousand cm^{-1}, based on a comparison with previously identified levels.[8] Also, because of good agreement of calculated and measured oscillator strengths in singly excited Na I, we believe that our calculated oscillator strengths are accurate to about a factor of two.

By comparing the relative intensities and positions of emission lines predicted by calculation to those observed experimentally, tentative identifications of some of the transitions were made. We confirmed or, in some cases, denied these identifications using the technique of laser enhanced fluorescence.

LASER ENHANCED FLUORESCENCE EXPERIMENTS

The key to the laser enhanced fluorescence experiments are the large populations which are stored in the two lowest levels of the $2p^5 3s3p$ manifold. The lowest of these levels is $2p^5 3s3p\ ^4S_{3/2}$. Since it autoionizes quite slowly (approximately 10^{-6} s), but still radiates in the XUV, this level has been termed quasi-metastable.[9] In recent experiments in this same hollow cathode, we measured a population in this level of 2×10^{10} atoms/cm^3. The next higher level, $2p^5 3s3p\ ^4D_{7/2}$, is that of highest J in the configuration, and is metastable against both autoionization and radiation. Here we measure a population of 10^{11} atoms/cm^3.[6] In the laser enhanced experiments, these large stored populations are transferred to other quartet levels, using a laser tuned to the transition wavelength suggested by the calculations and emission spectra. If the transfer wavelength assignment is correct, fluorescence from the upper quartet level is enhanced.

As an example of the laser enhancement technique, a Quanta Ray pulsed dye laser was tuned through the predicted $2p^5 3s3p\ ^4D_{7/2}$ — $2p^5 3s3d\ ^4F_{7/2}$ transition at 3865.5 Å; thereby transferring stored population from the metastable $^4D_{7/2}$ level to the $^4F_{7/2}$ level. During the time the transfer laser was on, the intensities of all lines which may originate from this upper level were observed. Those whose intensity increased (usually by a factor of 2 to 4) were taken to be correctly assigned. A check was made to see that lines that were not assigned this upper level were not enhanced.

RESULTS

The results of this study are summarized in Table I and in the Grotrian diagram of Fig. 1.

Table I shows the observed wavelengths grouped in series characterized by a common upper level. We have labeled levels by giving the largest LS-coupled component of calculated eigenvectors. All identifications shown in this table were verified by the laser enhanced fluorescence method. That is, each upper level was populated by transfer from $2p^5 3s3p\ ^4D_{7/2}$ or $^4S_{3/2}$, as allowed by the selection rule $\Delta J = 0, \pm 1$. All possible transitions from each upper level were observed as listed in Table I.

Table I also gives the calculated Einstein A coefficient and the calculated branching ratio for each of the observed transitions. The

Table I Na I quartet transition wavelengths, decay rates, branching ratios (B.R.), and observed line intensities

Upper Level	Lower Level	Wavelength (Air, ±.2Å)	A_{ik} (10^8 s)	g_i x B.R.	Obs. Intensity (arb. units)
$2p^5 3s3d\ ^4P_{1/2}$	$2p^5 3s3p\ ^4S_{3/2}$	3511.0	1.15	1.3	1.3
	$^4D_{3/2}$	4008.8	.15	.18	.27
	$^4P_{3/2}$	4286.7	.28	.32	.46
$^4P_{3/2}$	$^4S_{3/2}$	3502.5	.88	1.0	1.6
	$^4D_{5/2}$	3942.6	.21	.24	.32
	$^4D_{3/2}$	3997.7	.08	.09	.14
	$^4P_{5/2}$	4204.9	.25	.28	.46
	$^4P_{3/2}$	4273.9	.16	.18	.23
$^4D_{3/2}$	$^4S_{3/2}$	3416.2	.15	.03	*
	$^4D_{3/2}$	3885.7	.62	.12	.36
	$^4D_{1/2}$	3930.6	.25	.05	.15
$^4D_{5/2}$	$^4S_{3/2}$	3489.0	.42	1.5	1.3
	$^4D_{7/2}$	3872.9	.12	.45	.19
	$^4D_{5/2}$	3925.6	.45	1.6	.85
	$^4D_{3/2}$	3980.3	.05	.18	.14
	$^4P_{5/2}$	4185.5	.49	1.8	1.5
	$^4P_{3/2}$	4253.8	.05	.18	.17
$^4F_{5/2}$	$^4S_{3/2}$	3427.3	.28	.11	.09
	$^4D_{5/2}$	3848.0	.28	.03	.05
	$^4D_{3/2}$	3900.4	.77	.08	.15
$^4F_{7/2}$	$^4D_{7/2}$	3865.5	.58	1.4	1.3
	$^4D_{5/2}$	3917.9	.66	1.5	1.6
	$^4P_{5/2}$	4176.7	.35	.83	.82
$^4F_{9/2}$	$^4D_{7/2}$	3881.8	1.63	10.0	4.2
$2p^5 3s4s\ ^4P_{5/2}$	$^4S_{3/2}$	4432.3	.14	.91	2.8
	$^4D_{7/2}$	5071.2	.41	2.7	2.4
	$^4D_{5/2}$	5162.5	.15	1.0	.74
	$^4D_{3/2}$	5256.4	.02	.16	.06
	$^4P_{5/2}$	5621.0	.14	.91	.37
	$^4P_{3/2}$	5744.2	.04	.23	.11

*Not observed in emission scans; transition observed in laser transfer $2p^5 3s3p\ ^4S_{3/2} \rightarrow 2p^5 3s3d\ ^4D_{3/2}$.

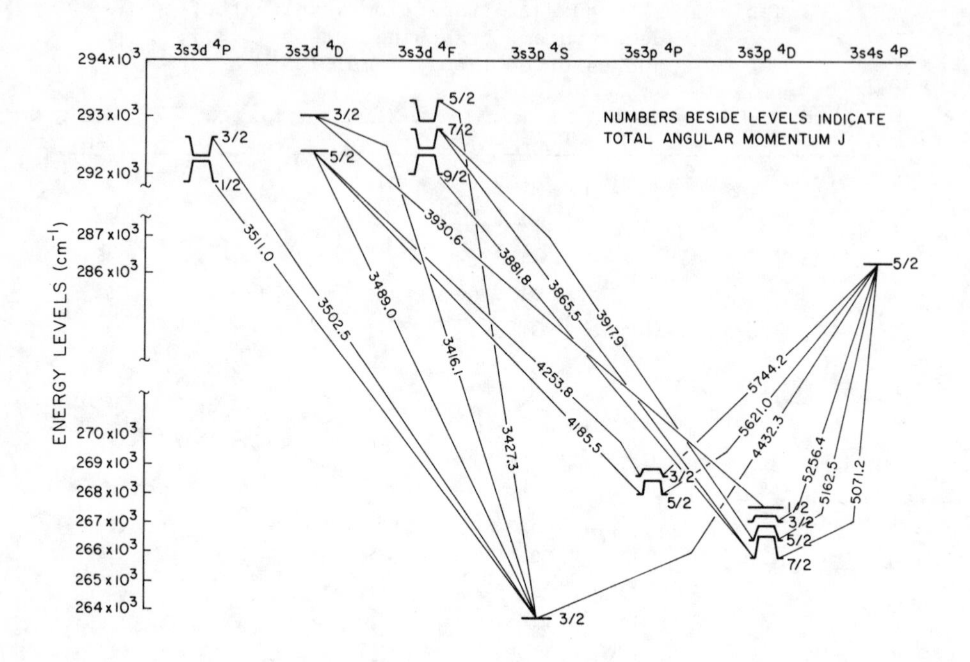

Fig. 1. Na I core-excited Grotrian diagram (configurations $1s^2 2s^2 2p^5$ $n\ell n'\ell'$ quartet system).

last column gives the experimental relative intensities of each of the observed lines. (Lines from $2p^5 3s3d$ and $2p^5 3s4s$ upper states have been normalized separately, since the cross sections for electron excitation of these two configurations are not necessarily the same.) In all cases, predicted and observed relative line intensities agree to within a factor of 3, consistent with the accuracy expected of our calculated oscillator strengths and branching ratios.

Figure 1 is a Grotrian diagram showing many of the brighter transitions. Based upon the photoabsorption measurements of Ref. 7, the energy of the lowest level in the quartet manifold, $2p^5 3s3p\ ^4S_{3/2}$, should be set at $263765 \pm 15\ cm^{-1}$.

In summary, this work establishes a Grotrian diagram interrelating the three lowest configurations of the quartet manifold of Na I, and shows the utility of laser enhanced fluorescence for ascertaining and verifying transitions among core-excited states.

ACKNOWLEDGEMENTS

The authors gratefully acknowledge helpful discussions with R. D. Cowan, A. Weiss, T. Lucatorto, R. W. Falcone, and P. J. K. Wisoff.

The work described here was supported by the U.S. Air Force Office of Scientific Research and the U.S. Army Research Office.

REFERENCES

1. S. E. Harris, Opt. Lett. 5, 1 (1980).
2. Joshua E. Rothenberg and Stephen E. Harris, IEEE J. Quant. Elect. QE-17, 418 (1981).
3. Robert D. Cowan, The Theory of Atomic Structure and Spectra (University of California Press, Berkeley, 1981), Secs. 8-1, 16-1, and 18-7.
4. H. G. Berry, R. Hallin, R. Sjodin, and M. Gaillard, Phys. Lett. 50A, 191 (1974).
5. R. Frohling and H. J. Andra, Phys. Lett. 97A, 375 (1983).
6. D. E. Holmgren, R. W. Falcone, D. J. Walker, and S. E. Harris, Opt. Lett. 9, 85 (1984); R. W. Falcone and K. D. Pedrotti, Opt. Lett. 7, 74 (1982).
7. J. Sugar, T. B. Lucatorto, T. J. McIlrath, and A. W. Weiss, Opt. Lett. 4, 109 (1979).
8. W. C. Martin and Romuald Zalubras, J. Phys. Chem. Data 10, 153 (1981).
9. S. E. Harris, D. J. Walker, R. G. Caro, A. J. Mendelsohn, and R. D. Cowan, "Quasi-Metastable Quartet Levels in Alkali-Like Atoms and Ions," Opt. Lett. (to be published).

MEASURED RADIATIVE LIFETIMES OF ROVIBRONIC LEVELS IN THE $A^1\Pi(v = 0)$ STATE OF CO AND COMPARISON WITH THEORY*

M. Maeda† and B. P. Stoicheff
Department of Physics, University of Toronto,
Toronto, Ontario, M5S 1A7, Canada.

ABSTRACT

Tunable VUV radiation was generated in the region of 154.5 nm by resonant 4-wave mixing of N_2-pumped dye-laser radiation in Mg vapor. Individual rotational levels (J' = 1 to 29) of the CO $A^1\Pi(v = 0)$ state were excited, and radiative lifetimes were determined by fluorescence decay measurements. The increased lifetimes of levels perturbed by nearby $d^3\Delta$ and $e^3\Sigma^-$ states were found to be in good agreement with theoretical values.

INTRODUCTION

The recent extension of tunable coherent sources to VUV wavelengths has provided the possibility of high-resolution spectroscopy in this region[1]. Moreover, since these sources generate short pulses, of the order of nanoseconds, measurement of radiative lifetimes of specific rovibronic levels is readily achieved. Such measurements have already been demonstrated in preliminary studies of CO[2] and NO[3]. Here we wish to report on recent extensive measurements of lifetimes of individual rotational levels (from J' = 1 to 29) of the $A^1\Pi(v = 0)$ state of $^{12}C^{16}O$. These experimental values provide a comparison with values calculated by Field[4]. It is well-known that the lowest electronic state, $A^1\Pi$, of CO has large spin-orbit interaction with nearby triplet states $d^3\Delta$ and $e^3\Sigma^-$, and that the effects of degeneracies can be calculated by perturbation theory[5]. The present lifetime measurements indicate overall good agreement with the theoretical values of Field[4].

EXPERIMENT

A fast, dual-N_2 laser system of the Blumlein-type was designed and built for pumping two dye lasers. It has two transversely excited discharge tubes connected to flat Mylar capacitors, with firing controlled by a common spark-gap switch. The dual-N_2 lasers each emitted 400 KW in pulses of 2.5 to 3 ns at a repetition frequency of 10 Hz, and with a jitter of less than 1 ns. The two dye lasers used the simple grazing incidence configuration with gratings of 3400 lines/mm, and a short cavity length of ∿7 cm in order to generate short pulses. With coumarin dyes, these lasers each produced ∿5 KW pulses with a duration of 1.5 to 2 ns, in a bandwidth of 0.3 cm^{-1}.

Resonantly enhanced 4-wave sum mixing (4-WSM) in Mg vapor was used

*Research supported by the Natural Sciences and Engineering Research Council of Canada, and the University of Toronto.

†Visiting Scientist, on leave from Department of Electrical Engineering, Kyushu University, Fukuoka 812, Japan.

to generate VUV radiation over the range 140 to 160 nm. One dye laser (coumarin 440) was tuned to the 2-photon transition $3s3d\,^1D_2 \leftarrow 3s^2\,^1S_0$ ($2\nu_1$ = 46403 cm^{-1}) in Mg. The other (coumarin 540) was scanned over a wide range ($\equiv \nu_2$). When the two beams at ν_1 and ν_2 were superimposed and focused in the heat-pipe containing Mg vapor at ~20 Torr and He buffer gas at ~200 Torr, 4-WSM at $2\nu_1 + \nu_2$ was obtained having a flux of ~10^7 photons per pulse.

The resulting VUV radiation was focussed in the centre of a small, stainless-steel, sample cell into which CO gas (Matheson, research grade) slowly flowed. This cell was fitted with an entrance, exit, and a side window, all of LiF. A solar-blind photo-multiplier tube (EMR 510G-08-13) was positioned at the side-window, at a distance of 1.5 cm from the incident beam, for collection of the fluorescence radiation. The signal was processed by a boxcar-averager and gated-integrator system (PAR 162/165).

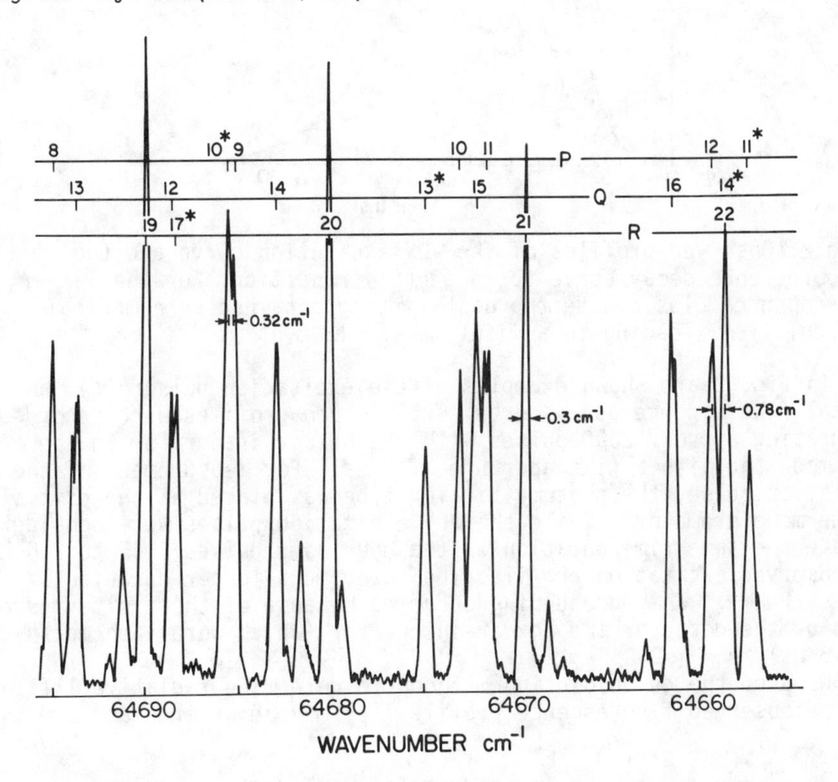

Fig.1. A part of the fluorescence excitation spectrum $A^1\Pi(v=0) \rightarrow X'\Sigma(v=0)$ of CO in the region of strong perturbations. The J' values labelled * indicate extra, triplet-like lines due to the interaction of the $A^1\Pi(v=0)$ and $e^3\Sigma^-(v=1)$ states.

Initially, the fluorescence excitation spectrum of the (0,0) band of the transition $A^1\Pi \rightarrow X^1\Sigma$ was recorded by tuning the VUV radiation over the wavelength range 154.0 to 155.5 nm, with an effective resolution of ~0.3 cm^{-1}. The observed lines were then identified according to the analysis of Simmons, Bass, and Tilford[6]. A small part of the spectrum, showing a region of maximum perturbation is produced in Fig. 1. Following this identification, the exciting radiation was tuned to each of the unblended rovibronic lines, and the decay of fluorescence intensity with time was measured.

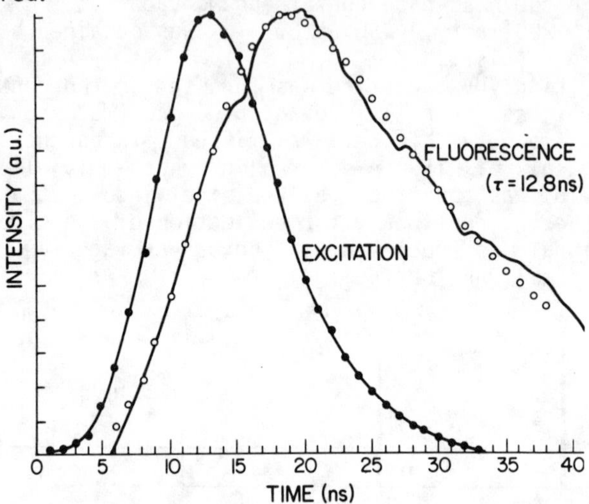

Fig.2. Observed profiles of the VUV excitation curve and the fluorescence decay curve for the P(8) transition. For the latter, the open circles denote the best fitting obtained by numerical calculation, leading to a lifetime τ = 12.8 ns.

In Fig. 2 are shown examples of the excitation pulse g(t), and of a typical fluorescence decay curve, f(t). Such profiles were recorded by integrating approx. 2000 pulses with the boxcar-integrator in the scanning mode and with a time aperture of 2 ns. For measurement of the excitation pulse, the photomultiplier tube was placed at the exit slit of the monochromator. While the VUV excitation pulses were considered to have the same time duration as the dye-laser pulses (1.5 to 2.0 ns), the observed excitation profiles indicated much longer duration of ∿10 ns (FWHM). This broadening occurred because of the finite response times of the detector and boxcar-integrator (which were each measured to be 4-5 ns).

We used the deconvolution method[7] to determine radiative lifetimes τ. The observed fluorescence profile f(t) may be written as

$$f(t) = \int_0^t dt' g(t') \exp[-\frac{t-t'}{\tau}]$$

Here g(t') is the overall instrumental response, including VUV pulse duration. Values of τ were obtained by least-squares fitting with a computer. With this method it was possible to determine τ-values which were the same as, or even smaller than, the instrumental response time. However, to determine τ-values down to 1-2 ns or less, faster detectors are necessary.

RESULTS AND DISCUSSIONS

The radiative lifetimes of rovibronic levels are normally expected to be independent of rotational quantum number, except of course, in the case of interaction with nearby levels, as in the v = 0 levels of the $A^1\Pi$ state of CO. For this state, mean values of radiative lifetimes over a vibrational level have been obtained by electron excitation[8,9] and by synchrotron radiation excitation[10]. High-resolution laser excitation such as performed in the present experiments, however, permits the evaluation of lifetimes for individual rovibronic levels within a vibrational level.

Radiation lifetimes for levels J' = 1 to 29 of the v' = 0 level of the $A^1\Pi$ state were obtained by measuring the fluorescence decay rates for transitions in the P, Q, and R branches. The results are shown in Figs. 3 and 4. In each graph, the lifetimes τ are normalized by the unperturbed lifetime τ_u, and plotted as a function of the rotational quantum number J'. All measurements were made at 50 mtorr of CO. A value of $\tau_u = 10.3 \pm 0.5$ ns for the lifetime of unperturbed levels was also obtained at 50 mtorr pressure by averaging the fluorescence decay rates of 26 unblended P, Q, and R transitions from unperturbed levels. This value was used in the graphs of τ/τ_u vs J' in Figs. 3 and 4. Finally, in order to assess the effect of radiation trapping, values of τ for the P(6) and P(20) transitions were measured for CO pressures as low as 8 mtorr. The zero-pressure extrapolations gave a value of τ_o = 9.4 ± 0.5 ns.

Also shown in Figs. 3 and 4, are the theoretical values of Field[4], normalized by the deperturbed lifetime $\tau_o = 9.9 \pm 0.1$ ns based on the measurements with synchrotron radiation[10]. (Values of $\tau_o = 10.69 \pm 0.3$ ns[8] and 11.9 ± 0.7 ns[9] based on electron excitation are not deperturbed values.) At high J'-values, the maximum perturbation occurs at J' = 27, as determined from the Q(27) and P(28) transitions. The P(28)

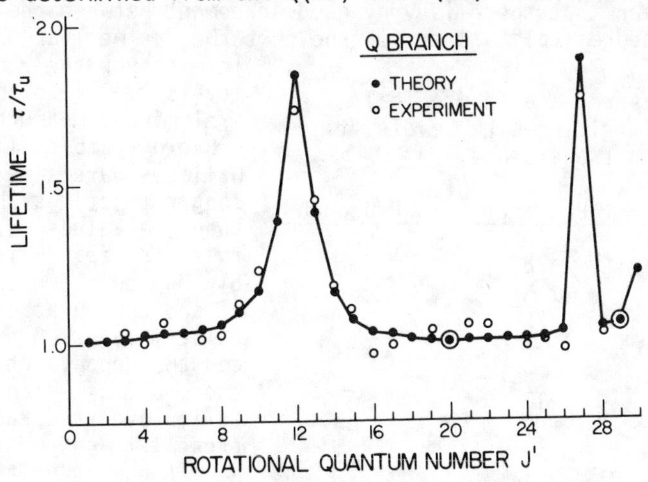

Fig.3. A graph of lifetimes as a function of rotational quantum number J', obtained from Q branch lines. The lifetimes are normalized by the unperturbed decay time τ_u. The theoretical values are those of Field[4].

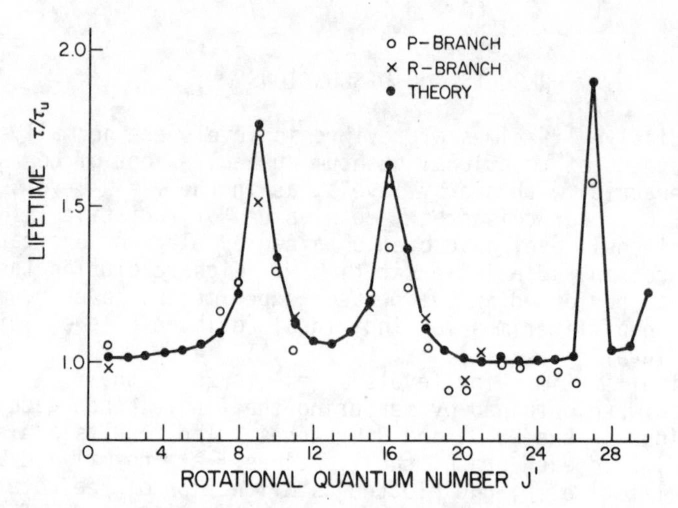

Fig.4. A graph of lifetimes as a function of rotational quantum numbers J', obtained from P and R branch lines. Lifetimes are normalized by τ_u, and theoretical values are those given by Field[4].

transition has rather low intensity and the R(26) transition is overlapped by that of P(17). This J' = 27 perturbation has been attributed[6] to interaction with the $d^3\Delta_i(v = 4)$ state, with all three triplet components affecting the $A^1\Pi$ levels of high-J'. At lower J'-values, the observed perturbations differ for Q branch transitions from those of P and R branches. These have been explained by the selection rules for interaction with the $e^3\Sigma^-(v = 1)$ state which provide for the $F_2(J = N)$ component affecting the Π^- component and thus perturbing the Q branch, while the F_1 and $F_3(J = N \pm 1)$ components each affect the Π^+ components resulting in perturbations of the P and R branches.

It is seen that there is very good agreement between the experimental and theoretical[4] values for the perturbed as well as unperturbed levels over the range of J' levels from 1 to 29. The preliminary measurements by Provorov, Stoicheff, and Wallace[2] gave somewhat longer lifetimes (by ∿1 ns) than the values given here. This discrepancy is probably due to the present use of shorter duration laser pulses, lower CO pressure, and the deconvolution method, all of which have led to improved accuracy of the present measurements.

A number of triplet-like lines resulting from the perturbations were also observed, but only a few of the P and R branch lines were sufficiently intense

Table 1. Measured and Theoretical Lifetimes (ns) of Triplet-Like Levels from P and R Branch Transitions.

J'	Theory	P Branch	R Branch
8	56.9	96	-
9	23.5	-	33
10	42.1	72	-
11	111	-	-
...			
15	127	113	-
16	62.3	97	89
17	25.7	31	35
18	37.4	-	-
19	100	-	-

- indicates overlapped lines

and free from overlap to use in the measurement of lifetimes. These values are given in Table 1 along with predicted values for comparison. As before, all of the experimental measurements were made at a CO pressure of 50 mtorr. While the perturbations lengthen lifetimes of singlet-like levels (Figs. 3 and 4), those of triplet-like levels are shortened[11] from the unperturbed $e^3\Sigma^-$ lifetimes of ∿3 μs and $d^3\Delta$ lifetimes of 4.2 μs. For some of the triplet-like transitions which exhibited relatively long fluorescence decays, we observed double-exponential decays. For example, the P(10) fluorescence decay had a long-lived component of ∿75 ns and a short-lived component of <20 ns. Undoubtedly, the longer lifetime corresponds to that of the triplet-like level, and the existence of a shorter lifetime suggests excitation transfer to a close-lying singlet state. We also note that each singlet-like line exhibited a weak, long-lived decay (out to ∿400 ns), presumably caused by excitation transfer to nearby states. A better understanding of the relevant transfer processes will require more detailed experimentation.

CONCLUSION

We have used a high-resolution, tunable and coherent VUV source producing short pulses (∿ 2 ns) to excite CO molecules to specific rotational levels of the $A^1\Pi(v = 0)$ state. The fluorescence decay was measured, and a deconvolution method was used to determine radiative lifetimes of these levels from J' = 1 to 29. A comparison with theoretical values calculated by Field[4] shows very good overall agreement.

The selective VUV excitation technique described here can be applied to lifetime measurements of a wide variety of atomic and molecular systems. We measured lifetimes down to 5 ns, but the finite response times of the photo-multiplier and data acquisition system precluded accurate measurement of shorter lifetimes. The development of fast detectors for the VUV region is important for extension of these measurement techniques to nanosecond and sub-nanosecond lifetimes.

We are grateful to R. W. Field for sending us his calculated data prior to publication, and for helpful discussions.

REFERENCES

1. B. P. Stoicheff, J. R. Banic, P. Herman, W. Jamroz, P. E. LaRocque, and R. H. Lipson, in Proc. Topical Meeting on Laser Techniques in Extreme Ultraviolet Spectroscopy (T. J. McIlrath and R. R. Freeman, eds.), Amer. Inst. Phys. New York (1982), pp. 19-31.
2. A. C. Provorov, B. P. Stoicheff, and S. C. Wallace, J. Chem. Phys. 67, 5393 (1977).
3. J. R. Banic, R. H. Lipson, T. Efthimiopoulos, and B. P. Stoicheff, Opt. Lett. 6, 461 (1981).
4. R. W. Field, private communication (1983).
5. R. W. Field, B. G. Wicke, J. D. Simmons, and S. G. Tilford, J. Mol. Spectrosc. 44, 383 (1972).
6. J. D. Simmons, A. M. Bass, and S. G. Tilford, Astrophys. J. 155, 345 (1969).
7. L. Hundley, T. Coburn, E. Garwin, and L. Stryer, Rev. Sci. Instrum. 38, 488 (1967).

8. R. E. Imhof and F. H. Read, Chem. Phys. Lett. 11, 326 (1971).
9. T. A. Carlson, N. Duric, P. Erman, and M. Larsson, Z. Physik A287, 123 (1978).
10. R. W. Field, O. Benoist d'Azy, M. Lavolleé, R. Lopez-Delgado, and A. Tramer, J. Chem. Phys. 98, 1839 (1983).
11. T. G. Slanger and G. Black, J. Chem. Phys. 58, 3121 (1973).

Chapter IV. Instrumentation

Variable Line Space Slitless Spectrometers at Grazing Incidence,

M.C. HETTRICK

U.C. Berkeley, Space Sciences Laboratory, Berkeley, CA 94720.

A plano grating consisting of variably-spaced grooves concentric about a common point (fig. 1) provides $\lambda/\Delta\lambda \sim 3000$ over a simultaneous factor two in soft x-ray wavelength (λ = 10-100 Å). The complete instrument contains a 1 arc sec 1-meter diameter collecting mirror whose 10 meter focal length includes a 1.5 meter spectrometer with an imaging detector having 15 μm pixels. The end-to-end efficiency is approximately 4%, and the grating has a factor 4 variation in line spacing across a 50 cm ruled width.

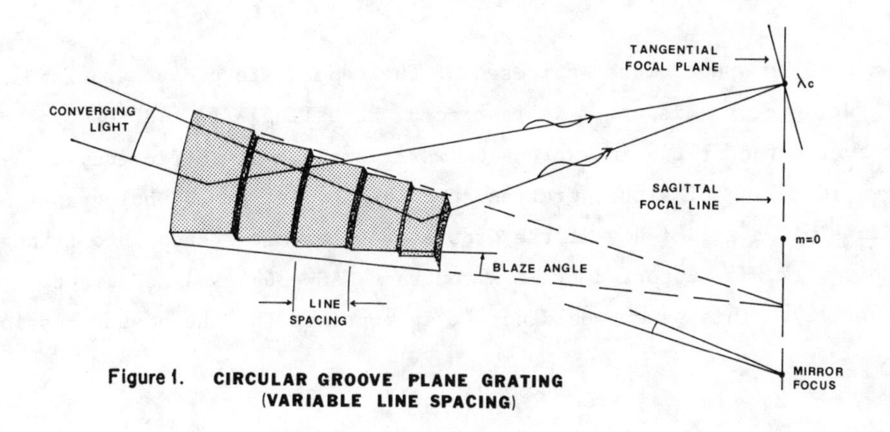

Figure 1. CIRCULAR GROOVE PLANE GRATING (VARIABLE LINE SPACING)

A plano grating with grooves which radiate from a common point (fig. 2), i.e. an oriental fan, operates in the conical diffraction mount and is shown to have much lower spectral resolution due to dispersion limits imposed by geometric effects and attainable line densities. However, in the extreme UV (λ = 100-1000 Å), raytraces treating both equal and varied angular spacings indicate $\lambda/\Delta\lambda \sim 2000$-3000 is attainable over a factor two in wavelength with high efficiency ($\sim$ 10%).

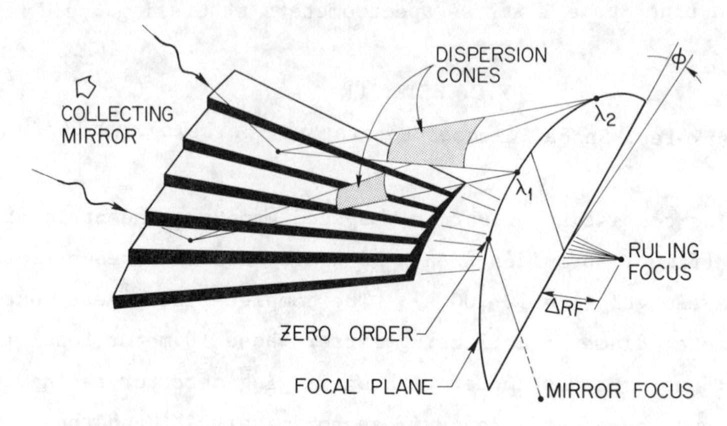

Figure 2. ORIENTAL FAN GRATING

An echelle spectrometer composed of the concentric groove grating and a fan grating cross-disperser is raytraced, revealing $\lambda/\Delta\lambda \sim 20{,}000$ is achievable over a factor two in extreme UV wavelengths. The efficiency is $\sim 2\%$. Stigmatic ruling of either grating, provided by extreme UV holography, can obtain similar efficiency if the groove profiles are etched into triangles.

This work was supported by NASA contract NASW-3636 and NSF grant INT-8116729. This paper would not be appropriate for the poster session.

VUV INSTRUMENT CALIBRATIONS FOR LASER-PLASMA STUDIES

P. D. Rockett, C. R. Bird, and R. Shepherd
KMS Fusion, Inc., Ann Arbor, Michigan 48106

ABSTRACT

Techniques for spectral and intensity calibrations of laser-plasma VUV instrumentation are described. A flexible laboratory x-ray station is shown to yield creditable results.

INTRODUCTION

Spectroscopy in the VUV region between .15 keV and 10 keV has been a major component of laser-plasma characterization for many years. High resolution spectroscopy has employed natural and artificial crystal instrumentation[1-4] as well as gratings[5,6], while low resolution spectroscopy has utilized active instruments, such as filtered x-ray diodes[7] and PIN diodes.[8] Several of these systems depended upon film as the direct recording material, and others utilized microchannel plate detectors for signal intensification prior to recording. Of the above mentioned instrumentation only x-ray diodes have been systematically calibrated for absolute sensitivity over a wide range of wavelengths.[9] Film calibration has been performed only sporadically[10,11], as has crystal calibration. Calibration of grating instruments is virtually unheard of in laser experiments.

These apparent lapses are due to the lack of time and/or funds available to contribute towards calibration of x-ray equipment. Many measurements after all only require relative data, and spectral characterization can often be performed in situ with known laser targets. Some experiments can, however, benefit greatly from a knowledge of the absolute spectral sensitivity of an x-ray instrument. Furthermore, some instrument capabilities are simply unknown until a full calibration is performed.

To support our needs for characterized instrumentation, we have fabricated an inexpensive soft x-ray calibration system called XCAL, which enables a quick response to the needs of laser-plasma experiments. A hot-filament point x-ray source sits within a ten inch i.d. vacuum chamber operated at 10^{-6} Torr and can be simultaneously viewed by an instrument under test and by a photon-counting detector. This small system has been supported by one full time person and has provided invaluable data on the sensivity of two crystal spectrographs, a grazing incidence spectrograph, and a new x-ray film. This permitted the performance of certain experiments, which required absolute data for meaningful comparison

0094-243X/84/1190171-09 $3.00

to theory. In addition unexpected capabilities were revealed in the calibrations, which resulted in the design of experiments previously considered unworkable.

X-RAY CALIBRATION FACILITY (XCAL)

The keystone of the KMSF, Inc. x-ray calibration facility (XCAL) is a J. E. Manson, Inc. radiatively-cooled point x-ray source. The electrostatically-focussed source is viewed at two orthogonal directions (each seeing a 45 degree take-off angle) by the test instrument and a photon counting detector. Measurements of incident flux are thus made in real time as can be seen in Figure 1. Operation with a modest emission current results in a photon flux which varies from 10^7 to 10^9 photons/sec-sr. This is approximately 100 times less bright than a Henke tube, but is still more than adequate for most film-based instrument calibrations. Source size varies from 300 μm to 700 μm in diameter by varying the filament emission current from 0.1 mA to .6 mA, respectively.

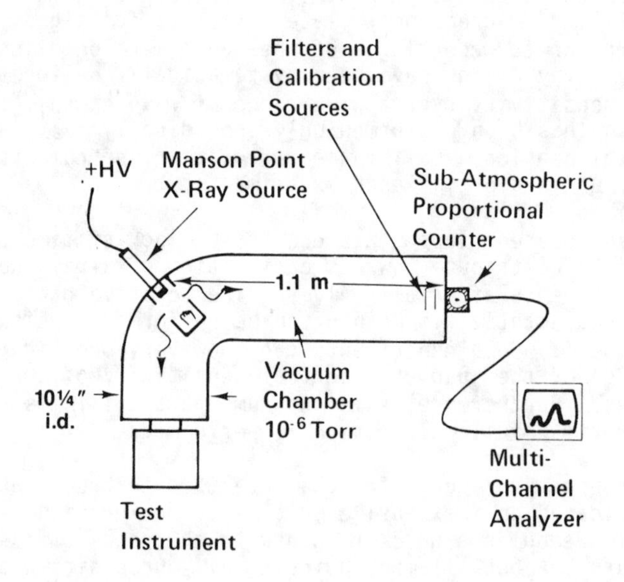

Figure 1. Schematic of XCAL, a point-source-based soft x-ray calibration system.

The wavelength range of XCAL presently extends from 67.4Å to 2.52Å using the characteristic lines listed in Figure 2. A change of anodes is a ten minute operation. Our photon counting is performed with a 50/50 argon-methane sub-atmospheric proportional counter. Ion generation and counting statistics limit resolution to approximately $12/\sqrt{E(eV)}$ where E is the detected photon energy.[12] Thus, closely spaced spectral lines may be unresolved, and MCA spectral calibration becomes important. Figure 3 describes the two types of isotope sources chosen to calibrate the multichannel analyzer energy scale. ^{55}Fe provides a Mn Kα/β line, while a ^{210}Po alpha source excites low energy calibration lines in C, F, O, or Al.

Anode	λ(Å)
Bo on Mg:	67.4
C	44.7
O on C, Mg	23.6
Cu (L)	13.3
Mg	9.89
Ti	2.76, 2.52

(Other users of Manson sources go to Cu Kα at 1.54 Å)

Figure 2. Present wavelength coverage of the KMSF x-ray calibration system (XCAL).

Sources:

^{55}Fe 2.11 Å

^{210}Po - alpha source

- C 44.5 Å
- F 18.4 Å
- Al 8.3 Å

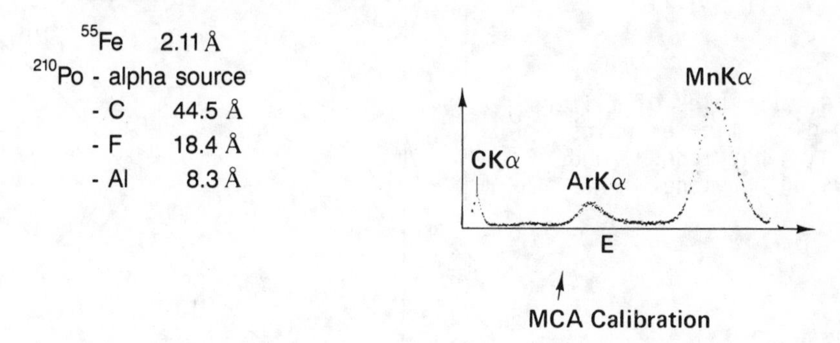

Figure 3. The multichannel analyzer is spectrally calibrated with isotope sources.

Equally important is the filtering of a given anode spectrum to isolate desired lines. A graphite anode for instance generates both a strong C Kα line and a strong O Kα line. As shown in Figure 4 a thick 4 μm parylene N filter will discriminate against oxygen while passing C Kα . A parylene D filter uses the chlorine L-edge to discriminate against carbon, but pass O Kα . In general K-, or L- edge filters are required to further reduce bremsstrahlung from higher Z anodes, improving line purity.

Oxygen
Discriminate against
with Parylene N, C_8H_8
TRANSMISSION
0.30
0.20
0.10
0.0
0.2
0.4
0.6
0.8
1.0
ENERGY (keV)

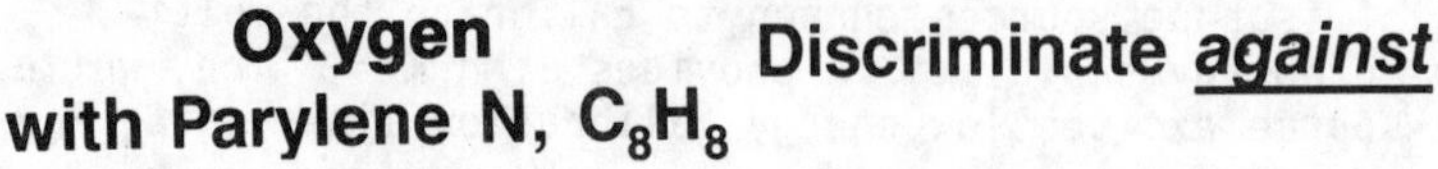

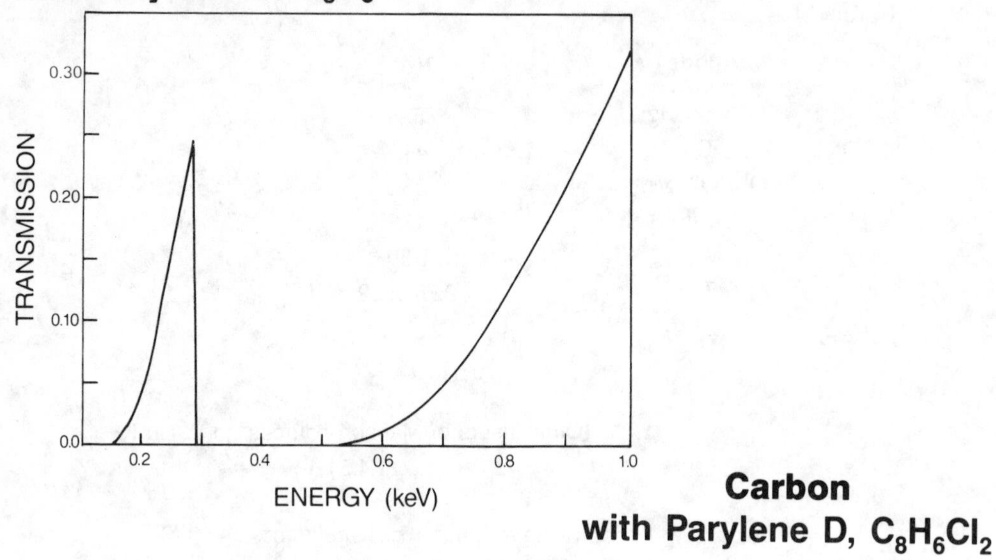

Carbon
with Parylene D, $C_8H_6Cl_2$
TRANSMISSION
0.8
0.6
0.4
0.2
0.0
0.2
0.4
0.6
0.8
1.0
ENERGY (keV)

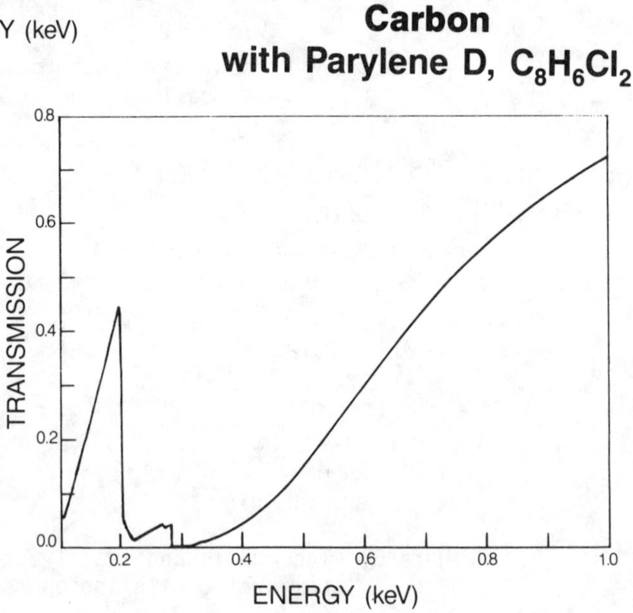

Figure 4. Examples of filters used to discriminate against O Kα (using parylene N) and C Kα (using parylene D)

DETECTOR CALIBRATIONS

Photographic film is the final element in most x-ray instrumentation, and a calibration of the instrument is incomplete without an understanding of the film response. XCAL was used to calibrate a new x-ray film from KODAK designated Direct Exposure Film or DEF. DEF was advertised as KODAK No-screen's replacement, and we therefore made a comparison to a credible No-screen calibration from NRL.[11] The resulting H&D curves appear in Figure 5. They showed DEF to be 2.5x more sensitive than No-screen at 1.25 keV and 5x more sensitive than No-screen at Ti $K\alpha/\beta$. This film was then applied to laser spectroscopic experiments with remarkable success.

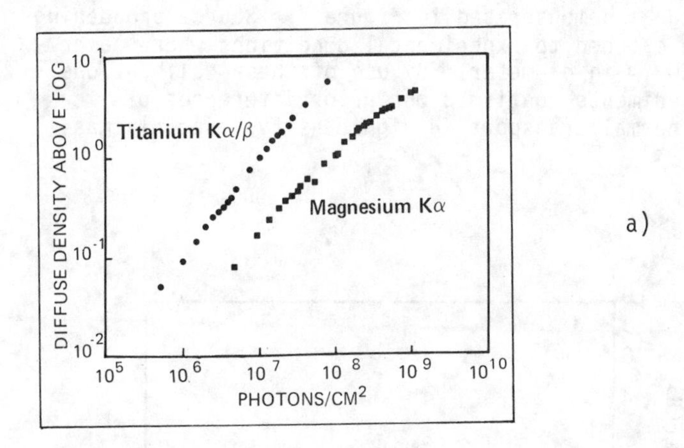

a)

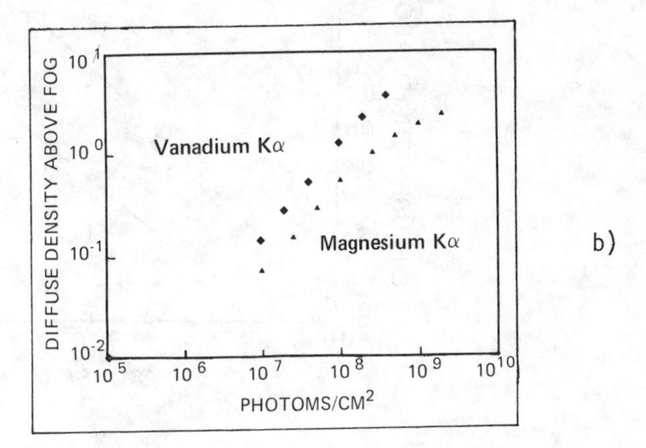

b)

Figure 5. a) Calibration of Kodak Direct Exposure Film performed at KMSF: b) calibration of Kodak No-screen film performed at NRL and LASL.

Several x-ray spectrographs were also calibrated with XCAL, permitting the inference of absolute flux levels in given spectral lines. A preliminary calibration of a 0.5 m grazing incidence spectrograph appears in Figure 6. The large range of angles of incidence across the grating in the Rowland circle mount make it impossible to calculate spectral sensitivity. Only an x-ray calibration of this sort can eventually yield a curve of sensitivity vs. wavelength. Two convex curved crystal spectrographs (LiF and PET) were calibrated with XCAL at Ti $K\alpha/K\beta$ (K-edge filtering will not exclude the $K\beta$ line until Z=23). Absolute reflectivities were obtained, which were used to interpret experimental data from earlier electron superthermal transport experiments. The small source diameter of XCAL resulted in minimal source broadening as demonstrated in Figure 7. Source broadening could in fact be matched to experimental conditions where laser targets are ~ 120 μm in diameter. By use of these calibrations, $K\alpha$ emission experiments confirmed our prior inferences of inhibited superthermal transport in high density laser plasmas.

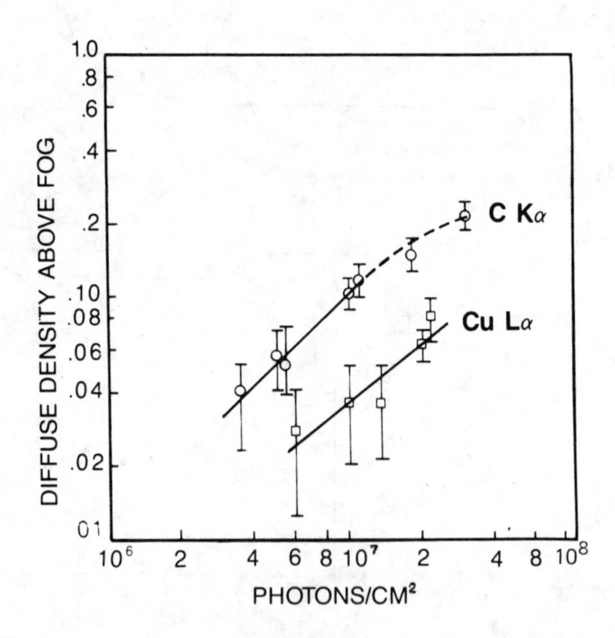

Figure 6. Preliminary calibration of a 0.5m grazing incidence spectrograph at C $K\alpha$ and Cu $L\alpha$.

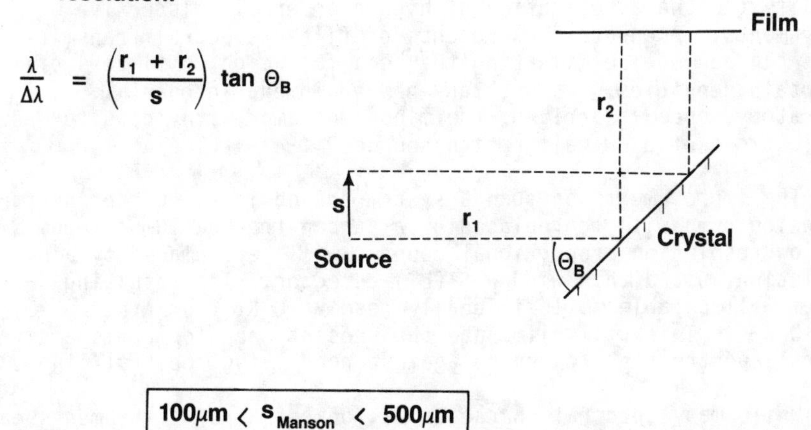

Figure 7. The small diameter (s_{Manson}) of the Manson point x-ray source has permitted calibration of curved. crystal spectrographs with minimal source broadening.

CONCLUSIONS

Many small laboratories operate extensive x-ray instrumentation, but avoid establishing a calibration facility for fear of the cost in manpower and dollars. Our experience has shown that the dedication of one full-time individual and very limited expenses can yield a productive, flexible facility for the calibration of most all x-ray equipment. Source brightness permits adequate exposures in minutes to hours. Care, however, must be taken to adequately characterize the x-ray source and the photon-counting detectors.

The rewards of calibration are copious. Meaningful transfer curves for film-based instruments enable accurate estimates of source spectral brightness and true FWHM's. Measured integrated crystal reflectivities can aid in the choice of optimum elements for specific experiments. Isolated characteristic x-ray lines can simplify the characterization of higher orders in dispersive instruments. Accurate measurements of filter spectral transmittence can be made, eliminating the need for weighing filters of uncertain densities. As with any new equipment in ones' laboratory, once it arrives, the mind runs amok with ideas for new applications of a CW calibration source.

The improvements on such a system are obvious. Our present diffusion pump will be replaced by a carbon-free helium cryopump. The low resolution proportional counter will be replaced by a high resolution, ultra-thin window Si(Li) detector. The resulting system will be able to individually resolve C $K\alpha$ from N $K\alpha$ from O $K\alpha$. This will eliminate the present need for a dispersive crystal spectrograph to insure against near-lying spectral lines.

While many spectral characterizations of x-ray instruments can be done with a laser-produced plasma, there is nothing quite so unambiguous and well-defined as a laboratory x-ray calibration facility.

ACKNOWLEDGEMENTS

This work was performed under the auspices of the Department of Energy under contract #DE-AC08-82DP-40152.

REFERENCES

1. M. H. Key, J. G. Lunney, J. M. Ward, R. G. Evans, and P. T. Rumsby, J. Phys. B 12, L213 (1979).

2. B. Yaakobi, D. Steel, E. Thorsos, A. Hauer, and B. Perry, Phys. Rev. Lett. 39, 1526 (1977).

3. K. B. Mitchell, D. B. Van Hulsteyn, G. McCall, and P. Lee, Phys. Rev. Lett. 42, 232 (1979).

4. P. G. Burkhalter, G. Charatis, and P. D. Rockett, J. Appl. Phys. 54, 6138 (1983).

5. P. G. Burkhalter, V. Feldman, and R. D. Cowan, J. Opt. Soc. Am. 64, 1058 (1974).

6. P. G. Burkhalter, J. Reader, and R. D. Cowan, J. Opt. Soc. Am. 70, 912 (1980).

REFERENCES (Cont.)

7. P. D. Rockett, W. C. Priedhorsky, and D. V. Giovanielli, Phys. Fluids 25, 1286 (1982).

8. G. Enright, M. C. Richardson, and N. H. Burnett, J. Appl. Phys. 50, 3909 (1979).

9. R. H. Day, P. Lee, E. B. Saloman, and D. J. Nagel, J. Appl. Phys. 52, 6965 (1981).

10. R. F. Benjamin, P. B. Lyons, and R. H. Day, Appl. Opt. 16, 393 (1977).

11. C. M. Dozier, D. B. Brown, L. S. Birks, P. B. Lyons and R. F. Benjamin, J. Appl. Phys. 47, 3732 (1976).

12. B. L. Henke and M. A. Tester, Advances in X-ray Analysis, (Plenum, New YOrk, 1975) Vol. 18, p. 76.

A High-Resolution VUV Spectrometer with Electronic Parallel Spectral Detector

C. L. Cromer, J.M. Bridges, T.B. Lucatorto, and J.R. Roberts
National Bureau of Standards, Washington, DC 20234

I. Introduction

We have recently constructed what we believe is a unique VUV spectrometer designed for photoabsorption studies of both stable and transient species. The instrument is comprised of three elements: a background continuum source which is provided by a laser-produced plasma, a commercially available 1.5M grazing incidence spectrometer, and a detector which consists of a high-resolution image intensifier coupled to an optical multichannel analyzer. The instrumental goal was to develop an instrument which would incorporate a continuum source with extremely short duration and a detector with high sensitivity and linear response over a large dynamic range in order to optimize VUV observations on transient laser-excited and laser-ionized vapors. Previous VUV absorption spectra of laser-produced species such as Li(2p) [1] and Ba^{++} [2] used a BRV source[3] and photographic detection; the disadvantages of the former system were the relatively long duration of the BRV pulse (~60ns)[4] and the difficulty in obtaining reliable photometric data from Shumann-type VUV photographic emulsions.

In an effort to provide a more sensitive, reliable, and convenient alternative to the photographic plate as detector of VUV spectral images, several researchers have developed VUV instruments which have utilized electronic recordings of intensified images. Timothy and Bybee[5] have constructed detectors for astronomical observations at UV and VUV wavelengths based on their Multi-Anode Microchannel Array (MAMA) concept of photon-counting array detectors. In their VUV devices, photons impinge directly on the front surface of the microchannel array (or Channel Electron Multiplier Array-CEMA). The ejected photoelectrons start a multiplicative cascade resulting in a pulse of $\sim 10^6$ electrons at the output surface for each initial photoelectron. These electron pulses are collected by an anode array which provides an electronic "image" of the VUV

0094-243X/84/1190180-13 $3.00 American Institute of Physics

spectra having 25 μm resolution. While these MAMA-based devices have single photoelectron sensitivity and excellent spatial resolution, the photon count rate is limited to about 1 MHz in each pixel, a factor which precludes their use in pulsed experiments such as ours where up to 10^4 photons/pixel per pulse (~20ns) are expected.

Another previous design concept which has been developed by members of the Tokamak plasma community is more suitable for experiments with high counting rates. This design again uses the front surface of the CEMA directly as the primary photosensitive element but then has an acceleration of the CEMA output electron bunches into a phosphor to create an intensified optical image of the original VUV spectrum. The phosphor is deposited onto a fiber optic image conduit which is coupled to a 2.56 cm linear self-scanned diode array having 1024 pixels of dimensions 25 μm x 2.5 mm (the Reticon 1024S [6] which is specially designed for spectrographic applications). Because electrons must have at least 4keV energy for efficient conversion in the phosphor, the spacing between the CEMA output suface and the phosphor screen (which form the planar boundaries of the high field region) must be significantly greater than in the MAMA design; this increased spacing results in a significant loss of spatial resolution. To optimally utilize the 25 μm resolution of the diode array given the reduced resolution of the intensified phosphor image, a fiber optic reducer is incorporated into the conduit. VUV instruments utilizing such detectors to observe the emission line spectra of Tokamak plasmas have been built by Fonck et al.[7], by H.W. Moos and his collaborators[8], by Schwob et al.[9], and by H. Manning et al. [10]

Recently Tondello and his collaborators have constructed an instrument suitable for photoabsorption measurements on transient species.[11,12]. As in our instrument, their spectrometer has a laser-produced plasma for its continuum source. The detector, however, is of a unique design. The VUV spectral image impinges on a scintillator-coated glass plate coincident with the Rowland cylinder of a vacuum spectrograph. The visible image thus formed is focussed by a lens system onto a self-contained intensified diode array. The system has fairly good spatial resolution (~45μm), but only moderate sensitivity.[12]

II. Instrumental

Ideally, photoabsorption studies of transient species should be performed with a continuum source of picosecond duration and high enough flux to produce a measureable spectrum with a single shot. Presently, a laser-produced plasma (LPP) is the most promising candidate for such a source. Preliminary studies have shown that LPP sources generally have the temporal characteristics of the exciting laser [13], [14] with pulse durations as short as ~25ps having been observed.[15] P.K. Carroll, et al.[13] have also made a qualitative study of the suitability of various target materials for producing continua in the range 4-200 nm, which covers the region of interest for the present spectrometer (4-90nm). They claimed that the rare earths from Sm to Yb produced fairly clean, strong continua in a set of overlapping intervals which covered the whole range 4 to 200nm, but they did not provide detailed spectral intensity measurements. Kühne [16] has subsequently made a radiometric comparison of a LPP source and a BRV source [3] in the range $40 \leq \lambda \leq 80$nm, and he showed the LPP source to be superior in overall reproducibility for all wavelengths and in intensity for $\lambda < 64$ nm. O'Sullivan et al.[17] measured the irradiance from LPP sources with Yb and Gd targets in the 115-220 nm range, and Mahajon et al[18], have made measurements with a Yb target at 122 nm. A definitive radiometric study of the spectral characteristics of the various rare earth LPP sources below 115 nm does not yet exist, however.

The spectrometer is a commercially available grazing incidence instrument (Acton Research Corporation model GIMS 551.5)[6] which is similar to a conventional monochromator except that the exit slit assembly is replaced by a specially designed bakeable chamber to house the CEMA assembly. (See Fig. 1). The instrument is supplied with three gratings, a 600 ℓ/mm, a 1200 ℓ/mm, and a 2400 ℓ/mm, which give it coverage in the wavelength range 4-90 nm. The specified wavelength resolution when used as a conventional monochromator with the 1200 ℓ/mm grating and 10 μm slits is better than 0.015 nm.

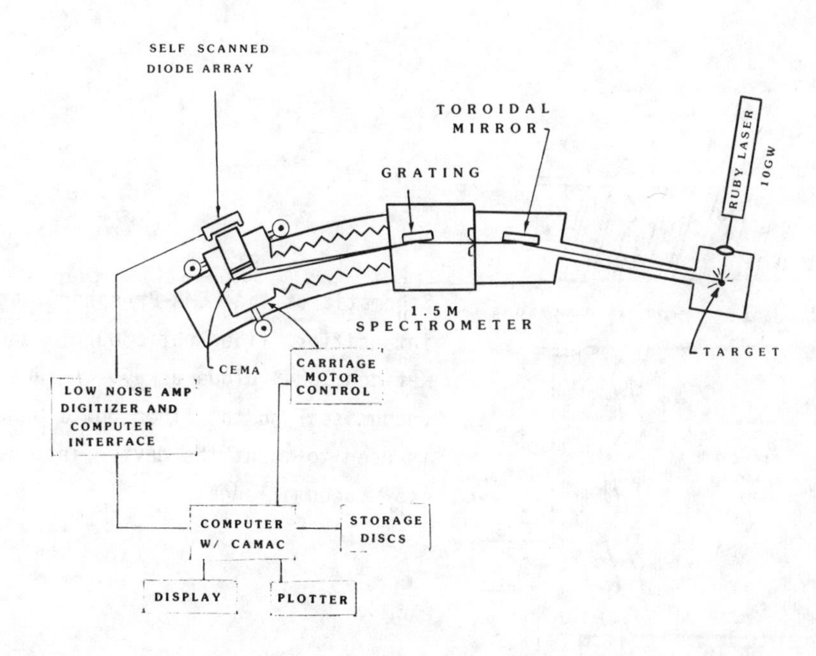

Fig. 1
Layout of a VUV Absorption Spectrometer

Fig. 1 shows the overall layout of the apparatus. A tube which constitutes the atomic absorption cell (e.g. for metal vapors this would be a heat-pipe oven) is placed between the LPP source and the slit. The ruby laser, capable of delivering up to 12J in a 25 ns pulse, is focussed onto the target. Synchronization of the laser pulse with the detector scan and data acquisition is controlled by a CAMAC-based LSI 11/2 computer.[6]

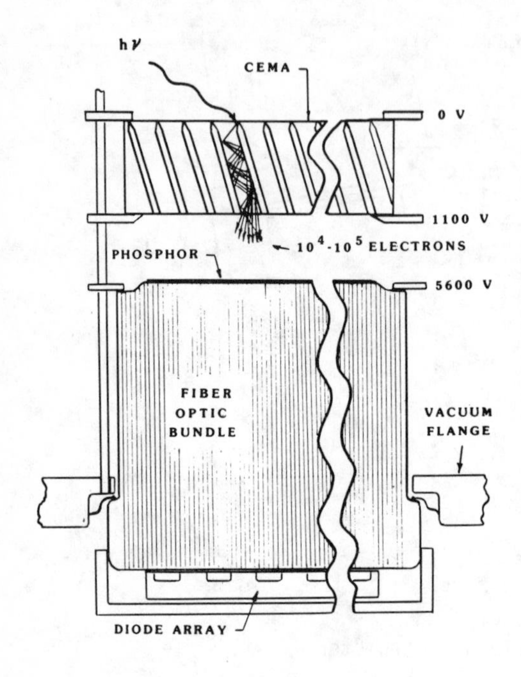

Fig. 2
Schematic of the CEMA-Phosphor image intensifier, fiber optic bundle and Reticon 1024S diode array. A bakeable vacuum seal on the fiber optic bundle is used to mount the device into a 4.5" vacuum flange.

Fig. 2 is a schematic of the detector assembly. While conceptually similar to the type developed for the Tokamak observations [7,8,9,10], ours differs from those previously reported on in two aspects: we have employed a special resolution-enhanced CEMA plate which is claimed to produce an approximately 20% resolution improvement,[19] and we have not used a fiber optic reducer. The CEMA has 12 μm channels on 15 μ centers biased at 15° to the plate normal; the input side is funneled and MgF_2 overcoated to improve sensitivity in the VUV range.[20] The design goal was to achieve a spectral resolution with the CEMA detector which was not much less than that achievable with a conventional single exit slit arrangement.

The CEMA-phosphor image intensifier section has two sources of resolution loss: the deviation from the Rowland circle of the flat front surface of the

CEMA, and the spread of the electron bunch as it travels from the output side of the CEMA to the phosphor. (The construction of a CEMA plate with front surface curvature to match the Rowland circle could have been attempted, but the advantages to our planned experiments were not worth the added effort at this time.) A 2.54 cm linear segment on a flat CEMA plate adjusted for minimum deviation from the Rowland circle departs from the circle by 50μm at the center and at each end; this deviation causes a (calculated) increase in image size (due to defocussing) from a minimum of 9 μm at λ=90 nm up to 40 μm at λ=4 nm. The latter figure represents a value which is about 2/3 of the expected minimum image size at focus on the Rowland circle and as large as could be tolerated. The fact that a flat CEMA plate with diameter larger than 2.54 cm would produce unacceptably large resolution loss was one of the reasons for choosing a straight fiber optic coupler to the commonly used 2.56 cm long Reticon 1024 S rather than a fiber optic reducer.

The spread of the electron bunch is caused by space charge repulsion and by the random non-axial (radial) velocity components of electrons as they exit the channels. Space charge should not cause significant effects for an electron bunch of less than 10^4 electrons exiting a single tube.[7] With a phosphor gain of 10, which occurs for a phosphor voltage (Vph) in the range $4kV \leqslant Vph \leqslant 5kV$,[7] a 10^4-electron bunch will generate 10^5 photons, about one-third of which enter a diode in the array (see Fig. 2) and result in an easily detectible signal. Thus considering the fact that a spectral line will fill the entire 2.5 mm height of the diode one can set the gain of the CEMA such that the single channel bunches do not exceed 10^4 electrons without much compromise in the dynamic range.

Assuming the CEMA gain adjusted below the space-charge effect limit indicated above, we must still deal with the effects of the non-axial velocity components. Three approaches can be used to reduce this effect:

1. A tailoring of the exit region of the channels to reduce the electron velocity (such as in "end spoiling"[7] and as in our present resolution-enhanced variety)[19]
2. An increasing of the phosphor voltage (Vph) to increase the ratio of the (average) axial to radial velocity component
3. A decreasing of the CEMA-phosphor distance (s)

Approaches 2 and 3 are readily applicable. The dependance of the diameter of the electron-bunch spot on the phosphor (d) to Vph and s is given to good approximation by $d \propto s/\sqrt{V_{ph}}$.[7] Clearly optimum resolution occurs at the smallest s compatible with maintaining a breakdown-free gap under the conditions of Vph > 4kV for adequate phosphor gain.

To minimize breakdown probability we used a high smoothness selected CEMA which reduced the chance for field-emission initiated breakdown and a bakeout procedure which reduced the chance of ion-feedback-stimulated runaway discharge. Adjusting s=0.4 mm and $V_{ph} \lesssim 4.5$ kV we have obtained reliable operation at background pressures up to 5×10^{-7} torr. Based on the experience of Fonck et al.[7] and our own prelimiary data (see next section) we expect that the phosphor image of a single initial event is less than 40 μm in diameter.

The reading out and digitizing of the signal stored on the diode array was performed by a commercially available unit (DARRS TN-6100 from Tracor Northern)[6] which was interfaced to the CAMAC-based LSI-11/2 system by a DARRS TN-6200.[6] (The combination of diode array, digitizer, interface, and computer constitute what is generally referred to as an Optical Multichannel Analyzer or "OMA".) The specified sensitivity is approximately 5000 diode electron-hole pairs for each count and the dark noise is 1.6 counts rms when operated at -20°C. The maximum number of counts per channel is 1.64×10^4, giving a dynamic range of 10^4. The minimum sweep time involving the readout of all 1024 pixels is 16 ms.

III. Preliminary Results

Initial tests consisted of observing the emission spectrum of a Ne hollow cathode source, observing the emission spectrum of a LPP on an aluminum target, and observing the absorption spectrum of an Al film and some Ne autoionizing resonances with an LPP continuum source.

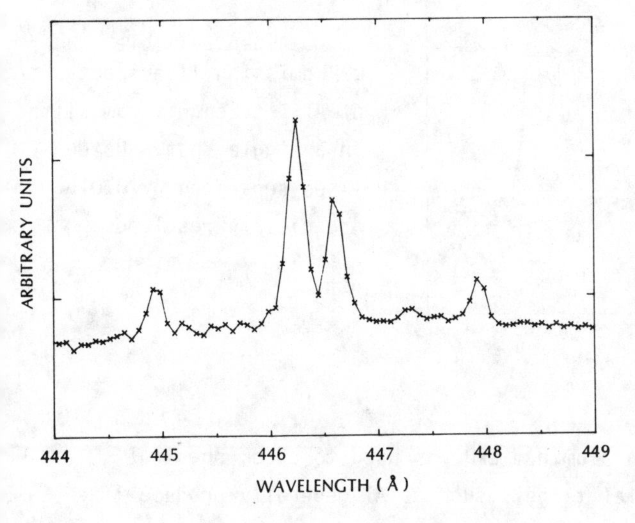

Fig. 3

Spectrum of Ne+ excited in a hollow cathode discharge. Each data point represents a pixel.

The Ne^+ lines in a low pressure (~0.1 torr) Ne hollow cathode discharge are practically free from self absorption and other signficant broadening effects. Therefore, they were used for the focussing and alignment of the CEMA plate relative to the Rowland circle. Fig. 3 shows the spectrum of the Ne^+: $2s^2 2p^4 3s\ ^2P \rightarrow 2s^2 2p^5\ ^2P$ multiplet in the region of 46.6 nm, the output of the OMA being photographed directly from the screen of an oscilloscope. Each data point represents a single diode or pixel. One observes that the spectral lines have a FWHM of about 3 pixels which translates into a spatial extent of 75 μm. Considering that the spectrometer image of the spectral line has an estimated width of about 60 μm, we see that the spatial resolution of the CEMA-phosphor image intensifier is approximately 40 μm; further tests of this resolution are in progress.

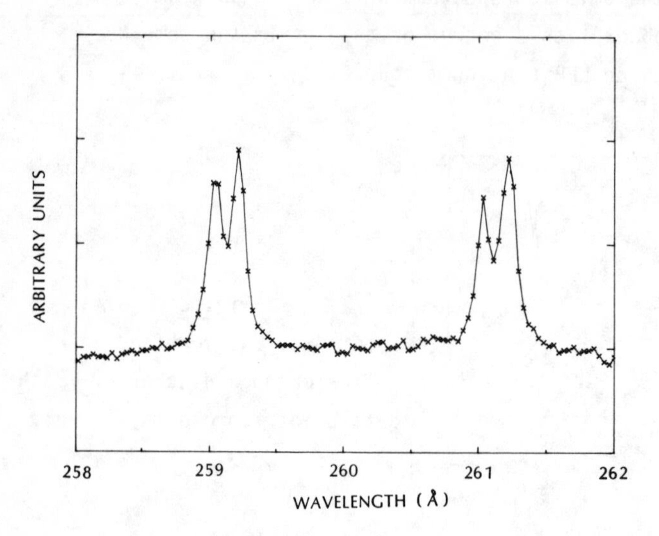

Fig. 4
LPP emission line spectrum of Al^{3+} taken with Nm slit in a single shot. Note lines separated by 0.0166 nm are clearly resolved.

The emission spectrum of an aluminum LPP was used to check spectral resolution. LPP's of light metal targets such as Al generally produce line emission from several ionization stages. Figure 4 is the part of the Al spectrum obtained which lies between 25.6 and 26.4 nm. The two sets of closely spaced Al^{6+} lines: 25.9035-25.9219 nm and 26.1053-26.1219 nm,[21] which are clearly resolved, demonstrate that a resolution of at least 0.016 nm has been achieved.

The most important test related to our intended use of the instrument is the performance on absorption spectra. Since a typical absorption spectrum is created by a relatively intense broadband continuum interrupted by the sample-produced small regions of decreased flux, (i.e. absorption features and lines), instrumental scattering and higher order diffraction are much more serious in absorption measurements than in emission spectroscopy. Indeed our inital attempts to obtain absorption spectra showed that the scatter of the zero order image in the bellows connecting the grating chamber to the CEMA chamber (see Fig. 1) overwhelmed the sample (Al and Ne) absorption. (This scatter was also present in the emission measurements, Figs. 3 and 4, but produced no noticeable interference.) To reduce the scatter, a special highly absorbent zero order trap and two long λ baffles were installed in the bellows.[22]

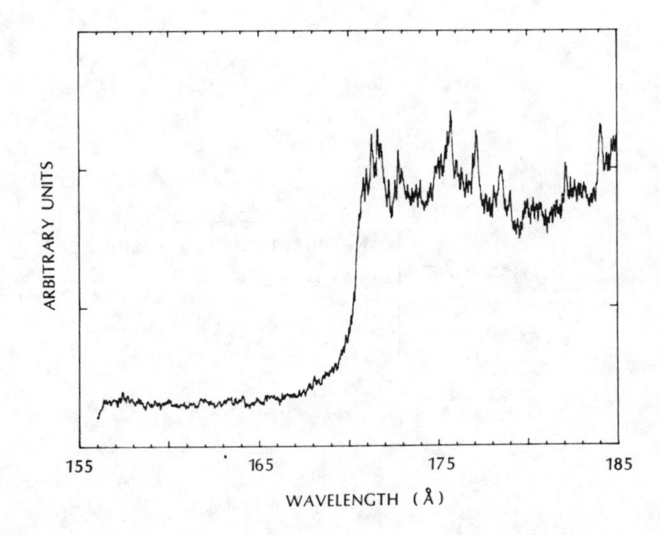

Fig. 5

An L_{23} edge taken with LPP tantalum background. Twenty shots with 5 μm slit. Note background at 16.5 nm is less than 15% intensity at 18.0 nm.

An Al thin film (~100 nm thick) filter was installed in the mirror chamber to eliminate second order interference in the region of the 27 nm Ne autoionizing resonance. Figure 5 is the recorded spectrum of the Al $L_{2,3}$ edge at 17.0 nm. Note that the observed flux for $\lambda < 16.5$ nm is less than 15% of that observed above the edge, which providds a measure of the scattering background in this spectral region.

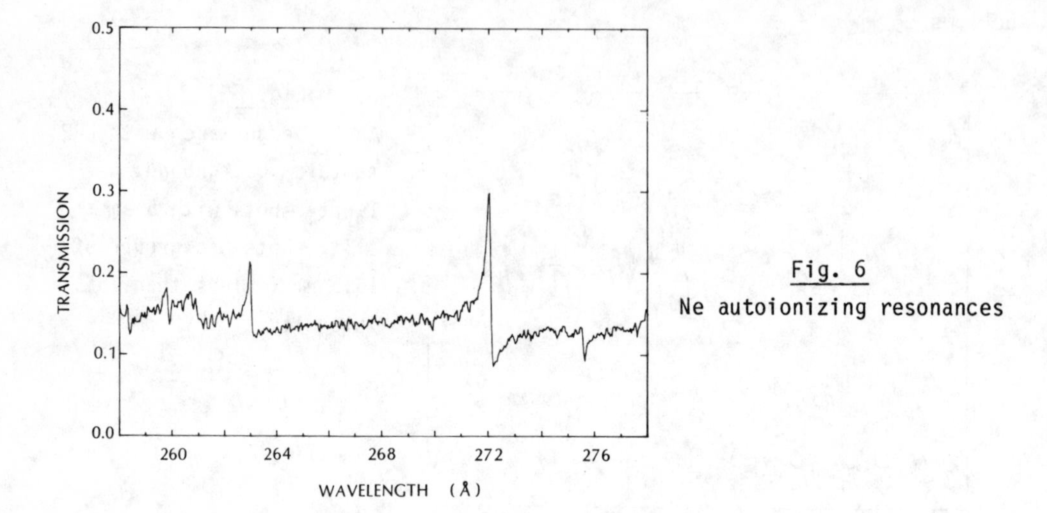

Fig. 6
Ne autoionizing resonances

In Fig. 6, we have the Ne $2s^22p^6\ {}^1S_0 \rightarrow 2s2p^6\ np\ {}^1P_1$ resonances first studied by Codling et al.[23] The observation of these lines provides a good instrumental test since their measurement requires both low scatter and fairly high resolution.

Acknowledgments

We wish to acknowledge the creative technical support of David C. Morgan who helped design and then skillfully constructed the zero order trap and the gimbal mount for the diode array. Many helpful discussions with W.T. Hill III, R.P. Madden, D.L. Ederer, T.J. McIlrath, A.C. Parr and B. Laprade are much appreciated. Partial support of this work is from AFOSR, Contract No. ISSA.84-0020 for which we are grateful.

References

1. T.J. McIlrath and T.B. Lucatorto, Phys. Rev. Lett. 38, 1390 (1977).

2. T.B. Lucatorto, T.J. McIlrath, J. Sugar, and S.M. Younger, Physical Rev. Lett. 47, 1124 (1981).

3. G. Balloffet, J. Romand and B. Vodar, C. R. Academy of Science 252, 4139 (1961).

4. T.B. Lucatorto, T.J. McIlrath, and G. Mehlman, Appl. Opt. 18, 2916 (1979).

5. J.G. Timothy and R.L. Bybee, Ultraviolet and Vacuum Ultraviolet Systems, SPIE Vol. 279, 129 (1981); J.G. Timothy and R.L. Bybee, Appl. Opt. 14, 1632 (1975).

6. The manufacturer or brand name is used for purposes of identification only; the authors' choice of a product is not to be construed as an endorsement of that product in any way nor be constituted as any kind of judgement on the merit of that product.

7. R.J. Fonck, A.T. Ramsey, and R.V. Yelle, Appl. Opt. 21, 2115 (1982).

8. W.L. Hodge, B.C. Stratton, and H.W. Moos, Rev. Sci. Instrum. 55, 16 (1984).

9. J.L. Schwob, M. Finkenthal, and S. Suckewer, Abstracts for the Seventh International Conference on Vacuum Ultraviolet Physics.

10. H.L Manning, J.L. Terry, and E.S. Marmar, Bull. Am. Phys. Soc. 28, 1250 (1983).

11. E. Jannitti, P. Nicolosi, G. Tondello, and Wang Yongchang, this Proceedings; E. Jannitti, P. Nicolosi, and G. Tondello, submitted to Opt. Commun.

12. G. Tondello, personal communication.

13. P.K. Carroll, E.T. Kennedy and G. O'Sullivan, Appl. Opt. 19, 1454 (1980).

14. D.J. Nagel, C.M. Brown, M.C. Perrar, M.L. Ginter, T.J. McIlrath, and P.K. Carrol, Appl. Opt. 23, 1428 (1984).

15. D.J. Bradley, A.G. Roddie, W. Sibbett, M.H. Key, M.J. Lamb, C.L.S. Lewis, and P. Sachsenmaier, Opt. Commun. 15, 231 (1975).

16. M. Kuhne, Appl. Opt. 21, 2124 (1982).

17. G.O'Sullivan, J.R. Roberts, W.R. Ott, J.M. Bridges, T.L. Pittman, and M.L. Ginter, 1982 Opt. Lett. 7 31.

18. C.G. Mahajon, E.A.M. Baker, and D.D. Burgess, Opt. Lett. 4, 283 (1979).

19. Specifications provided by the Galileo Electro-Optic Co. (See Ref. 6 above).

20. L.B. Lapson and J.G. Timothy, Appl. Opt. 15, 1218 (1976); R. Malina, Ph.D. thesis, Univ. of California, Berkeley, unpublished.

21. R.L. Kelly and L.J. Palumbo, Atomic and Ionic Emission Lines Below 2000 Å, NRL Report 7599 (1973).

22. C.L. Cromer, J.M. Bridges, and T.B. Lucatorto, to be published.

23. K. Codling, R.P. Madden, and D.L. Ederer, Phys. Rev. 155, 26 (1967).

PHOTOEMISSIVE MATERIALS FOR 0.35μm LASER FIDUCIALS IN X-RAY STREAK CAMERAS*

C. P. Hale, H. Medecki, P. H. Y. Lee
Lawrence Livermore National Laboratory, Livermore, CA 94550

ABSTRACT

Using a soft x-ray streak camera, materials are tested for suitability as transmission photocathodes when irradiated by 0.35μm laser pulses. Preliminary measurements of sensitivity, dynamic range, and temporal resolution are reported. A practical fiber-optic fiducial under development for laser fusion x-ray diagnostics on the LLNL Nova laser system is described.

INTRODUCTION

Prompted by the need for laser driver-related, synchronized fiducials on laser fusion x-ray streak camera target diagnostics, we conducted an investigation of materials which efficiently photoemit when irradiated by light pulses of 0.35μm wavelength. Commonly used x-ray photocathode materials such as gold and cesium iodide which we have not characterized at this wavelength were studied; we also investigated other materials which were expected to be more effective 0.35μm photoemitters.

An LLNL Model 2 x-ray streak camera was used to obtain the data on photoemissive characteristics. The camera is electronically optimized for recording soft x-rays with transmission-mode, thin film photocathodes of a demountable, atmosphere-compatible design[1]; successful cathode materials for 0.35μm (photon energy 3.5eV) therefore had to operate under similar restrictions.

In this report the photocathode candidate materials and experimental apparatus are described, and results of this study are presented. A description of a practical fiber optic fiducial cathode now being tested for implementation on the Nova laser system will be followed by a description of the work to be completed.

CATHODE MATERIALS FOR 0.35μm WAVELENGTH

Cathode materials in this study must be compatible with the soft x-ray streak camera used to record the laser pulses. The materials must be thin-film deposited on a transparent substrate for transmission-mode photocathode use, and the material must withstand occasional exposure to atmosphere and a typical working vacuum of 10^{-6} Torr. High stability of the thin film and its oxides and low hygroscopicity are essential requirements.

Additionally, the candidate materials must have work functions near or below the 3.5eV photo energy. The materials tested in this

*Work performed under the auspices of the U.S. Department of Energy by the Lawrence Livermore National Laboratory under contract number W-7405-ENG-48.

Summary of Photoemissive Candidates to Date		
Photocathode, thickness	Work function (eV)	Substrate, thickness
Nb, 100-300-500Å*	2.29	SiO_2, 1 mm
Al, 100-300-500Å*	2.98	SiO_2, 1 mm
CsI, 1000Å**	6.00	CH, 1000Å
Au, 300Å**	4.73	CH, 1000Å
Al, 200Å	2.98	CH, 1000Å
Al/Au, 200Å†	2.98/4.73	CH, 1000Å
Inconel, 200Å	—	SiO_2, 1 mm
Ba, 200Å	2.70	SiO_2, 1 mm

*Stepped thickness for comparative sensitivity measurement
†200Å maximum on either edge of slit, decreasing thickness toward center
**Soft x-ray photocathodes

Table 1. A summary of materials and respective substrates used in this study. Cathode thicknesses are all ±30A.

study are shown in Table 1. Cesium iodide and gold, both ordinarily used in soft x-ray applications, were studied here for their sensitivity to 0.35μm light pulses and to give possible multiphoton process comparison with more efficient photoemitters.

Thicknesses of cathode and substrate chosen also depended on practical considerations. If cathode thickness exceeds the escape depth of the electrons, then sensitivity is reduced because photoelectrons originating from photons absorbed beyond the escape depth will not be emitted into the vacuum. Conversely, if the cathode is thinner than the escape depth, sensitivity is reduced due to light being transmitted rather than absorbed. Furthermore, the technique to uniformly deposit very thin (10-100Å) layers of material under practical conditions is difficult and results in non-uniform photoemission.[2] After an analysis of the case for thin gold cathodes, Stradling, et al., recommended a 200Å gold layer for optical performance.[2] Cathodes for this study were deposited at a variety of thicknesses at or near 200Å with a typical uncertainty of ±30Å. The 0.35μm transmission of 1mm-thick quartz substrate exceeds 90%, and 1000Å parylene transmits 70% at 0.35μm as measured with a Cary 14 spectrophotometer.

EXPERIMENTAL APPARATUS

Using the LLNL Monojoule laser facility, a 1-50μJ, 40 picosecond 0.35μm light pulse is generated by frequency tripling the output of an Nd:glass laser system operating at 1.06μm fundamental wavelength (Figure 1). A 58° prism is used to refract out the

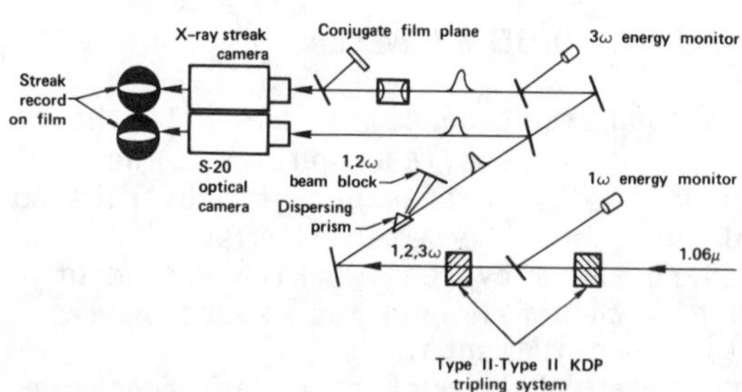

Figure 1. Experimental apparatus for studying photoemissive materials with 0.35μm light pulses and an x-ray streak camera.

Figure 2. Conjugate film plane showing cylindrical lens focus and unfocused Fresnel structure.

unused 1.06μm and 0.53μm components of the tripler output; conversion efficiency of third harmonic production was typically 0.5-1.0%. PIN diodes calibrated for 1.06μm and 0.35μm were used to characterize the system prior to cathode tests to determine 0.35μm intensity at the cathode surface. A conjugate-plane film diagnostic monitors the cylindrical lens line focus incident on the photocathode. An LLNL optical streak camera fitted with an S-20 photocathode registers pulse width and relative energy in parallel with the Model 2 LLNL x-ray streak camera fitted with the demountable cathode under analysis.

Multiple data shots were taken from each candidate material over a range of incident intensities to measure threshold for usable signals (corresponding to approximately 10 photoelectrons emitted from a 100μm-long, 100μm-wide portion of the cathode slit), temporal resolution, and estimation of dynamic range (in this study considered to be 0.35μm intensity at threshold compared to intensity at the onset of blooming on film) using Polaroid 107 film as the streak recording medium. Fluorescent material sensitive to UV light was positioned above and below the cathode slit to insure accurate beam alignment. As can be seen in Figure 2, the line-focus was surrounded by considerable Fresnel ring structure due to poor spatial filtering; the variation in intensity across the beam was judged to be of limited importance in the 10^5-10^6 W/cm^2 range covered in this study.

EXPERIMENTAL RESULTS

Initial experiments were designed to test cathode materials for their ability to photoemit when irradiated with 0.35μm laser pulses. Threshold sensitivities and saturation intensities as defined earlier were recorded and preliminary estimates of dynamic range were plotted (Figure 3).

The most sensitive material studied was 200Å Al on 1000Å parylene. While no distinct threshold effect was observed, usable signals were recorded at approximately 0.15 MW/cm^2 and saturation began at approximately 1.5 MW/cm^2. It is well known that thin-film Al under the conditions of this experiment (exposure to atmosphere, 10^{-6} Torr working vacuum) absorbs oxygen quickly and leaves essentially a layer of aluminum oxide only; however, there is evidence that suggests that this diffusion of oxygen into the electropositive metals may not adversely affect the work function as much as previously believed.[3] This is also true of Barium, which in this study was seen to exhibit an efficient threshold intensity (approximately 6.7 MW/cm^2) but lower dynamic range than aluminum.

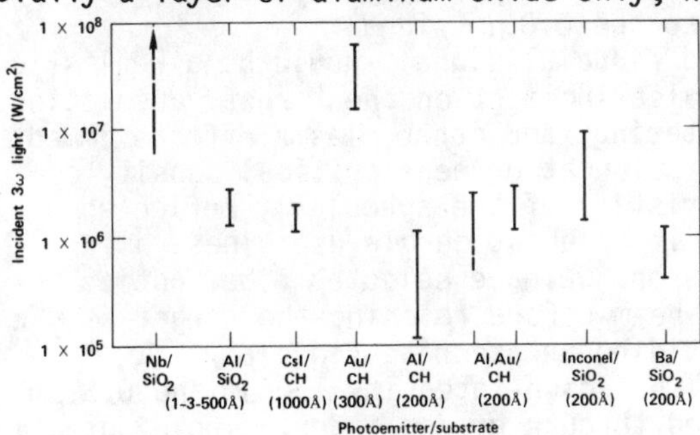

Figure 3. Illustration of sensitivity range for photoemissive materials studied. Lower and upper bounds are defined in text; dashed lines represent inconsistent signal strength and in the case of Nb, eventual surface damage.

Aluminum on quartz was seen to be less sensitive than its counterpart on parylene; this may be accounted for by some undetermined difference in attenuation between the substrate materials or some difference between the two Al deposits. The Al on quartz was a stepped-thickness deposit, as was the niobium candidate. The hope was to define changes in photoemission for discreet thicknesses of cathode material, but clear indication of a variation in signal level from one segment of the cathode to another could not be detected with the present experimental apparatus. Niobium was expected to photoemit well at 0.35μm, but no thickness segment produced streak signals until the incident 0.35μm energy exceeded the damage threshold of the material. Electron microscopy of the Nb surface revealed a peeling, blistered structure; streaked signals recorded after damage occurred may have been due to field emission of electrons from the damaged surface.

Two Au cathodes of different thicknesses were studied. One cathode was of dual Al/Au construction, 200Å maximum thickness respectively. The Au segment of this cathode demonstrated a threshold sensitivity of approximately 1.5 MW/cm^2, in good agreement with an estimation of 1.0 MW/cm^2 by Marjoribanks, et al., of 100Å-thick Au on 25μm polyethylene substrate.[4] The 300Å-thick Au cathode studied showed a much higher threshold of approximately 25 MW/cm^2, possibly due to greater 0.35μm absorption.

As can be seen in Figure 3, sensitivities of other candidate materials fell into a wide range recorded between 0.1 MW/cm^2 and 100 MW/cm^2 of incident 0.35μm light.

DEVELOPMENT OF A PRACTICAL FIDUCIAL CATHODE

The programmatic goal of this project has been the ability to provide x-ray streaked diagnostics with a temporally accurate, synchronized representation of the laser driver pulse for comparison with x-ray channels. Three methods of realizing this goal are:

1. Provide a photocathode sensitive to both x-rays and 0.35μm, and propagate these two radiations as spatially separated beams;
2. Provide a bifurcated photocathode with separate channels for x-ray and 0.35μm signals;
3. Utilize existing gold or cesium iodide x-ray photocathodes, but modified mechanically to accept a spatially separated fiber optic cathode sensitive to the 0.35μm signal.

Ideally, a driver-related fiducial signal should be a realistic representation of the light pulse incident on the target; absorption in the resultant plasma, scattering, and other plasma effects immediately following target irradiation at or near critical density can seriously modify the characteristics of the specularly reflected light from the target, making it of questionable usefulness as a driver fiducial. For this reason, we have selected fiber optic transport of a portion of the beam before reaching the target as the most attractive method of providing an accurate fiducial.

It has been demonstrated that high-intensity 1.06μm and 0.53μm light pulses can be transmitted through graded-index, single optical fibers efficiently and without observable temporal distortion.[5] Accordingly, 50μm core diameter graded-index glass fibers were deposited on one end with aluminum approximately 200Å thick to act as spatially confined 0.35μm photocathodes (Figure 4).

The cathode end of the fiber is mounted in a small hole drilled through the x-ray cathode cup; care is taken to keep the aluminized tip close to the surface of the cup to prevent field emission. The 0.35μm laser pulse is focused down and aligned on the fiber by maximizing transmitted light through a glass window on the camera with a photodiode. Once the fiber is aligned for maximum signal, the demountable cathode cup is levered into place inside the camera.

Tests indicate that a usable fiducial streak signal can be obtained (Figure 5) with 0.35μm intensities of approximately 8×10^6 W/cm^2 with no fiber damage. Improvements in the deposition process and characterization of aluminized fibers are expected to lower the input intensity to the 10^5-10^6 W/cm^2 range. No temporal distortion has been observed.

By providing dual fibers adjacent to the x-ray cathode slit, the fiducial streaks will precisely define the slit orientation for accurate photo-digitization. Additionally, timing of the cameras and proper operation can now be easily verified by recording the fiducial only, without firing actual laser shots in the target chamber.

CONCLUSIONS

Of the candidate materials surveyed, 200Å Aluminum on parylene stands out a good photocathode for 0.35μm wavelength light. There are clear indications of wide dynamic range and accurate temporal resolution for this material. Other adequate materials were surveyed; data on gold and cesium iodide photocathodes at 0.35μm wavelengths were generated. An optical fiber cathode utilizing a thin (200Å) Al layer has been developed and is now being tested for practicality. Tests indicate that this is is a feasible method of delivering a time-resolved, driver-synchronized 0.35μm fiducial to x-ray diagnostics in the LLNL ICF program and similar facilities. This approach provides not just an optical benchmark but a well characterized streak signal of laser driver pulse shape. It also

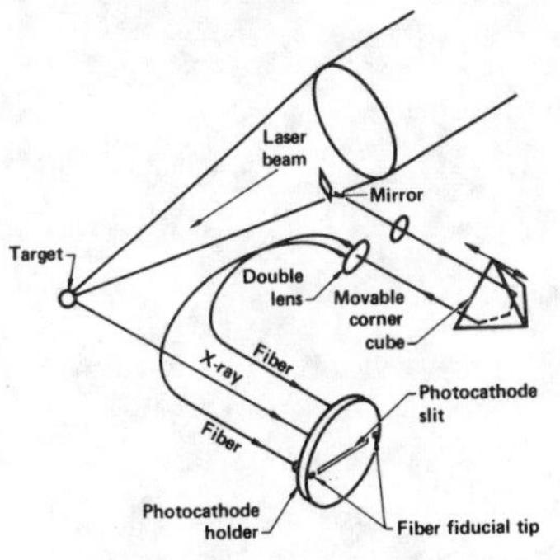

Figure 4. Implementation of A1 fiber fiducials. In actual facility use, a mirror before target will supply 0.35μm light to fibers.

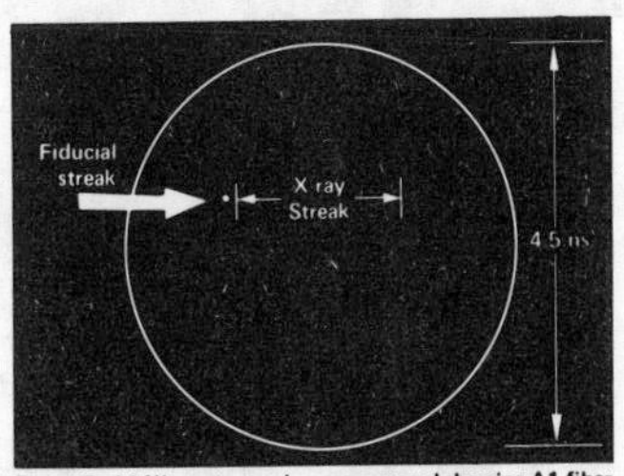

Figure 5. LLNL x-ray streak camera record showing A1 fiber optic fiducial signal, ~6μJ of 0.35μm light incident on fiber. Time axis is pointed upward; pulsewidth is ~40ps.

improves testing and timing of the diagnostic at the laser facility, and the capability of testing the x-ray camera using a small laser system operating at 0.35μm wavelength.

Future work will involve detailed measurements of dynamic range, threshold sensitivity, and temporal accuracy of candidate materials using a 50%, 0.35μm transmission etalon and shot-to-shot optical streak camera comparison. A technique is under development to determine optimum thicknesses of cathode materials and deposition procedures. Concurrent with this effort will be the development of a similar fiber optic system for the fourth harmonic at 0.26μm wavelength on the Novette laser system in the near future.

ACKNOWLEDGEMENTS

The authors wish to acknowledge the guidance and suggestions of their colleagues R. E. Turner, E. M. Campbell, and D. E. Campbell through the course of this work. G. Stone provided help in preparing a number of the experimental cathodes studied in this experiment. Thanks are also due to W. G. Halsey and C. L. McCaffrey for their expert assistance with work performed on the scanning electron microscope.

REFERENCES

1. G. L. Stradling, Lawrence Livermore National Laboratory Report, Livermore, Calif., UCRL-53271 (April 19, 1982).
2. Laser Program Annual Report-1979, Lawrence Livermore National Laboratory, Livermore, Calif., UCRL-50021-79 (1980) pp. 5 to 16.
3. A. H. Sommer, Photoemissive Materials (Robert E. Krieger Publishing Company), p. 20, pp. 27-28.
4. R. S. Marjoribanks, M. C. Richardson, J. Delettrez, S. Letzring, W. Seka and D. M. Villeneuve, Opt. Comm. 44, No. 2, 133 (1982).
5. R. A. Lerche and G. E. Phillips, Lawrence Livermore National Laboratory Report, Livermore, Calif., UCID-19176 (1981).

X-UV GAIN AMPLIFICATION STUDIES IN LASER PLASMA USING NORMAL INCIDENCE MULTILAYERS MIRRORS

P.DHEZ[+,++] G.JAMELOT, A.CARILLON, and P.JAEGLE.
ERA 719 ,CNRS.Batiment 350 Universite Paris Sud (91405) ORSAY (FRANCE)

P.PARDO and D.NACCACHE.
Institut d'Optique Applique. Batiment 503. BP 43 (91406) ORSAY Cedex (FRANCE)

ABSTRACT

Present state of art in multilayers mirrors and in population inversion in laser plasma permits to test possible laser cavity.

INTRODUCTION

The recent measurements in Orsay on X-UV Amplification of Spontaneous Emission (ASE) observed with different length of plasma, up to about 10 mm. has enabled us to plan experiments with still longer medium [1]. The progress achieved in X-UV ASE has led us to look, as soon as possible, for the feasibility and the direct test of mirrors for an X-UV laser cavity.

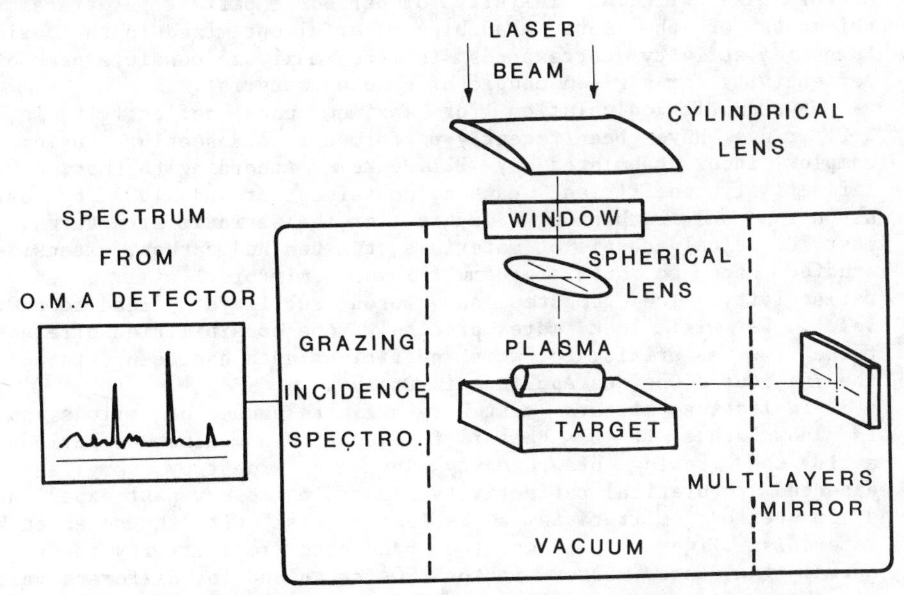

Fig.1. Scheme of the existing set up to measure Amplification of Spontaneous Emission (ASE) by using different lenght of plasmas. The added normal incidence mirror should permit new tests of the amplification gain by plasma.

++ and LURE Batiment 209C Université Paris Sud 91405 ORSAY

For an X-UV laser cavity the normal incidence multililayered mirrors[2,3] especially optimized for the desired range, are most promising. But before to solve entirely all the tecnical problems, and in particular for the semi-transparent exit mirror, it seems important to test it's principle, the mirror feasibility and the possible damaging effect on the mirror. This first step can be achieved by using a mirror providing just a single pass through an elongated plasma. Such a plasma has been previously studied[4].

The scheme of a possible experiment is shown in figure 1. The Optical Multichanel Analyser detector, with a scintillator, permits to get limited spectral range for each laser shoot. So a direct comparaison between the spectra obtained with and without mirror can reveal the enhancement of amplification provided by the mirror, if it has a sufficient reflectivity in the considered spectral range.

Let us look first , what normal incidence reflectivity coefficient can be obtained in the 100Å range in which we are interested. Next by using reflectivity coefficient as well as previously measured plasma coefficent in this range, we will calculate the observable effect on the spectrum.

NORMAL INCIDENCE MULTILAYERS MIRRORS FOR THE 100 Å RANGE

The use of the Vinogradov- Zeldovich's transcendantal equation is the easiest way to compute the peak reflectivity for multilayer mirrors[5]. The obtained peak reflectivity corresponds to the ideal mirror case with an infinity of periods , perfect interfaces and thickness of the more absorbing material optimized in the period. Such reflectivity corresponds to the maximum possible peak of reflectivity for a given couple of choosen material.

Systematic calculation for maximum peak reflectivity in the X-UV range have been recently performed by A.Rosenbluth[6] using the complex index tabulated by B.L.Henke[7]. According to that, a 0.70 reflectivity coefficient can be obtained around 100Å by using strontium and ruthenium bilayers for the periodic structures. The most classical couple of materials, tugsten and carbon, extensively studied up to now for multilayer mirrors, will give 0.17 reflectivity. The tungsten and boron multilayer lead to a 0.23 value. We will look more precisely the possibilities offered by these two materials, for which sufficient data has been obtained at the Institut d'Optique Applique in Orsay[8].

Realistics mirrors must have a finite number of periods and it is known that it is better to evaporate a minimum of periods to avoid the growing of rugosity which can destroy very fast the expected theoretical reflectivity. According to our past experience, a 30 periods mirror is a realistic task with the selected W/B materials. Figure 2 shows the predicted reflectivity for such a normal incidence mirror in the 100Å range and for different values of the repartition coefficient β.This parameter β is the ratio of the thickness of the more absorbing material (in our case W) to the period length. In this calculation the period has been adjusted, according to the index values in this range to get the maximum reflectivity at the desired wavelength corresponding to the Al^{+10} line at 106Å.

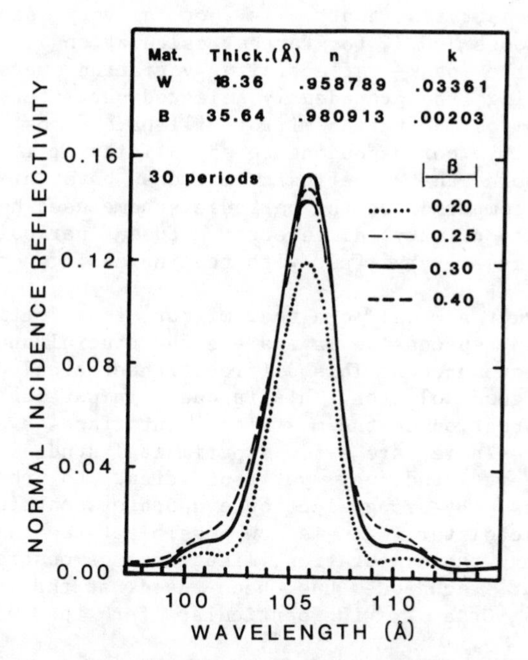

Fig.2. Normal incidence reflectivity calculated for a 30 periods W / B multilayer mirror with different tungsten thickness (β is the ratio of the tungsten thickness to the period lenght).

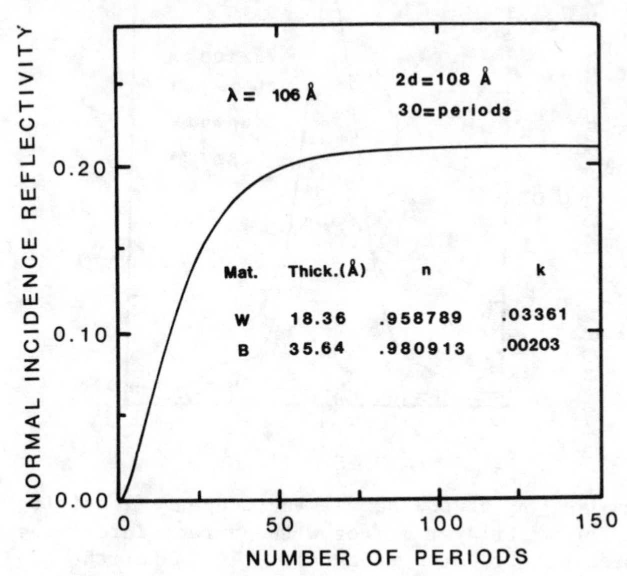

Fig.3. Calculated peak reflectivity variation at 106Å versus number of W/B periods with $\beta = 0.34$.

This calculation demonstrates that β is not a very sensitive parameter so we will choose β=0.34 for further calculations.

Figure 3 shows the peak reflectivity variation versus the number of periods using the precedently selected parameters. From such a curve we know our 30 period mirror will have almost 3/4 of the maximum reflectivity corresponding to an infinity of identical periods. It is known that aperiodic mirrors have a better reflectivity . But, compared to the periodic scheme used here for calculation, the reflectivity difference between periodic and aperiodic mirror decrease very fast with the increasing number of their periods.

In view to construct a model of a real mirror, imperfections of the interfaces between successive layers are the crucial and final point to be taken into account.This difficult theoretical problem has not yet found a good solution. This is due , in particular, to the lack of real information on the multilayer interfaces available at the present time. There are a few experimental studies on the qualities of multilayers[9] and not yet sufficient to check the theoretical predictions. By using electron evaporator containing an in situ X ray reflectometer, it is now possible to evaluate the rugosity growing during the evaporation, like first demonstrated by E.Spiller . A similar technique has been used at the Institut d'Optique Applique in Orsav , in particular for the W and B materials choosen here[10].

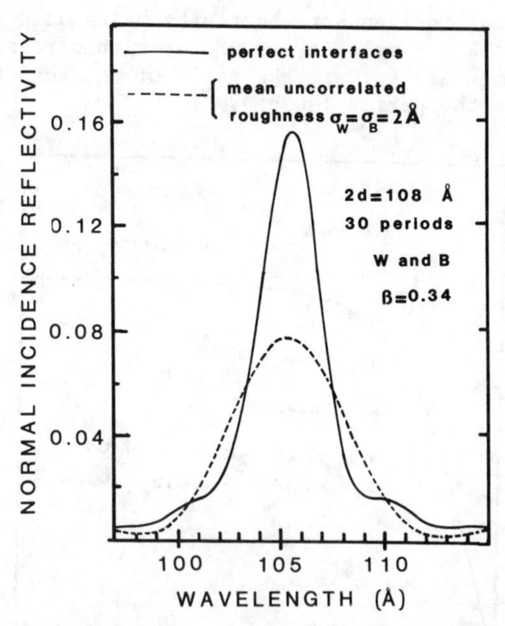

Fig.4. Comparison of the normal incidence reflectivity calculated for multilayer mirror with perfect interfaces and with mean uncorrelated roughness σ = 2Å for the two layering materials used.

Figure 4 shows the importance of the rugosity for the considered 30 periods normal incidence mirror. With the perfect interfaces the reflectivity is about 0.15 at the peak, and the half width is about 3.5Å. With the uncorrelated rugosities of equal values σ=2Å on the two materials the reflectivity will drop down to its half value and the band pass is almost doubled. Such uncorrelated defect have the worst effect on the reflectivity and gives a lowest pessimistic reflectivity limit. Such calculations are also the easiest to be performed. They are equivalent to the use of the well known Debye-Waller factor in X ray to take into account the atomic thermal motion.

PREDICTED EFFECT FOR A SINGLE PASS AMPLIFICATION ENHANCEMENT THROUGH AN ELONGATED PLASMA

Absorption coefficientdetermination in laser plasma can be obtained by comparaison of the emitted intensity from two plasmas with different lengths. Figure 5 is an example of spectre registrated with 7.1 mm. and 4.5 mm. long aluminum plasmas previously studied , by using a set up similar to that described in figure 1, but without the proposed mirror. Figure 6 displays corresponding absorption coefficient spectrum with a negative absorption resulting from ASE at 106Å.

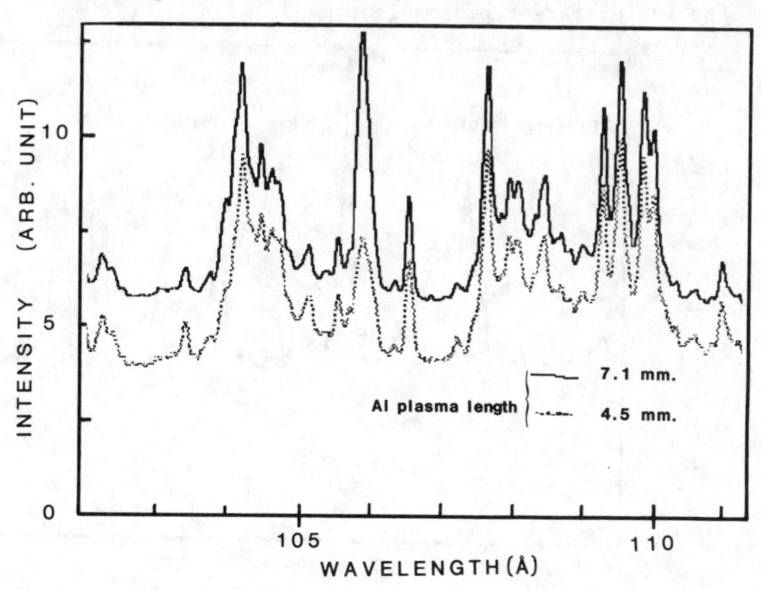

Fig.5. Spectra registred with plasma lenght 7.1mm. and 4.5mm. used to calculate absorption coefficient of the plasma.

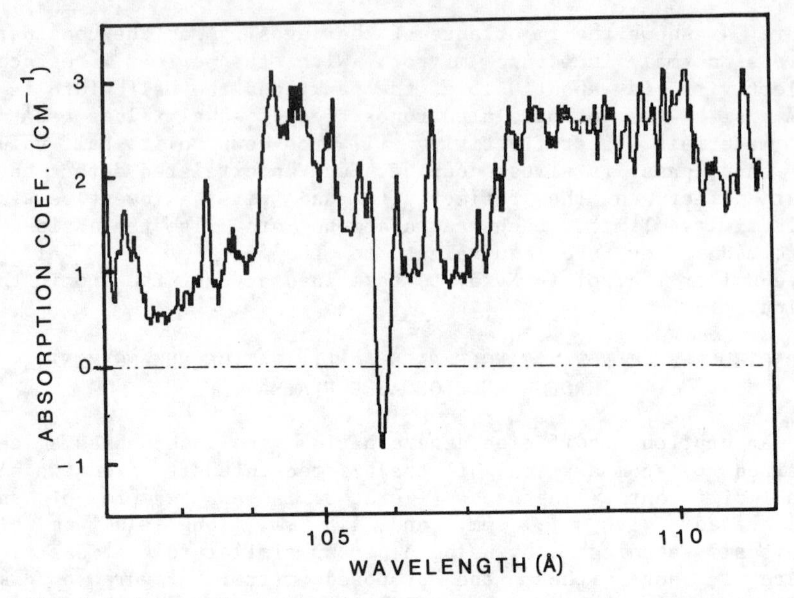

Fig.6. Absorption coefficient deduced from spectra emitted by two different plasma length.

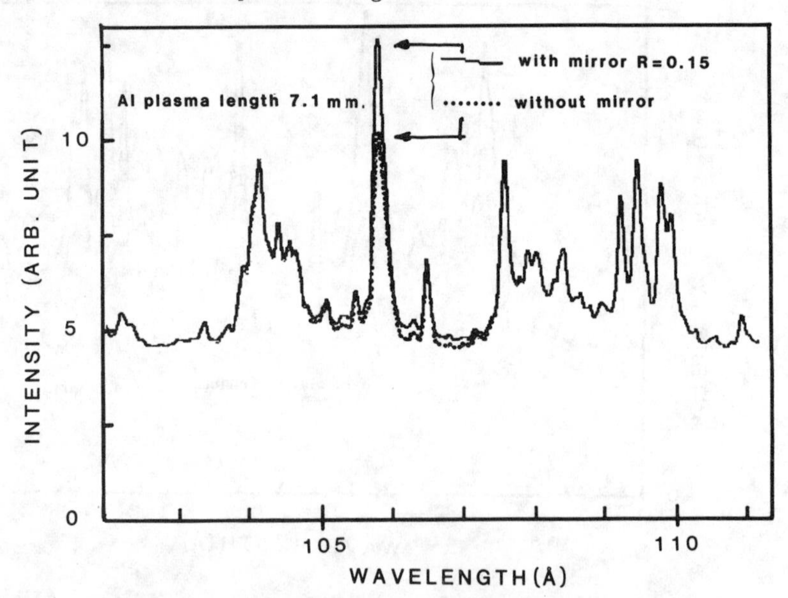

Fig.7. Comparaison of the spectrum emitted by a 7.1mm. plasma and the perdicted spectrum obtained by adding a normal incidence multilayer mirror of 0.15 reflectivity.

Such absorption coefficients can now be used to calculate the undergoing amplification of the reflected band pass returned by a given mirror. Reflectivity curve for multilayers mirror can be computed by using a simple gaussian function with peak and half values corresponding to the calculated reflectivity curve. Figure 7 gives an example of calculation with a 7.1 mm. plasma length and the same spectrum as in figure 5 where a 0.15 reflectivity and 3.5A band pass have been used. In spite of the relatively low reflectivity used, a very clear effect appears at 106A, around the Al^{+10} line . With this 7.1 mm. plasma length and the 1 cm^{-1}. amplification coefficient, the minimum reflectivity needed to observe such an effect is about 0.10.

Evidently the importance of this effect with a single pass amplification will increase accordingly to the mirror reflectivity. But, the use of longer plasma, with a given mirror, must also have a very important effect.
To test that point, several calculations have been done with different plasma lengths. Figure 8 shows a comparaison between the calculated spectra obtained from a 20 mm. length plasma with and without the mirror. In this figure the multilayers mirror has the same characteristics as used in figure 7. Owing to the plasma length, the ratio between the amplified line and the absorbing part of the spectrum is enhanced. Under these condition, a 0.05 reflectivity mirror must be sufficient to observe a noticeable effect in line intensity.

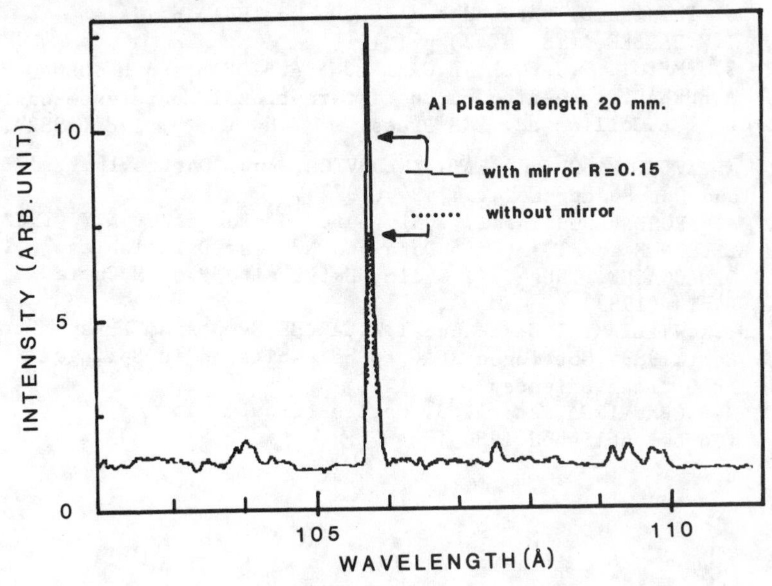

Fig.8. Calculated spectrum emitted by a 20mm. length of plasma alone and comparaison with the predicted spectrum obtained by adding a normal incidence mirror of 0.15 reflectivity.

The use of the mirror appreciably increases the possibilities of detecting features like ASE in plasma emission. In the experiment with mirrors, only a limited range of the spectrum within the band pass is affected, so this will also permit an easier and more efficient statistical treatment of the spectra.

CONCLUSION

New experiments for ASE measurements can be fruitfully performed by using reflecting mirror of the type needed for a laser cavity.

An increase of plasma length and the use of normal incidence mirror are the two main technical objectives to promote. The present qualities of multilayers mirrors should allow a direct observation of the predicted effect.

ACKNOWLEDGEMENTS

The authors wish to thank R.Karnatak for hepful comments.

REFERENCES

1. P.JAEGLE et al., Invited paper at this conference.
2. E.SPILLER, AIP Conference Proceedings Nb.175 (American Institut of Physics,D.T.Atwood and B.L.Henke 1981) p.124.
3. T.W.BARBEE, (id ref 2) p.131.
4. G.JAMELOT, P.JAEGLE, A.CARILLON, A.BIDEAU and H.GUENNOU, A.SUREAU.Proceeding of the International Conference on Laser'81. Carl B.Collins ed. STS Press. Mc Lean VA, p. 178 (1982).
5. A.V.VINIGRADOV and B.Ya.ZELDOVICH, App. Optics,16,1, 89 (1977) and Opt.Spectrosc.42,4,404 (1977).
6. A.E.ROSENBLUTH,Thesis .University of Rochester,N.Y (1982).
7. B.L.HENKE et al, Atomic Data and Nuclear Data Tables 27,1(1982).
8. D.NACCACHE. These III Cycle, Universite P.et M.Curie. Paris (1983).
9. W.K.WILLIAM, Z.U.REK and T.W.BARBEE,Symposium X-Ray Microscopy, Sept.1983. Gottingen RFA(to be published in Springer Series in Optical Sciences).
10. J.P.CHAUVINEAU et al. J. Optics (France) 15. (To be published 1984).

PERFORMANCE OF TRANSITION METAL--CARBON MULTILAYER MIRRORS FROM 50 to 350 eV

D. R. Kania, R. J. Bartlett, W. J. Trela
Los Alamos National Laboratory, Los Alamos, NM 87545

E. Spiller
IBM Corporation, Yorktown Heights, NY 10598

L. Golub
Harvard Smithsonian Center for Astrophysics, Cambridge, MA

ABSTRACT

We report measurements and theoretical calculations of the reflectivity and resolving power of multilayer mirrors made of alternate layers of a transition metal (Co, Fe, V, and Cr) and carbon (2d ≈ 140 A) from 80 to 350 eV.

INTRODUCTION

Recent developments in thin film technology have made it possible to fabricate coatings, multilayer mirrors, that enhance surface reflectivity in the vacuum ultraviolet and soft x-ray region.[1,2] Multilayer mirrors form an artificial crystal lattice consisting of alternate layers of high and low atomic number (Z) materials. The high Z material acts as a scattering plane while the low Z material acts as a spacer between the high Z planes. Like a natural crystal these coatings obey Bragg's law, $\lambda/2d=\sin\theta$, i.e., the ratio of the incident wavelength, λ, to the 2d spacing of the multilayer equals the sine of the incident angle, θ, measured from the mirror surface. We have measured the reflectivities of four transition metal (Co, Fe, Cr, and V)--carbon multilayer mirrors between 80 and 350 eV. The 2d spacing of the mirrors was ≈ 140 A. The angular range examined was 15° to 80°.

Calculations of the multilayer mirrors performance may be made using the equations of classical electrodynamics[3] and compilations of the optical constants of the relevant materials.[4] Peak reflectivity calculations were performed and compared to the measured peak reflectivities.

Extrapolation of the calculated reflectivity was required because of a lack of optical constant data in the region below 100 eV. Inclusion of the effects of interfacial roughness which reduces the multilayer mirror reflectivity yields excellent agreement between the calculated and measured values. It is important to note that other factors, such as uncertainties in the optical constants and diffuse boundaries may also contribute to the reduction in the reflectivity.

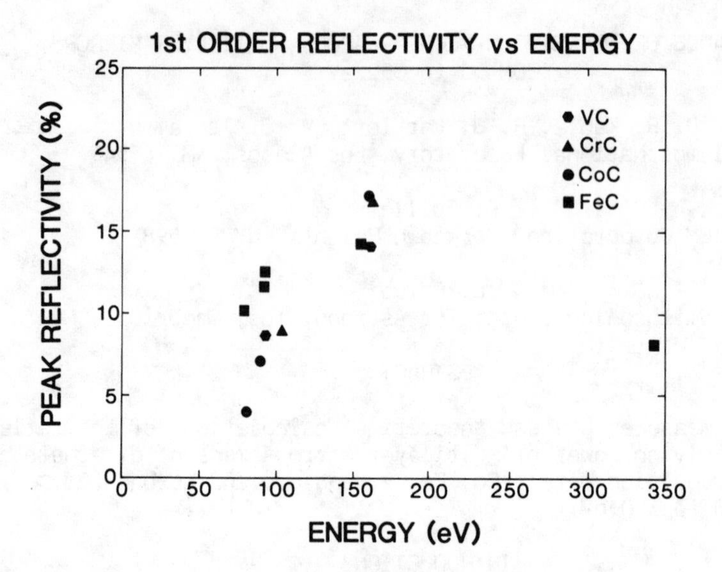

Fig. 1. Measured peak reflectivity vs. energy for several transition metal-carbon multilayer mirrors. The effective 2d spacing for the mirrors was approximately 140A. The angular range was 15° to 80°.

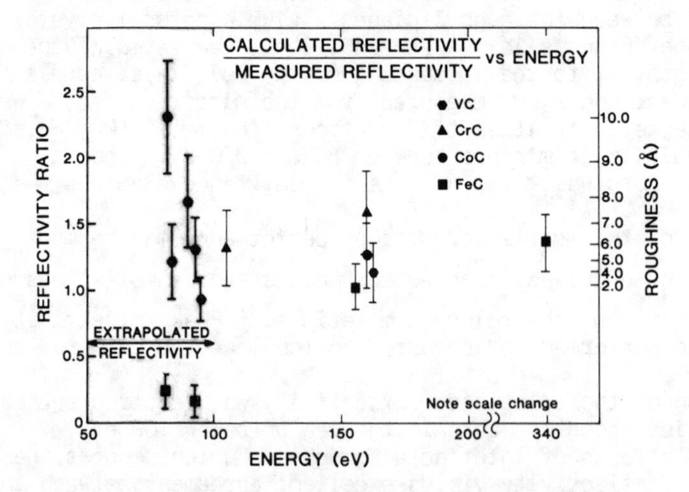

Fig. 2. The reflectivity ratio (calculated/ measured) vs. energy for several transition metal-carbon multilayers. The calculated reflectivies are extrapolations below 100 eV. The roughness for a given reflectivity ratio is shown on the right hand scale.

EXPERIMENT

The multilayer mirrors used in the present investigation were fabricated by electron beam evaporation.[1] An <u>in situ</u> soft x-ray (γ = 31.6 or 67.6 A) monitor was used to maximize the reflectivity of the multilayer during fabrication. The structure which results is not a regular lattice with constant layer thickness throughout, rather the thickness ratio of the low Z to high Z material increases towards the surface of the multilayer mirror. Table I includes the average characteristics of the multilayers studied in this experiment.

Table I Multilayer Characteristics

	V-C	Cr-C	Co-C	Fe-C
Average 2d spacing (A)	134	134	143	143
Number of layer pairs	15	14	20	20
Average thickness ratio high Z/low Z	0.7	0.7	0.4	0.4

The reflectivity measurements were performed at the Stanford Synchrotron Radiation Laboratory. The photon beam from the synchrotron was monochromatized by a "grasshopper type" (Rowland circle grazing incidence) monochromator with a 1200 l/mm grating. The samples could be rotated (θ) independent of the detector (2θ). A single channetron electron multiplier with a micromachined aluminum photocathode was used to measure the reflected, I_R, and incident, I_o, S-polarized photon beams. Data was collected by fixing the sample and detector angles and scanning the photon energy. The errors in the reflectivity $R = I_R/I_o$ were approximately 20%.

DATA AND ANALYSIS

Figure 1 shows the measured peak reflectivity vs. energy for the multilayer mirrors listed in Table I. This may be compared to calculations of the peak reflectivity based on the method of P. Lee[5] and the optical constant compilations of Henke, et. al.[4] Unfortunately, the optical constant tabulations are incomplete below 100 eV, therefore the calculated reflectivities between 80 and 100 eV are linear extrapolations of the reflectivity above 100 eV. It is reasonable to expect this extrapolation to be accurate for all the materials except iron which has a 3s electron binding energy of 92 eV. Changes in the optical constants associated with this resonance may make the extrapolation less accurate. Figure 2 shows the reflectivity ratio (calculated/measured), R_R, vs. energy for all the samples listed in Table I.

The error bars are representative of the experiment and do not contain the uncertainties in the optical constants or extrapolations.

We note that within experimental error nearly all of the multilayers perform below calculational levels, i.e., $R_R > 1.0$. The exceptions, the FeC data below 100 eV, are probably a result of the uncertainty introduced by the extrapolation of the reflectivity below 100 eV into a resonance region in iron. Many effects may cause this reduction: surface roughness, diffuse boundaries, and uncertainty in the multilayer parameters (optical constants, material density, and material distribution). We choose to assume that all of the discrepancy is due to surface and interfacial roughness. The reduction in reflectivity for a rough boundary between two media[6] coupled with the Bragg condition is

$$R_R = \exp\left[+ (2\pi\ \sigma/d)^2\right] \tag{1}$$

where σ is the root mean square roughness.

Using the average reflectivity ratio for each sample (we have left out the Fe-C samples below 100 eV) we have calculated a σ for each sample using equation 1. The calculated roughness and sample standard deviations are summarized in Table II.

TABLE II

Multilayer Mirror	Calculated Roughness (A)	Sample Deviation (%)
FeC*	4	20
CoC	8	30
V-C	4	15
CrC	6	12

*Excluding data below 100 eV.

The right hand scale of Fig. 2 provides an indication of the roughness associated for a given reflectivity ratio.

A complete diffraction profile of a V-C sample is shown in Fig. 3. The structure observed is typical of all of the samples. The central peak has a resolving power, the peak energy divided by the full width at half maximum, of 20 which is consistent with the theoretical expectation that the resolving power is nearly equal to the number of layer pairs contributing to the reflectivity which is 15 in this case.[1] The structure in the wings of the main peak is attributed to the aperiodicity of the multilayer structure, i.e., the ratio of high Z to low Z material in the multilayer is a function of depth.

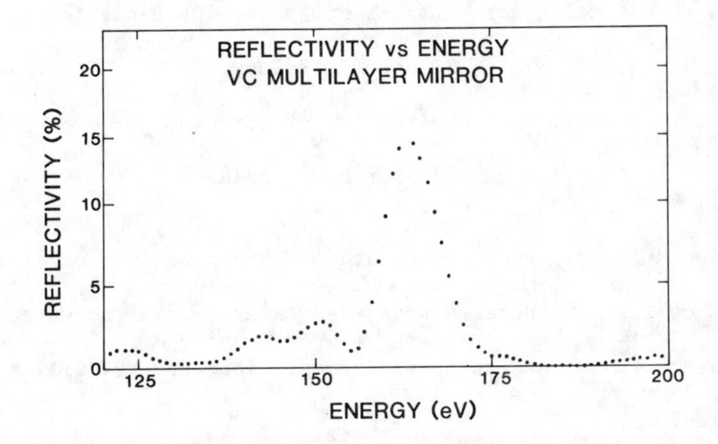

Fig. 3. The reflectivity in percent of a V-C multilayer mirror vs. photon energy.

CONCLUSION

We have demonstrated that multilayer mirrors can be used a efficient reflectors of soft x-rays for non-grazing incidence. The performance of these structures can be calculated with allowance for imperfections in the fabrication process and uncertainties in the optical constants.

REFERENCES

1. E. Spiller, "Evaporated Multilayer Dispersion Elements for Soft X-Rays," in Low Energy X-Ray Diagnostics, 1981, edited by D. T. Attwood and B. L. Henke, American Institute of Physics, New York, p. 124, 1981.

2. T. Barbee, "Sputtered Layered Synthetic Microstructures (LSM) Dispersion Elements," in Low Energy X-Ray Diagnostics, 1981, edited by D. T. Attwood and B. L. Henke, American Institute of Physics, New York, p. 131, 1981.

3. M. Born and E. Wolf, "Principles of Optics," Pergamon Press, New York, pp. 51-70, 1970.

4. B. L. Henke, P. Lee, T. J. Tanaka, R. L. Shimabukuro, B. K. Fujikawa, Atomic Data and Nuclear Data Tables, 27, (1982).

5. P. Lee, Opt. Commun., 43, 237 (1982).

6. H. E. Bennett and J. O. Porteus, J. Opt. Soc. Am., 51, 123 (1961).

Molecular Dynamics Probed by VUV Radiation

John C. Polanyi
University of Toronto
Toronto, Canada

Laser-induced fluorescence has proved to be a powerful tool for the study of inelastic collisions in gases and at surfaces. Joys and sorrows experienced in the attempt to extend such measurements into the VUV will be discussed and exemplified.

0094-243X/84/1190212-01 $3.00

PHOTOIONIZATION OF THE H ATOM IN STRONG ELECTRIC FIELDS BY RESONANT TWO-PHOTON EXCITATION

Karl H. Welge[*†‡] and H. Rottke[*]
[*]Fakultaet fur Physik, Universitaet Bielefeld, FRG
[†]Joint Institute for Laboratory Astrophysics, University of Colorado and National Bureau of Standards, Boulder, Colorado 80309

ABSTRACT

The photoionization of the H atom in strong electric fields, F, by resonant two-photon excitation, H(1) + VUV → H(2) + UV → H^+ + e, has been investigated at energies from the classical field ionization saddle point, $E_{sp} = -2\sqrt{F}$ a.u., through the zero field ionization limit, E = 0, into the continuum, E > 0. The atoms have been excited to single Stark levels in n = 2 with tunable pulsed VUV laser light around the Lyman-α line in an atomic beam with sub-Doppler resolution. The ionization from selected Stark levels by the UV was observed as a function of the UV wavelength. The VUV and UV were linearly polarized either perpendicular, σ, or parallel, π, to the electric field. Ionization spectra have been taken with (π,π), (π,σ), (σ,π), and (σ,σ) polarization and investigated at field strengths from 2-7 KV/cm. All spectra exhibit quasi-stable state structure superimposed on a continuum at energies $0 > E > -|E_{sp}|$. For the first time, also oscillatory structures have been observed with the H atom, extending well into the continuum region, E > 0, and below the zero field limit, E < 0. Their field dependence follows $\propto F^{3/4}$, with a deviation of a few percent. The oscillations appeared pronounced in (π,π) excited spectra, while not observable with the other polarization combinations.

INTRODUCTION

The interaction of atoms with strong external electric and magnetic fields, that is, where the field strength is comparable with, or larger than, the internal binding forces, has gained considerable attention in recent years.[1] At laboratory field strengths the strong interaction situation is always given in the vicinity of ionization limits, that is, in high Rydberg states and the adjacent continuum. While the extensive theoretical work in this field is based largely on the hydrogen atom, experimental studies have been performed almost always with non-hydrogen atoms.[1] In fact, no experiments are known with the H atom in strong magnetic fields (with dominating diamagnetic interaction), and the only ones in electric fields have been carried out by Koch and collaborators who used a method of fast atomic beam laser spectroscopy combined with Stark tuning levels into resonance with fixed frequency CO_2 laser lines.[2,3] They have

---

[‡]JILA Visiting Fellow, 1983-84. Permanent address: Universitaet Bielefeld, FRG.

performed precise measurements on the energy and ionization rate of discrete, quasi-stable states at energies above the classical field ionization saddle point, $E_{sp} = -2\sqrt{F}$ a.u., and well below the zero field ionization limit, $E < 0$.

A conceptually straightforward and versatile method, widely used in studies on high Rydberg atoms in strong fields, is the state selective excitation with field ionization in atom beam arrangements.[1] Experimental obstacles have prevented so far the application of this technique to the hydrogen atom. In this work we have essentially overcome these obstacles by two-step excitation of the atoms using tunable VUV laser light at the Lyman-α line for the first, resonant step and tunable UV laser light for the second, ionizing step:

$$H(n=1) + VUV \rightarrow H(n=2) + UV \rightarrow H^+ + e \quad . \qquad (1)$$

EXPERIMENTAL

The VUV was generated by frequency tripling in krypton, $h\nu_{VUV} = 3\,h\nu_{UV'}$. The fundamental, UV', was obtained directly from a tunable dye laser in the ultraviolet pumped by an excimer laser at 308 nm. The UV' fundamental leaving the krypton cell was separated from the VUV by a 30° LiF prism. This was necessary in order to prevent the UV' from entering the excitation-ionization arrangement and thus to avoid ionization of the H(n=2) atoms by it.[4] The excimer laser also pumped another dye laser to produce the tunable UV light for the second-step excitation.

The experimental arrangement consisted essentially of two parallel field electrodes and an electron multiplier placed behind the positive electrode which carried a wire mesh through which electrons from the field region could pass and be detected by the multiplier. The VUV and UV laser beams passed, anti-collinearly, through the field, such that the pulses (~5 ns duration, 10 Hz repetition rate) arrived simultaneously at the center of the field. Hydrogen atoms, produced in a microwave discharge, entered the field region in an atomic beam, well collimated to an effective, perpendicular Doppler width of ~2 GHz at 121.6 nm. T
he axis of the atomic beam, the laser beams and the electric field intersected each other at right angles. The field electrode arrangement was enclosed in a cryogenetically cooled (~20 K) metal housing provided with the appropriate openings for the atomic and laser beams, as well as for the electrons to reach the multiplier, positioned outside the housing. More detail of the experimental setup and procedure will be published elsewhere.[5]

EXPERIMENTS AND RESULTS

The experiments have been carried out in the energy range from the saddle point, E_{sp}, to the zero field ionization limit, $E = 0$, ($IP_o = 109678.77\ cm^{-1}$) and into the continuum region, $E > 0$. While the VUV wavelength was kept at the (n=1) → (n=2) transition, the UV laser was scanned and the electron signal recorded as a function of the UV wavelength, yielding the ionization spectrum from the respective Stark level in the n = 2 state.

As the excitation-ionization cross section essentially depends on the spatial quantization,[6,7] the polarization of both laser beams had to be well defined with respect to the electric field direction. In these experiments both parallel (π) and perpendicular (σ) polarization has been applied, yielding the four two-photon combinations: (π,π), (π,σ), (σ,π), and (σ,σ). By the same token, the n=1 → n=2 transition had to be selectively excited with respect to each of the four Stark sublevels in the n=2 manifold, requiring sufficient spectral resolution in the VUV transition. Figure 1 shows of the Stark splitting measured at a field strength F = 5465 V/cm; once for σ and once for π polarization. The lines are labeled by the low-field

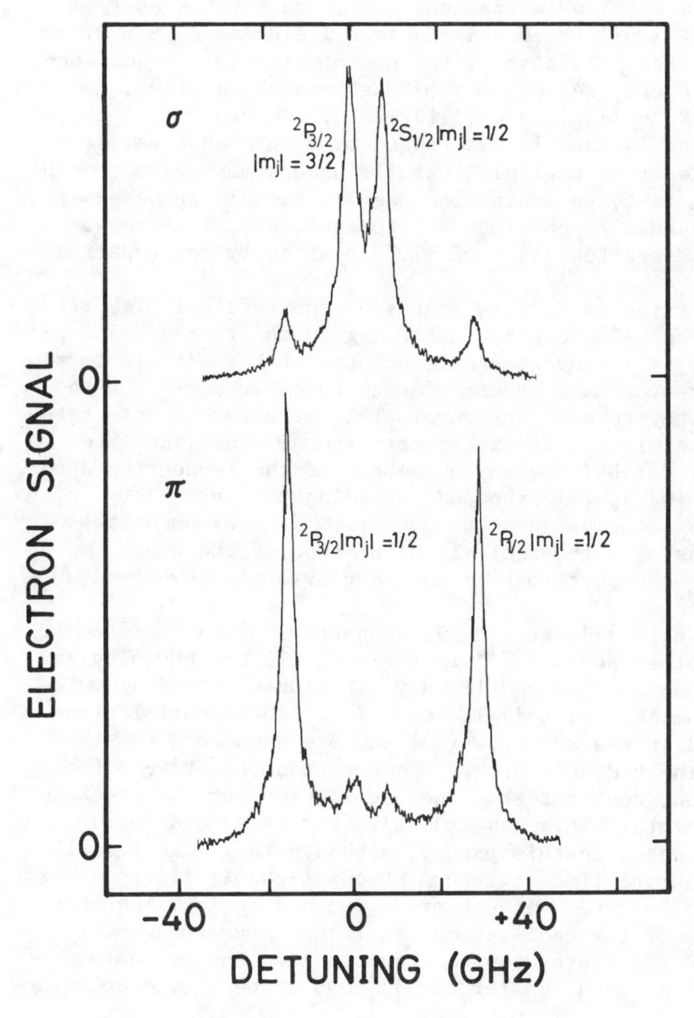

Fig. 1. Stark spectra of the H(n=1 → n=2) transition at the Lyman-α line, ~121.6 nm. Field strength F = 5665 V/cm. VUV laser polarization: upper spectrum perpendicular (σ), lower spectrum parallel (π).

quantum number notation in the n = 2 manifold. The effective resolution of the Stark lines of ~3 GHz width at half maximum results from the combined contributions of the laser bandwidth at 121.6 nm and the perpendicular Doppler line width of the atomic beam. This measurement has been made with the resonant two-step photoionization technique previously employed by us.[4] However, different from the previous experiments, the ultraviolet radiation, obtained here from the second laser, had a wavelength such that the photon energy was well above (~3000 cm^{-1}) the zero field ionization limit. At F = 5465 V/cm only the two strongest lines ought to be present in the respective σ and π spectra. The appearance of two low-intensity components is due to imperfect linear polarization of the VUV laser light.

Ionization spectra have been taken and investigated at different field strengths and with all four polarization combinations. Figure 2 shows two examples obtained at F = 5715 V/cm: (a) with (π,π) and (b) with (π,σ) polarization. The saddle point energy corresponding to F = 5715 V/cm and the zero field limit, E = 0, are marked in the figures. To discuss the photoionization in an electric field three energy regions may be distinguished: (I) $E < -|E_{sp}|$, (II) $0 \geq E \geq -|E_{sp}|$, and (III) $E > 0$. In region (I) the lifetime of a state is largely determined by spontaneous emission, the tunneling rate being negligibly small, except very close to the saddle point. In fact, no ionization was observed in these experiments below the saddle point since the ionization rate there was smaller than the detection limit of ~10^{-5} s given by the experimental conditions.

The strong mixing of Rydberg states by the external electric field in the region (II) results, according to theory[1,6-10] and previous experiments,[2,3] in quasi-stable quantum states with ionization lifetimes ranging over many orders of magnitude, from ~10^{-5} s to <10^{-12} s, where they form a continuum. This structure of discrete lines is clearly exhibited by the spectra shown. The lines are identified by the parabolic quantum numbers of the respective upper, ionizing states, $|n_1,n_2,|m|\rangle$ grouped according to $n = n_1 + n_2\ |m| + 1$, where n is the principal quantum number. The assignment has been made by means of a theoretical calculation of the state energy, following the perturbation method previously developed and employed.[2,8,9]

As has been observed first by Freeman et al. with Rb [11] and later also with other atoms,[12-14] in region (III) the photoionization cross section can show oscillatory structures extending well into the continuum, E > 0, and also to E < 0. These oscillations clearly developed in the spectrum (a), and are seen here for the first time with the hydrogen atom. The spectrum (b), taken with (π,σ) polarization, does not show the oscillations at the precision of these measurements. This was true also for (σ,π) and (σ,σ) spectra (not presented in this paper), although they also clearly exhibited the discrete line structure of quasi-stable states. These findings essentially agree with theory[6,7] according to which the modulation degree of the oscillations is relatively strongest for the excitation of final states with the magnetic quantum number $|m| = 0$, while it is much smaller for $|m| > 0$ states, evidently too

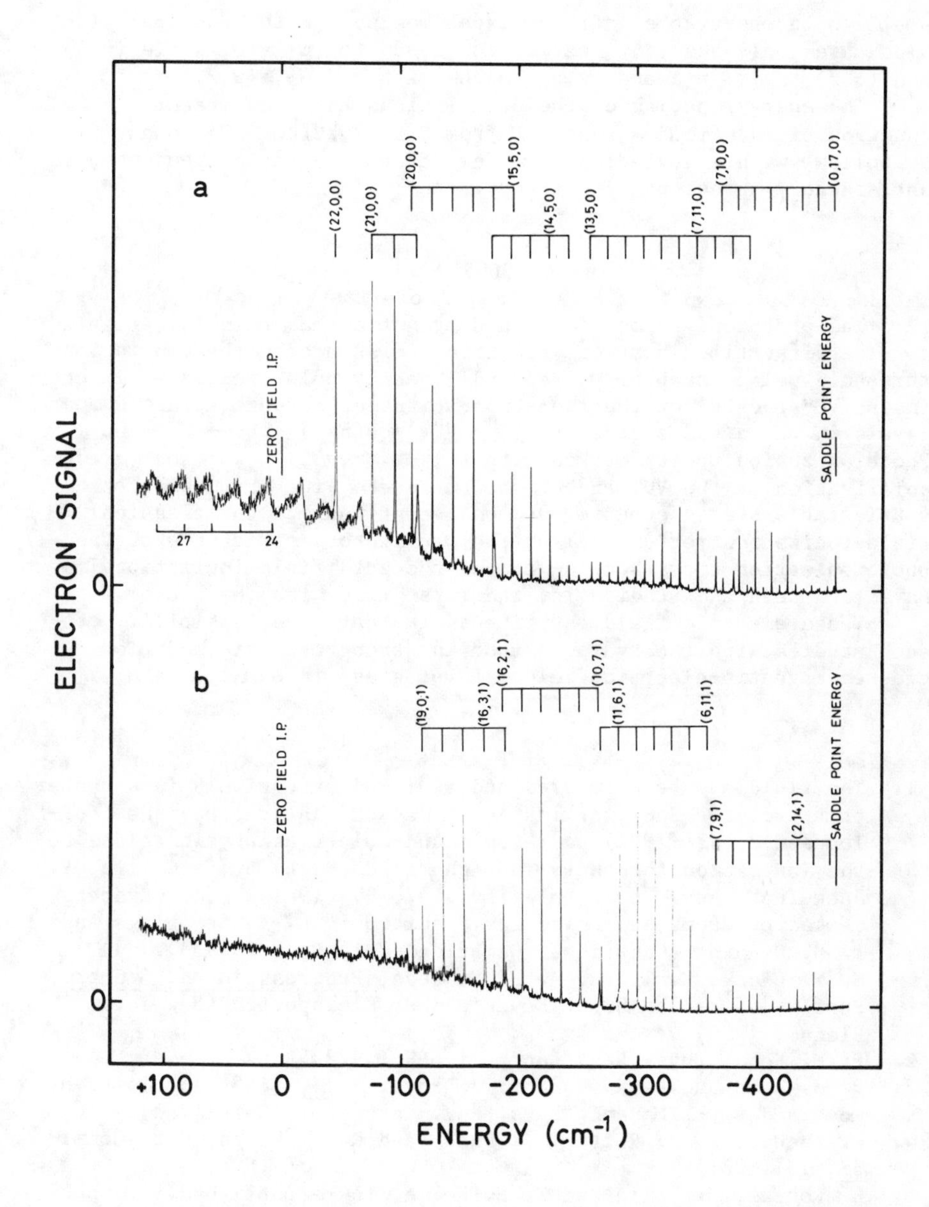

Fig. 2. Photoionization spectra from the excited state $2^2P_{3/2}|m_j| = 1/2$ (see Fig. 1) of the H atom in an electric field F = 5714 V/cm. Polarization: (π,π) in spectrum (a) and (σ,σ) in (b). Lines labeled by parabolic quantum numbers (n_1,n_2,$|m|$) grouped according to $n = n_1 + n_2 + |m| + 1$.

small to be observable in these experiments. Of the four polarization cases only the (π,π) excitation leads to $|m| = 0$, while (π,σ) and (σ,π) to $|m| = 1$ and (σ,σ) to the mixture $|m| = 0,2$.

The energy spacing of the oscillations has been measured as a function of the field strength F from 2 to 7 kV/cm. We found it to follow, with a deviation of a few percent, a $F^{3/4}$ dependence, in satisfactory agreement with theory.[15]

CONCLUSIONS

The photoionization of the hydrogen atom has been investigated for the first time in strong electric fields around the ionization threshold using tunable, pulsed and linearly polarized laser light in the VUV and UV for the two-step excitation through selected Stark levels in the n = 2 state; $H(n=1) + VUV \rightarrow H(n=2) + UV \rightarrow H^+ + e$. Photoionization spectra taken with (π,π), (π,σ), (σ,π), and (σ,σ), polarization of the VUV and UV show discrete line structures of quasi-stable states superimposed on a continuum in the classical field-ionization region. Also, the oscillatory structures of the photoionization cross section around the zero field ionization limit have been seen and studied for the first time with the H atom.

These electric field experiments indicate the feasibility of such studies with the hydrogen atom in strong magnetic and also crossed magnetic-electric fields, in progress in our laboratory.

REFERENCES

1. The field has been covered and reviewed extensively in a number of articles. See, for instance: a) Journal de Physique, Colloque C-2, 43 (1982) on "Atomic and molecular physics close to the ionization threshold in high fields." b) S. Feneuille, P. Jacquinot, Adv. Atom. Mol. Phys. 17, 99 (1981). (c) "Rydberg states of atoms and molecules," edited by R. F. Stebbings and F. B. Dunning (Cambridge, Cambridge Uni. Press, 1983). d) C. W. Clark, K. T. Lu, A. F. Starace, Progress in At. Spectr., Part C, edited by H. J. Beyer and H. Kleinpoppen (New York, Plenum, 1984).
2. P. M. Koch, Phys. Rev. Lett. 41, 99 (1978).
3. P. M. Koch and D. R. Manani, J. Phys. B 13, L645 (1980); P. M. Koch and D. R. Manani, Phys. Rev. Lett. 46, 1275 (1981).
4. H. Zacharias, H. Rottke, J. Danon and K. H. Welge, Opt. Comm. 37, 15 (1981).
5. H. Rottke, H. Holle and K. H. Welge, to be published.
6. E. Luc-Koenig and A. Bachelier, J. Phys. B 13, 1743 (1980); ibid., 1769 (1980).
7. D. A. Harmin, Phys. Rev. A 24, 2491 (1981); ibid., A26, 2656 (1982).
8. H. J. Silverstone and P. M. Koch, J. Phys. B 12, L557 (1979).
9. R. J. Damburg and V. V. Kolosov, J. Phys. B 12, 2637 (1979); ibid., 9, 3149 (1976); ibid., 14, 829 (1981).
10. V. D. Kondratovich, V. N. Ostroviskii, JETP 52, 198 (1980).

11. R. R. Freeman, N. P. Economou, G. C. Bjorklund and K. T. Lu, Phys. Rev. Lett. 41, 1463 (1978); R. R. Freeman and N. P. Enonomou, Phys. Rev. A 20, 2356 (1979).
12. W. Sandner, K. A. Safinya and T. F. Gallagher, Phys. Rev. A 23, 2448 (1981).
13. T. S. Luk, L. DiMauro, T. Bergemann and H. Metcalf, Phys. Rev. Lett. 47, 83 (1981).
14. B. E. Cole, J. W. Cooper and E. B. Saloman, Phys. Rev. Lett. 45, 887 (1980).
15. A. R. P. Rau, J. Phys. B 12, L193 (1979); A. R. P. Rau and K. T. Lu, Phys. Rev. A 21, 1057 (1981).

USE OF SYNCHROTRON AND LASER RADIATIONS FOR PRESENT AND FUTURE PHOTOIONIZATION STUDIES IN EXCITED ATOMS AND IONS

François J. Wuilleumier
Laboratoire de Spectroscopie Atomique et Ionique and LURE, Université Paris-Sud, B.209c, 91405-Orsay Cedex, France

ABSTRACT

The combined use of synchrotron and laser radiations allows inner-shell and outer-shell photoionization experiments to be made on atoms in excited states. The present status of these experiments, carried out in Orsay, is presented. The opportunities opened up by future electron/positron -storage ring based sources of VUV radiation are discussed.

INTRODUCTION

Until a recent past, the vast majority of photoionization measurements has been performed on atoms and molecules in their ground state. Very valuable information has been gained, especially with the expanding availability of high flux of monochromatized synchrotron radiation over a broad energy range (at present, from a few eV to a few hundreds eV).[1] The importance of photon impact experiments as a key to our understanding of atomic structure is illustrated by the long list of parameters they permit to obtain, such as binding energies of atomic electrons, oscillator strengths of discrete transitions, partial subshell photoionization cross sections, angular distribution asymmetry parameters of Auger and photoelectrons, spin polarization of the photoelectrons.

This field is limited, however, in some ways: dipole selection rules confine the class of final states that can be reached, and the initial states often are ensembles of near-degenerate levels (particularly in open-shell atoms), which complicates the interpretation. If atoms can be prepared in an excited initial state for photoionization, these limitations can be circumvented and, furthermore, knowledge can be gained on the behavior of excited atoms and molecules, which is of great potential importance in such fields as plasma physics, astrophysics and photochemistry.

The study of single photoionization processes in excited atoms requires the use of two different photon sources: the one, to prepare the atoms in specific initial states, and the other one to photoionize these excited states. With the advent of high-power frequency-tunable lasers, it has become possible to prepare a sizable stationary fraction of valence-excited atoms and to laser-ionize these laser excited atoms. However, available laser energies and tunability ranges have restricted these experiments to outer electrons of alkali- and alkaline-earth atoms.[2] In a few cases, metastable excited states of rare gas atoms have also been prepared by electron impact[2] or by using a cold cathode dc discharge source,[3] further photoionize with the beam of a dye laser. Finally, in some alkali-and alkaline earth atoms placed onto excited states by means of a pulsed laser, inner-shell photoabsorption has been studied, using ultraviolet continuum radiation from a pulsed BRV source, whose repetition rate was synchronized with that of the laser.[4]

The broad energy spectrum of synchrotron radiation was an obvious invitation to extend these earlier measurements. The first observation of inner-shell photoionization in laser-excited atoms by synchrotron radiation was made in Orsay, in 1981, using electron spectrometry, and was reported on at the first topical meeting of this type, already held in Boulder, two years ago.[5] Since that time, extensive studies have been carried out, using a cw ring dye laser to create the excited states and the synchrotron radiation from the ACO storage ring to probe inner- and outer-shell photoionization, as well as autoionization in excited sodium and barium atoms. The present status of these experiments will be presented in the first part of this paper.

The intensity of synchrotron radiation can be greatly enhanced and its spectral distribution modified through the insertion of undulators in a straight section of a storage ring. The generation of stimulated radiation has recently been achieved at LURE, in the first successful operation of a storage ring-free electron laser.[6] Other experiments are in preparation, such as the production of coherent VUV radiation by bunching the electron beam in an optical klystron.[7] A new generation of storage rings, starting with Super ACO in Orsay, is being built, primarily for the use of these new insertion devices. The opportunities opened up, for the experiments on excited atoms, molecules and ions, by these future electron/positron storage ring based sources of VUV and soft x-ray radiation will be discussed in the second part of this review.

EXPERIMENTAL

The main characteristics of laser and synchrotron radiation sources are quite different. While an electron storage ring is able to provide synchrotron radiation up to a rather high photon energy (typically 1000 eV for a small ring, 10 000 eV for a large ring), photons available from laser systems are restricted to very low energies (typically 1 eV to 4 eV for cw dye lasers, up to 10-15 eV for pulsed lasers). On the other hand, monochromatized beams of synchrotron radiation are less intense and spectrally broader by orders of magnitude than those available from continuously tunable lasers. In the VUV region, 10^{12} to 10^{13} photons/sec in a band width of 1% is the highest flux obtainable with a good toroïdal grating monochromator illuminated by the radiation emitted from the bending magnet of a modern storage ring;[8] in addition, a resolution of 1000 can almost be considered as ultimate. Compared to these numbers, 10^{18} photons/sec in a 20 MHz band width (relative width of 10^{-7}) is the ordinary intensity that a modest cw laser can deliver to an experiment in the narrow energy range it is tunable. Moreover, the emission of synchrotron radiation is pulsed, with typical repetition frequency of 10 MHz. Thus, the number of photons in one pulse is extremely low, 10^5 to 10^6 photons after monochromatization, which means that pulsed lasers are poorly suited for the type of experiments described here. Even if one

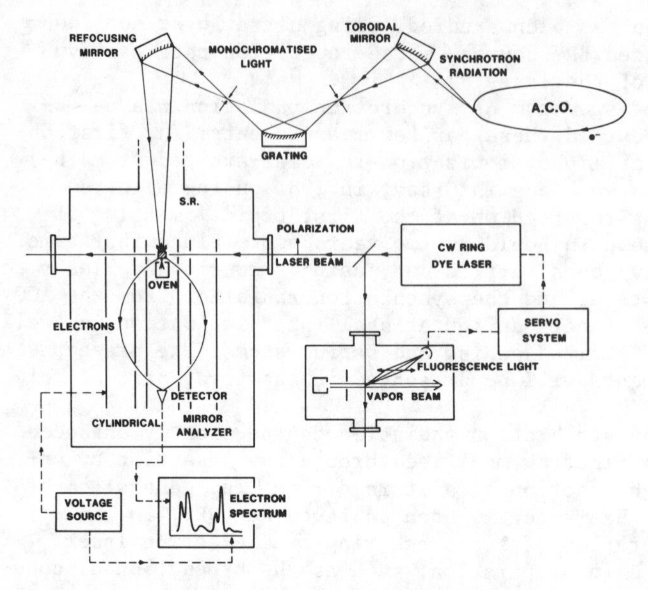

Fig.1.- Experimental set up. See detailed description in the text. (from Ref. 10).

is able to synchronize a 1μs-wide, 10 Hz pulsed laser pumped source with a 1ns-wide, 10 MHz pulsed synchrotron radiation source, each laser pulse would shine on only 10 synchrotron pulses and the combined laser-synchrotron duty factor would be at most $1:10^5$. Thus, only cw dye lasers are used presently for these photoionization studies on free atoms. In addition to that, since no laser laboratory has yet been developed in a synchrotron radiation center, only commercially available systems can be used at the present time in combination with synchrotron radiation. Some laboratory built lasers are, of course, able to provide larger powers and to cover a more extended energy range, but they are hardly portable.

Fig.1 shows the experimental set up used in these experiments. The synchrotron radiation from the electron storage ring ACO (536 MeV, 150 mA) is monochromatized with a toroïdal grating monochromator efficient in the 15-170 eV photon energy range,[8] and is refocussed into the source volume of a cylindrical mirror electron analyzer whose axis is collinear with the photon beam.[1] A weakly collimated effusive beam of atoms in the vapor phase is produced by a resistivically heated oven mounted on the axis of the CMA. The laser beam traverses the electron spectrometer in a direction perpendicular to the CMA axis. The electrons produced in the source volume of the CMA, at the intersection of the three beams, are energy analyzed at the magic angle of 54°44'. The laser is a cw ring dye laser (Spectra Physics 380A) operating in the monomode regime. It is pumped by an Ar ion laser (maximum power 18 Watts). In the case of sodium, the laser is locked to the $3\ ^2S_{1/2}$ (F=2) $\rightarrow 3\ ^2P_{3/2}$ (F=3) transition, at 5890 Å; for the experiments on barium, it is locked to the $6s^2\ ^1S_o \rightarrow 6s6p\ ^1P_1$ resonance line, at 5535 Å. The laser beam is linearly polarized in the horizontal plane. However, since the density of atoms in the interaction zone is high (10^{12} to 10^{13} atoms/cm^3), the alignment of the intermediate state is lost by radiation trapping.

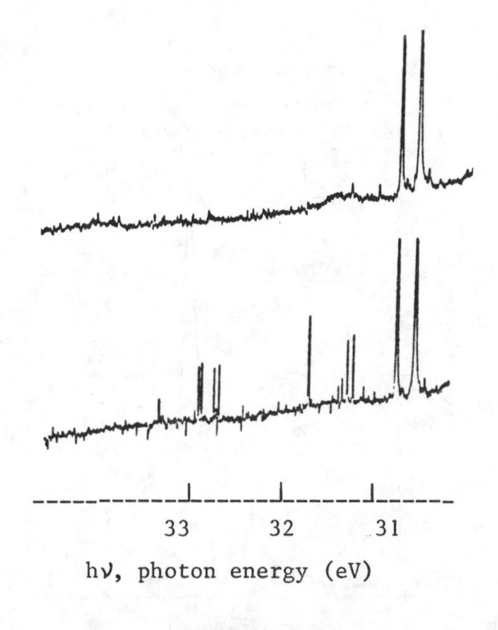

Fig.2.- Photoabsorption spectrum of Na atoms in the ground state (upper trace) and in the excited 3p $^2P_{3/2}$ state (lower trace). (from Ref.4).

Energies of the main transitions to doubly excited states of the $2p^5 3s3p$ configuration (Ref.4):

$2p^5(^2P)3s3p(^3P)\,^4P_{5/2}$:	31.19 eV
$^4P_{3/2}$:	31.24 eV
$^2D_{3/2}$:	31.34 eV
$^2D_{5/2}$:	31.40 eV
$^2S_{1/2}$:	31.78 eV
$2p^5(^2P)3s3p(^1P)\,^2D_{5/2}$:	32.68 eV

This experimental apparatus differs in two main aspects from that used in the preliminary experiment.[5] First, an auxiliary collimated metallic vapor beam, placed outside of the CMA, is used to stabilize and to lock the laser to the resonance line. Second, a liquid nitrogen trap, placed in front of the oven, contributes to significantly reduce the background in the electron spectra.

EXPERIMENTAL RESULTS ON SODIUM

To make easier the description of the experimental results obtained in the photoionization and autoionization of laser excited Na atoms, we recall, in Fig.2, part of the absorption spectrum of Na atoms excited in the $2p^6 3p\ ^2P_{3/2}$ state.[4] The structures observed in the lower trace are due to discrete excitations of a 2p electron to autoionizing states of the $2p^5 3s3p$ type in the $2p^6 3p$ atoms. In the experiments with synchrotron radiation, two processes have been studied by electron spectrometry: the direct ionization of a 2p electron into the continuum and the autoionization of these doubly excited states.

Fig.3 shows a photoelectron spectrum of Na atoms at 60 eV photon energy, in a flat region of the continuum absorption. Part a) was obtained during the feasibility experiments of 1981,[5] part b) is an example of the results nowadays obtainable under improved laser and sodium beam conditions.[2] In the absence of laser radiation, one observes photolines corresponding to the inner 2p shell ionization of Na atoms in the $2p^6 3s\ ^2S$ ground state. When the laser is on, new photoelectron lines appear around 40 eV binding energy, corresponding to the direct ionization of a 2p electron in the laser-excited atoms. Since the screening of the 2p electrons by the outer electron is reduced in the excited state, the binding energy of a 2p electron is greater in the excited state than in the ground state. Because of the coupling of the

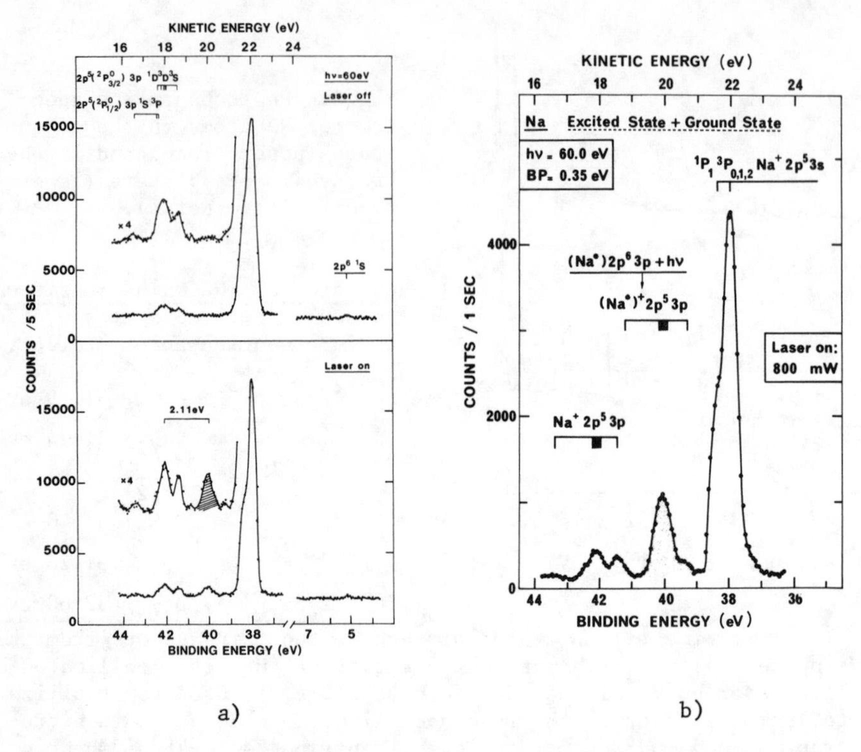

Fig.3.- Photoelectron spectra of Na taken at 60 eV photon energy. a) = preliminary results, obtained in the feasibility experiment, Ref.5 (note that, in this reference, Fig.2, that is similar to a) has been interchanged with Fig.4); b) = present results (from Ref.2)

two open shells in the final $2p^5 3p$ state, the removal of a 2p electron from an excited atom gives rise to a multiplet structure whose energies can be deduced from the $2p^5 3p$ final states, produced by ionization of the ground state via electron correlations, by a simple translation of 2.11 eV. It is worthwhile to notice that the various final states of the $2p^5 3p$ configuration are not populated in the same way via the one-photon and the two-photon routes. Experiments with higher resolution would be necessary to study in more details this fine structure effect.

In an atom like Na, where the outer electron is weakly coupled to the core electrons, one has predicted that the photoionization of a 2p electron at a fixed photon energy is practically independent of the atomic orbit that the valence electron occupies.[9] The relative population of atoms in the excited state can be deduced from the reduction in intensity of a photoelectron peak corresponding to 2p ionization in the ground state atoms. Then, by simply comparing the ratio of the area under the peaks at 40 eV and 38 eV binding energy, respectively, it has been possible to check that, within an experimental uncertainty of 15-20%, the variation of the 2p cross section is, indeed, negligible. This provides an easy way to determine the relative population

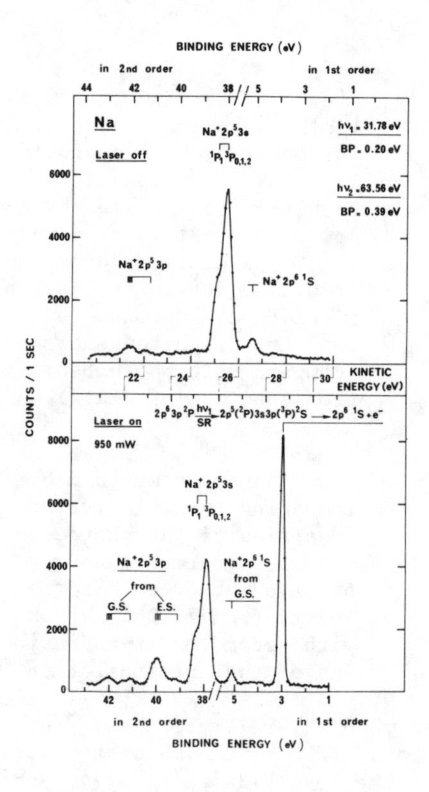

Fig.4.- Electron spectra ejected from Na atoms in the ground state (top, laser off) and from a mixed medium partly formed of atoms in the $2p^63p\ ^2P_{3/2}$ state (bottom, laser on) by 31.78 eV and 63.56 eV photons simultaneously transmitted by the monochromator. See text for explanation (from Ref.10). (The lower part of this figure is similar to Fig.4 of Ref.5 that had been interchanged with Fig.2).

of atoms in the excited state. Comparison of Fig.3a and 3b illustrates also the drastic reduction of the background level, once the liquid nitrogen trap has been placed in the experiment,and the large increase in the population of excited atoms: in the original experiment, the relative density of atoms in the excited state was, at most, 10% (note, for reference, that the relative intensity of correlation satellites at 60 eV is also about 10%), while, in the improved experimental conditions, this relative density appears to be close to 30%.

For the study of the even-parity autoionizing resonances excited by step-wise laser-VUV two-photon absorption, the synchrotron radiation is tuned to the 30-32 eV region indicated by the earlier photoabsorption measurements (Fig.2).[4] Fig.4 shows a typical electron spectrum obtained in this energy range, here at 31.78 eV. Since the monochromator transmits also radiation diffracted in second order by the grating, at 63.56 eV, one observes again the photoionization into the continuum of 2p electrons, as in Fig.3. One sees also a peak due to photoionization of a 3s electron (binding energy 5.14 eV) with 31.78 eV photons. The second feature in the lower frame of Fig.4 is the intense electron line at 3.03 eV binding energy, that is the binding energy of the 3p excited electron. This line arises from the decay of the autoionizing state $2p^5(^2P)3s3p(^3P)^2S$ produced by photoabsorption of 31.78 eV photons in the excited atom. When the monochromator is tuned slightly off the 31.78 eV energy, this peak disappears, because the nonresonant photoemission of 3p electrons is very weak at this photon energy; moreover, since this doubly excited state is far in energy from the other terms of the $2p^53s3p$ configuration, no other transition is excited. To the extent that the radiative decay of the doubly excited state is negligible, the oscillator strength for the excitation of such even-parity resonances via inner-shell photoabsorp-

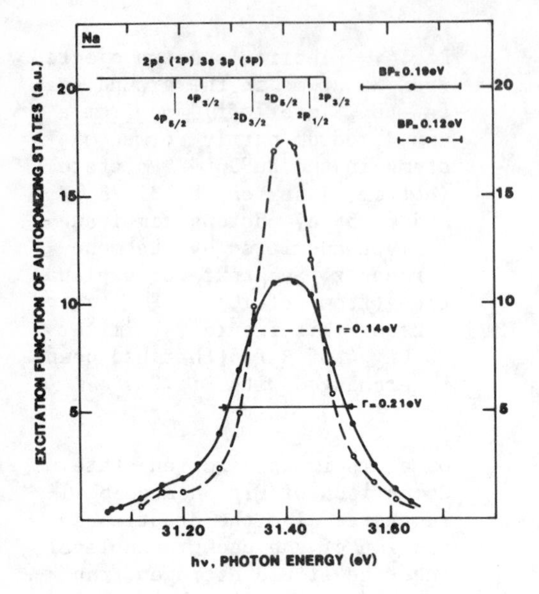

Fig.5.- Variation of the ratio between the area of the autoionization line, following the formation of the doubly excited states by absorption of 31.40 eV photons, and of the photoelectron line due to ionization by 62.80 eV photons, as a function of photon energy (compare with photoabsorption spectrum of Fig.2).

tion can be determined by comparing the relative intensities of the photoelectron peak, simultaneously observed at 40 eV binding energy from 2p ionization with second order radiation, and of the autoionization peak.[10]

The situation is more complicated when several resonances are excited simultaneously in the finite band width of the monochromator, as it is the case around 31.40 eV, for the $2p^5(^2P)3s3p(^3P)^2D_{3/2,5/2}$-$^2P_{1/2,3/2}$ excited states. Fig.5 shows the excitation function of the autoionizing states determined by measuring the relative intensity of the autoionization line as a function of the photon energy, for two different values of the band width of the monochromator.[10] Both excitation functions peak at the same photon energy of 31.40 eV, that is the excitation of the $^2D_{5/2}$ state.From this result, from the accurate measurement of the kinetic energy and of the width of the autoionization line, and from the FWHM of the excitation function, one can conclude that the transition to this $^2D_{5/2}$ excited state is largely dominant.[10]

The detailed values of the oscillator strengths for each transition or group of transitions will be published elsewhere.[10] Here, we would like only to mention that the sum over all excited states of the $2p^53s3p$ configuration gives an overall f-value of 0.23(4). There is no calculation to be compared with this experimental value. However, one can note that it is not far from the f-values calculated[11] for the $2p^6 \rightarrow 2p^53s$ transitions in the isoelectronic sequence of Ne-like ions: 0.17 for NeI, 0.20 for NaII, 0.21 for AlIV and SiV. As for the 2p photoionization cross section, the coupling with the 3p electron does not seem to modify strongly the transition probability.

One can see, on the better resolution results of Fig.5, that the $^4P_{5/2}$ term shows up with some intensity, confirming that the representation of this state must include some dipole components.[12] The relative intensity of the autoionization line due to the decay of this excited state is about 3% of the line observed at 31.40 eV.

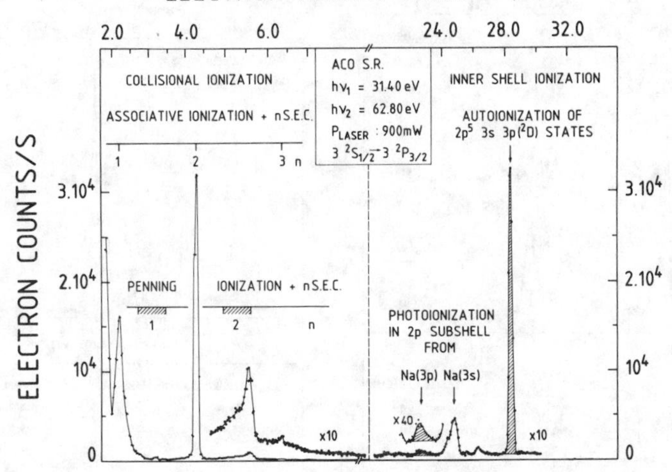

Fig.6.- Electron energy spectrum from Na atoms simultaneously laser-excited to the 3p $^2P_{3/2}$ state and irradiated with monochromatized ACO synchrotron radiation (31.40 eV and 62.80 eV photons). High energy electrons (right part) comes from the photoionization and autoionization processes produced by the photons. Low energy electrons (left part) are formed in collisional ionization and subsequently heated by superelastic collisions.

It can be noted, in Fig.4, that the kinetic energy of all electrons is reduced when the laser is on. This effect is due to a plasma potential drop that develops in the laser-excited medium through purely collisional processes - associative ionization and Penning ionization- that produce molecular and atomic ions.[13,14] These processes involve only collisions between Na atoms in excited states, including the formation of higher Na(nl) states via energy-pooling collisions. The energies of all primary electrons produced by these mechanisms are lower than 1.2 eV, but they are observed at higher kinetic energies, as shown in Fig.6, because they undergo superelastic collisions with Na(3p) atoms that increase their energies by steps of 2.11 eV. In Fig.6, electrons produced by associative ionization are observed after 1, 2 or 3 superelastic collisions at 2.1 eV, 4.2 eV and 6.3 eV, respectively, electrons produced by Penning ionization of higher excited states are seen, after 1 or 2 superelastic collisions , at 3.3 eV and 5.4eV. From the simultaneous measurement of the inner-shell photoionization processes, as shown in Fig.6, important parameters for the interpretation of the collisional processes, such as the kinetic energy of the electrons and the density of atoms can be deduced. In particular, the intensity of the autoionization line, excited at 31.40 eV photons, is a very sensitive indicator of the density of excited states in the medium, once the determination of the oscillator strength has been made.

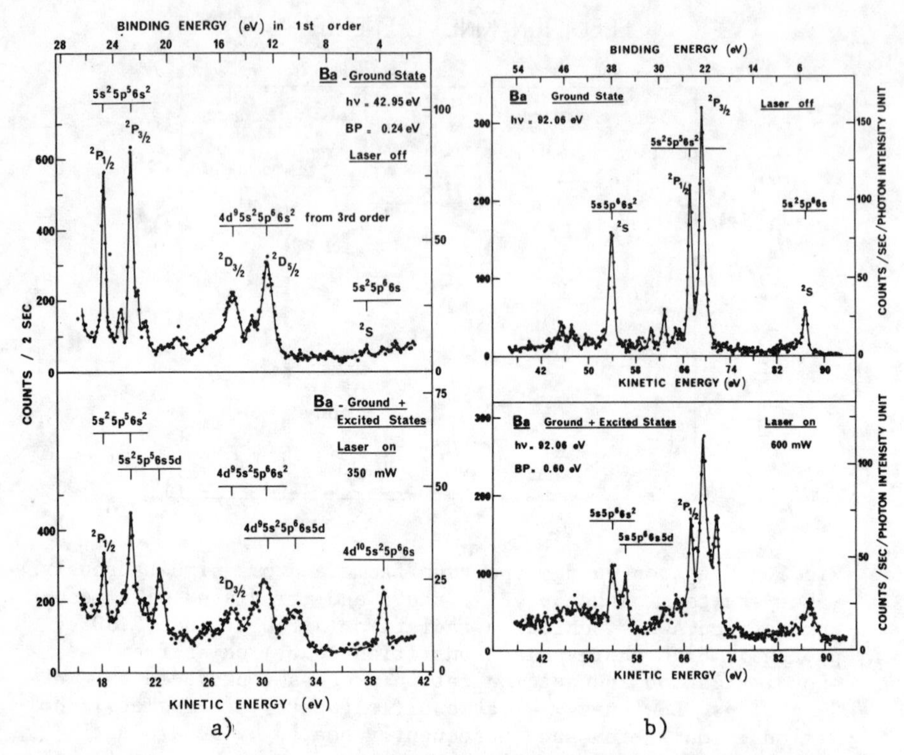

Fig.7.- Photoelectron spectra ejected from Ba atoms in the ground state and in the excited $6s5d^{1,3}D$ states by 43 eV and 129 eV photons (a) and by 92 eV photons (b). (a from Ref.15, b from Ref.16).

PRELIMINARY RESULTS ON BARIUM

Although the laser was tuned to the $6s^2\ ^1S_0 \rightarrow 6s6p\ ^1P_1$ transition, at 2.24 eV photon energy, most of the excited atoms decay rapidly by radiative transition to the metastable 6s5d $^1D_2$ state at the high laser power needed to produce a sizable density of excited atoms. In fact, very recent measurements[17] of the 5d photoionization cross section, carried out at high resolution between 15 eV and 25 eV photon energy, have shown that the $^3D_{1,2,3}$ states are also equally populated, likely by collisional transfer from the $^1D_2$ state.

Fig.7a shows a typical example of photoelectrons ejected from Ba atoms in ground and excited states. With the monochromator set at 42.95 eV, but transmitting second and third orders as well, the 5p and 4d subshells are ionized by 42.95 eV and 128.95 eV photons, respectively. In the spectrum recorded without the laser, the corresponding photolines appear with a relatively high intensity, while the 6s cross section is evidently much weaker at this photon energy. When the laser is turned on, additional lines appear, from 5p and 4d ionization in the excited atom; in addition, an intense peak appears at 3.8 eV binding energy, due to photoionization of the excited 5d electron. One notes that the new structures in the 5p and 4d photolines are on the low binding energy side: the binding energies of these electrons are lowered by about 2 eV. Since the mean radius of the 5d and 6p orbitals are respectively smaller and larger than that of a 6s orbital,[18] this shift of binding energy to lower values confirmed that most of the excited atoms are in the metastable $6s5d\,^{1,3}D$ states. For high laser powers, no 5p or 4d photoline has ever appeared at higher binding energy than in the ground state. In addition, no photoelectron line originating from the 6p subshell is present in the spectra, while one can see distinctly, in the high resolution-low photon energy spectra, a photoline due to ionization of a 6s electron in the excited atom.[17] Thus, the absence of a 6p line is not due to a low value of the 6p cross section, but is a clear indication that the density of Ba atoms in the $6s6p\,^{1}P_{1}$ state is negligible in the vapor in these experimental conditions.

A simpler spectrum is presented in Fig.7b, at 92 eV photon energy, below the 4d ionization thresholds. Here, only first order photons are transmitted and ionize Ba atoms in the 5s, 5p and 6s subshells. Again, when the laser is on, new photoelectron lines are visible, due to ionization of excited Ba atoms. Since there is only one line due to ionization of the 5s electron in the ground state, the observation of the 5s line in the excited state shows clearly that the binding energy of the 5s electrons is also shifted, by about 2 eV, to lower value. Comparing Fig.7a and Fig.7b, one notes also that, at high photon energy, the 5d cross section is not anymore enhanced, as it is at lower photon energy.

The relative density of Ba atoms in the excited states can be deduced from the reduction in intensity of a photoelectron line clearly due only to ionization of Ba atoms in the ground state (5s, $5p_{1/2}$ or $4d_{3/2}$ subshells) or from a comparison of the intensities of some lines due to ionization in the ground and excited states, respectively, assuming little variation of the corresponding photoionization cross sections. One may expect that it should be the case for the 5s subshell. Both independent sources of data indicate that more than 50% of Ba atoms are transfered in the metastable excited $6s5d\,^{1,3}D$ states.

Because of the presence of higher orders, there is some overlap between photoelectron lines originating from the various subshells, when the photon energy is varied. This makes the analysis of the data more difficult or, at certain photon energies, even impossible. However, the variation of the 5d cross section has been obtained over a large photon energy range, and the 5p and 4d cross sections at some photon energies.[17]

FUTURE OF THESE EXPERIMENTS

The results described in the preceeding paragraphs illustrate the level of experiments that can be presently made with the flux emitted from the bending magnet of a storage ring. However, such experiments are still difficult, because of the number of experimental techniques that must be worked out simultaneously when the synchrotron radiation beam is on. Large progress could be made in optimizing the various parameters involved in these experiments, in particular

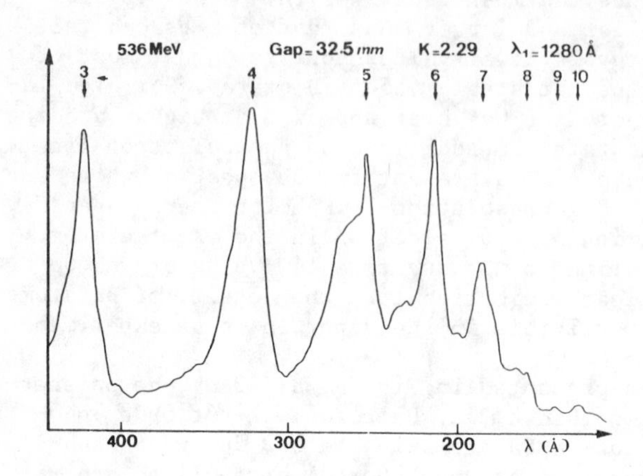

Fig.8.- Spectral distribution of the ACO undulator at 536 MeV in the VUV range. The spectrum is not corrected for the monochromator response. (from Ref.19).

in improving the electron detection and choosing another geometry to allow angular distribution studies to be made in relation with the polarization of the laser and the alignment of the excited states. But , in the excitation range where the cw dye lasers are nowadays tunable, the main limitations come from the low intensity of synchrotron radiation. Wide possibilities are open up for the future, since this intensity can be greatly enhanced through the insertion of undulators in the straight section of a storage ring.[19]

An undulator is a periodic array of N dipole magnets placed in such a section. The spatially alternating magnetic field causes the electrons to oscillate about their straight path. The emitted radiation is concentrated in a very small opening angle. Interference effects in the radiation produced by the same electron at essentially collinear source points result in a spectral brightness increased by N^2 and a spectrum with quasi-monochromatic peaks.[19] The emission wavelength of the fundamental is in the UV or the VUV region, depending on the electron energy and on the magnetic field in the undulator.

The emission of a permanent magnet undulator has been measured in the VUV spectral range with ACO operating at 536 MeV.[19] A spectrum of the harmonics 3 to 10 is shown in Fig.8. The gain in brightness is about 250 (for N=17).

Undulators can also be used to produce coherent radiation in the VUV. The generation of stimulated radiation from interaction of the circulating electron beam with the spatially oscillating magnetic field has recently been achieved, in the first successful operation of a storage ring-free electron laser in the visible.[6] To produce coherent radiation in the VUV range, it is also possible to bunch the electron beam on an optical wavelength scale, using a powerful external laser in connection with an optical klystron. Experiments of this type are underway in Orsay.[7]

In the new Super ACO (800 MeV, 400 mA, critical wavelength = 19 Å) positron storage ring being built in Orsay, 6 straight sections will be available for undulators. The combined use of a cw dye laser beam and of an undulator will allow large progress to be made in photoionization studies of excited atoms and molecules. In addition to make easier the measurements mentioned above, the following experiments may become feasible:

- measurements at lower atomic densities, allowing to study of metals more difficult to vaporize (e.g. transition metals, rare earths) and to reduce the intensity of low energy electrons produced by collisional ionization; this source of electrons could be a troublesome source of background masking the photoelectrons in the particularly interesting energy range just above the ionization threshold.
- measurements at lower laser powers, leading to an extension of the tunability range in which cw dye lasers are presently available; this means also a possibility to study other metals, requiring higher photon energies for the production of the first excited state, and to excite directly high lying excited states in the samples presently under investigation.
- production of excited atoms with an hole in an innershell by promoting, with the undulator beam, an inner electron onto some optical orbitals and the beam of a cw laser to create a second excitation or to photoionize the excited electron.

Moreover, a new class of experiments will be extensively developed, using an undulator beam in connection with a pulsed dye laser. The number of photons emitted in one pulse of a 30-pole undulator in Super ACO should be about 10^8 to 10^9 in a band width of 1%. Either the undulator beam or the pulsed laser could be used to prepare an excited system and the other to probe it as a function of time delay between both pulses. This will allow dynamics of the excited states to be studied. In addition, any combination of several pulsed lasers used in cascade would give access to almost any excited state, since pulsed lasers are, and will become more and more, tunable over a large energy range extending into the VUV region. Many new experiments will be feasible in atoms, but the largest field of investigation will be the excited states of molecules that are exclusively studied now via multiphoton excitation-ionization.

Ultimately, photoionization experiments on multiply charged positive ions may become possible. Two new ion sources have been built in France. In the Electron Cyclotron Resonance Ion Source (ECRIS),[20] the source plasma is located in a magnetic well allowing long exposure time to electron bombardment for efficient ionization; the electrons are accelerated by an oscillating electromagnetic wave at the frequency of the electronic cyclotron resonance. In the Electron Beam Ion Source (EBIS and CRYEBIS),[21] an intense beam of electrons is accelerated by potential differences and magnetically confined on the axis of a solenoid; they serve to successively ionize the ions electrostatically trapped by the electric field created by the space charge of the electron beam. With an ECR source, taking as an example Ar^{10+} ions, it seems possible to obtain, after extraction, in the source volume of an electron spectrometer, ion densities in the order of 10^9 cm^{-3}. Many metallic ions can also be produced with a relatively high current in the microampere region.[20] The mounting of either one (or both) sources in the immediate vicinity of the beam emitted by an undulator of Super ACO would open up a whole new field, completely unexplored, to the experimental investigation.

ACKNOWLEDGEMENTS

The success of these new experiments has been made possible by a close cooperation with J.M. Bizau, B. Carré, D. Ederer, J.L. Picqué, P. Koch, J.C. Keller, J.L. LeGouët, P. Dhez. I would like also to thank J.P. Briand for helpful discussions

REFERENCES

1. F. Wuilleumier, Atomic Physics, vol.7 , 1981, p.491-527.
2. F. Wuilleumier, X-ray and Inner-Shell Physics, ed. B. Crasemann (American Institute of Physics, Proceeding Series n°94, New-York, 1982) p. 615-632. (review of previous work on excited states).
3. J. Ganz, B. Lewandowski, A. Siegel, W. Bussert, H. Worbel, M.W. Ruf and H. Hotop, J. Phys. B 15, L485(1982).
4. J. Sugar, T.B. Lucatorto, T.J. McIlrath and A.W. Weiss, Opt. Lett. 4, 109(1979).
5. J.M. Bizau, F. Wuilleumier, P. Dhez, D. Ederer, J.L. LeGouët, J.L. Picqué and P. Koch, Laser Techniques for Extreme Ultraviolet Spectroscopy, ed. T.J. McIlrath and R.J. Freeman (American Institute of Physics, Proceeding Series n°90, New York, 1982) p.331-343.
6. M. Billardon, P. Elleaume, J.M. Ortega, C. Bazin, M. Bergher, M. Velghe, Y. Petroff, D. Deacon, K.Robinson and J.M. Madey, Phys. Rev. Lett. 51, 1652 (1983).
7. Y. Petroff, private communication.
8. P.F. Larsen, W.A.M. van Bers, J.M. Bizau, F. Wuilleumier, S. Krummacher, V. Schmidt and D. Ederer, Nucl. Instr. Meth.195,245(1982).
9. T.N. Chang and Y.S. Kim, J. Phys.B 15, L835 (1982).
10. J.M. Bizau, F. Wuilleumier, P. Dhez, D. Ederer, J.C. Keller, J.L. LeGouët, J.L. Picqué, P. Koch and B. Carré, to be published.
11. W.L. Wiese, M.W. Smith and B.M. Miles, NSRDS-NBS 22, vol.II, 1969.
12. S. Harris, paper in this volume.
13. J.L. LeGouët, J.L. Picqué, F. Wuilleumier, J.M. Bizau, P. Dhez, P. Koch and D. Ederer, Phys. Rev. Lett. 48, 600(1982).
14. B. Carré, J.M. Bizau, P. Dhez, D.L. Ederer, P. Gerard, J.C. Keller, P.M. Koch, J.L. LeGouët, J.L. Picqué, G. Spiess and F. Wuilleumier, submitted to Optics Communications, 1984.
15. J.M. Bizau, B. Carré, P. Dhez, D. Ederer, P. Gerard, J.C. Keller, P. Koch, J.L. LeGouët, J.L. Picqué, G. Wendin and F. Wuilleumier, XIII ICPEAC, Abtr. of Contributed Papers, ed. S. Eicher, W. Fritsch, I.V. Hertel, N. Stolterfoht, U. Wille, Berlin, 1983, p.27.
16. J.M. Bizau, B. Carré, P. Dhez, D. Ederer, P. Gerard, J.C. Keller, P. Koch, J.L. LeGouët, J.L. Picqué, G. Wendin and F. Wuilleumier, 15th EGAS, Madrid, Abstract of Contributed Papers, 1983, p.89.
17. J.M. Bizau , D. Cubaynes, P. Gerard et al., to be published.
18. G. Wendin, private communication.
19. M. Billardon, D.A.G. Deacon, P. Elleaume, J.M. Ortega, K.E. Robinson, G. Bazin, M. Bergher, J.M.J. Madey, Y. Petroff and M. Velghe, J. Physique 44, C1 (1983).
20. F. Bourg, R. Geller, B. Jacquot, J. Lamy, M. Pontonnier, J.C. Rocco, Nucl. Instr. Meth. 193, 325(1982).
21. J. Arianer et al. Nucl. Instr. Meth. 193, 401 (1982).

STATE-SELECTIVE VACUUM UV SPECTROSCOPY OF SMALL MOLECULES

C.R.Vidal, P.Klopotek and H.Scheingraber
MPI fuer Extraterrestrische Physik, 8046 Garching, FRG

ABSTRACT

State-selective spectroscopy of the CO and NO molecule is demonstrated using the methods of frequency selective excitation spectroscopy, of fluorescence spectroscopy, of vacuum uv visible two step excitation spectroscopy and of state selective life time measurements.

INTRODUCTION

Coherent vacuum uv sources based on the methods of four wave mixing in gases [1,2] have been developed in recent years to a point where these light sources have become attractive for laser spectroscopy in the vacuum uv and efforts have been started in several laboratories to explore the different possibilities of using these light sources under the aggravating boundary conditions of the vacuum uv spectral region.

For the purpose of analyzing unidentified molecular spectra of astrophysical interest we have initiated a program for testing novel techniques of state-selective spectroscopy in the vacuum uv. So far we succeeded to demonstrate four different techniques of state-selective spectroscopy:

(1) frequency selective excitation spectroscopy,
(2) fluorescence spectroscopy,
(3) vacuum uv visible two step excitation spectroscopy,
(4) state selective life time measurements.

These methods have been tested in performing spectroscopy of the CO and NO molecule yielding numerous new spectroscopic results.

THE LASER SYSTEM

The experiments to be presented have been carried out using an excimer laser and a nitrogen laser pumped dye laser system. Both systems generate coherent vacuum uv radiation with the method of two-photon resonant sum frequency mixing in Mg-Kr and Sr-Xe systems which are generated in a modified concentric heat pipe oven [3]. Both laser systems are designed to cover a spectral region from about 120 to 200 nm depending on the dyes used. The line width is typically 0.1 - 0.3 cm^{-1}. The number of photons are about 10^{11} to 10^{12} per shot at a repetition rate of up

0094-243X/84/1190233-13 $3.00

to 20 sec^{-1}. The wavelength calibration was done on the fundamental waves in the visible part of the spectrum using the iodine spectrum of Gerstenkorn and Luc [4]. As an example Fig.1 shows a small portion of the (1,0) band of the NO D-X electronic system together with the corresponding portion of the iodine spectrum. These spectra were taken in a simple two-photon resonant third harmonic experiment near the $5s^2\ ^1S$ - $5p^2\ ^1S$ two-photon resonance of strontium which requires the calibration of only one laser. The dispersion was determined from the iodine spectrum in a least squares fit. The standard error was typically 0.015 cm^{-1} in the vacuum uv which corresponds to an accuracy of 2 parts in 10^7.

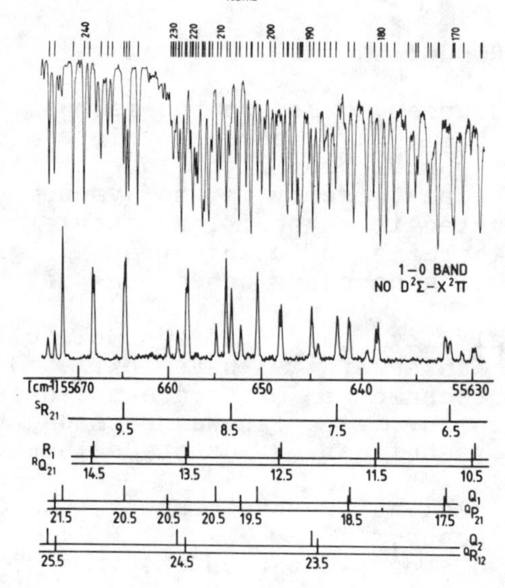

Fig.1: Excitation spectrum of the NO D-X electronic system showing a small portion of the (1,0) band. The spectrum was taken with the third harmonic of a laser, the fundamental wave of which is calibrated by the iodine spectrum shown above.

FREQUENCY SELECTIVE EXCITATION SPECTROSCOPY

In normal excitation spectroscopy individual states of a molecule are excited by a laser and a photodetector collects all the photons of the subsequent fluorescence. Spectra of this kind generally resemble absorption spectra if the quantum efficiency is similar for all excited states. This requires that no anomalies occur in the excited state such as a strong predissociation. The strongest absorption bands of the CO molecule in the spectral region from 110 to 160 nm are due to transitions of the A-X system and the upper half of Fig.2 shows an excitation spectrum of the (4,0) band for which the fluorescence occurs in the vacuum uv back down to the ground state. Spectra of this kind with a lower resolution have already been taken on the CO molecule [5,6] and on the NO molecule [7]. Simultaneously with the A-X system the intercombination bands can be excited in the same spectral region. These bands originate from a transition from the singlet ground state to one of the excited triplet states and they have a transition moment which is typically three orders of magnitude smaller than the transition moment of the allowed A-X system.

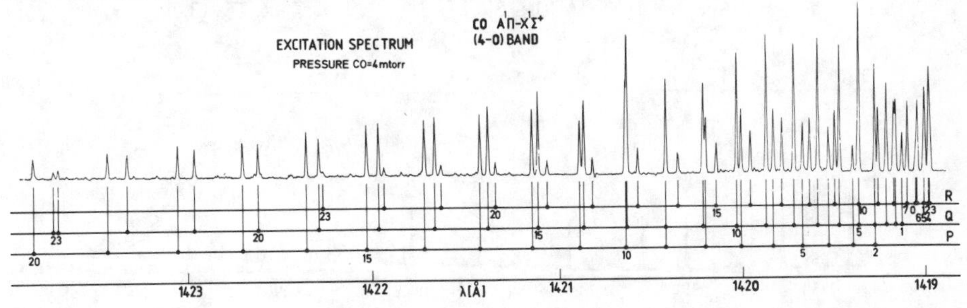

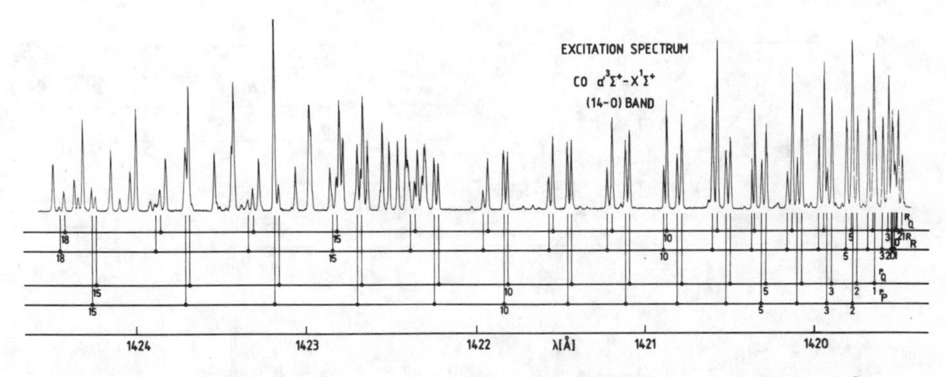

Fig.2: Excitation spectra of two overlapping bands in the CO spectrum. The upper half shows the (4,0) band of the A-X system obtained from the fluorescence in the vacuum uv. The lower half shows the underlying weak (14,0) band of the $a'^3\Sigma^+$ - $X^1\Sigma^+$ intercombination system which was obtained from the fluorescence in the infrared.

These intercombination bands have already been observed and identified by Herzberg and coworkers[8-11]. However several bands which are particularly interesting for an analysis of the perturbations, have not yet been observed because they overlap with one of the strong bands of the A-X system. Since the excited triplet states will predominantly fluoresce into the $a^3\Pi$ state, the corresponding fluorescence occurs primarily in the infrared. Using an infrared sensitive photomultiplier together with a cut off filter we have been able to completely separate the intercombination bands from the very much stronger bands of the A-X system. The lower half of Fig.2 shows the (14,0) band of the $a'^3\Sigma^+$ - $X^1\Sigma^+$ system which occurs in the same spectral region as the (4,0) band of the A-X system in the upper half of Fig.2 and which shows no trace of the allowed (4,0) band any more. In the same manner several other bands of the $a'^3\Sigma^+ - X^1\Sigma^+$, of the $e^3\Sigma^- - X^1\Sigma^+$ and of the $d^3\Delta - X^1\Sigma^+$ system have been analyzed and measured for the first time.

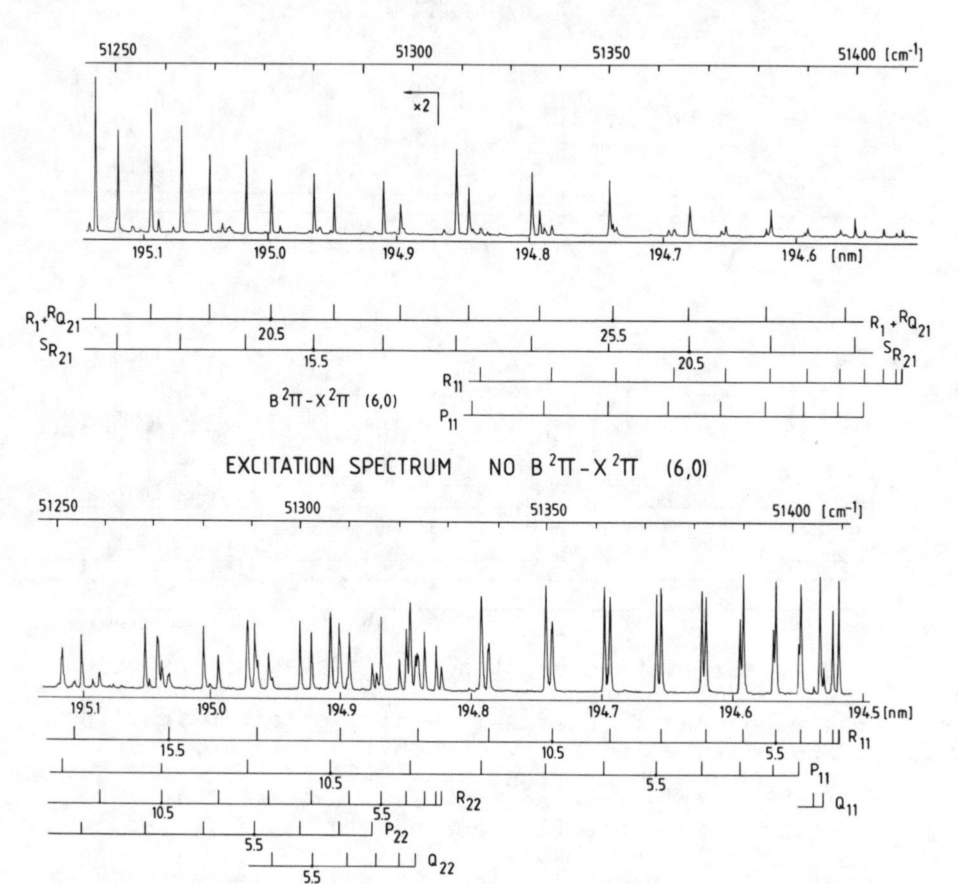

Fig.3: Small portions of the excitation spectra of two overlapping bands in the NO spectrum. The upper half shows the (3,0) band of the A-X system as obtained from the fluorescence of the (3,1) band. The lower half shows the underlying (6,0) band of the B-X system as obtained from the fluorescence of the (6,2) band.

Similar to the intercombination bands of the CO molecule we have also been able to separate overlapping bands of the NO molecule. Since in this case one has a situation in which the different electronic systems fluoresce in the same spectral region, we had to use a spectrometer which was alternately adjusted to one of the strongest vibrational bands of a particular system in a spectral region where the other system has no significant fluorescence. By looking at the (3,1) band of the A-X system and independently at the (6,2) band of the B-X system we have been able to separate the (3,0) band of the A-X system almost completely from the overlapping (6,0) band of the B-X system. Fig.3 shows a small portion of the two overlapping bands. In this manner our knowledge of the term values of the NO molecule can be significantly extended.

FLUORESCENCE SPECTROSCOPY

In the previous excitation spectra the laser was scanned and the photodetector collected the laser induced radiation in a particular spectral region. In case of the fluorescence spectroscopy the laser frequency is fixed to a particular molecular transition and the laser induced

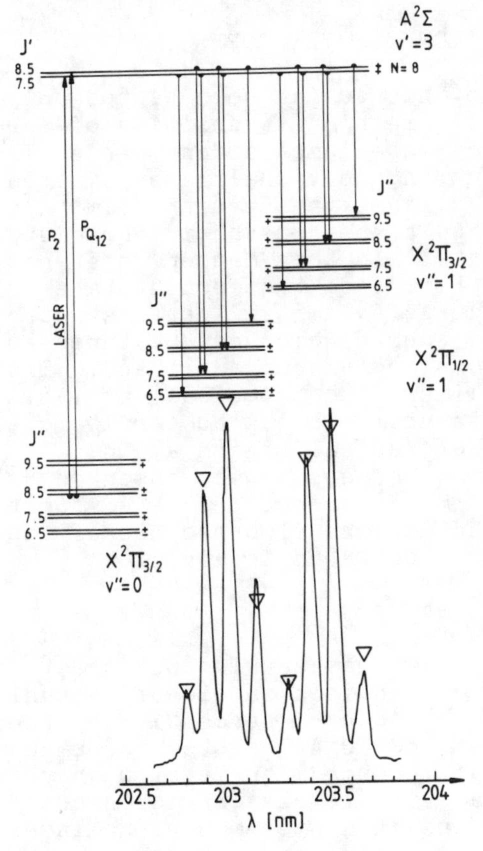

Fig.4: Fluorescence spectrum of the (3,1) band of the NO A-X system. The partial energy level diagram indicates the pumping and the fluorescence transitions. The open triangles are the results of calculations based on relations given by Earls [13].

fluorescnce is spectrally resolved. In this manner the Franck Condon factors of the NO A-X system [12] have already been measured. Similarly we have also taken fluorescence spectra of the C-X and the D-X system of the NO molecule and have determined the Franck Condon factors.

In high resolution scans of the laser induced fluorescence we have also been able to resolve the rotational structure and the fine structure splitting of individual vibrational bands. Fig.4 shows an example of the (3,1) band of the NO A-X system. The energy level diagram indicates the pumping transition and the fluorescence transitions. The open triangles show the results of calculations using relations given by Earls [13] which are based on a theory of Hill and Van Vleck [14]. The agreement is generally very good although we noticed small systematic discrepancies of so far unknown origin. The measurements show a strong J-dependence and the relative intensities clearly indicate the transition from Hund's coupling case (a) to case (b) with increasing values of J [15].

TWO STEP EXCITATION SPECTROSCOPY

This kind of spectroscopy has so far been tested only on the CO molecule. In this method a transition of the A-X system or of one of the intercombination bands is pumped in the vacuum uv. Using an additional laser excited by the same pump laser as the vacuum uv system, the molecule is excited in a second step from the intermediate state to a higher excited state. In this manner one has access to states whose absorption from the ground state is beyond the lithium fluoride cut off limit. The first step requires a coherent vacuum uv source which can be tuned to a particular line of a desired vibrational band. The second step greatly simplifies the spectral analysis because most spectroscopic measurements can be carried out in the more convenient visible spectral region again.

The two step excitation process can be observed by means of the laser induced fluorescence of the first pumping step and also by means of the fluorescence of the second pumping step. This is illustrated in Fig.5.

(1) Isotope shift of the $E^1\Pi$ -state

The experiment shown in Fig.5 was carried out on the $^{13}C\ ^{16}O$ isotope combination which in a sample of natural abundance has a concentration of about 1 percent. In the first step the (4,0) band of the CO A-X system is pumped on the P(2)-line populating an intermediate level with v=4 and J=1. In the second step a visible laser is scanned across the (0,4) band of the E-A system and a laser induced fluorescence (LIF) is observed whenever the visible laser hits a transition from the intermediate level. For a $^1\Pi - {}^1\Pi$ transition one generally expects three lines, a P-, a Q- and an R-line. Since for J=1 the P-line cannot exist only two lines show up in Fig.5 demonstrating immediately the isotope shift of the band origin. Simultaneously we observed the laser induced fluorescence due to the initial pumping process of the vacuum uv laser. Since the laser is fixed on a particular transition, the amplitude of the fluorescence is constant until the visible laser hits in the second pumping step a transition from the intermediate level which reduces the population density and hence also the fluorescence associated with this level. As a result a laser reduced fluorescence (LRF) can be observed which is shown on the top of Fig.5. The experiment in Fig.5 also illustrates the sensitivity of the method because it was carried out on a weak transition of a rare isotope combination.

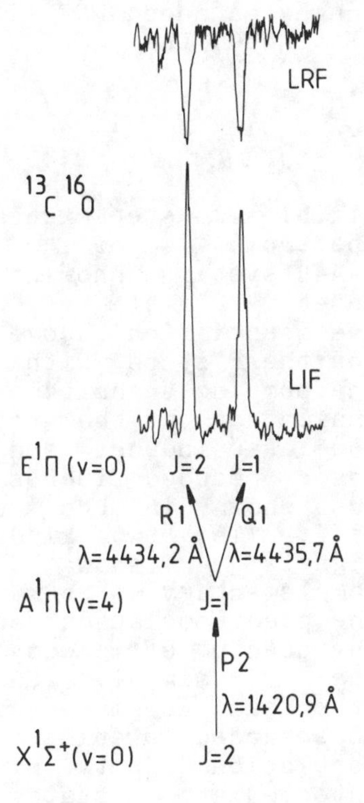

Fig.5: Two step excitation spectrum of the X-A-E system of the $^{13}C\,^{16}O$ isotope combination. The pumping transitions employed are shown on the bottom. The laser reduced fluorescence from the intermediate state and the laser induced fluorescence of the final state are shown in the upper half.

(2) Predissociation of the $B^1\Sigma^+$ and the $C^1\Sigma^+$ states

A similar double resonance experiment revealing a strong predissociation is shown in Fig.6 where the (2,0) band of the CO A-X system was pumped by the vaccum uv laser either on the R(16)- or on the R(17)-line. In the second step the visible laser is scanned across the (1,2) band of the

B-A system. In this case for a $^1\Sigma$ - $^1\Pi$ transition one expects either two lines to show up, a P- and an R-line, or one line, a Q-line. In Fig.6 we note that for both initial pumping transitions the LIF signal of the R-line is very much weaker than the LIF signal of the P-line. This is a clear manifestation of the predissociation of the B state which occurs for v=1 above J=17. The predissociation drastically reduces the quantum efficiency of the B state and hence the amplitude of the LIF signal. The LRF signal due to the first pumping laser, however, shows the opposite behavior. Below the predissociation limit the LRF signal is rather small because the second pump transition is quickly saturated. Above the threshold the population density of the upper level in the B state is reduced so fast due to predissociation that the visible transition does not saturate as easily and the population density of the intermediate level can be reduced more effectively giving rise to a strongly enhanced LRF signal. In this manner the LIF signal and the LRF signal supplement each other.

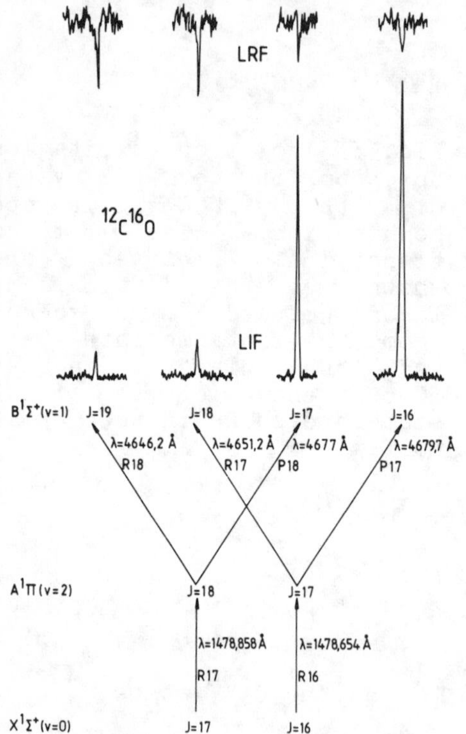

Fig.6: Two step excitation spectrum of the X-A-B system showing the onset of the predissociation above J=17 for the B state with v=1. The pumping transitions are indicated on the bottom. The laser induced and the laser reduced fluorescence are shown in the upper half. The two kinds of signals show a complementary behavior. The predissociation reduces the quantum efficiency for the LIF signal. On the other hand it gives rise to an enhanced lowering of the population density of the intermediate state and hence causes an increased LRF signal.

Using the same double resonance technique we also investigated a possible predissociation of the $C^1\Sigma^+$ state which turned out to be neglibible and which did not show the behavior observed by Schmid and Geroe [16] supporting instead results of Tilford and Vanderslice [17].

(3) Collision induced transitions in the $A^1\Pi$ state

In another double resonance experiment where a similar two-step excitation was performed as in Fig.6, we were able to study collision induced transitions in the A state. By pumping in the first step either through the P(3) or the Q(2) line of the (3,0) band of the A-X system, a significant population density was alternately established in either one of the two Λ-components of the intermediate level with J=2 which differ with respect to the parity. At low pressures of a few mTorr very simple double resonance spectra on the (0,3) band of the B-A system are obtained which are shown in the upper half of Fig.7 and which consist either of a Q-line or of a P- and R-line.

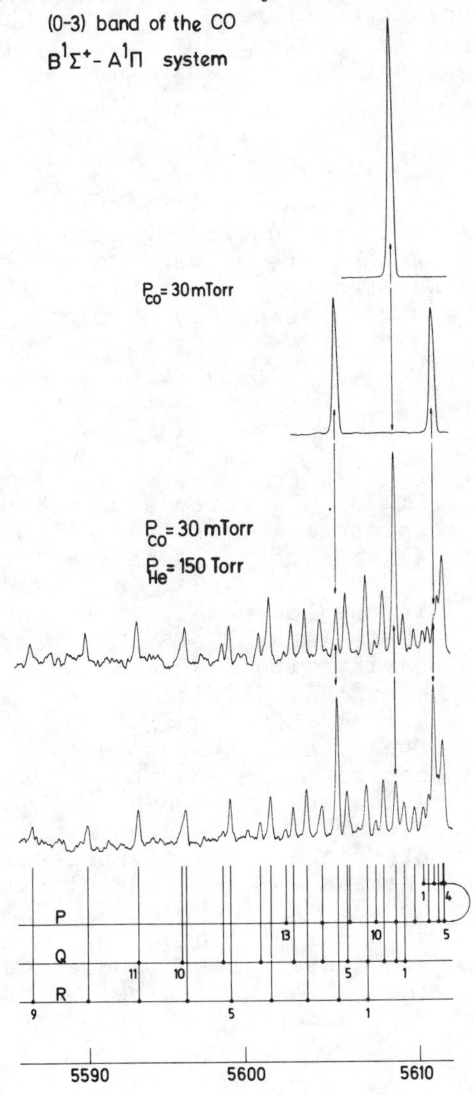

Fig.7: Two step excitation of the X-A-B system for the low pressure, collision free case (upper half) and for the high pressure case (lower half) showing collision induced satellite lines.

Adding typically 150 Torr of helium additional lines are observed as shown in the lower half of Fig.7. These lines originate from collision induced transitions in the A state which populate rovibronic levels in the vicinity of the initial laser populated level. From measurements of this kind the relative population densities of the intermediate state are obtained which provide collisional cross sections after solving the rate equations associated with the different levels of the A state. An analysis of this kind is shown, for example, in reference 18 for the Li_2 molecule. Using then H_2 as a collision partner dramatically different spectra were observed which show a different structure and an underlying continuum and which are not yet understood. We suspect that in this case some photochemistry is going on.

(4) Perturbations of the $E\,^1\Pi$ state

In a different study we have investigated perturbations of the $E\,^1\Pi$ state. By pumping through the R(29) line and the Q(30) line of the (4,0) band of the A-X system, the two Λ-components of opposite parity of the level with J=30 in the A state were alternately populated.

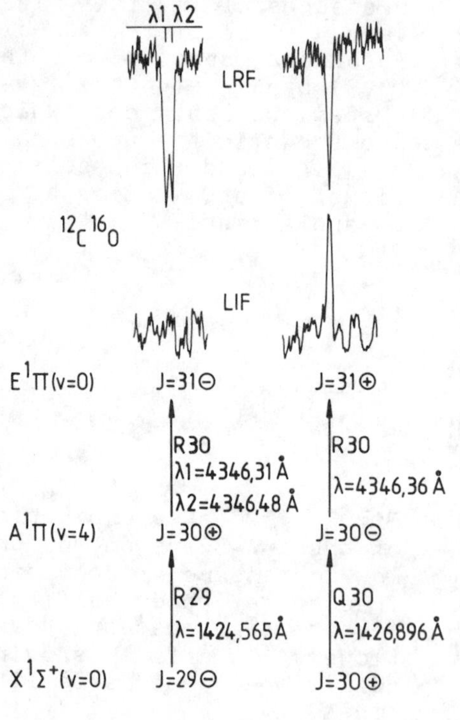

Fig.8: Two step excitation spectrum of the X-A-E system. The pumping transitions are indicated in the lower half. Pumping the level of negative parity with J=30 results in a normal LIF signal and a normal LRF signal. Pumping the other Λ-component of opposite parity an anomalous spectrum is observed where the LIF signal disappears, whereas the LRF signal shows a doublet. The spectrum indicates an accidental predissociation into a state of $^1\Sigma^+$ symmetry.

In a two step excitation experiment shown in Fig.8 the laser induced fluorescence which results from a pumping process from the intermediate level of negative parity, shows a normal behavior. Pumping from the other parity level a very different spectrum is observed. The LIF signal disappears indicating a strong predissociation of the E state. The LRF signal, on the other hand, shows a splitting where the two lines indicate a transition to a perturbed and a perturbing level. Because of the selection rules the perturbing state has to have a $^1\Sigma^+$ symmetry. This is a nice example of an "accidental predissociation" as first reported by Simmons et al.[19] and by Tilford et al.[20].

(5) Fine structure of the $c^3\Pi$ state

As a last example demonstrating the power of the two step excitation spectroscopy Fig.9 shows a double resonance experiment in which the vacuum uv system pumps in the first step a line of an intercombination band. In this manner a level with v=14 and K=1 of the $a'\,^3\Sigma^+$ state is populated. In the second excitation step a transition into the $c^3\Pi$ state is pumped and the LIF signal shows for the first time the fine structure splitting of the

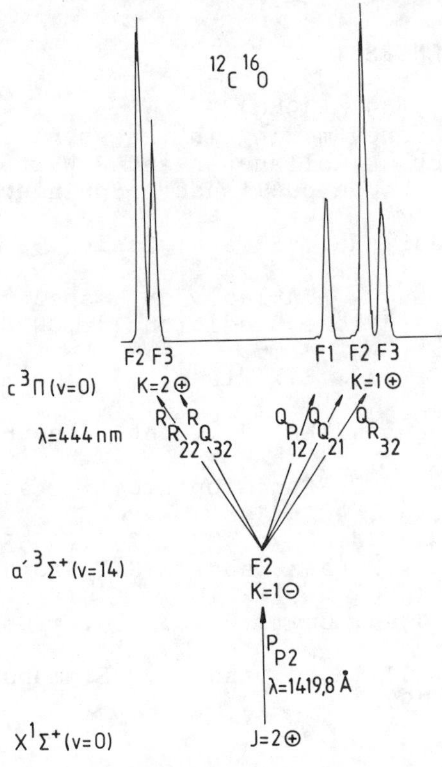

Fig.9: Two step excitation of the X-a'-c system showing the fine structure splitting of the $c^3\Pi$ state in the LIF signal. The lower half shows schematically the different transitions used in the pumping scheme.

c $^3\Pi$ state. The K value was selected to obtain the largest possible fine structure splitting and a detailed analysis[21] gave a fine structure constant A=1.49 cm^{-1}.

The preceding experiments have in some cases been repeated on the rare isotope combination $^{13}C\,^{16}O$. Together with two step excitation experiments employing the intercombination bands, the large sensitivity and the power of the vacuum uv visible double resonance spectroscopy has been demonstrated.

LIFE TIME MEASUREMENTS

Life time measurements on individual rotational levels of the (0,0) band of the CO A-X system were first reported by Provorov et al.[5]. They show a strong J-dependence of the fluorescence life times due to strong singlet triplet perturbations. We have carried out complementary life time measurements on individual rovibronic levels of the excited triplet states. In this case the life times are rather large and perturbations have significantly more dramatic effects on the life times of individual rovibronic levels. Our results differ significantly from recent measurements of Shadfar et al.[22] who used pulsed low energy electron impact excitation.

REFERENCES

1. C.R.Vidal: Appl. Opt. 19 , 3897 (1980).
2. C.R.Vidal, "Four wave frequency mixing in gases," in Tunable Lasers edited by L.Mollenauer and J.White, Topics in Applied Physics, to be published, Springer, Heidelberg, 1984.
3. H.Scheingraber and C.R.Vidal, Rev. Scient. Instr. 52 , 1010 (1981)
4. S.Gerstenkorn and P.Luc, "Atlas du spectre d'absorption de la molecule d'iode," editions du CNRS, Paris (1978)
5. A.C.Provorov, B.P.Stoicheff and S.Wallace, J. Chem. Phys. 67 , 5393 (1977)
6. R.Hilbig and R.Wallenstein, IEEE J. Quant. Electr. QE-17 , 1566 (1981)
7. J.R.Banic, R.H.Lipson, T.Efthimiopoulos and B.P.Stoicheff, Opt. Lett. 6 , 461 (1981)
8. G.Herzberg and T.J.Hugo, Can. J. Phys. 33 , 757 (1955)
9. G.Herzberg, J.D.Simmons, A.M.Bass and S.G.Tilford, Can. J. Phys. 44 , 3039 (1966)
10. J.D.Simmons and S.G.Tilford, J. Chem. Phys. 45 , 2965 (1966)
11. G.Herzberg, T.J.Hugo, S.G.Tilford and J.D.Simmons, Can. J. Phys. 48 , 3004 (1970)

12. H.Scheingraber and C.R.Vidal, AIP Conf. Proc. 90 , 95 (1982)
13. L.T.Earls, Phys. Rev. 48 , 423 (1935)
14. E.Hill and J.H.Van Vleck, Phys. Rev. 32 , 250 (1928)
15. J.T.Hougen, "The Calculation of rotational line intensities in diatomic molecules," NBS Monograph 115 (1970)
16. R.Schmid and L.Geroe, Z. Phys. 96 , 546 (1935)
17. S.G.Tilford and J.T.Vanderslice, J. Mol. Spectr. 26 , 419 (1968)
18. C.R.Vidal, Chem. Phys. 35 , 215 (1978)
19. J.D.Simmons and S.G.Tilford, J. Mol. Spectrosc. 53 , 436 (1974)
20. S.G.Tilford, J.T.Vanderslice and P.G.Wilkinsonm, Can. J. Phys. 43 , 450 (1965)
21. P.Klopotek and C.R.Vidal, J. Chem. Phys. to be published
22. S.Shadfar, S.R.Lorentz, W.C.Paske and D.E.Golden, J. Chem. Phys. 76 , 5838 (1982)

^{81}Kr DETECTION USING RESONANCE IONIZATION SPECTROSCOPY

S. D. Kramer, C. H. Chen, S. L. Allman, and G. S. Hurst
Oak Ridge National Laboratory, Oak Ridge, Tennessee 37831

B. E. Lehmann
University of Bern, Bern, Switzerland

ABSTRACT

A resonance ionization ion source, in conjunction with a mass spectrometer, was used to count 1000 individual ^{81}Kr atoms. In this method, a vacuum ultraviolet beam was used to resonantly excite krypton atoms in order to obtain atomic species selectivity and a small quadrupole mass spectrometer provided isotopic selectivity. After selection, the krypton ions were counted as they were implanted into a target. The method will also work for other isotopes.

INTRODUCTION

Resonance ionization spectroscopy (RIS) has proved to be a very sensitive atomic and molecular technique[1,2] The use of a resonance ionization source in a mass spectrometer can provide selective detection based on both atomic weight and number.[3-6] Recently, we have developed a facility using RIS and a small mass spectrometer for direct counting of small numbers of krypton atoms of a selected isotope.[7] In this work, we demonstrate that 1000 atoms of ^{81}Kr in a sample can be directly counted. For long-lived isotopes, such as ^{81}Kr which has a 200,000-yr half-life, this is a substantial improvement over radioactive counting techniques. The detection of a small number of ^{81}Kr atoms (~1000) is very important for groundwater and polar ice cap dating,[8] as well as for solar neutrino research.[9] The detection of small samples of ^{82}Kr atoms can be used in the study of double beta decay.

EXPERIMENTAL PROCEDURE

The experimental detector chamber is shown in Fig. 1. The method utilizes a pulsed coherent light source for resonantly ionizing krypton atoms and a quadrupole mass spectrometer for isotope separation.[2,8,10] An "atom buncher" was developed to enhance the probability that krypton atoms will be in the ionization light beam at the time of the light pulse. After krypton atoms have been ionized and isotopically selected, the ions are accelerated through a potential of 10 kV and implanted into a Be-Cu target. An electron multiplier is used to count each implanted ion by detecting the secondary electrons emitted by the target. The krypton density is kept low enough so that each electron multiplier pulse corresponds to the implantation of a single ion. Since the implanted krypton atoms will stay in the room-temperature Be-Cu target indefinitely, every isotopically selected krypton atom will eventually be ionized, implanted, and counted. The number of

0094-243X/84/1190246-07 $3.00

isotopically selected krypton atoms can then be determined by summing all of the counts.

The laser scheme[11] for excitation and ionization of krypton is shown in Fig. 2. In the RIS scheme 116.5 nm radiation (500 nJ/pulse)

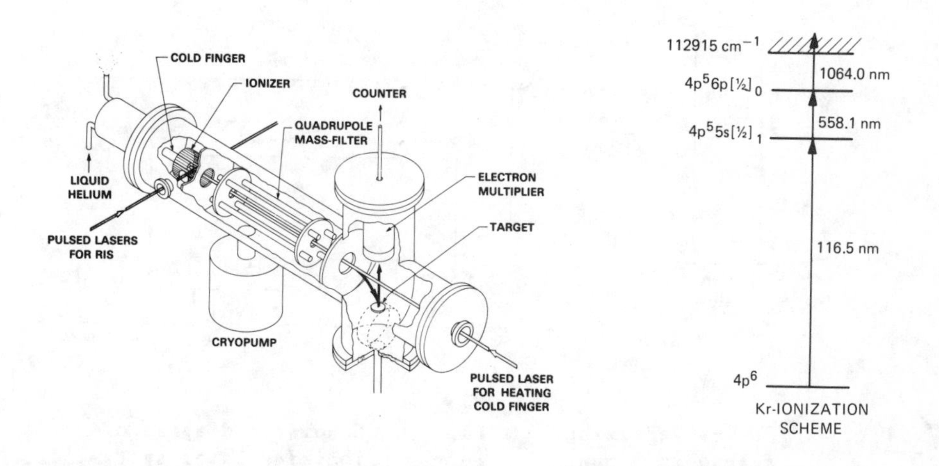

Fig. 1. Noble gas atom detection chamber.

Fig. 2. Krypton ionization scheme.

excites krypton from the ground state to the $5s'[1/2]_1$ state. Then photons at 558.1 nm (2 mJ/pulse) further excite it to the $6p[1/2]_0$ state that can be ionized with the 1.06-µm fundamental radiation (70 mJ/pulse) of a Nd:YAG laser.

Figure 3 shows the four-wave sum mixing process in xenon which was used to produce the required VUV light. Figure 4 shows the way the 0.2 mJ/pulse input beams at 252.5 nm and 1507 nm which were used to produce the VUV light were generated. As shown here, the 200-mJ/pulse, second-harmonic output of a 10-Hz, Q-switched Nd:YAG laser (Quanta-Ray DCR-1A) was split in half and used to pump two dye lasers. The wavelength at 252.5 nm, which was tuned to excite the $5p^5 6p[3/2]_2$ two-photon allowed transition in xenon, was generated by mixing the doubled output of one visible dye laser (Quanta-Ray PDL) with the residual 1.06-µm pump laser output. The second visible dye laser (Quanta-Ray PDL) pumped a hydrogen Raman cell whose second Stokes-shifted output produced the 1507-nm radiation. The input beams were focused with separate 40-cm focal length lenses and made coaxial by use of a dichroic beam splitter before entering the 15-cm long xenon gas cell where VUV generation occurred. The VUV output was optimized by using argon to achieve phase matching. A separate nitric oxide ionization chamber was used to monitor the VUV output. With a xenon pressure of 40 Torr and an argon pressure of 364 Torr, a VUV output of 500 nJ at 116.5 nm was produced. The calculated VUV bandwidth is about 1.5 cm^{-1}. To avoid self absorption by krypton in the generation cell, high purity gases (<2 ppm krypton) were used.

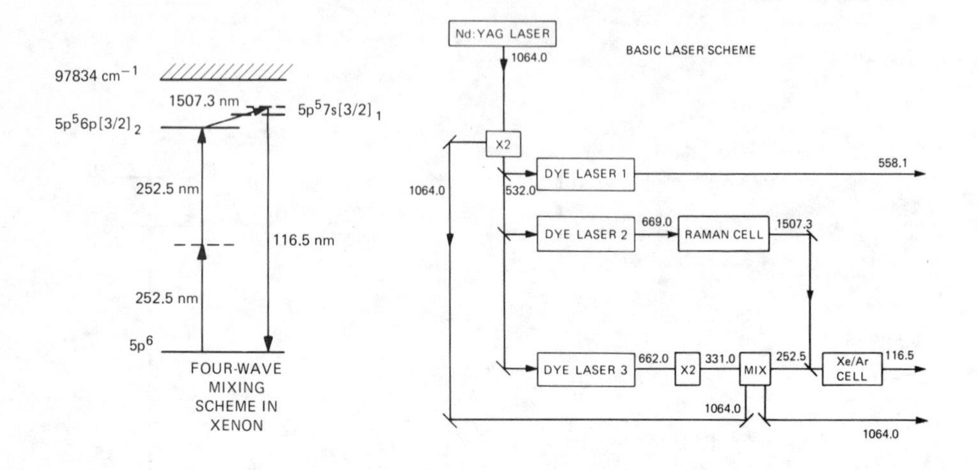

Fig. 3. Four-wave mixing scheme in xenon.

Fig. 4. Schematic diagram of ionizing laser system.

After passing through a LiF window, the VUV light and the other detector laser beams entered the actual krypton detection chamber where they passed through the ionization region of a quadrupole mass spectrometer (Extranuclear Corporation). It was not necessary to control the phase relationship between the laser pulses and the mass spectrometer radio frequency field.[11] Ions created in this region were injected into the spectrometer. Detuning the lasers provided verification that the krypton was the species being ionized. By putting a known amount of krypton into the 4-liter detector chamber, it was found that the detection probability was 5×10^{-8}/pulse. With this detection efficiency and a laser repetition rate of 10 Hz, about 500 h would be required to count all of the atoms of the selected isotope in the chamber.

To reduce the time required to ionize all of the krypton atoms, a cold finger, atom buncher[12] was developed. The atom buncher was designed so that krypton atoms could be condensed on a well-defined 0.4-cm diameter cold (15 K) spot. The trapped atoms were released by heating the cold spot with a 1-μs long evaporation laser pulse produced by a flashlamp pumped dye laser.[13] With a 7-μs time delay between this evaporation laser pulse and the ionization laser beams, the probability of having a krypton atom in the ionization laser beam can be enhanced dramatically. In this time the krypton atoms have traveled about 0.1 cm from the surface of the cold finger. The 0.1-cm diameter ionization detection laser beams are positioned so that they intercept the krypton atoms at this point. With this system, ~1% of the krypton on the cold finger can be ionized and detected by a single laser pulse.

The buncher surface cools back to ~15 K within 100 μsec. Thus, any krypton atoms that were not ionized by the laser beams return to the cold spot with a characteristic recurrence time of about 1 min and remain there until the next laser pulse. With a 10-Hz laser repetition rate, the overall detection probability in one laser pulse was measured to be 3×10^{-5}. This is the probability of detecting a single krypton atom in the chamber with a single laser pulse. Thus, the buncher used at 10 Hz increases by a factor of 600 the gas density in the interaction region. A slower repetition rate would increase it even more. So, for certain applications, such as VUV generation, an atom buncher might be used in the same manner as a pulsed nozzle jet. By using the buncher the time to count all the isotopically selected krypton atoms in the chamber is reduced to about 1 h.

Since the krypton atoms must be counted in a closed chamber to avoid pumping of the krypton, vacuum conditions are critical. During the counting period the chamber pressure must be kept below 10^{-5} Torr to operate the quadrupole mass filter and the ion detector. The outgassing rate of krypton has to be kept extremely low ($<10^3$ Kr/s) for accurate counting of 1000 atoms of ^{81}Kr because the abundance sensitivity of the mass spectrometer for adjacent masses is only about 10^4. In order to achieve this outgassing level, the chamber was roughed down to a few microns by a liquid nitrogen sorption pump and subsequently pumped by a closed-cycle helium cryopump to obtain a pressure of about 10^{-8} Torr. A thermostatically controlled oven was used to bake the chamber at 250°C while pumping for two to three days. After baking, the chamber was cooled down and evacuated by a cryopump for another two days. The total chamber pressure then achieved was ~10^{-10} Torr. When counting atoms, the main chamber is isolated from the cryopump and is pumped only with a titanium-zirconium getter which does not remove noble gases. The total outgassing rate of the most abundant krypton isotope, ^{84}Kr, was measured to be ~30 atoms/sec, and the total pressure rise in one week in the isolated chamber was less than 10^{-7} Torr.

A sample containing enriched ^{81}Kr was obtained from the National Bureau of Standards.[14] After mixing the sample with helium, 1000 atoms (±20%) of ^{81}Kr, about 5×10^4 atoms of other stable krypton isotopes, and 1×10^{10} helium atoms were placed in the chamber. Figure 5 shows the experimental data obtained when counting the 1000 atoms of ^{81}Kr after subtraction of a small background signal. The experimental conditions were such that the probability of ionizing and counting one ^{81}Kr atom by a single laser pulse was less than 5% even at the beginning of the run. This procedure insured digital counting in which each atom is counted one at a time. The exponential decrease in the signal with time is consistent with implantation of the ^{81}Kr in the target. Checks of the ^{80}Kr signal before and after the ^{81}Kr implantation revealed, as expected, no change in the ^{80}Kr signal. The results in Fig. 5 suggest that ~2100 ^{81}Kr atoms were counted. The number obtained is about a factor of 2 higher than the number of ^{81}Kr atoms initially in the chamber. This overestimate is due primarily to the low sticking coefficient of the imperfectly activated Be-Cu target which was measured to be 0.6 for 10-kV krypton ions.[15] A calibration at the 10,000-atom level using ^{78}Kr from natural krypton gas was made. The number of counts was ~23,000, which is also about a factor of 2 larger than the number of ^{78}Kr atoms put into the chamber. Our estimated detection limit is about 300 ^{81}Kr atoms. The accuracy at the 1000-atom level is about 30%.

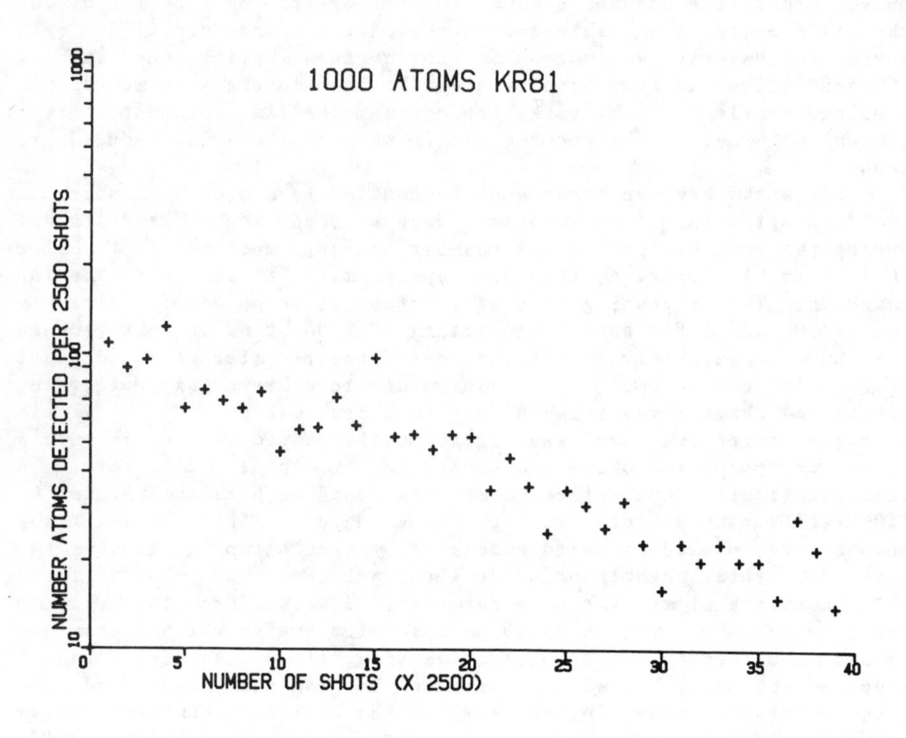

Fig. 5. Data obtained with a sample of 1000 atoms of ^{81}Kr.

CONCLUSION

We have shown that 10^3 atoms of ^{81}Kr can be counted in a sample containing a large number of other species. This accomplishment differs significantly from earlier atom counting.[16,17] In the present method, atoms of a selected type are ionized, removed from the sample, and detected until all atoms have been counted individually. Even in a volume of several liters, this sorting and counting process can be accomplished in a few hours.

Developments are planned which should significantly improve the abundance sensitivity and decrease the counting time. Other work is in progress for isotopically preenriching the samples before they are introduced to the apparatus described above.[15,18] This technique can be used to count ultralow-levels of isotopically selected xenon atoms.[19] Provided that light outputs of about 100 nJ/pulse at 107 nm can be generated for one-photon excitation of argon, isotopically selected argon atoms could also be counted.

The authors acknowledge the valuable help of M. G. Payne, R. D. Willis, and R. C. Phillips. This research sponsored by the Office of Health and Environmental Research, U.S. Department of Energy under contract W-7405-eng-26 with the Union Carbide Corporation and by the Nationale Genossenschaft fur die Lagerung radioaktiver Abfalle (Switzerland).

REFERENCES

1. C. H. Chen, G. S. Hurst, and M. G. Payne, in PROGRESS IN ATOMIC SPECTROSCOPY, edited by H. F. Beyer and H. Kleinpoppen (Plenum Press, New York, 1984), pp. 115-150.

2. G. S. Hurst, M. G. Payne, S. D. Kramer, and J. P. Young, Rev. Mod. Phys. 51, 767 (1979).

3. S. D. Kramer, G. S. Hurst, J. P. Young, M. G. Payne, M. K. Kopp, T. A. Callcott, E. T. Arakawa, and D. W. Beekman, Radiocarbon 22, 428 (1980).

4. D. L. Donohue, J. P. Young, and D. H. Smith, Int. J. Mass Spectrom. Ion Phys. 43, 293 (1982).

5. C. M. Miller, N. S. Nogar, A. J. Gancarz, and W. R. Shield, Anal. Chem. 54, 2377 (1982).

6. J. D. Fassett, J. C. Travis, L. J. Moore, and F. E. Lytle, Anal. Chem. 55, 765 (1983).

7. C. H. Chen, S. D. Kramer, S. L. Allman, and G. S. Hurst, Appl. Phys. Lett. (in press).

8. Private communication with H. Oeschger, H. Loosli, and B. E. Lehmann at the University of Bern (Switzerland).

9. G. S. Hurst, C. H. Chen, S. D. Kramer, M. G. Payne, and R. D. Willis, in SCIENCE UNDERGROUND, edited by Michael Martin Nieto, W. C. Haxton, C. M. Hoffman, E. W. Kolb, V. D. Sandberg, and J. W. Toevs (AIP Conference Proceedings No. 96, American Institute of Physics, New York, 1983), pp. 96-104.

10. G. S. Hurst, M. G. Payne, S. D. Kramer, and C. H. Chen, Phys. Today 33(9), 24 (1980).

11. S. D. Kramer, C. H. Chen, M. G. Payne, and G. S. Hurst, Appl. Optics 22, 3271 (1983).

12. G. S. Hurst, M. G. Payne, R. C. Phillips, J.W.T. Dabbs, and B. E. Lehmann, "Development of an Atom Buncher," J. Appl. Phys. (in press).

13. J. F. Ready, J. Appl. Phys. 36, 462 (1965).

14. This sample was prepared by a gaseous diffusion step followed by neutron activation of ^{80}Kr. Thus, the ratio of $^{81}Kr/^{80}Kr$ was about 0.03, and the only other major isotope was ^{78}Kr in the ratio $^{78}Kr/^{80}Kr = 0.7$. (Courtesy of Dr. F. J. Schima, National Bureau of Standards.)

15. C. H. Chen, R. D. Willis, and G. S. Hurst, "Enrichment of Rare Gas Isotopes Using a Quadrupole Mass Spectrometer," Vacuum (in press).

16. G. S. Hurst, M. H. Nayfeh, and J. P. Young, Appl. Phys. Lett. 30, 229 (1977).

17. S. D. Kramer, C. E. Bemis, Jr., J. P. Young, and G. S. Hurst, Opt. Lett. 3, 16 (1978).

18. R. D. Willis, S. L. Allman, C. H. Chen, G. D. Alton, and G. S. Hurst, Vac. Sci. Technol. (in press).

19. C. H. Chen, G. S. Hurst, and M. G. Payne, Chem. Phys. Lett. 75, 473 (1980).

SPECTROSCOPIC STUDY OF Xe_2 USING VUV LASER EXCITATION*

R. H. Lipson, P. E. LaRocque, and B. P. Stoicheff
Department of Physics, University of Toronto
Toronto, Ontario M5S 1A7, Canada.

ABSTRACT

Fluorescence excitation of Xe_2 at 149.0 nm has been observed using a pulsed supersonic jet for dimer formation and cooling, and tunable coherent VUV radiation (from 4-wave mixing in Mg) for excitation. The resolved vibronic structure due to the many isotopic Xe_2 dimers yielded unambiguous vibrational numbering (v = 37 to 46) of the vibrational levels. Analysis of the spectrum gave values of T_e = 63,802(6) cm^{-1}, ω_e = 124.5(3) cm^{-1}, $\omega_e x_e$ = 0.934(3) cm^{-1}, and D_e = 4,243(6) cm^{-1}, for the constants of the first excited state. Preliminary investigations are underway of a second band system of Xe_2 at 129.5 nm using VUV radiation from 4-wave mixing in Zn vapor, and of a Kr_2 band system at 124.0 nm using Hg vapor for VUV generation.

INTRODUCTION

The electronic spectra of rare gas dimers have been a subject of interest for many years, mainly because these dimers are model systems for studying Van der Waals interactions, and because of their potential as media for VUV and XUV lasers. Many absorption and some emission and fluorescence band systems have been investigated, and a great deal of information has been derived about ground state molecular constants and potential curves. Much of what is known spectroscopically about Xe_2 to-day is based on the experimental work of Freeman, Yoshino, and Tanaka[1], and of Castex and her colleagues[2-4], as well as on the theoretical work of Mulliken[5]. The ground state molecular constants are accurately known[1], and good estimates of the constants and potential curves of the lowest excited states have been obtained[3] from a thorough quantitative study[4] of absorption by Xe_2 dimers in the wings of the atomic Xe resonance lines at 147 and 130 nm. Here, we wish to report on recent results obtained from fluorescence excitation spectra of Xe_2 in the same wavelength region. These new spectra have permitted the unambiguous numbering of upper-state vibrational levels, and therefore, the evaluation of spectroscopic constants for the first excited state (0_u^+).

EXPERIMENT

Two experimental techniques were combined in our laboratory for this investigation: four-wave sum-mixing (4-WSM) to generate tunable

*Research supported by the Natural Sciences and Engineering Research Council of Canada, and the University of Toronto.

0094-243X/84/1190253-06 $3.00

and monochromatic VUV radiation[6], and a pulsed supersonic jet to produce large numbers of rotationally and vibrationally cold dimer molecules[7]. A schematic diagram of the experimental apparatus is shown in Fig. 1. The XeCl excimer laser (Lumonics TE-861M) with power output of 6 MW in a 7 ns pulse, pumped two dye-oscillator-amplifier systems. These had outputs of ∿30 KW in a 0.1 cm^{-1} line-width. The two beams were spatially overlapped in a Glan-Thompson prism, and focussed into a vertical heat pipe[8] containing either Mg or Zn vapor as the nonlinear medium. As shown earlier in this laboratory, with Mg vapor[9,10], VUV radiation having ∿10^8 photons/pulse in a 0.2 cm^{-1} linewidth is readily generated over the range 174 to 135 nm, and with Zn vapor[10,11] ∿10^7 photons/pulse in a line-width of ∿0.5 cm^{-1}, from 140 to 106 nm. The pulsed supersonic jet (Lasertechnics - LPV valve) was synchronized to the excimer laser at 10 Hz. When triggered, the jet valve emitted a 300 μs gas pulse through a 1 mm diameter pinhole. Typically, a back pressure of ∿5 atm was used with a 5% Xe-He mixture. The gas expanded into a small vacuum chamber which was kept at ∿10^{-4} torr by a 650 ℓ/s diffusion pump.

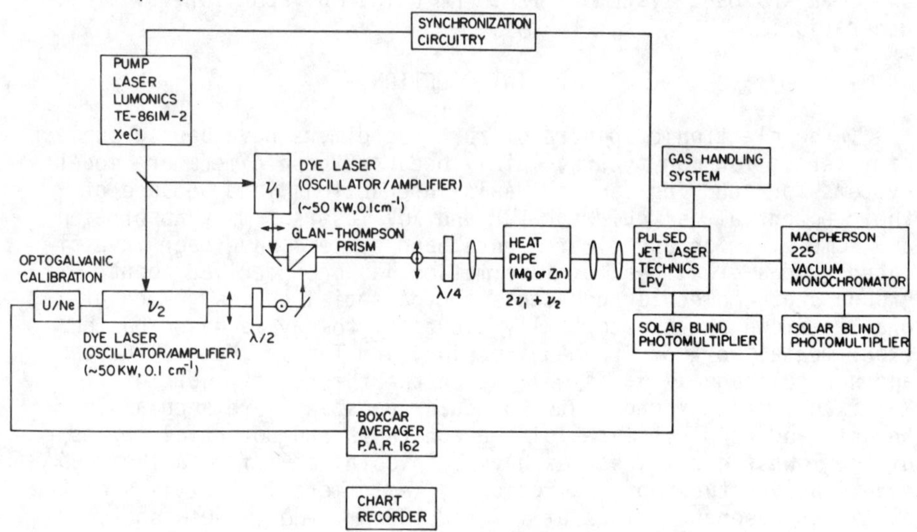

Fig. 1. Experimental arrangement for generation of tunable coherent VUV radiation, and for fluorescence excitation of Xe_2 dimers produced in a pulsed supersonic jet.

Fluorescence was excited by focussing the tunable, coherent VUV beam, on axis, ∿10 mm from the pinhole. The fluorescence radiation was collected by a LiF lens (f = 3 cm, d = 2 cm) and detected by a solar-blind photomultiplier (EMR-510G-08-13). The signal was processed by an integrator/boxcar averager (PAR M164/M62) and recorded. An estimate of the system sensitivity from the observed signal strength indicated that ∿10^5 Xe_2 molecules were excited per pulse. Wavelength calibration of the observed spectrum was made by exciting simultaneously the optogalvanic spectrum of uranium[12], in a

U-Ne hollow cathode lamp (Westinghouse WL-36077), with a small portion of the tunable dye-laser radiation used in the 4-WSM generation. The optogalvanic signal was also averaged by the integrator/boxcar electronics (PAR M165/M62) and displayed with the Xe_2 spectrum on a two-pen chart recorder. In this way it was possible to determine an unknown VUV wavelength to an accuracy of 2 parts in 10^6 (corresponding to 0.15 cm^{-1} at 70,000 cm^{-1}).

RESULTS AND DISCUSSION

In Fig. 2 is shown the fluorescence excitation spectrum observed near the first resonance line of Xe at 146.96 nm, under the conditions of our experiment. It is ascribed to Xe_2 since the signal intensity increased quadratically as a function of Xe back pressure, as expected for dimer formation. This progression of 10 highly-structured bands is assigned to transitions $0_u^+(v') \leftarrow 0_g^+(v'' = 0)$ based on the work of Castex. The bands have overall widths of $\sim$30 cm^{-1}, and are separated by $\sim$50 cm^{-1}. Each band appears to have similar structure, consisting of 10 or more features degraded to the blue, with widths of $\sim$1.5 cm^{-1} and spacings of $\sim$4 cm^{-1}. Such resolved features within each band have not been observed previously. Here, we attribute them to vibronic bands of the various isotopes of Xe_2, and the ensuing analysis (Fig. 3) confirms this assignment. The nine stable isotopes of Xe lead to a possible 45 dimers with natural abundances ranging from 8 x 10^{-5}% ($^{126}Xe_2$) to 14%($^{129,132}Xe_2$). The sharp vibronic bands of 16 dimers having abundances >1% were observed in the most intense spectra.

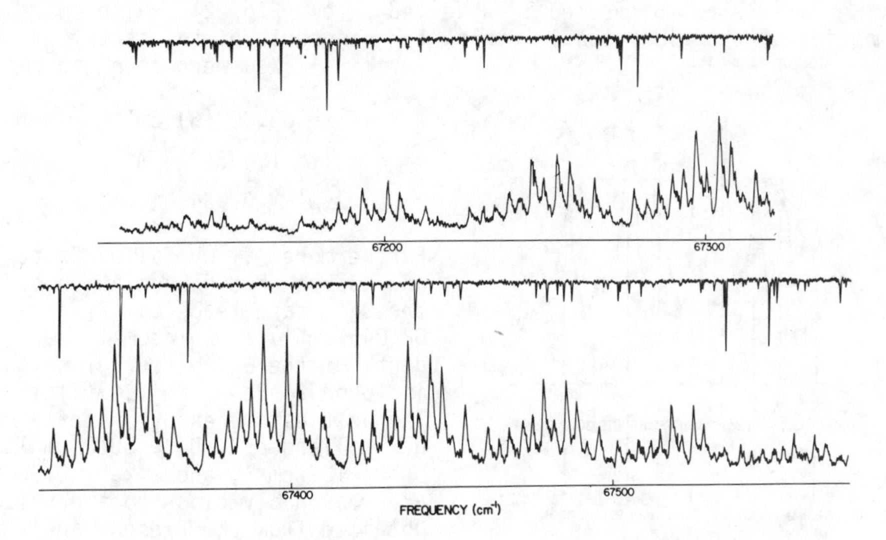

Fig. 2. Fluorescence excitation spectrum of Xe_2, near 149 nm, showing 10 vibronic bands with sub-bands due to isotopes of Xe_2. The upper traces are opto-galvanic spectra of U used for wavelength calibration.

The observation of resolved isotopic vibronic bands immediately suggested that vibrational numbering of the observed bands could be determined readily. Firstly, it was found that using theoretical rotational constants[13] each isotopic band envelope could only be fitted for rotational temperatures <1°K. While the vibrational temperature could have been somewhat higher, it seemed reasonable to assume that only the $v'' = 0$ vibrational level was populated. Thus observed frequencies of the fluorescence excitation spectra could be described by the following relation (in cm^{-1}):

$$\nu_i = T_e + \rho_i\omega_e'(v'+\tfrac{1}{2}) - \rho_i^2\omega_e x_e'(v'+\tfrac{1}{2})^2 - \rho_i \times \frac{21.12}{2} + \rho_i^2 \times \frac{0.65}{4} - \rho_i^3 \times \frac{0.003}{8} \qquad (1)$$

Here T_e is the electronic energy of the upper state, ω_e' and $\omega_e x_e'$ are the vibrational frequency and anharmonic constant, respectively, and ρ_i is the ratio $[\mu(^{129,132}Xe_2)/\mu_i]^{1/2}$ where μ_i is the reduced mass of a particular dimer isotope, i. The ground state constants ($\omega_e'' = 21.12$, $\omega_e x_e'' = 0.65$, and $\omega_e y_e'' = 0.003$ cm^{-1}) in Eq.(1) are those of Freeman et al.[1] derived for $^{129,132}Xe_2$.

In the present analysis, measured frequencies of 10 isotopic vibronic bands in each of the 8 most intense bands of the progression shown in Fig. 2, were averaged for two recorded spectra, and fitted to Eq.(1) by the method of least-squares. Since the quantum number v' is an integer, it was used as a running parameter and was varied until the chi-squared parameter (χ^2) of the fit was minimized. This occurred unambiguously for one set of quantum numbers (with $\chi^2 = 0.67$) giving v' = 37 to 46 for the 10 observed bands (Fig. 2). With this numbering of the vibrational levels, the constants of the state 0_u^+ for $^{129,132}Xe_2$ were found to be:

$$T_e = 63{,}802(6)\ cm^{-1}$$

$$\omega_e' = 124.5(3)\ cm^{-1}$$

$$\omega_e x_e' = 0.934(3)\ cm^{-1}$$

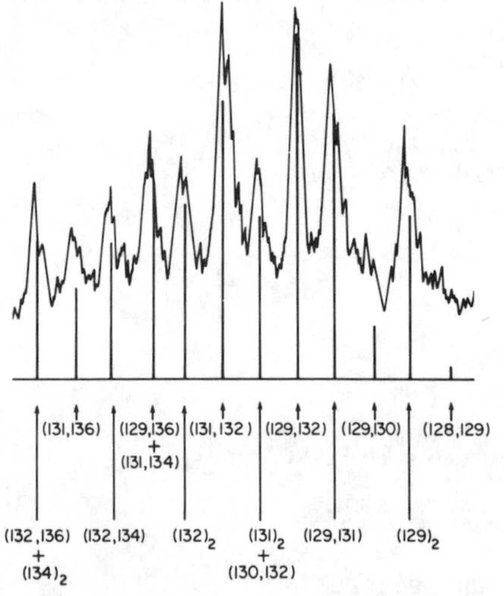

Fig. 3. One of the vibronic bands and the calculated "stick spectrum" for 15 isotopic species of Xe_2.

Furthermore, if we follow Castex and assume that dissociation of the 0_u^+ state leads to $^3P_1$ (at 68,045 cm^{-1}) + $^1S_0$ atoms, the depth of the 0_u^+ potential well is found to be $D_e = 4{,}243(6) cm^{-1}$. The best estimates by Castex and her colleagues[3], $D_e \approx 5000\ cm^{-1}$, $\omega_e' \approx 137\ cm^{-1}$, and $\omega_e x_e' \approx 0.5 cm^{-1}$, are reasonably close to the values obtained from the present analysis. In Fig. 3 is shown one band of the progression from Fig. 2, together with the calculated "stick" spectrum. The agreement between the observed and calculated spectra with respect to isotopic intensities and band positions is excellent. The root-mean-square deviation between observed and calculated vibronic frequencies based on Eq.(1) and the derived constants is $\pm 0.13\ cm^{-1}$, well within the error of measurement.

CONCLUSION

The present investigation has yielded the first experimental determination of the spectroscopic constants of an excited state of Xe_2. Preliminary studies of the fluorescence excitation spectrum of Xe_2 near 130 nm, have produced the spectrum shown in Fig. 4. Excitation was by VUV radiation generated by 4-WSM in Zn. In this spectrum, the isotopic pattern of vibronic bands is also visible. However, the isotopic structure is more closely spaced, indicating that transitions to lower vibronic levels than in the 149 nm band system, are occurring. In addition, we have resolved isotopic structure in 10 vibronic bands of Kr_2, in the region 124.0 to 125.0 nm (referred to as System II, by Tanaka, Yoshino, and Freeman[14]). Fig. 5 shows the isotopic structure of one of the bands, with rotational structure evident in sub-bands of the two most abundant isotopes. Analysis of these spectra is in progress.

Clearly, the experimental techniques used for these spectroscopic studies will be applicable to the whole family of rare gas dimers (including mixtures), as suitable tunable laser sources are developed in the VUV and XUV regions.

We wish to acknowledge the technical expertise of D. Warwick in building the jet apparatus, and to thank P. Herman for the design and construction of the dye amplifiers.

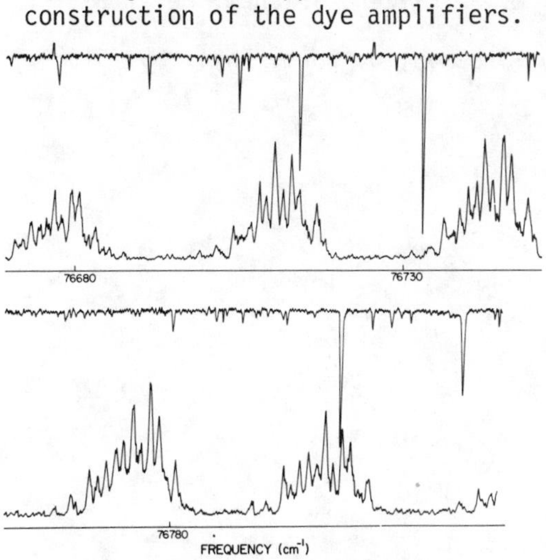

Fig. 4. Fluorescence excitation spectrum of Xe_2 near 130nm. Note that the sub-band isotopic structure is the same as that of Fig. 3.

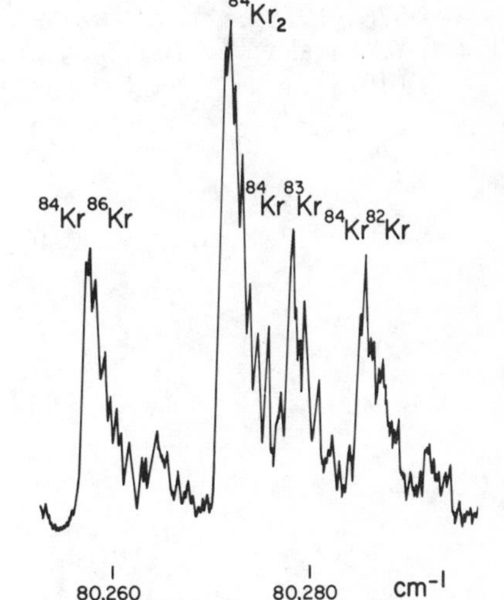

Fig. 5. One of the 10 observed bands of Kr_2 near 124.5nm showing isotopic sub-bands and resolved rotational structure.

REFERENCES

1. D. E. Freeman, K. Yoshino, and Y. Tanaka, J. Chem. Phys. 61, 4880 (1974).
2. M.-C. Castex and N. Damany, Chem. Phys. Lett. 13, 158 (1972); 24, 437 (1974).
3. O. Dutuit, M.-C. Castex, J. Le Calvé, and M. Lavollée, J. Chem. Phys. 73, 3107 (1980).
4. M.-C. Castex, J. Chem. Phys. 74, 759 (1981).
5. R. S. Mulliken, J. Chem. Phys. 52, 5170 (1970).
6. W. Jamroz and B. P. Stoicheff in Progress in Optics XX, E. Wolf ed. (North-Holland, Amsterdam, 1983), pp. 325-379.
7. P. M. Dehmer and J. L. Dehmer, J. Chem. Phys. 68, 3462 (1978).
8. H. Scheingraber and C. R. Vidal, Rev. Sci. Instr. 52, 1010 (1981).
9. S. C. Wallace and G. Zdasiuk, Appl. Phys. Lett. 28, 449 (1976).
10. B. P. Stoicheff, J. R. Banic, P. Herman, W. Jamroz, P. E. LaRocque, and R. H. Lipson, in Proc. Topical Meeting on Laser Techniques in Extreme Ultraviolet Spectroscopy (T. J. McIlrath and R. R. Freeman, eds.) Amer. Inst. Phys. New York (1982), pp. 19-31.
11. W. Jamroz, P. E. LaRocque, and B. P. Stoicheff, Opt. Lett. 7, 617 (1982).
12. B. A. Palmer, R. A. Keller, and R. Engleman, Jr., An Atlas of Uranium Emission Intensities in a Hollow-Cathode Discharge, Los Alamos Sci. Lab., University of California, Report LA-8251-MS, UC34a, (1980).
13. W. C. Ermier, Y. S. Lee, and K. S. Pitzer, J. Chem. Phys. 69, 976 (1978).
14. Y. Tanaka, K. Yoshino, and D. Freeman, J. Chem. Phys. 59, 5160 (1973).

VACUUM ULTRAVIOLET PHOTOIONIZATION OF COLD MOLECULAR BEAMS*

John C. Miller and R. N. Compton
Oak Ridge National Laboratory, Oak Ridge, Tennessee 37831

ABSTRACT

Experiments will be described which combine nonlinear optics, mass- and energy-resolved ion detection, energy-resolved electron detection, and a supersonic nozzle to study vacuum ultraviolet photoionization and photoelectron spectroscopy of ultracold molecules.

INTRODUCTION

The vacuum ultraviolet (VUV) spectral region ($\leq$200 nm) has traditionally been a difficult region in which to perform spectroscopic experiments on molecules and to interpret the resulting data. Experimental difficulties inherent in this region include low reflectivities, low window transmission (below 100 nm, no windows are available), and the need for a vacuum environment. Traditional monochromatic light sources for the VUV are weak, difficult to operate, and require vacuum monochromators which trade resolution for photon flux. Finally, the photophysics and photochemistry of high-lying states of molecules are complicated by their high density of states and distributed oscillator strengths as well as the numerous channels available to the molecule for the disposition of excess energy. Predissociation, ionization, and autoionization broaden spectral lines and complicate interpretation. Nevertheless, VUV spectroscopy has developed into a mature field over the last 50 years. A large body of reliable data has been acquired and an accompanying framework of theoretical explanations has been developed.

Very recently, a renaissance in the field of molecular spectroscopy in the VUV region has occurred. This resurgence has been driven by the technological advances provided by synchrotron light sources, tunable lasers, and supersonic molecular beams, each of which alleviates or eliminates some of the experimental and interpretational problems mentioned above. The synchrotron has made available an intense, broadly tunable, moderate resolution light source for the VUV and x-ray regions which also requires a monochromator. The well known attributes of tunable dye lasers such as high-intensity, very narrow bandwidth, spatial and temporal coherence, and short pulse times are being translated into the UV and VUV spectral regions. Coherent excitation in the VUV is now possible due to the development of new lasers in this spectral region and by the emergence of the complementary techniques of multiphoton excitation and nonlinear frequency conversion. Finally,

* Research sponsored by the Office of Health and Environmental Research, U.S. Department of Energy, under contract DE-AC05-84OR21400 with Martin Marietta Energy Systems, Inc.

supersonic molecular beam sources allow rotational and vibrational cooling which is imperative in order to simplify conjested spectra. The renaissance of VUV spectroscopy was amply detailed in the first topical meeting on "Laser Techniques in the Extreme Ultraviolet" and will be expanded by the present meeting.

We have developed a versatile apparatus, based on the above cited technologies and techniques, for the study of VUV spectroscopy and photoionization of molecules. A pulsed, supersonic molecular beam is crossed with the focused beam of a pulsed dye laser or its doubled or tripled output. Ions are created by multiphoton ionization (MPI) with the fundamental beam (three to five photons), two-photon ionization using the doubled or tripled beam, or one-photon ionization with VUV light. Use of a second laser allows for two-color experiments.

The ionization event is monitored by time-of-flight mass spectrometry and by photoelectron energy analysis. This apparatus thus combines our previous work on VUV spectroscopy of high-pressure gases using nonlinear optical techniques[1] with fragment ion and photoelectron analysis extensively used for MPI experiments on effusive beams.[2-4] The addition of the supersonic molecular beam allows examination of vibrationally and rotationally cold monomer beams as well as molecular clusters.

This paper will describe experiments designed to demonstrate and test the various capabilities of this new apparatus. These will include one-, two-, and four-photon ionization experiments on NO emphasizing photoelectron spectroscopy.

EXPERIMENTAL

The experimental arrangement is shown schematically in Fig. 1. The excitation source is an excimer-pumped dye laser (Lambda Physik EMG101/FL2000E) or its doubled or tripled output. Frequency conversion is achieved with an interactive frequency doubling system (Inrad) or by third-harmonic generation (THG) in rare gases. The THG cell is equipped with quartz and MgF_2 optics for the input and output, respectively, and contains a single electrode for monitoring MPI events in the tripling medium.[5] A second, independent, nitrogen-pumped dye laser (Molectron UV-400/DL400) provides two-color capability with the addition of a second lens to the chamber. The molecular beam/cluster source is a pulsed, supersonic nozzle (Quanta-Ray PSV-2) capable of delivering gas pulses of 60-100 μs duration at backing pressures up to 10 atm and a nozzle temperature of 70°C. The nozzle is currently operated as a free jet expansion although a skimmer will be installed in the future. Translation in the x, y, and z directions are possible. The source is typically operated at ~10 Hz with a 0.5 mm aperture. The laser is triggered externally after a variable delay with respect to the valve opening. Following either single- or multiphoton ionization the ions and electrons are detected and characterized independently and simultaneously. Ions are extracted with a small electric field and mass-analyzed by time-of-flight techniques with a resolution ($m/\Delta m$)

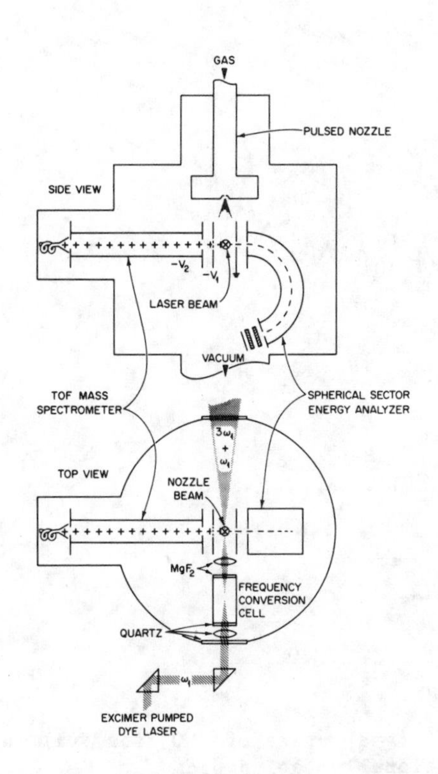

Fig. 1. Schematic of experimental apparatus.

of about 50. The electron kinetic energy distribution (i.e., the photoelectron spectrum) is determined with a 160° spherical sector energy analyzer. Energy resolution ranges from 50 to 200 meV, depending on experimental conditions (ion intensity, analyzer pass energy, etc.). Photoelectron angular distributions may be determined by eliminating the extraction fields and rotating the polarization of the laser beam about the acceptance direction of the analyzer with a double Fresnel rhomb. Further details of the ion and electron analysis may be found in previous papers.[2-4] Signal processing can be in either an analog mode using boxcar integrators or in digital mode using single pulse counting techniques and a multichannel analyzer.

BOUND STATE SPECTROSCOPY

We have used the MPI technique to investigate the ro-vibronic spectroscopy of a number of Rydberg states of NO in supersonic beams following the pioneering work of Johnson et al.[6,7] The states studied thus far (with the MPI scheme used given in parenthesis) include the A $^2\pi$ (v=0) state, (2+2 and 1+1), A $^2\pi$ (v=1) state (2+2), the C $^2\pi$ (v=0) state (2+1), the F $^2\Delta$ (v=3) state (1+1), and the N $^2\Delta$ (v=1) state (1+1). Studies of the latter two states employed VUV light generated by THG in xenon as the first photon and the blue dye laser fundamental as the ionizing photon as described previously for high pressure studies.[1] Studies of the A $^2\pi$ (v=0) state by a (1+1) scheme employed frequency doubling in a KPB crystal.

As an example, the spectrum of the F $^2\Delta$ (v=3) state is shown in Fig. 2 for both a room temperature gas beam (trace a) and a supersonically cooled beam (trace b). Optimum phase-matching for the THG at these wavelengths occurred at about 50 torr of xenon. The beam is a 5% NO/He mixture expanded through a 0.5-mm hole at a backing pressure of ~2 atm. Because of rotational warming of the

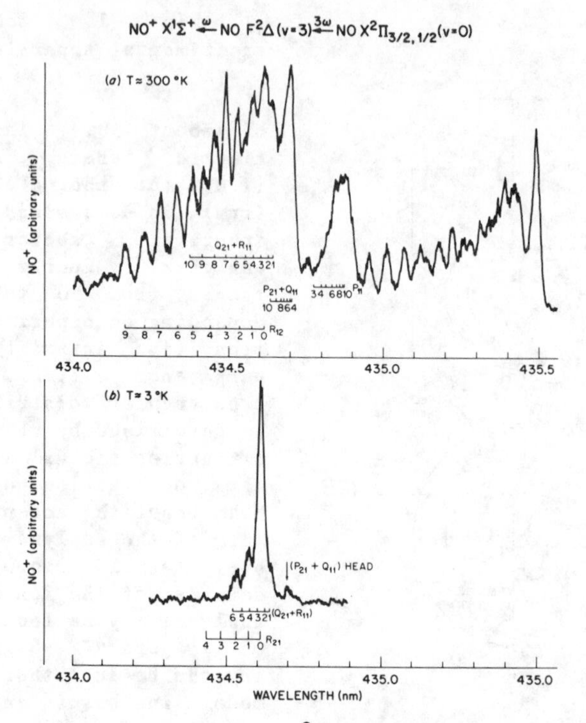

Fig. 2. MPI spectra of the $F^2\Delta$ (v=3) state of NO for (a) a room temperature beam, and (b) rotationally cooled beam.

gas due to collisions with the extraction plates (see Fig. 1), the coldest spectra were obtained at laser delay times corresponding to the arrival of the earliest molecules from the gas pulse. At later delay times the temperature was considerably warmer or essentially room temperature. The rotational temperature of Fig. 2(b) corresponds to a temperature of about 3 K if only the J=1/2 and 3/2 levels are used to characterize a Boltzmann distribution. The higher J levels occur with larger populations than expected from a 3 K Boltzmann distribution, however, and may have some contribution from the warm NO present as background gas in the chamber (average background pressure was typically 10^{-5} to 10^{-6} torr). For the more intense transitions to the A $^2\pi$ and C $^2\pi$ states temperatures below 1°K could be obtained. The assignments of the warm spectra of Fig. 2 are from Jungen[8] and are confirmed by the rotationally cold spectra. Cooling of high vibrational levels (ω_e = 1904 cm^{-1}) is complete as is the electronic cooling of the $^2\pi_{3/2}$ state (T_e = 120 cm^{-1}). At equivalent backing pressures, cooling of pure NO was less complete (T $\simeq$ 10-25 K) than that for 5% NO/He and 1% NO/Ar mixtures which were essentially the same. AC Stark broadening was only observed for the A $^2\pi$ (v=0) state as previously discussed by Otis and Johnson.[7]

PHOTOIONIZATION SPECTROSCOPY

The relatively low ionization potential of NO allows for the use of frequency tripling techniques to perform one-photon ionization experiments. The emphasis of these experiments thus shifts from bound states below the ionization threshold to autoionizing states coupled to the continuum. Figure 3 shows the spectrum of NO in the region of 380 to 386 nm obtained by frequency tripling the dye laser in xenon gas near its $6s'[1/2]^0J=1$ resonance.[5] The rotational temperature of the gas is unknown but is probably in the range of 5-10 K based on rotationally resolved spectra taken under similar conditions. The positions of the absorption lines in this region and their assignment[9] as listed by Ono et al.[10] (including some calculated line positions) are given above the spectrum. This spectrum should represent the highest resolution spectrum (0.007 nm) obtained for the coldest NO sample (5-10 K) reported to date. The previous work of Ono et al.[10] used conventional light sources with a resolution of 0.014 nm and obtained a gas temperature estimated to be 20 K. Although there is a general correspondence between the spectrum of Fig. 3 and the previous absorption[9] and photoionization spectra[10] a satisfactory line-by-line match is not obtained with either. The present spectrum should, however, be considered preliminary and work is continuing in this area. More careful calibration and systematic variation in the degree of rotational cooling are necessary to resolve the discrepancies. Studies of other spectral regions using THG near the 5s and 5s' resonances of Kr are also possible.

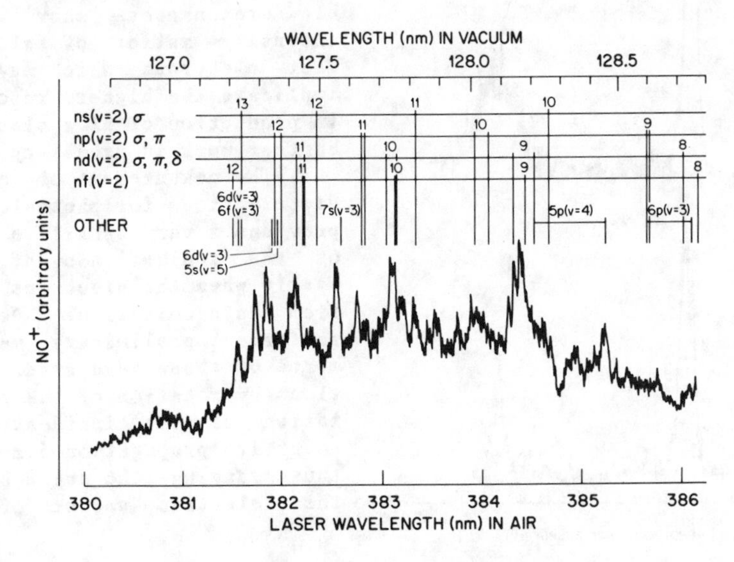

Fig. 3. One photon photoionization spectra of a rotationally cold beam of NO excited by VUV light produced by THG.

PHOTOELECTRON SPECTROSCOPY

We have extended our previous interest in MPI-PES[3,4] by including the new capabilities of the present apparatus for studies of NO. For instance, studies of the photoelectron spectrum of ultracold NO (~1°K) following MPI (2+2) via the A $^2\Sigma$ (v=0, J=1/2) state were performed under moderate resolution (see Fig. 4b). The appearance of the spectrum is independent of rotational temperature and initial and final state rotational quantum number. Similar spectra for NO have been described and discussed in detail previously.[3,11-13] Briefly, a higher energy electron peak near 1.7 eV is due to direct two-photon ionization of the A state. Near zero-energy electrons are not yet completely understood but have been attributed to vibrational autoionization.[3] Electrons due to electronic autoionization have also been observed.[3,11] The new configuration of the apparatus allows several additional experiments to be performed which further clarify the photoelectron spectra.

The spectrum of Fig. 4(a) represents the MPI-PES via the same state as in trace (b) but following frequency doubling of the input laser; i.e., the two spectra represent (1+1) and (2+2) MPI-PES via the same state. We have argued previously that the (2+2) process allows near resonances in the third photon region which ultimately may be responsible for the slow electrons via vibrational autoionization. The similarity of the spectra of Fig. 4(a) and (b) seems to rule out this possibility. However, other experiments involving accidental double resonances show varying intensity ratios of slow and fast electrons which may again implicate the higher resonances in production of slow electrons. Further work is necessary.

The measurement of angular distributions for photoelectrons provides a very sensitive probe of the ionization event, especially when the electrons result from autoionization. We have performed preliminary measurements of these angular distributions by rotation of the polarization of the incident laser about its propagation direction, thus varying the angle between the electric vector of the

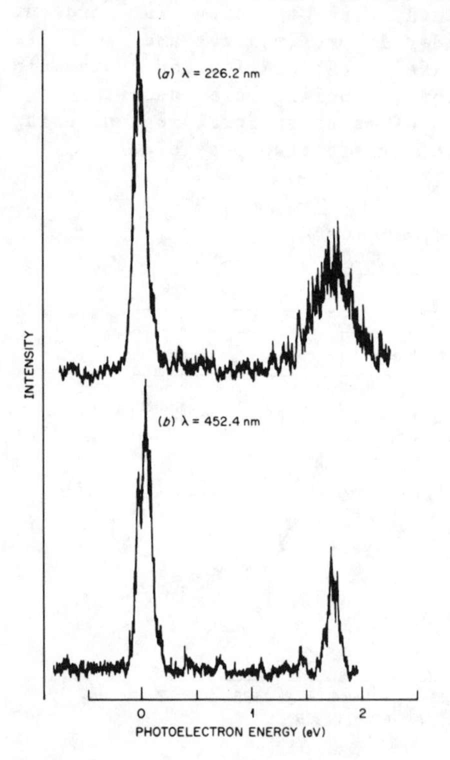

Fig. 4. The photoelectron spectra of supersonically cooled NO following (a) UV MPI (1+1) and (b) visible MPI (2+2).

light and the acceptance direction of the analyzer. The angular dependence of the slow and fast electrons following MPI (2+2) via the A $^2\Sigma^+$ (v=1) state of NO are observed to be quite different. In general, the angular dependence can be expressed as

$$I(\theta) = \sum_{N=0}^{K} B(N) \cos^{2N}\theta \,, \tag{1}$$

where K is the order of the ionization process (i.e., K = 2 for a two-photon ionization).

For the distributions of the fast electrons, the following equation results from fitting the angular distribution to an Nth order regression formula

$$I(0) = 1 + 7.0 \cos^2\theta - 10.2 \cos^4\theta - 7.6 \cos^6\theta - 3.3 \cos^8\theta \quad . \tag{2}$$

The slow electrons exhibit a nearly isotropic distribution. Such a distribution implies that the molecule has "lost its memory" concerning the original orientation. The process of autoionization can lead to such a distribution. Although still considered preliminary, these distributions contain the most detailed clues so far available for understanding the origin of the slow electrons.

Finally, we have measured MPI-PES and angular distributions for the (1+1) MPI via the F and N states of NO using THG as described in a previous section. Only electrons from direct ionization are observed and their energy and angular dependences are similar to those measured for direct ionization in previous experiments.

TWO-COLOR EXPERIMENTS

Two-color MPI experiments (i.e., use of two independently tunable dye lasers) offer a number of significant advantages over those using a single laser. The major such advantage is the spectral simplification which arises from selecting a single rovibronic level with the first laser. The second laser will then only interact with these excited molecules and a simple PR or PQR structure will emerge in the probe laser spectrum. This simplification is complementary to but perhaps more versatile than that provided by supersonic cooling. Furthermore, by defining additional resonances in the MPI scheme (i.e., replacing virtual states with real states) the ionization cross section is increased and the experiment is better defined. Also, different selection rules may allow access to different states with sequential two-color excitation than with simultaneous multiphoton excitation. Finally, varying the relative polarization of the two lasers provides additional information about assignments. Several groups have previously published two-color spectra of NO.[15,16] We have observed several additional two-color MPI pathways and plan to use this very powerful excitation technique to simplify and characterize photoelectron spectra. In addition, we hope to demonstrate two-color MPI where one excitation involves VUV light from THG.

REFERENCES

[1] John C. Miller, R. N. Compton, and C. D. Cooper, J. Chem. Phys. 76, 3967 (1982).

[2] C. D. Cooper, A. D. Williamson, John C. Miller, and R. N. Compton, J. Chem. Phys. 73, 1527 (1980).

[3] John C. Miller and R. N. Compton, J. Chem. Phys. 75, 22 (1981); 75, 2020 (1981); and Chem. Phys. Lett. 93, 453 (1982).

[4] J. H. Glownia, S. J. Riley, S. D. Colson, J. C. Miller, and R. N. Compton, J. Chem. Phys. 77, 68 (1982); J. C. Miller, R. N. Compton, R. E. Carney, and T. Baer, J. Chem. Phys. 76, 5648 (1982).

[5] John C. Miller, R. N. Compton, M. G. Payne, and W. R. Garrett, Phys. Rev. Lett. 45, 114 (1980); John C. Miller and R. N. Compton, Phys. Rev. A 25, 2056 (1982).

[6] D. Zakheim and P. Johnson, J. Chem. Phys. 68, 3644 (1978).

[7] C. E. Otis and P. M. Johnson, Chem. Phys. Lett. 83, 73 (1981).

[8] C. H. Jungen, Can. J. Phys. 44, 3197 (1966).

[9] E. Miescher and F. Alberti, J. Chem. Phys. Ref. Data 5, 309 (1976).

[10] Y. Ono, S. H. Lin, H. F. Prest, C. Y. Ng, and E. Miescher, J. Chem. Phys. 73, 4855 (1980).

[11] J. Kimman, P. Kruit, and M. J. van der Wiel, Chem. Phys. Lett. 88, 576 (1982).

[12] M. G. White, M. Seaver, W. A. Chupka, and S. D. Colson, Phys. Rev. Lett. 49, 28 (1982); M. G. White, W. A. Chupka, M. Seaver, A. Woodward, and S. D. Colson, J. Chem. Phys. 80, 678 (1984).

[13] Y. Achiba, K. Sato, K. Shobatake, and K. Kimura, J. Chem. Phys. 78, 5474 (1983).

[14] J.Berkowitz, Chap. VII in PHOTOABSORPTION, PHOTOIONIZATION, AND PHOTOELECTRON SPECTROSCOPY (Academic Press, 1979).

[15] T. Ebata, T. Imajo, N. Mikami, and M. Ito, Chem. Phys. Lett. 89, 45 (1982); T. Ebata, N. Mikami, and M. Ito, J. Chem. Phys. 78, 1132 (1983); T. Ebata, Y. Anezaki, M. Fujii, N. M. Kami, and M. Ito, J. Phys. Chem. 87, 4773 (1983).

[16] W. Y. Cheung, W. A. Chupka, S. D. Colson, D. Gauyacq, P. Avouri, and J. J. Wynne, J. Chem. Phys. 78, 3625 (1983); M. Seaver, W. A. Chupka, S. D. Colson, and D. Gauyacq, J. Phys. Chem. 87, 2226 (1983).

COHERENT VUV AND SOFT X-RAY RADIATION FROM UNDULATORS IN MODERN STORAGE RINGS

Kwang-Je Kim, Klaus Halbach, David Attwood
Lawrence Berkeley Laboratory, One Cyclotron Road, Berkeley, CA 94720

ABSTRACT

Magnetic structures in modern storage rings provide an assured route to fundamentally new opportunities for extending coherent radiation experiments to the vacuum ultraviolet and soft x-ray spectral regions. Coherent power levels of order 10 milliwatts are anticipated, in a fully spatially coherent beam, with a longitudinal coherence length of order 1μ m. In addition to broad tuneability and polarization control, the radiation would occur in 20 psec pulses, at 500 MHz repetition rate.

I. INTRODUCTION

Undulators in modern storage rings are capable of generating substantial coherent power in the heretofore inaccessible spectral region extending from photon energies of a few eV to several thousand eV. In addition to providing an **assured route** to coherent techniques in the important soft x-ray and vacuum ultraviolet spectral regions (collectively referred to here as the XUV), the radiation is also **broadly tuneable**. The latter feature is a significant advantage with respect to potential x-ray lasers that might become available in the next decade or so, in that it permits imaging and spectroscopic probe studies to be conducted around the important atomic and molecular resonances so prominent in that region. The purpose of this paper is to give an introductory review of the coherence characteristics of undulator radiation.

II. MODERN SYNCHROTRON RADIATION SOURCES

An extremely relativistic electron experiencing bending motion produces a narrow cone of synchrotron radiation[1] directed along its instantaneous direction. The angular width of the radiation cone, γ^{-1}, is typically several hundred microradians, where γ is the ratio of total electron energy to its rest energy.

Historically, the primary source of synchrotron radiation in storage rings has been at the bending magnets, as illustrated at the top of Figure 1. However, bending magnet radiation is incoherent, and spread across a wide horizontal fan. A more efficient manner of producing and collecting synchrotron radiation involves utilization of periodic magnet structures, in particular permanent magnet which permits combination of high field strengths with short periods.[2] This general scheme is illustrated in Figures 1 and 2. These many period, N, magnetic structures produce an N-fold increase in radiated flux and offer options for spectrally shifting to higher photon energies with a strong field "wiggler", or coherent intensification on axis with a moderate field "undulator". The parameter determining this option is the deflection parameter K = .934 x magnetic field strength in Tesla x period length in cm. K is the ratio of the maximum electron deflection angle to the natural radiation angle γ^{-1}. For K>>1, the wiggler case, the electron experiences large magnetic deflections, thus producing more flux at higher photon energies, and radiating into a proportionately larger radiation cone. Although incoherent in nature, wigglers are currently the most powerful source of laboratory x-rays.[3]

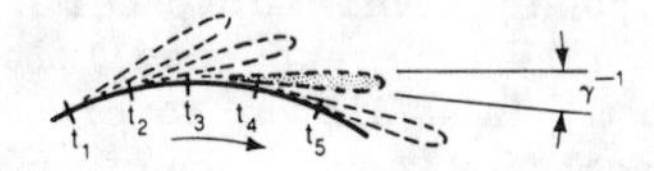

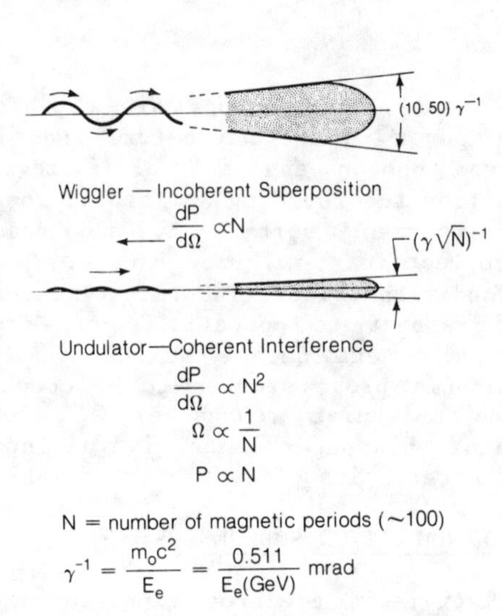

Figure 1. Synchrotron radiation sources include (a) historically significant bending magnets, with relativistically narrowed radiation patterns, (b) multiperiod strong field wigglers, and (c) many period coherent undulators.

Undulators are fundamentally different. As structures for which K $\lesssim$1, the radiated fields associated with various magnetic periods all fall within the primary radiation cone of the individual relativistic electrons. Consequently, interference effects occur between radiation from the various periods. This results, for instance, in N-fold field additions on axis, or N^2 increases of power density (intensity) on axis. To the present synchrotron radiation community this would be described as greater source "brightness". More significantly, undulators have introduced substantial coherent radiation opportunities to important regions of the spectrum.

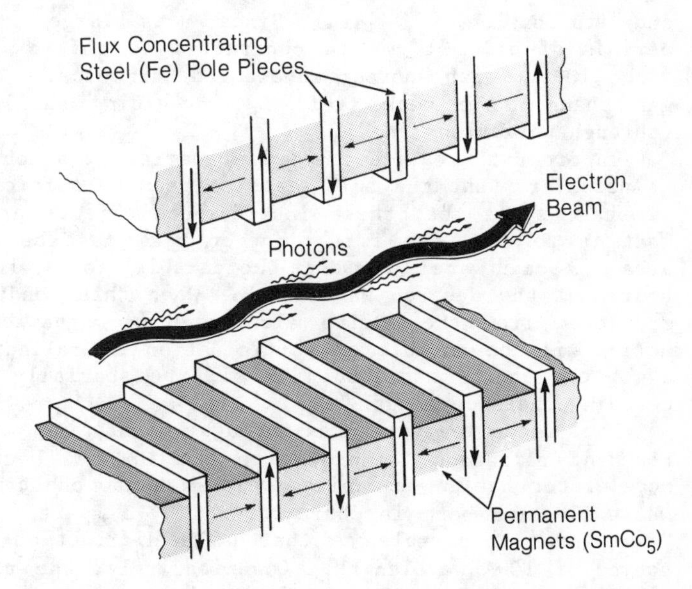

Figure 2. Many period magnetic insertion devices generate an N-fold increase of radiated power. Wigglers are strong field structures, driving electron trajectories of large angle, leading to an incoherent superposition of fields. Undulators employ electron deflections within the nominal radiation cone, such that the coherent superposition of fields dominate - leading to on axis field intensification, radiation cone narrowing, and substantial coherence properties in modern low-emittance storage rings.

III COHERENT UNDULATOR RADIATION

The spectral characteristics of undulator radiation can be understood by considering the radiation seen by an observer fixed in the laboratory frame, looking directly at the relativistic electron as it passes through the magnetic structure. Electromagnetic waves, emitted as the electron passes successive periods of the structure, will arrive at the fixed observer at times seperated by $\Delta t = \lambda_u (1-\beta_z)/c$, where λ_u is the magnetic period length, c is the velocity of light in vacuum, and $c\beta_z$ is the electron's average velocity in the z-direction. In the extreme relativistic case $\beta_z = 1 - (1 + K^2/2)/2\gamma^2$, from which it follows that the radiated fields are relativistically contracted from the undulator period λ_u to an electromagnetic wavelength

$$\lambda_x = c\Delta t = \frac{\lambda_u(1 + K^2/2)}{2\gamma^2} \tag{2}$$

and its harmonics, λ_x/n. Since is large (~1,000), undulator periods of order 1 cm are contracted to soft x-ray wavelengths of order 100 Å, with shorter wavelength harmonics. Changing the magnet gap changes the peak field B_o, providing easily tuned radiation (through variations in K).

In order to realize the full benefits of a coherent undulator it is necessary that the full electron beam be sufficiently constrained in phase space that these interference effects are not washed out. That is to say, the full electron beam must be constrained to an area . solid angle product comparable to a diffraction limited source at the desired wavelength. When this condition is achieved, radiation from the entire electron beam-magnetic structure interaction will be capable of participation in collective interference, e.g., the entire radiated flux will be spatially coherent within a certain coherence length.[4]

The technological breakthrough permitting this extension of coherent radiation techniques to the XUV has been provided by the accelerator design community, which can now construct a storage ring whose electron beam is characterized by a space . angle phase space "emittance" comparable to that of a diffraction limited radiation source of 100Å wavelength. Quantitatively, the resulting radiation will be spatially coherent if the phase space of the electron beam, described by the horizontal and vertical emittances ε_x and ε_y, are equal to or less than the diffraction limited radiation phase space defined by $d(2\theta) = 2.4\lambda_x$, where λ_x is the wavelength. When this condition prevails, that is when

$$\varepsilon_x \varepsilon_y \leqq (2.4\lambda_x)^2, \tag{3}$$

the resulting radiation will be fully spatially coherent. The spatial radiation pattern of an undulator designed for Berkeley's proposed ALS synchrotron radiation facility[5] ("Advanced Light Source") is shown in Figure 3.

Undulator radiation is both spatially and longitudinally coherent, displaying spectral peaks* of width $\lambda_x/\Delta\lambda_x \sim N$ as illustrated in figure 4. The coherence length, for

---

* As pointed out by B. Kincaid of Bell Laboratories, the higher harmonics, as in his 20th harmonic optical klystron, are threatened by period to period random variations in magnetic field strength. However, present fabrication tolerances provide sufficient fidelity to assure performance of the fundamental and low order harmonics of interest in this paper.

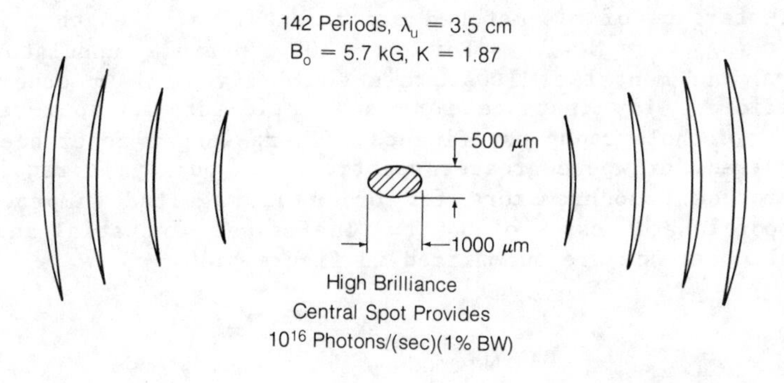

Figure 3. The spatial distribution of undulator radiation, demonstrating coherent interference at 500 eV photon energy. Calculations are for undulator D, of the proposed ALS synchrotron facility at Berkeley (Ref. 5).

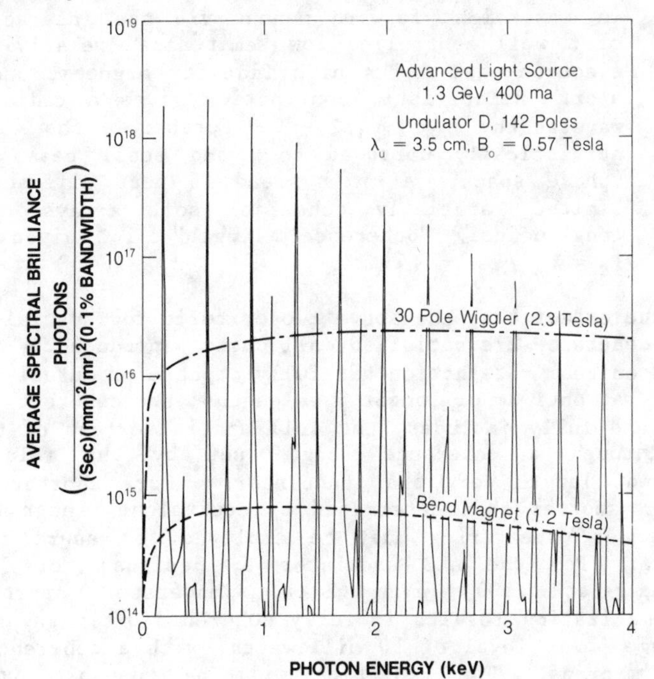

Figure 4. Spectral Characteristics (harmonic structure) of undulator D radiation is compared with that of a wiggler and a bend magnet at the ALS.

interference of mis-matched experimental path lengths, is defined as $l_c = \lambda_x^2/\Delta\lambda_x = N\lambda_x$. For a 100 period undulator radiating at a fundamental of 100Å, this would give a 1 μm coherence length, sufficient for instance for many anticipated soft x-ray microprobe and microholography experiments. Where longer coherence lengths are required, or where greater spectroscopic purity is required, one can introduce monochromators for order of magnitude improvements, with proportionate losses of power. The issues of spatial and longitudinal coherence are summarized in figure 5.

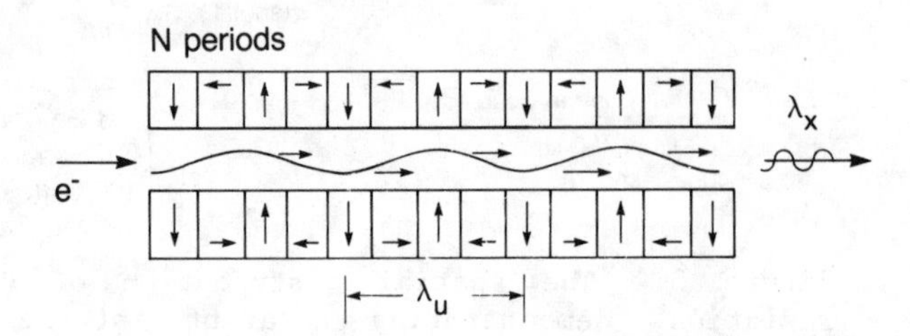

Figure 5. Quasi-coherent soft x-rays are produced in a straight forward manner via the interaction of a well controlled low "emittance" relativistic electron beam and a high fidelity magnetic undulator. Relativistic contraction gives a radiation wavelength $\lambda_x = \lambda_u/2\gamma^2$. Further, the relativistically narrowed cone and small beam size (phase space, $\Delta A \cdot \Delta\Omega$), lead to near diffraction limited (spatially coherent) soft x-rays. The longitudinal "coherence length" is given by $l_c = \lambda_x^2/\Delta\lambda_x = N\lambda_x$.

To the extent that the above two criteria for spatial and longitudinal coherence are satisfied, all photons produced by the undulator-electron beam interaction are fully capable of mutual interference. At optimum or longer wavelengths, the coherent volume is set by, and fully includes, the full spatial extent of the electron beam, through a coherence length set by the relativistically contracted length of the undulator. For shorter wavelength radiation shorter than determined by equivalent electron and photon phase spaces, aperturing must be employed to assure full spatial coherence. For the ALS undulator D, designed for 3rd harmonic radiation at 500 eV (24Å), moderate aperturing and monochromatization results in fully coherent soft x-ray radiation at an average power level of 10 milliwatts, with a coherence length of several microns. The radiation will be tuneable over a broad spectral range including the K-edges of such important elements as carbon, oxygen and nitrogen. Further, the radiation will occur in 20 psec pulses at a 500 MHz repetition rate.[6] Similar performance will be available in the VUV utilizing separate undulators specifically designed for that purpose, and placed elsewhere in the ring. The proposed ALS has twelve long straight sections to accomodate different undulators optimized for use in different spectral ranges. The projected ALS capabilities are summarized in Figure 6 for four of its undulators (A, B, C, and D).

For comparative purposes, Figure 6 also shows the results of recent non-linear laser harmonic and mixing experiments.[7] Producing radiation in 10 psec to several nsec bursts, at typically 10 Hz repetition rates, these radiation sources fall off very rapidly in power as they extend to photon energies beyond ten eV, typically as wavelength to an exponent of order the non-linearity involved, e.g., λ^{9}, etc.

In addition, we note that if available, true lasers would be characterized by narrower spectral content and higher peak power, thus complimenting tuneable coherent undulator radiation capabilities as described here. At the present, however, high reflectivity mirrors do not exist for the XUV, although progress is being made.[8] If such optics did become available, the undulator/storage ring techniques described here could be operated as XUV Free Electron Lasers (FELs), also broadly tuneable, but of greatly increased spectral purity and power. Other proposed schemes for (soft) x-ray lasing are likely to be of high spectral purity, perhaps involve a few discrete spectral lines, but not be broadly tuneable, across absorption features of interest to the experimenter. In addition, such potential lasing schemes would likely be single pass super-radiant schemes, also quasi-coherent in nature. They are also likely to be driven, at least in their early stages, by large pulsed energy facilities, for example, high power lasers originally developed for inertial fusion.

Table I. Typical parameters for a soft x-ray undulator.

• Electron Beam Energy	1 GeV ($\gamma = \frac{E}{m_o c^2} = 2{,}000$)
• Beam Current	0.4 amps
• Magnetic Wavelength (λ_u)	3 cm
• Magnetic Periods (N)	100
• Photon Wavelength (λ_x)	$\frac{\lambda_u}{2\gamma^2} = 25\text{Å}$ (500 eV) (Broadly tuneable 100eV to few keV)
• Bandwidth ($\frac{\lambda_x}{\Delta\lambda_x}$)	N = 100
• Angular Width (2θ)	$\frac{1}{\gamma\sqrt{N}} \sim 50\ \mu\text{rad}$
• Polarization	Linear, circular, . . .
• Radiated Power	1 watt cw in 20 psec bursts
• Coherent Power	10 milliwatt at 500 eV, tuneable (spatially coherent, $\ell_c = \frac{\lambda^2}{\Delta\lambda} \sim$ few microns)

IV. POLARIZATION CONTROL

In addition to the space-time characteristics of electromagnetic radiation, there is also polarization. In this section we describe a method[9] which utilizes a sequential pair of crossed undulators to obtain complete polarization control. The idea is illustrated in Figure 7. Each undulator produces radiation on axis of linear polarization. By properly overlapping the two cross-polarized wave trains it is possible to produce linear or circular polarization, depending on their relative phases. Phase variation in this scheme is controlled by an electron path modulator between the two linear undulators - effectively using modulated electron transit time to the second undulator as a means of varying phase between the two wave

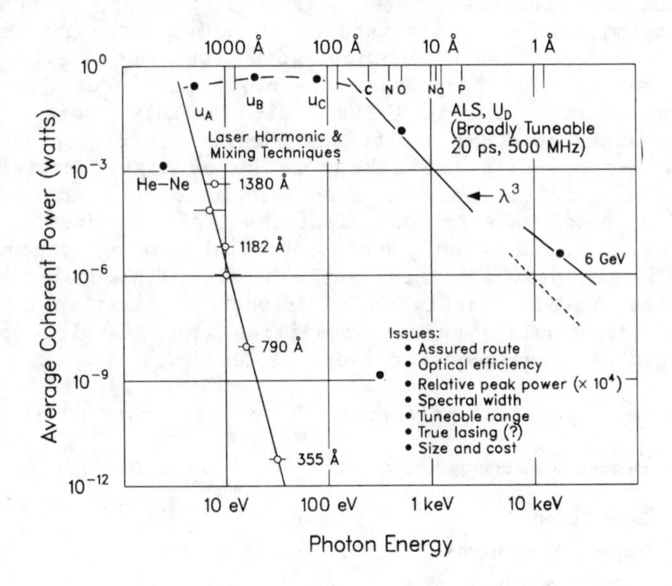

Figure 6. Broadly tuneable coherent radiation will be available - via an assured route - at interesting power levels throughout the VUV and soft x-ray spectral ranges. Laser harmonic and mixing techniques are not competitive in the region beyond 10 eV. Potential XUV lasers may someday provide narrower spectral features and higher peak powers, but will not likely provide broad tuneability, and thus are complimentary to coherent undulator techniques. Undulators may also enjoy additional capabilities as bunching techniques and mirrors (FELs) become available.

trains - each of which is driven by the electron beam. In order to maintain well defined polarization and phase control it is again necessary that the electron beam emittance be small. Variable polarization control is expected to play an important role in future experiments, such as in probes of bio-chemical structures in which image contrast or differential scattering signals are provided by polarization sensitive scattering from structures in which the diameter or helical pitch may be of sizes comparable to the probing VUV or soft x-ray wavelength.

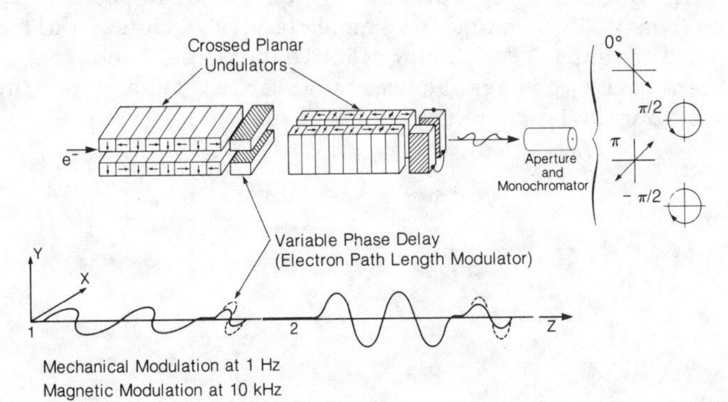

Figure 7. Variably polarized radiation can be generated by using crossed undulators in low emittance storage rings.

Table II summarizes parameters which characterize variable polarization capabilities that could be provided at two potential sites, Brookhaven's present VUV ring, and Berkeley's proposed ALS.

Table II Two potential sources of variable polarization radiation are described.

Ring	VUV (NSLS)	ALS
E_e	750 MeV	1.3 GeV
σ_θ	9×10^{-2} m.r.	2×10^{-2} m.r.
N	5	30
λ_u	10 cm	4 cm
λ_1 (ϵ_1)	470 Å (26 eV)	47 Å (260 eV)
$\Delta\lambda/\lambda$	4%	0.14%
P	86%	84%
Flux at Sample (1% Optical Eff)	$\sim 10^{12}$ Photons/sec	$\sim 10^{12}$ Photons/sec

V. CONCLUSION

Coherent VUV and soft x-ray radiation is predictably available with modern low emittance storage rings and permanent magnet undulators. While covering a wide spectral region, with broadly tuneable radiation, these facilities will provide pulses typically of 20 psec duration, at a 500 MHz repetition rate. In Table II, we have summarized typical parameters for a soft x-ray undulator. In addition to providing an assured route to the earliest coherent XUV radiation experiments, these multi-user facilities will eventually be upgradable to Free Electron Lasers as high reflectivity mirrors become available, thus assuring the facilities of long and prosperous lifetimes.

REFERENCES

1. H. Winick and S. Doniach, Synchrotron Radiation Research (Plenum Press, New York, 1980).

2. K. Halbach, Nucl. Instr. Meth. 187, 109 (1981), Journal de Physique Colloque C1 - 211, Tome 44, (1983).

3. E. Hoyer et al., Nucl. Instr. Meth. 208, 117 (1983)

4. The coherence properties of Synchrotron Radiation were first described by A.M. Krondratenko and A.N. Skrinsky (Opt. Spectrosc. 42, 189, Feb. 1977) using intuitive arguments. A rigorous description of undulator coherence properties in terms of correlated field quantities is given by K-J. Kim (to be published)

5. Advanced Light Source Conceptual Design Report, LBL Pub. 5084, Lawrence Berkeley Laboratory (1983); see also R.C. Sah, IEEE, NS-30, 3100 (1983).

6. R.C. Sah, A. P. Sabersky, and D.T. Attwood, "Picosecond Pulses for Future Synchrotron Radiation Sources", submitted to Topical Meeting on Ultrafast Phenomena, Monterey, CA, June, 1984.

7. See papers on the subject in the Proceedings of this conference.

8. D.T. Attwood et al., Report of the Working Group on Short Wavelength Optics, Proceedings of the Conference on Free Electron Generation of Extreme Ultraviolet Coherent Radiation, Optical Society of America, (Brookhaven, September 1983), edited by J.M.J. Madey and C. Pellegrini.

9. K-J. Kim, "A Synchrotron Radiation Source with Arbitrarily Adjustable Elliptical Polarization", New Rings Workshop, SSRL Report (Stanford, August 1983); also Nucl. Instr. Meth. 219, 425 (1984).

This work was supported by the director, Office of Energy Research, Office of Basic Energy Science, Division of Material Sciences.

PRODUCTION OF COHERENT XUV AND SOFT X-RAYS USING A TRANSVERSE OPTICAL KLYSTRON

R. R. Freeman and B. M. Kincaid
AT&T Bell Laboratories, Murray Hill, NJ 07974

ABSTRACT

This paper describes the theory of the production of coherent XUV and soft X-rays using a Transverse Optical Klystron (TOK). A TOK uses a high powered laser in conjunction with an undulator magnet to produce laser-like output of XUV radiation from a relativistic electron beam.

INTRODUCTION

With the recent demonstration of a storage ring-based free electron laser operating in the visible, the future of the FEL laser sources of coherent radiation appears particularly bright. While the advances in FEL physics will be substantial in the demonstration of FEL oscillation in the near UV, there is, however, a real need in the laser/spectroscopy community for a tunable laser source between 500 Å and 2000Å. In this region, no laboratory source of widely tunable, high-powered coherent radiation is available. Pushing FEL technology into the wavelength region below 2000Å requires a new and dedicated storage ring. If, instead, one uses the transverse optical klystron (TOK) harmonic technique, it should be possible to reach this wavelength region using the existing 700MeV, NSLS storage ring, although at a lower average power than could be obtained with a dedicated FEL storage ring. Since the TOK does not use an optical cavity, but rather uses the electrons in the storage ring as a kind of non-linear optical medium, operation is not limited by mirror losses, as in the FEL. As a result, the TOK does not produce a high average power but can provide an useful ($\sim 10^{-3}$ to 10^{-6}) conversion of the input laser peak power to much higher frequencies. We are currently building a TOK device for use on the NSLS storage ring in collaboration with members of the NSLS scientific staff.[1]

An optical klystron is a device in which a relativistic electron beam produces coherent electromagnetic radiation by interacting with an external laser beam in an undulator magnetic field. This device is the relativistic generalization of the microwave klystron.[2] Because the energy exchange between the electrons and the light in this case is due to the transverse electric field of the laser, the device is called a transverse optical klystron. Just as in an ordinary klystron, coherent radiation is produced at the harmonics as well as at the fundamental of the input laser frequency, resulting in a relatively efficient conversion of the input visible laser to output wavelengths in the extreme ultraviolet.

The generation of coherent light by the TOK can be described as a three step process: energy modulation, compaction or bunching, and radiation. A schematic representation of the TOK is shown in Fig. 1. The electron beam is energy modulated in the modulator by the combined action of the laser and the undulator magnet, the energy modulation is changed into a spatial modulation or bunching at the input laser's wavelength in the compactor section, and the bunched electron beam radiates coherent light at the fundamental and odd harmonics of the input laser in the radiator section.

The use of a compactor section to enhance bunching has found application in FEL design. The idea was first proposed by K. Robinson in 1960,[3] although it was not published until 1977 by Vinokurov and Skrinsky,[4] and has been theoretically examined by several authors.[5] Csonka[6] first proposed creating the energy modulation using an external laser in 1978, and there has been extensive theoretical work on the production of harmonics using the TOK since then.[7] There exists at least two other proposals to implement a TOK on a storage ring to create higher harmonics of a powerful input laser.[8] The only reported experimental work on the TOK concept has been Novosibirsk[9] Orsay.[10] The frequency variation of the output intensity as a function of the magnetic field strength in the compactor compares extremely well with theory, and the interference pattern in the emission from two undulators separated by a compactor has been observed. Although no experimental work using a laser to generate coherent harmonics was reported, the Orsay group has observed a gain enhancement in the FEL, as expected. (Indeed, the successful operation of the FEL at Orsay in the visible depended critically upon the gain enhancement due to the compactor).[11] These experimental studies show that the central principle of the TOK, enhanced bunching of the electron beam at optical wavelengths using a special magnetic compactor section, is in accord with theory.

PHYSICAL PRINCIPLES OF OPERATION

In this paper we explain the general physical principles behind the operation of each of the three sections of the TOK, and derive expressions valid in the plane wave limit for energy modulation, bunch compaction and radiation of coherent light at the harmonics of the laser. More detailed calculations of conversion efficiencies using Gaussian laser modes have been performed, but are beyond the scope of this report and will be reported elsewhere.

1. Modulator

The interaction of the electron beam with the laser beam in the periodic field of the undulator produces an energy modulation. The first point to be addressed is how a transverse laser electric field can actually do this, since the electron's velocity is perpendicular to the laser field, at least to a first approximation. The answer is, of course, that the undulator magnet produces a transverse velocity component in the electron's motion that can couple to the laser light. The matching of the laser wavelength, the electron energy, and the undulator period must be correct, or no resonant energy transfer between the laser and the electron takes place. In an undulator, a relativistic electron emits synchrotron radiation at a fundamental wavelength

$$\lambda = \frac{\lambda_0}{2\gamma^2}(1+K^2)\ , \tag{1}$$

where γ is the electron energy in rest mass units, λ_0 is the magnet period and

$$K = \frac{\lambda_0 e B_0}{2\sqrt{2}\pi mc^2}\ . \tag{2}$$

Here B_0 is the peak value of the magnetic field.

If one views the motion of the electron from an inertial frame moving with the average forward velocity of the electron, $v=\beta^* c$, the laser is Doppler shifted to longer wavelengths, and the undulator period is compressed by the Lorentz contraction. The magnetic field of the the undulator is boosted by a factor of γ^*, and the moving magnetic field produces a strong electric field, so, in the moving frame, the undulator magnet looks like a very strong electromagnetic wave, with a period of λ_0/γ^*. The laser wavelength in the moving frame is

$$\gamma^*(1+\beta^*)\ \lambda\ . \tag{3}$$

Here

$$\gamma^* = \frac{1}{\sqrt{1-\beta^{*2}}} . \tag{4}$$

Due to the magnetic field, the average longitudinal beta is reduced to

$$\beta^* = 1-\frac{1+K^2}{2\gamma^2}, \tag{5}$$

yielding

$$\gamma^{*2} = \frac{\gamma^2}{1+K^2} . \tag{6}$$

In the moving frame, the electron is driven by the sum of the laser field and the transformed undulator field. For small fields, the electron executes simple periodic motion under the influence of the electric field of the undulator wave, since the laser field is very small compared to the boosted undulator field. For large undulator fields, however, corresponding to $K > 1$, the electron is driven to relativistic transverse velocities. The transformed undulator magnetic field then bends the simple harmonic motion into a "figure eight" curve.[12] It is this motion that is responsible for the emission of harmonics of the fundamental frequency given by (1), since, in the moving frame, the electron is actually radiating synchrotron radiation.

When the wavelength of the laser, λ_L, is equal to the spontaneous emission wavelength, λ, given by (1), a resonance condition exists. In the moving frame, the laser and undulator wave have the same wavelength at resonance, λ_0/γ^*. Therefore, a weak stationary standing wave envelope modulation exists, with a period of $\lambda_0/2\gamma^*$. The amplitude of the standing wave is proportional to the product of the laser electric field and the undulator field. The electron's figure eight motion causes it to sample different parts of the standing wave field. This produces a net average force on the electron along the z direction, proportional to the gradient in the electric field intensity, the ponderomotive force. The electron accelerates or decelerates under the influence of this force, and can execute bound energy oscillations (synchrotron oscillations) in the ponderomotive potential. If electrons are uniformly distributed along the beam, for a short interaction time in the undulator-laser standing wave field, electrons gain or lose energy according to their initial phase relative to the standing wave pattern.

In the moving frame, the electric field of the laser $E_L' = E_L/2\gamma^*$ and the electric field, E_0', of the undulator wave is, to order $\frac{1}{\gamma^2}$, equal to $\gamma^* B_0$. Then the envelope of the standing wave can be written as

$$E_0'(1+(E_L'/E_0')\cos 2k'z') , \tag{7}$$

where $k' = \gamma k \equiv 2\pi\gamma/\lambda_0$. The ponderomotive force[13] on the electron is given by

$$F' = -\frac{e^2}{4m\omega^2}\nabla E'^{\,2} . \tag{8}$$

Making use of the the fact that $E_L' << E_0'$ for all realistic lasers, one readily finds

$$F' = -\frac{e^2\lambda_0' E_0' E_L'}{2\pi mc^2}\sin 2k'z' , \tag{9}$$

or, writing the variables in terms of their laboratory values, the maximum value of F' is

$$F'_{max} = \frac{e^2\lambda_0 B_0 E_L}{4\pi\gamma mc^2} . \tag{10}$$

Since F' is parallel to the direction of motion, the force in the lab frame is the same, F'. Thus, the maximum change of energy, measured in the lab frame for a modulator of length $L_0 = N\lambda_0$, is given by

$$\Delta E_{max} = \frac{e^2\lambda_0 B_0 E_L L_0}{4\pi\gamma mc^2} . \tag{11}$$

(In this discussion, we have neglected the effect that the figure eight motion has on the Fourier components of the z motion, and hence on the ponderomotive force. When this is included, expression (11) is multiplied by the usual Bessel function factor, which is of order unity.) The value of ΔE for a given electron depends on its initial phase, Φ, relative to the laser and undulator fields, so,

$$\Delta E = \Delta E_{max} \sin\Phi . \tag{12}$$

Notice that the energy modulation takes place on a length scale $\lambda_0/2\gamma^*$ in the moving frame, so, transforming back to the lab frame, the wavelength of the energy modulation is just

$$\frac{\lambda_0}{2\gamma^{*2}} . \tag{13}$$

At resonance (Eq (1)) this is just λ_L, the laser wavelength. Thus, the interaction of the laser and undulator fields produces a periodic energy modulation with the wavelength of the laser, and proportional to the product of the field strengths and the length of the interaction region.

Notice that the maximum energy modulation occurs when the laser is tuned to resonance. Since an equal number of electrons are accelerated and decelerated, there is no net change in the average energy of the electron beam or the laser field. This is also true in a conventional microwave klystron, where for a monoenergetic electron beam the energy modulation process is lossless, producing theoretically infinite power gain. In the free electron laser net energy gain or loss is achieved by detuning slightly away from resonance. This causes the standing wave envelope in the moving frame to drift, and this moving potential pattern can either accelerate or decelerate electrons in the beam, causing a net energy change. In general, the energy modulation effect we have been discussing is much larger than the average energy shift.

It is apparent that in order for the energy modulation of Eq. (11) to have any significance, it must exceed the natural energy spread of the storage ring, σ_ϵ. At the VUV ring at NSLS, $\sigma_\epsilon = 7.3\times10^4$ eV for γ=700.[14] To estimate the required laser power for a TOK, let us assume a modulator with length L_0=1 meter, K=2, and a laser at λ= 5320Å. focused such that the Rayleigh length of the focus equals the modulator length. Under these conditions, Eq. (11) gives

$$\Delta E_{max} = 7.5\times10^{1}(P_{L\ watts})^{1/2}\ \mathrm{eV}\ , \tag{14}$$

where P_L is the laser power in watts. For $\Delta E_{max}/\sigma_\epsilon$ to be greater than one requires a laser power greater than 325 KW. Such powers are only obtainable from pulsed lasers. Lasers producing pulses with peak power exceeding 100 MW and coherence lengths greater than the electron bunch length are available commercially, however. In practice, a 100 MW laser will produce a modulation $\Delta E_{max}/\sigma_\epsilon$ greater than 20, so, as we will see later, a significant power output up to about the 20th harmonic should be possible.

2. Compactor

The principle of the klystron is that a spatially bunched beam can radiate significantly more power than an unbunched one. It is the function of the compactor section to translate energy modulation into spatial bunching. This is accomplished by the longitudinal dispersion or momentum compaction factor of the bunching section, given by $\alpha = \frac{\gamma}{s}\frac{\partial s}{\partial \gamma}$, where s is distance travelled by the electron. This produces a path difference ΔS given by $\Delta S = \alpha L\ (\Delta E/E)$. For free space, $\alpha = -\frac{1}{\gamma^2}$. We may estimate what length of free space is required by assuming we have used the 100 MW, 5320Å laser described above to impress an energy modulation $\Delta E = 20\sigma_\epsilon$. Then, $\Delta s = 20\sigma_\epsilon \alpha L/\gamma mc^2$. Maximum bunching requires $\Delta s \cong \lambda/\pi$, so,

$$L \approx \left(\frac{\lambda}{\pi}\right)\left(\frac{mc^2}{20\sigma_\epsilon}\right)\gamma^3\ . \tag{15}$$

For the NSLS storage ring $\gamma = 700$ and $\sigma_\epsilon = 7.3\times10^4 eV$, so Eq. (15) yields $L \cong 18$ meters! Clearly, a field-free drift space which could fit in the space available on the NSLS ring ($\sim$2.5 meter) is not sufficient. There are other problems with using free space as a compactor, as have been discussed by others.

In an undulator magnet, however, the electrons do not travel in straight paths, and we must use $s = \beta^* ct$ to derive the momentum compaction.

$$\alpha = \frac{\gamma}{s}\frac{\partial s}{\partial \gamma} = \frac{\gamma}{s}\frac{\partial s}{\partial \beta^*}\frac{\partial \beta^*}{\partial \gamma} = \frac{\gamma}{\beta^*}\frac{\partial \beta^*}{\partial \gamma}\ . \tag{16}$$

Using the definitions of β^* and γ^*, (5) and (6), we get

$$\alpha = -\frac{1+K^2}{\gamma^2}\ . \tag{17}$$

Thus, the effect of the periodic magnetic field is to increase the effective length by a factor $(1+K^2)$. For $K=2$, the length is effectively 5 times longer, so an actual device would be 5 times shorter to achieve the same bunching. Therefore, in our example above where a free space length $L=18$ meters was required, approximately 4 meters with $K=2$ would achieve the same result.

Hence a compactor can be designed so that a given physical length will fully bunch a beam with a given energy modulation. If ΔE. is small, then significant compaction will be required, but if ΔE. is large, which, as will be seen later, is desirable for good conversion efficiency for the higher harmonics, then the natural α of the modulator and radiator undulators will bunch the beam. To reach the 100 Å range in the TOK using an extremely high power excimer laser with a short fundamental wavelength, no separate compactor would be required. Such devices are designed to be "integrated" rather than modular, with the energy modulation, bunching and radiation taking place in a distributed fashion in a single magnet, as shown in Figure 2. In this limit, the system becomes more like an optical traveling wave tube than a klystron.

1. Bunching We are now in a position to calculate the optimal bunching condition and the harmonic content of the bunched beam, following the derivations used by Webster and Slater[2]. When the beam exits the modulator at time t_0, there is very little bunching, and the current $I_0 = e \cdot dn/dt_0$, where n is number of electrons. The current at the exit of the compactor at time t is

$$I(t) = e \cdot dn/dt = I_0 \cdot dt_0/dt. \tag{18}$$

The time t is related to the time t_0 by the relation

$$t = t_0 + \frac{L}{\beta^* c} + \frac{\alpha L}{\beta^* c}\frac{\Delta \mathrm{E}}{\mathrm{E}}. \tag{19}$$

Here ΔE is the energy modulation. Using (12) with Φ written as ωt_0, we obtain

$$I(t) = \frac{I_0}{(1+\eta \cos \omega t_0)}. \tag{20}$$

Here $\eta = 2\pi\Delta s/\lambda$, and $\Delta s = \alpha L \Delta \mathrm{E}/\mathrm{E}$. Note that t_0 is given in terms of t implicitly by Eq. (19).

Eq. (20) describes a current at the entrance to the radiator which is bunched in space. In fact, Eq. (20) describes the adjustment of the compactor to obtain the maximum harmonic content in the beam: for some t, $\cos\omega t_0 = -1$ and $I(t) \rightarrow \infty$ for $\eta = 1$, i.e.,

$$\alpha = \frac{\lambda}{2\pi L}\frac{\mathrm{E}}{\Delta \mathrm{E}}. \tag{21}$$

Eq. (21) is the fundamental design equation for the operation of the compactor section.

Fourier expansion of Eq. (20) yields the charge density as a function of t and z.

$$\rho(z,t) = \rho_0 \left[1 + \sum_{n=1}^{\infty} a_n \cos\frac{2\pi n}{\lambda}\left(\frac{z}{\beta^*} - ct\right) \right], \tag{22}$$

where

$$a_n = 2J_n(n\eta) \tag{23}$$

Note that for large n the Bessel function $J_n(x)$ is a maximum when $x \cong n$, which is just the requirement discussed above to maximize harmonic output, $\eta = 1$.

2. Energy spread Up to this point the initial electron beam as been assumed to be monoenergetic, with no energy spread. Under this assumption, even a vanishingly small energy modulation imposed in the modulator can yield perfect beam bunching if there is sufficient α in the compactor. This is analogous to the fact that the power gain of an ordinary klystron is infinite for a perfectly monoenergetic electron beam. Of course, real electron beams have energy spread, and the higher harmonics in the charge density are reduced by a factor δ_ϵ,

$$\delta_\epsilon = \frac{1}{\sqrt{2\pi}\sigma_\epsilon} \int_{-\infty}^{\infty} \exp-(\frac{\Delta E^2}{2\sigma_\epsilon^2}) \cos(\frac{2\pi n \alpha L \Delta E}{\lambda E})\, d(\Delta E) .$$

$$= \exp-(\frac{1}{2}\frac{n^2}{n_c^2}) . \tag{24}$$

Here n_c is the cut-off harmonic given by

$$n_c = \frac{\Delta E_{max}}{\sigma_\epsilon} . \tag{25}$$

The factor δ_ϵ multiplies the Fourier coefficient a_n in (23). Hence, if the nth harmonic is desired, ΔE_{max} must be greater than $n\sigma_\epsilon$. This may be viewed either as a requirement on the laser or as a design constraint on the electron storage ring. In addition, if a large energy spread is produced by the laser beam, a significant amount of time must pass for the natural synchrotron radiation damping of the storage ring to cool the electron bunch before another laser shot is attempted. If this is not done, the factor δ_ϵ will reduce the higher harmonic output significantly.

3. Angular spread Another factor related to the properties of real electron beams spoils the harmonic output. Until now, we've assumed that the electron beam has no angular spread. If an electron moves through the compactor with an average angle ϕ with respect to the z direction, in a compactor section of length L it will cover an extra distance $L\phi^2/2$, and thus pick up an extra phase $\pi L\phi^2/\lambda$. If there is a Gaussian distribution of angles in the beam with width σ_ϕ,

$$\delta_\phi = \frac{1}{\sqrt{2\pi}\sigma_\phi} \int_{-\infty}^{+\infty} \exp-(\frac{\phi^2}{2\sigma_\phi^2}) \cos(\frac{2\pi n}{\lambda}\frac{L\phi^2}{2})\, d\phi$$

$$= [(1+b)/2b^2]^{1/2} , \tag{26}$$

where

$$b = [1+(\frac{2\pi n L\sigma_\phi^2}{\lambda})^2]^{1/2} .$$

The factor δ_ϕ also multiplies the a_n given by Eq. (23) and is a measure of how the higher harmonic content of the bunched beam is spoiled by the inherent angular divergence of the stored electron beam. In the NSLS VUV storage ring, $\sigma_\phi \cong 10^{-4}$ radian, and for $\lambda = 5320\text{Å}$, $L=1$ meter and $n=11$, Eq. (26) gives $\delta_\phi = .70$. For higher harmonics, $\delta_\phi \sim n^{-1/2}$. This is to be compared with the dependence of δ_ϵ on n for large n, $\delta_\epsilon \sim \exp(-n^2)$. Clearly the effect of energy spread in the storage ring on the high harmonics is much more severe than is the effect of angular spread.

The final expression for the Fourier coefficients of the bunched beam at the exit of the compactor is

$$a_n = 2J_n(n\eta)\delta_\epsilon\delta_\phi . \tag{27}$$

The design of a specific TOK device reduces essentially to maximizing a_n for the harmonic of interest as the beam enters the radiator.

3. Radiator

The normal incoherent radiation from an unbunched beam in enhanced by a large factor when the beam is bunched. Since incoherent radiation is emitted at the fundamental wavelength, given by (1), and its harmonics, the presence of spatial harmonics in the bunching, described by (22), produces an enhancement in the harmonic emission.

1. Coherent Enhancement The enhancement can be shown to be given by

$$\frac{d^2I}{d\omega d\Omega} = |F|^2 \frac{d^2I_0}{d\omega d\Omega} , \tag{28}$$

where the L.H.S. is the emitted energy per unit solid angle and frequency for the bunched beam, and $|F|^2$ is the coherent enhancement factor, multiplying the emitted energy for a single electron. A simplified view of the derivation of F is that since in the nth spatial harmonic in the beam there are $a_n Ne$ electrons distributed longitudinally in a phase coherent manner according to $\cos(\frac{2\pi z}{\lambda})$, then the coherent enhancement is just

$$|F|^2 = \frac{1}{2}(a_n Ne)^2 . \tag{29}$$

This is because these $a_n Ne$ electrons are all radiating in phase, and hence, their radiation adds as the square of the number of electrons. Since N_e is very large, on the order of 10^{11} electrons in a single bunch for NSLS-VUV, the coherent enhancement can be very large. The source brightness for the radiation emitted is proportional to (28), so one can see that optically bunching the electron beam increases source brightness by the coherent enhancement factor, producing laser-like angular divergence and frequency bandwidth.

2. Solid angle and bandwidth To get the total energy emitted one must multiply (28) by the solid angle and fractional linewidth into which the $a_n \cdot Ne$ electrons radiate. For our plane wave description, where the electron beam is uniformly illuminated by the laser,

$$d\Omega = \frac{(\frac{\lambda_L}{n})^2}{(radiating\ area)}$$

$$= \frac{\lambda_L^2}{n^2\, 2\pi\sigma_x\sigma_y} , \tag{30}$$

where σ_x and σ_y are the one sigma widths of the electron beam. If we assume that the coherence length of the laser in the modulator exceeds the storage ring bunch duration, then $d\omega/\omega$ is given by the transform limit of the Gaussian bunch duration,

$$\frac{d\omega}{\omega} = \frac{\lambda_L}{2n\sqrt{\pi}\sigma_z} . \tag{31}$$

Thus we have an expression for the total energy emitted for the nth harmonic,

$$W_n = \frac{d^2 I_0}{d\omega d\Omega}\Big|_{\substack{\omega=\omega_n \\ \theta=0}} \frac{N_e^2 \lambda_L^2 c}{8\sqrt{\pi}\, n^2 \sigma_x \sigma_y \sigma_z} a_n^2 . \tag{32}$$

3. Comparison with unbunched beam in undulator We can compare this to the total incoherent radiation of the unbunched beam passing through the same undulator by noting that

$$d\Omega = \frac{\pi}{nN\gamma^{*2}}$$

$$= \frac{2\pi\lambda}{Ln} , \tag{33}$$

where L is the length of the radiator and N is the number of radiator periods. Also, for the unbunched beam,

$$\frac{d\omega}{\omega} = \frac{1}{nN_r} . \tag{34}$$

Suppose, as a worst case comparision of power levels, that one is able to collect all of the power emitted in either the coherent or incoherent case. Then the ratio of the two is

$$R = \frac{1}{n} \cdot a_n^2 \cdot Ne \frac{\lambda^2}{16\pi^{5/2}} \cdot \frac{N_r L_r}{\sigma_x \sigma_y \sigma_z} \tag{35}$$

To get an estimate of this ratio, assume a fundamental λ of 5320Å, and a radiator of 1 meter, with 20 periods. Also we use the beam dimensions of the VUV ring at NSLS: σ_x=1.0mm, σ_y=.08mm, and σ_z=45mm. Then

$$R = \left(\frac{a_n^2}{n}\right) \cdot (8.4\times10^5) . \tag{36}$$

Further, if we were to measure the intensity through an aperture whose solid angle subtended at the undulator were less than that given by Eq. (30), the measured ratio would be

$$R = a_n^2 \cdot Ne \cdot \frac{N\lambda}{4\sqrt{\pi}\sigma_z}$$

$$= 5\times10^6 \times a_n^2 . \tag{37}$$

Finally, if the light through the small aperture is to be used in an experimental situation where the absorption linewidth is small compared to Eq. (31) (e.g., an atomic resonance), then the power ratio will be just the coherent enhancement factor, (29). The factor a_n may be estimated by assuming that the laser has imposed sufficient energy modulation so that $n_c > n$. Then, since $J_n(n) \sim n^{-1/3}$ for large n, we have $a_n^2 \approx 4n^{-2/3}$.

4. Power output We have calculated output powers from Eq. (32) in the plane wave limit assuming a $\lambda = 5320\text{Å}$ laser with 100 MW input power. These results show peak output powers of approximately 300 watts at 600Å, 3 KW at 1000Å and 30 KW at 1800Å.

CONCLUSION

The above discussion has presented the essential physics of the TOK in a simplified manner. A real device will use focused laser beams, and the energy gain, momentum compaction, and radiation will be at least partially distributed long the whole device. This calculation has been done, with the details to be reported elsewhere. Figures 3,4,5 show the efficiencies of harmonic conversion using all parameters of a real device.

References

[1] Our collaborators at NSLS are: A. van Steenbergen, C. Pellegini and A. Luccio. Professor F. DeMartini of the University of Roma is also a collaborator.

[2] D. L. Webster, J. Applied Phys. **10,** 501 (1939); J. C. Slater, *Microwave Electronics* (Van Nostrand, Princeton, NJ, 1950) p. 222.

[3] K. W. Robinson "Ultra Short Wave Generation", unpublished (1960).

[4] N. A. Vinokurov and A. N. Skrinsky, Preprint InP 77-59, Novosibirsk (1977); N. A. Vinokurov, Proc. 10th Int. Conf. on High Energy Charged Particle Accelerators, Serpukov, **2,** p. 454 (1977).

[5] See, for example, R. Coisson, Part. Acc., **11,** 245 (1981).

[6] P. L.Csonka, Part. Acc. **8,** 225 (1978).

[7] P. L. Csonka, Part. Acc. **11,** 45 (1981); V. Stagno, G. Brautti, T. Clauser, and I. Boscolo, Nuovo Cimento **56B,** 219 (1980); I. Bosocolo and V. Stagno, Nuovo Cimento **58B,** 267 (1980); I. Boscolo, M. Leo, R. A. Leo, G. Soliani and B. Stagno, Opt. Comm. **36,** 337 (1981); W. B. Colson, IEEE J. Quant. Elect. **QE-17,** 1417 (1981); F. DeMartini, *Physics of Quantum Electronics,* **7,** Chapter 32, Addison-Wesley (1980); R. Coisson and F. DeMartini, *Physics of Quantum Electronics,* **9,** Chapter 42, Addison-Wesley (1982); P. Elleaume, *Physics of Quantum Electronics,* **8,** Chapter 5, Addison-Wesley (1982).

[8] S. Baccaro, F. DeMartini, and A. Ghigo, Opt. Lett. **7,** 174 (1982).

[9] A. S. Artamonov, et. al. Nucl. Instr. and Meth., *177,* 174 (1980).

[10] C. Bazin, et al., *Physics of Quantum Electronics,* **8,** Chapter 4, (1982); J. M. J. Ortega, et al., Nucl. Inst. and Meth. (to be published); M. Bellardon, et al., 'Recent Results of the ACO Storage Ring FEL Experiment', presented at the Bendor 1982 FEL Conference; P. Elleaume, 'Optical Klystrons', unpublished paper presented at the 1982 Bendor FEL Conference.

[11] M. Billardon, et.al. Phys. Rev. Lett. *51,* 1652 (1983).

[12] E. M. Purcell, 'Production of Synchrotron Radiation by Wiggler Magnets', unpublished (1972); appears in Proceedings Wiggler Workshop, SLAC, Eds. H. Winick and T. Knight, SSRP Report 77/05 (1977).

[13] F. F. Chen, *Introduction to Plasma Physics,* p 256ff, Plenum Press, (1974).

[14] M. Sands, SLAC-121 (1970).

INTEGRATED DEVICE

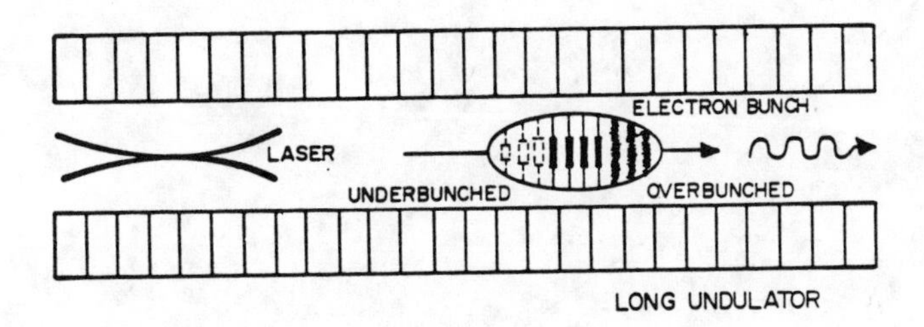

Figure 1

TRANSVERSE OPTICAL KLYSTRON (TOK)

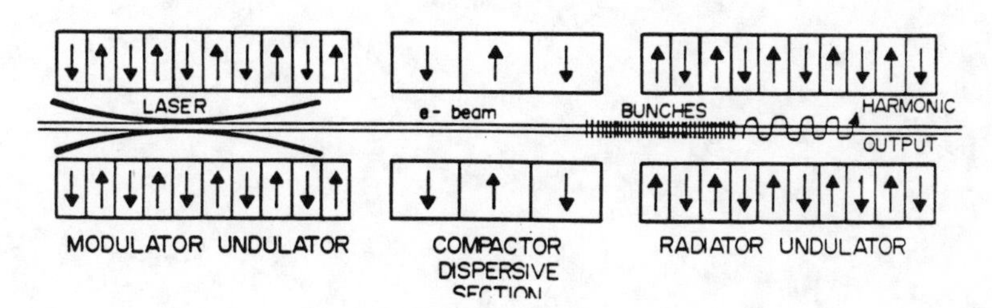

Figure 2

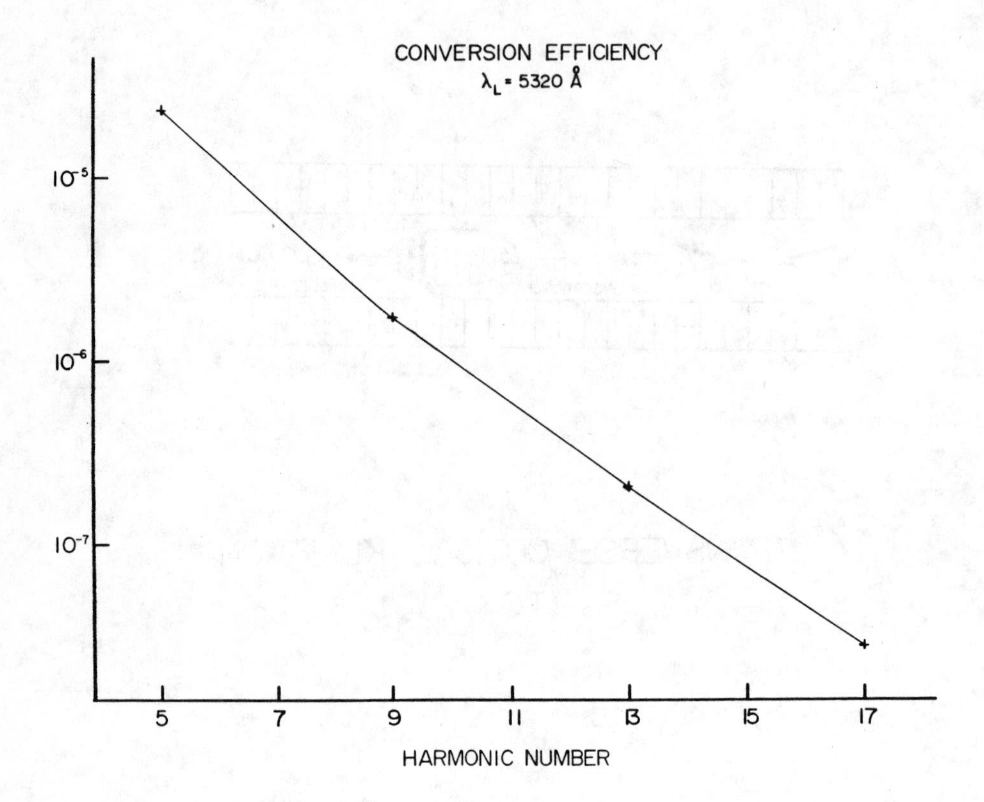
CONVERSION EFFICIENCY
λ_L = 5320 Å
10^{-5}
10^{-6}
10^{-7}
5
7
9
11
13
15
17
HARMONIC NUMBER

Figure 3

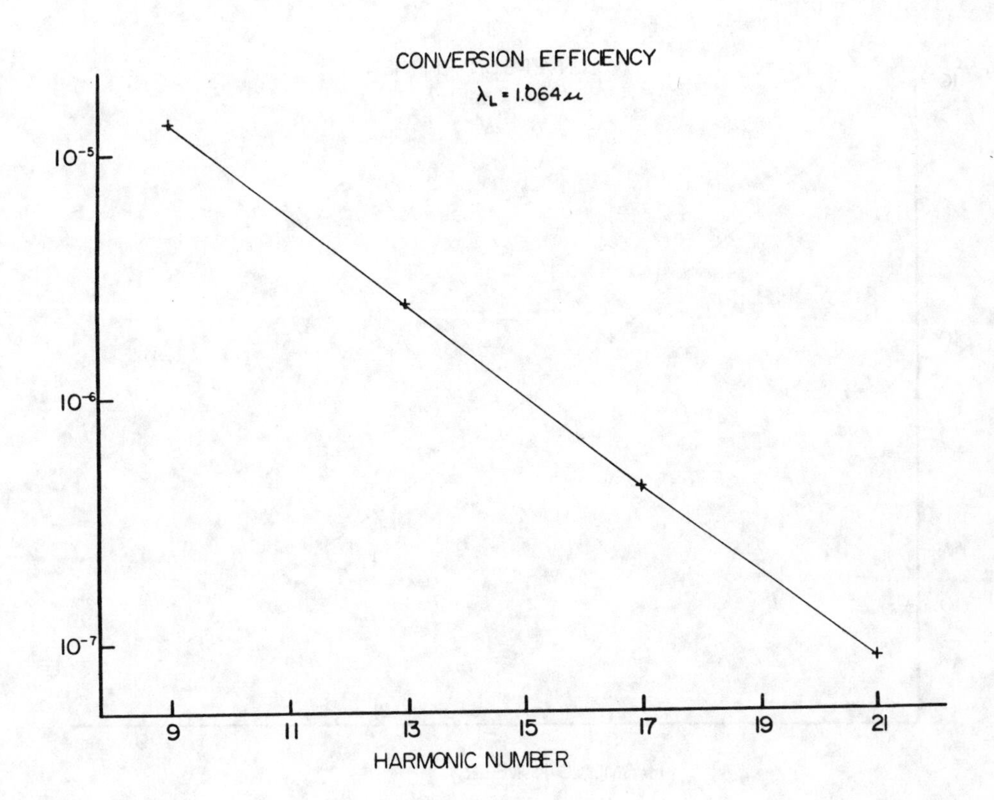

CONVERSION EFFICIENCY
λL = 1.064 μ
10^-5
10^-6
10^-7
9
11
13
15
17
19
21
HARMONIC NUMBER

Figure 4

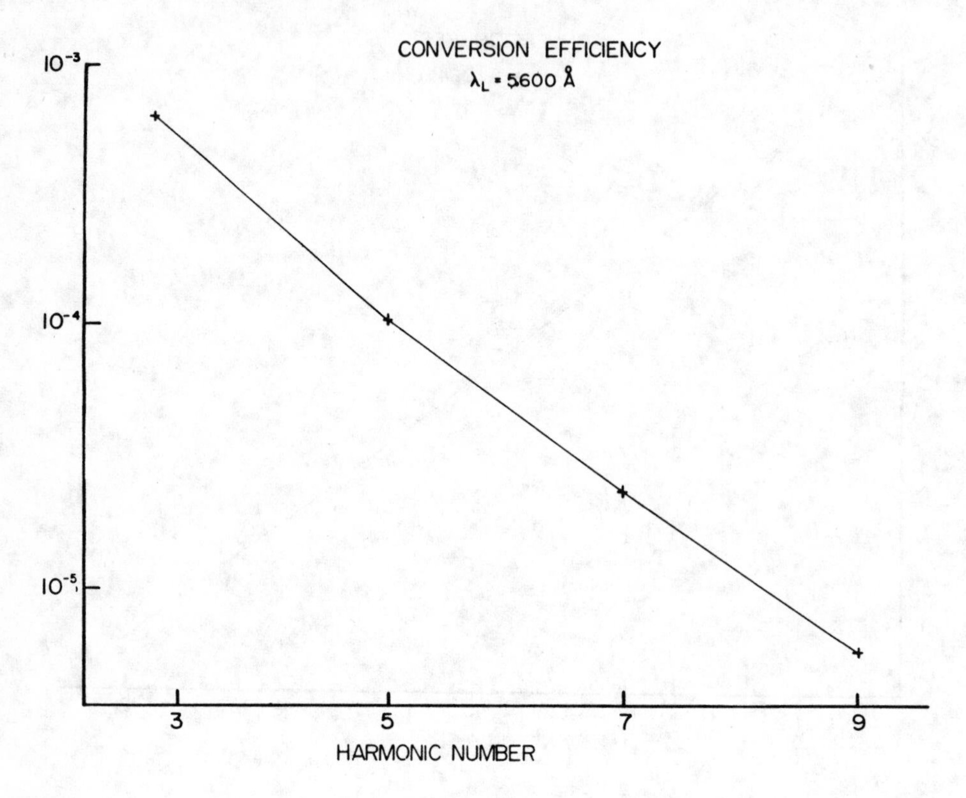
CONVERSION EFFICIENCY
λ_L = 5600 Å
10^{-3}
10^{-4}
10^{-5}
3
5
7
9
HARMONIC NUMBER

Figure 5

AN XUV/VUV FREE-ELECTRON LASER OSCILLATOR

John C. Goldstein, Brian E. Newnam,
Richard K. Cooper, and Jack C. Comly, Jr.
Los Alamos National Laboratory
Los Alamos, New Mexico 87545

ABSTRACT

It is shown, from computations based on a detailed theoretical model, that modest improvements in electron beam and optical mirror technologies will enable a free-electron laser, driven by an rf linear accelerator, to operate in the 50-200-nm range of optical wavelengths.

INTRODUCTION

The problems and prospects of extending free-electron laser technology from the visible and near infrared, where such devices are currently operating, to the ultraviolet were extensively discussed at a recent conference.[1,2] Although significant technical problems must be overcome before free-electron lasers (FELs) can be operated in the VUV (100-200 nm) and the XUV (50-100 nm), the present lack of other intense and tunable sources of coherent radiation at these wavelengths (see however Ref. 3 for a review of the generation of short wavelengths by frequency mixing processes) together with the intrinsic properties of FELs—continuous tunability in wavelength and output in the form of a train of picosecond pulses—make the development of such devices potentially very rewarding.

The basic problem to be solved in order to operate a free-electron laser oscillator in the XUV/VUV is unchanged:[1,2] because only low reflectivity mirrors are available at the shorter wavelengths, a laser must have a large weak-signal gain to overcome the losses of the optical resonator. Hence, we consider only uniform undulator magnets because other (tapered) magnet designs, which yield high efficiency operation at large optical power levels, suffer from proportionately lower small-signal gains. Although the optical klystron design has been successfully used to enhance the small-signal gain of an FEL,[4] this device is useful only if the quality of the driving electron beam is very good,[5] as in an electron storage ring.

The necessity of having a large small-signal gain forces one to consider long undulator magnets which in turn put increasingly severe constraints on allowable electron beam quality. The emittance and energy spread of the electron beam reduce the available small-signal gain from a given undulator from that which a perfectly monoenergetic electron beam would produce. Although the importance of these effects was recognized and treated in our earlier study of a linac-driven XUV FEL,[2] the present work introduces a more refined theoretical model which treats the effects of an imperfect electron beam in a more realistic way.

0094-243X/84/1190293-11 $3.00

In the following sections, the theoretical model is presented and is followed by a tabulation of XUV/VUV system parameters which are derived from the small-signal gain theory and realistic estimates of achievable mirror reflectivities and electron beam quality from rf linacs. Gain saturation curves at five different wavelengths are calculated for the given system parameters and are used to obtain laser output (peak) powers versus mirror reflectivities at optimized output coupling values. The designs developed in this work are refinements of the basic system design given in Ref. 2 to which the reader is referred for additional details.

ELECTRON DYNAMICS

Consider a beam of relativistic electrons propagating along the $\hat{z}$-axis of a plane-polarized uniform undulator magnet. The undulator is assumed to be coaxial with an optical resonator in which only the lowest-order Gaussian mode is excited. The vector potential for the combined static magnetic field and the optical mode is given by $\vec{A} = \hat{x}A_x = \hat{x}\,[A_w(z,y) + (c/\omega)E(x,y,z)\,\sin(kz-\omega t + \phi(x,y,z)]$ where

$$A_w = (B_w/k_w)\cosh(k_w y)\sin(k_w z) \tag{1}$$

and $E(x,y,z)$ and $\phi(x,y,z)$ are the electric-field amplitude and phase of the lowest-order Gaussian mode.[6] The vector potential of Eq. (1) gives rise to a static magnetic field primarily polarized in the $\hat{y}$-direction with a wavelength $\lambda_w = 2\pi/k_w$ and a peak amplitude B_w:

$$\vec{B} = \hat{y}\,B_w\cosh(k_w y)\cos(k_w z) - \hat{z}B_w\sinh(k_w y)\sin(k_w z) \quad . \tag{2}$$

The basic variables of the theory are the relativistic factor γ_o of an electron, whose total energy is $\gamma_o mc^2$, and its phase $\psi = k_w z + kz - \omega t + \phi(x,y,z)$. By differentiating ψ with respect to z (denoted by ψ'), one can easily show that for a strongly relativistic electron (γ_o>>1) moving primarily along the $\hat{z}$-axis (β_x, β_y << 1), ψ obeys the following equation of motion:

$$\psi' = k_w + \phi' - (k/2\gamma_o^2)\,[1 + \gamma_o^2\,(\beta_x^2 + \beta_y^2)] \quad . \tag{3}$$

The dependence of the transverse velocities ($\beta_x = v_x/c$, $\beta_y = v_y/c$) on the electron's z-coordinate will be derived next.

The motion of an electron in the $\hat{x}$-direction is determined from conservation of this component of canonical momentum:

$$\gamma_o m\, v_x = \frac{|e|}{c} A_x + v_{xi} \quad , \tag{4}$$

where v_{xi} is the electron's initial x-velocity upon entering the undulator. Note that A_x does depend on the coordinate x through E and ϕ, but that, for the magnetic-field strengths and optical intensities of interest in this work, the contribution of the optical field E to A_x is some six orders of magnitude smaller than that of the static magnetic field A_w. Hence, Eq. (4) is a good approximation, and leads to

$$\beta_x{}^2 \cong (a_w{}^2/2\gamma_o^2)[1 + (k_w y)^2] + \beta_{xi}{}^2 \tag{5}$$

where $a_w = |e|B_w\lambda_w/(2\pi mc^2)$ is the dimensionless vector potential of the static magnetic field (a_w ~1 usually), we have averaged $\beta_x{}^2$ over a undulator wavelength, and we have assumed that the electron's y-coordinate is small enough to take only the first order term in y of Eq. (1).

In the $\hat{y}$-direction, electrons execute betatron oscillations described by the following trajectory:

$$y(z) = y_i \cos(k_\beta z) + (\beta_{yi}/k_\beta) \sin (k_\beta z) \quad . \tag{6}$$

Here y_i (β_{yi}) is the electron's initial y-position (velocity) on entering the undulator, and the betatron wavelength λ_β is given by

$$\lambda_\beta = 2\pi/k_\beta = \sqrt{2}\gamma_o\lambda_w/a_w \quad . \tag{7}$$

Note that, for this simple harmonic motion, conservation of energy leads to the following relation:

$$\beta_y^2 (z) + (k_\beta y(z))^2 = \beta_{yi}^2 + (k_\beta y_i)^2 \quad . \tag{8}$$

Hence, from Eqs. (5), (6), and (8) we have

$$1 + \gamma_o^2 (\beta_x^2 + \beta_y^2) \cong 1 + \frac{a_w^2}{2} [1 + (k_w y)^2] + \gamma_o^2\beta_{xi}^2 + \gamma_o^2\beta_y^2$$

$$= 1 + \frac{a_w^2}{2} + \gamma_o^2 [\beta_{xi}^2 + \beta_{yi}^2 + (k_\beta y_i)^2] \tag{9}$$

We can now use (9) in (3). Note that the terms that depend on the electron's initial position and velocity are small, so that we write the phase equation as

$$\psi' = k_w + \phi' - (k/2\gamma^2)(1 + \tfrac{1}{2} a_w^2) \tag{10}$$

where γ is the effective value that is shifted slightly from the initial value γ_o by $\delta\gamma$:

$$\gamma = \gamma_o - \delta\gamma \tag{11}$$

$$\delta\gamma = \frac{0.5\gamma_o^{\ 3}}{1+\frac{1}{2}a_w^2}\left[\beta_{xi}^2 + \beta_{yi}^2 + (k_\beta y_i)^2\right] \tag{12}$$

Eqs. (11) and (12) are the central results of this theory: in the FEL, the motion of an electron which enters the undulator with initial transverse velocities (β_{xi}, β_{yi}) and positions (x_i, y_i) can be calculated as if its energy is slightly reduced by $\delta\gamma$ from its initial value γ_o. Accompanying Eqs. (10) - (12) is the usual equation of motion for γ in a plane-polarized undulator.[7] These equations, coupled to Maxwell's equations for the electromagnetic field, are solved numerically[7] to calculate the gain of the FEL.

The initial positions and velocities of electrons entering the undulator are determined from the transverse phase-space distribution of the beam. We shall take this distribution to be the product of uncorrelated Gaussians:

$$\rho(x_i,\beta_{xi};\, y_i,\beta_{yi})\; dx_i d\beta_{xi} dy_i d\beta_{yi}$$

$$= (\varepsilon_x \varepsilon_y)^{-1} \prod_{j=1}^{4} d\alpha_j \exp\left[-(\alpha_j/\bar{\alpha}_j)^2\right] \tag{13}$$

where $\{\alpha_1, \alpha_2, \alpha_3, \alpha_4\} = \{x_i, \beta_{xi}, y_i, \beta_{yi}\}$ and the barred quantities are the constant parameters of the distributions. The transverse emittances ε_x and ε_y are given by the corresponding phase space areas

$$\varepsilon_x = \pi\, \bar{x}_i \bar{\beta}_{xi};\; \varepsilon_y = \pi \bar{y}_i \bar{\beta}_{yi} \tag{14}$$

and we will assume that they are both equal to the emittance ε:

$$\varepsilon_x = \varepsilon_y = \varepsilon \quad . \tag{15}$$

According to Eqs. (11) and (12), the distribution of initial values corresponds to a distribution of γ values. Although we have not derived an analytical expression for this distribution starting from Eq. (13), we have numerically generated this distribution for a given emittance ε. Such a distribution is shown in Fig. 1 with parameters chosen for the 82-nm laser example. Note that the effective energy is always less than the nominal energy $\gamma_o mc^2$ which, in this case, equals 204.4 MeV. This is because γ reflects the reduction in the electron's $\hat{z}$-velocity as it propagates away from the undulator's axis. If the electron has a non-zero transverse initial velocity, clearly v_z will be reduced from what it would have been if no initial transverse velocity were present. Also, if the electron has an initial $\hat{y}$-displacement, it moves in a region of higher magnetic field which produces a larger undulator velocity (v_x) and a consequent reduction of v_z. The theory is solved[7] with a spread of electron energies generated numerically for each case and always qualitatively resembling the shape in Fig. 1.

A distribution of electron γ's corresponds, through the basic FEL resonance condition $\lambda = \lambda_w (1 + \frac{1}{2} a_w^2)/(2\gamma^2)$, to a distribution of optical wavelengths λ. An important point to note here is that the distribution of optical wavelengths corresponding to Fig. 1 is about ten times wider than the gain bandwidth (homogeneous width) due to a monoenergetic beam of electrons[8] which is determined solely by the length and wavelength of the undulator magnet. Hence, the FEL in the present study is strongly inhomogeneously broadened. Since the monoenergetic beam gain curve is antisymmetric about the resonance wavelength,[8] the net gain due to the energy spread distribution of Fig. 1 is approximately given by the derivative of that distribution. Hence, one qualitatively expects to see a gain vs wavelength that has a sharp absorption feature (negative gain) at short wavelengths followed by a low, broad region of positive gain to longer wavelengths. This shape is indeed seen in the numerical calculations. In what follows below, the reported values of gain will refer to the maximum gain calculated for any optical wavelength.

One thus sees that the effect of the beam emittance is to severely inhomogeneously broaden the gain and thus to lower the peak gain. If one crudely models the shape of Fig. 1 as a Gaussian, one finds that the peak small-signal gain varies strongly with ε as ε^{-3}: one power of ε comes from the electron beam density (ε determines the beam area), while ε^{-2} comes from a Lorentzian broadening factor. Hence, it is very important to minimize the emittance of the electron beam in order to maximize the gain.

Another source of energy spread is the finite temporal duration of the electron pulse which implies that some electrons arrive either earlier or later than the moment of peak rf field in the accelerating gaps of the linac. The shape of Fig. 1 has included in it this effect for a Gaussian electron pulse of 9.1 ps full width at half maximum and an rf frequency of 1.3 GHz.

TWO-DIMENSIONAL EFFECTS

Although the theoretical model presented above is essentially one-dimensional, there are two important two-dimensional effects that cannot be neglected: ε determines the transverse area of the electron beam, and the gain for a Gaussian resonator mode differs significantly from that of a plane optical wave.

In the $\hat{y}$-direction, the betatron motion keeps the electron beam confined. If the beam size *is* "matched" to the undulator at its entrance, that is the $\hat{y}$-radius y_i is chosen according to

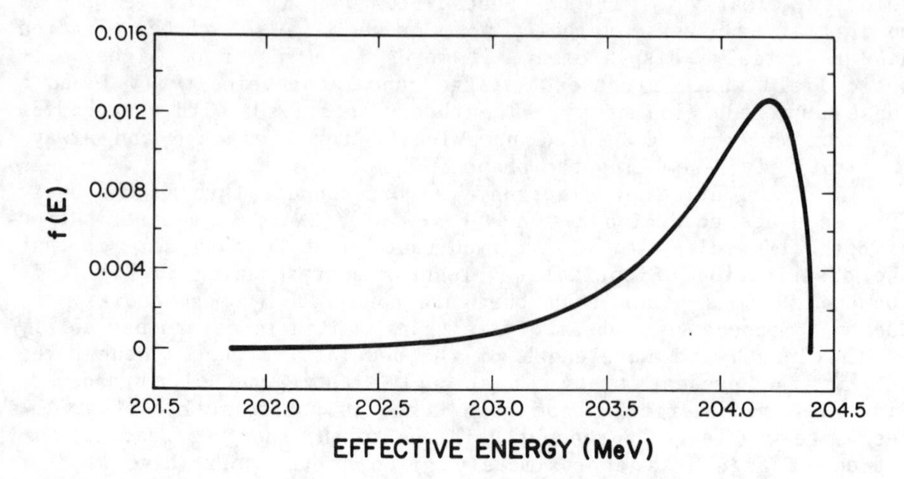

Fig. 1. Distribution of effective electron energy due to transverse emittance.

$$\bar{y}_i^{\,2} = \varepsilon\lambda_\beta/(2\pi^2) \tag{16}$$

then the motion of individual particles will keep this envelope radius constant as the beam propagates down the undulator. In the $\hat{x}$-direction, there is no constraining field, so we take the beam to be externally focused such that at the middle of the undulator of length L_w the beam is circular:

$$\bar{x}^2(z) = \bar{y}_i^{\,2}[1 + ((z-0.5\ L_w)/RR_e)^2]\ , \tag{17}$$

where

$$RR_e = \pi\bar{y}_i^{\,2}/\varepsilon \quad . \tag{18}$$

In general, the beam cross-section is elliptical. We further average $\bar{x}(z)$ over the length of the undulator to finally obtain an average electron beam area A_{eb}:

$$A_{eb} = 0.5\ \pi\bar{y}_i^{\,2} \left\{\sqrt{1+\mu^2} + \mu^{-1}\ \ln\,[\mu + \sqrt{1+\mu^2}]\right\}\ , \tag{19}$$

with

$$\mu \equiv 0.5\ L_w/RR_e \quad . \tag{20}$$

In our previous design of an XUV FEL,[2] we determined that a series of quadrupole doublets external to the undulator could maintain a small beam area. However, this beam focusing increases the electron's transverse velocities. The net result of these two effects on the FEL gain will be presented elsewhere.[9]

We use the results of an earlier study[10] to relate the one-dimensional plane-wave gain, G_{pl}, to the gain for a Gaussian mode, G_g:

$$G_g = G_{pl} \cdot f_g \cdot f_c \quad , \tag{21}$$

where G_{pl} would be the Colson formula[8] for a monoenergetic electron beam but here is obtained numerically according to the theory above, and the filling factor f_g is given by

$$f_g = 1/(1 + A_{lb}/A_{eb}) \quad . \tag{22}$$

The area A_{lb} of the light beam at the focus is equal to the Rayleigh range of the resonator RR_l multiplied by one-half the optical wavelength. The correction factor f_c, which depends on $\alpha = \sqrt{\lambda L_w / A_{eb}}$ and $q = L_w/RR_l$, is tabulated in Ref. 10. There is an optimum value of q for each α, but values of α greater than 5 do not raise further the maximum gain.

SYSTEM RESULTS

We have used the above theory with the previously specified undulator magnet[2] to determine those electron beam properties that lead to acceptable small-signal gains at XUV and VUV wavelengths. The undulator has B_w = 7500 g, λ_w = 1.6 cm, and a_w = 1.12. We find that a twelve-meter undulator is needed for the shorter wavelengths, but that a six-meter undulator is sufficient for the longer wavelengths. The system properties are specified below in Table I.

Table I. System parameters

λ, nm	50	82	100	150	200
γ_o	510.33	400.0	360.86	294.59	255.12
$\frac{\varepsilon}{\pi}$, cm-rad	3.25×10^{-6}	4.88×10^{-6}	5.40×10^{-6}	6.62×10^{-6}	7.64×10^{-6}
$\frac{\varepsilon_n}{\pi}$, cm-rad	1.66×10^{-3}	1.95×10^{-3}	1.95×10^{-3}	1.95×10^{-3}	1.95×10^{-3}
I, Amp	100	100	100	100	100
L_w, cm	1200	1200	1200/600	600	600

Can low emittance values be achieved with high currents? The present state of the art is summarized on Fig. 2 where one sees recent data plotted with the Lawson-Penner curves[2] which represent the empirically-observed performance of somewhat older linacs. The normalized emittance $\varepsilon_n = \gamma_o \varepsilon$ is usually a constant for a given linac current. The charge per bunch times the rf frequency is the average current in the linac which can be related to the peak current (listed in Table I) by assuming a pulse shape and duration. To achieve FELs at wavelengths ≧82 nm we require a normalized emittance (shown on Fig. 2 as the starred dot to which the arrow points) about a factor of nine below the LP curve. However, as can be seen from Fig. 2, several recent results already fall below the LP curve, so that we are here asking for slightly more improvement (~factor of three) over recent results. To achieve oscillation at 50 nm requires a 15% better ε_n than for the longer wavelengths. These improvements appear not to be excessive, and an approach currently under study[2] is to use a laser-irradiated photocathode electron gun that would replace the buncher sections of a linac in which most of the emittance growth is believed to occur.

Using the theory above with the parameters of Table I, we compute the gain saturation curves shown in Fig. 3 for three XUV wavelengths. Using the same resonator design as before,[2] we show in

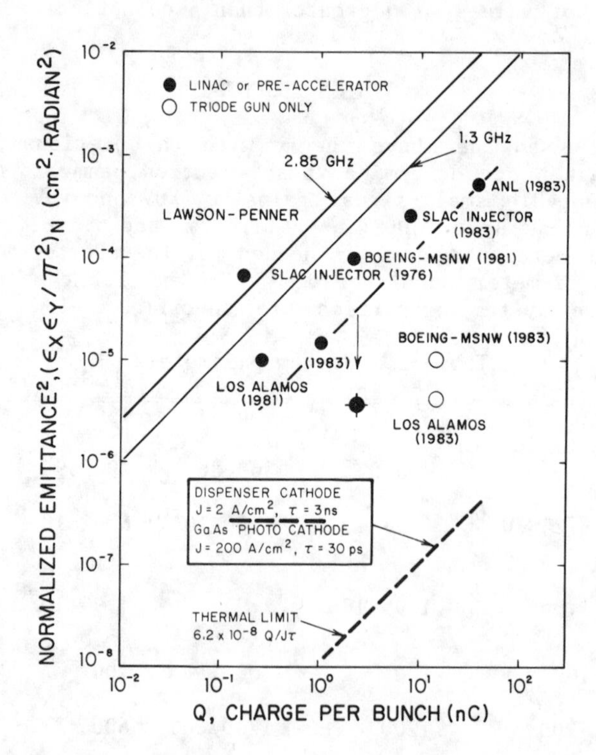

Fig. 2. Emittance-squared vs charge per bunch for linacs.

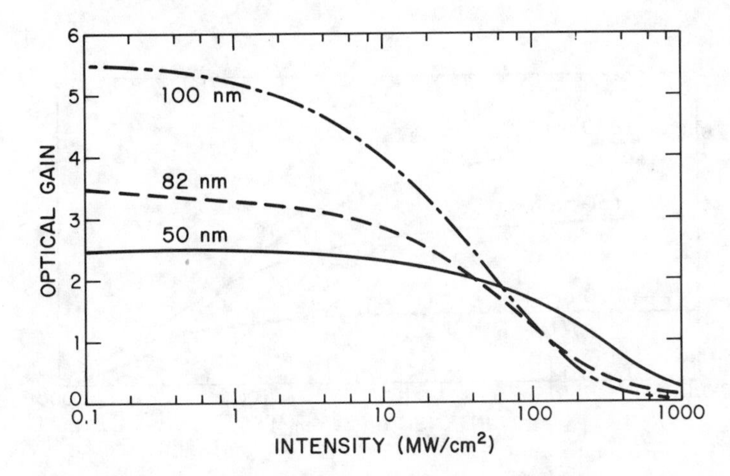

Fig. 3. XUV gain saturation curves.

Fig. 4 the expected peak output power as a function of mirror reflectivity for optimized output coupling at each reflectivity. One notes that peak powers of hundreds of kilowatts are possible if sufficient reflectivity is available; on the other hand, the present design will not oscillate at 50 nm unless a reflectivity of 60% is achieved.

Fig. 5 shows calculated gain saturation curves for the VUV wavelengths using a six-meter undulator. The peak output powers, for optimized output-coupling fractions, are shown for these cases in Fig. 6 where one sees that peak powers of a megawatt are achievable at longer wavelengths with good mirrors. In both wavelength ranges, thermal distortion of the grazing-incidence reflectors[2] limit the average power output to about 200 W (for $\lambda/20$ distortions).

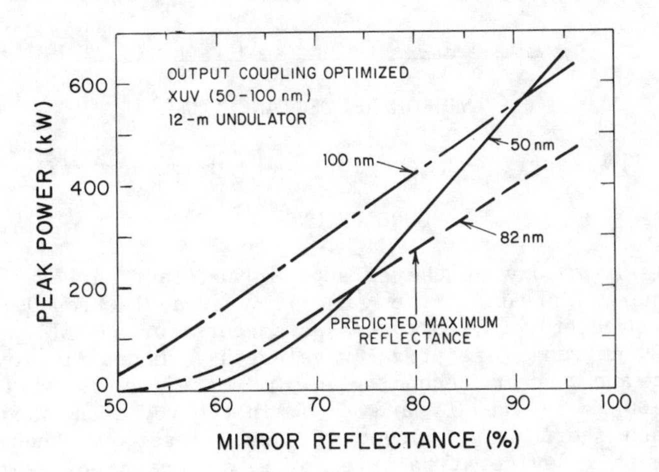

Fig. 4. XUV laser peak output power vs mirror reflectivity.

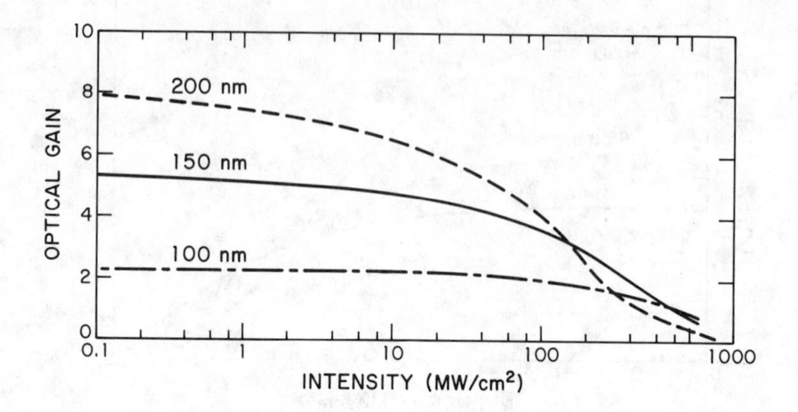

Fig. 5. VUV gain saturation curves.

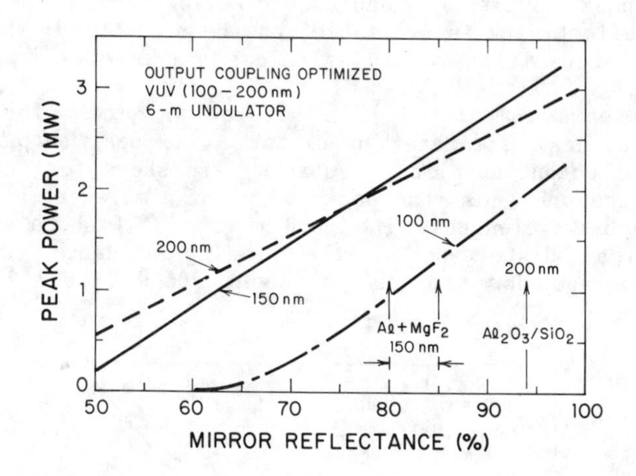

Fig. 6. VUV laser peak output power vs mirror reflectivity.

CONCLUSIONS

A realistic theory of the effects of an electron beam's emittance and energy spread on the gain of a free-electron laser has been used to predict required system parameters for an FEL operating in the 50-200 nm range of optical wavelengths. Modest improvements in rf linac and mirror technologies are required to yield peak powers of ~100 kW in the XUV and ~1 MW in the VUV. The wavelength tunability and short pulse output of such a laser, at these power levels, make this device attractive for a wide range of scientific studies.

ACKNOWLEDGMENTS

The authors are pleased to acknowledge helpful suggestions made by C. A. Brau, J. S. Fraser, R. W. Warren, and J. M. Watson.

REFERENCES

1. Proceedings of the Optical Society of America Topical Conference on Free Electron Generation of Extreme Ultraviolet Coherent Radiation, Brookhaven National Laboratory, Upton, NY, Sept. 19-22, 1983; to be published.
2. B. E. Newnam, J. C. Goldstein, J. S. Fraser, and Richard K. Cooper, ibid.
3. R. Hilbig, G. Hilber, A. Timmermann, and R. Wallenstein, paper MA1, these proceedings.
4. J. Billardon, P. Elleaume, J. Ortega, C. Bazin, M. Velghe, Y. Petroff, D. A. G. Deacon, K. E. Robinson, and J. M. J. Madey, Phys. Rev. Lett. 51, 1652 (1983).
5. P. Elleaume, in Free-Electron Generators of Coherent Radiation, Vol. 8 of the Physics of Quantum Electronics, S. F. Jacobs, G. T. Moore, H. S. Pilloff, M. Sargent III, M. O. Scully, and R. Spitzer, eds. (Addision-Wesley, 1982), p. 119.
6. H. Kogelnik and T. Li, Proc. IEEE 54, 1312 (1966).
7. J. C. Goldstein and W. B. Colson, in Proceedings of the International Conference on Lasers "LASERS '82", R. C. Powell, ed. (STS Press, 1982), p. 218.
8. W. B. Colson in Novel Sources of Coherent Radiation, vol. 5 of the Physics of Quantum Electronics, S. F. Jacobs, M. Sargent III, and M. O. Scully, eds. (Addision-Wesley, 1978), p. 157.
9. T. S. Wang, R. K. Cooper, and J. C. Goldstein, to be published.
10. W. B. Colson and P. Elleaume, Appl. Phys. B 29, 101 (1982).

Extreme-Ultraviolet and X-ray Emission and Amplification by Non-relativistic Electron Beams Traversing a Superlattice

A. E. Kaplan and S. Datta
School of Electrical Engineering
Purdue University
West Lafayette, IN 47907

Abstract

High-energy electrons emit resonant electromagnetic radiation when passing through a spatially periodic medium. It is conventionally assumed that ultra-relativistic electron beams are required to obtain significant emission. We demonstrate theoretically the feasibility of exploiting solid-state superlattices with short periods to obtain both spontaneous and stimulated emission in the far-ultraviolet and soft X-ray range using non-relativistic beams.

Introduction

Fast-moving electrons emit electromagnetic waves when moving from one medium into another with a different dielectric constant [1-3]. This is known as transition radiation and was predicted by Ginzburg and Frank [1]. In a spatially periodic medium the waves emitted at different interfaces interfere so that a resonant emission is obtained when the following condition is satisfied [2-4]:

$$\sqrt{\bar{\epsilon}}\cos\theta = c/v - n\lambda/\ell \tag{1}$$

where λ is the wavelength of radiation, ℓ is the period of the spatially varying dielectric constant ϵ of medium (it is conventionally assumed that the variations are very small), v is velocity of the electrons (assumed normal to the interfaces), θ is the angle between the direction of wave propagation and electron motion, n is an integer, and $\bar{\epsilon}$ is a "mean" ϵ. This condition is readily derived by requiring that the waves emitted at different interfaces interfere constructively at a distant point. Usually the period $\ell >> \lambda$, so that ultra-relativistic beams ($v/c \simeq 1$) are required in order to satisfy Eq(1) for real θ. Recently the possibility of stimulated resonance radiation of ultra-relativistic (~ 50 Gev) electrons traveling through a stack of metal foils was considered [4], with $\ell \sim 7$ cm.

In this paper, we demonstrate the feasibility of using non-relativistic electron beams in order to attain both spontaneous and stimulated emission in the ultraviolet and X-ray range using solid-state superlattices with $\ell \sim 100$ Å so that $\ell/n\lambda \sim 1$. We show that the wavelength of resonant radiation and the required energy of electrons are determined by the parameter $Q = n\lambda_p/\ell$, where λ_p is a "mean" plasma wavelength of the medium. If $n\lambda_p << \ell$ (i.e. $Q << 1$), as is assumed in all previous work [2-4], then the wavelength of the resonant radiation has an order of magnitude of $\lambda \sim \lambda_p Q$, and the kinetic (dimensionless) energy of the electrons eU/mc^2 must exceed the critical amount $\sim 1/\sqrt{Q^2-\theta^2} >> 1$ $(\theta < Q)$ which constitutes the use of ultra-relativistic beams. On the contrary, if the period of the spatially periodic medium is chosen so that $Q >> 1$ (i.e. period ℓ is much shorter than the plasma wavelength, e.g. $\ell \sim$ 50-200 Å), the wavelength of resonant radiation becomes of the order of $\sim \lambda_p/Q = \ell/n$, and the critical kinetic energy of the beam turns out to be extremely small: $eU/mc^2 \sim 1/2Q^2$, which may be less than a few kilovolts even for very short wavelength radiation. The advantages of the proposed method are: (1) the frequency of radiation can be easily tuned in a very wide range by simply varying accelerating potential of beam (which is very hard to do with ultra-relativistic beams) (2) the range of the possible angles of the wave propagation is almost unlimited, and (3) the cost of equipment and energy required for experiments with non-relativistic beams is insignificant compared to large accelerating machines. This last consideration is perhaps the most important.

The main requirements for non-relativistic short-wavelength resonant radiation is a periodic medium with a very short spatial period. Fortunately, the development of molecular beam epitaxy (MBE) and other techniques in recent years has made it possible to grow very thin films (~ 100 Å and less) with precise boundaries. Periodic structures composed of thin films of different materials, in particular superlattices, have also been fabricated [5]. Using these structures, non-relativistic electron beams with energies 20-200 KeV can be used to generate radiation of wavelength 100-200 Å and less. The concept of EM radiation of electron beam in spatially periodic structures in general is known since microwave traveling wave tube amplifiers (with nonrelativistic beams) and their recent optical modification - free-electron lasers (with relativistic beams). An important feature of transition radiation discussed in this paper is that the electron beam travels *through* the material structure rather than in vacuum *above* the structure (as in traveling wave tubes), or through a spatially modulated magnetic field (as in many free-electron lasers). This makes it possible to use very short spatial periods. The problem is the energy transfer from the electron beam to the material structure causing heating and possible damages. This problem will be briefly addressed at the end of this paper.

In this paper, we show that resonant spontaneous emission with a total power of 0.1 mW (around a wavelength ~ 200 Å) can be obtained with a 75 KeV electron beam carrying a current of only 1 mA. The spontaneous emission can be used as a narrowband source by selecting the radiation in a narrow angular range. To get stimulated emission with a gain of 5% per pass, however, requires a significantly larger current (5×10^{10} A/cm^2); to avoid sample burnout it will be necessary to use pulsed operation with pulse lengths ~0.1 ps, which is a difficult task. However, it should be noted that coherent sources are not available at these short wavelengths and getting significant stimulated emission is a difficult problem in general.

Spontaneous Transition Emission

We will first obtain the resonant wavelength of radiation from Eq(1) noting that [3] $\ell\sqrt{\overline{\epsilon}} = \ell_1\sqrt{\epsilon_1} + \ell_2\sqrt{\epsilon_2}$, where $\ell_{1,2}$ and $\epsilon_{1,2}$ are respectively the thicknesses and dielectric constants of alternating layers forming the superlattice, $\ell = \ell_1 + \ell_2$ is its spatial period. For short wavelengths ($\lambda^2 << \lambda_{1,2}^2$), $\epsilon_{1,2} = 1 - \lambda^2/\lambda_{1,2}^2$ where $\lambda_{1,2}$ are the plasma wavelengths of the two materials forming the superlattice, ($\lambda_{1,2}^2 = 4\pi mc^2/e^2N_{1,2}^e$ where N^e is the density of electrons). Thus, the mean dielectric constant may be written as $\overline{\epsilon} \simeq 1 - \lambda^2/\lambda_p^2$, where $\lambda_p^{-2} = \ell^{-1}(\ell_1\lambda_1^{-2} + \ell_2\lambda_2^{-2})]$. Substituting this into (1) and solving it for λ, one gets the resonant wavelengths:

$$\lambda = \lambda_p \frac{Q}{Q^2 + \cos^2\theta} \left[\frac{1}{\beta} \pm \frac{\cos\theta}{Q} \sqrt{(\gamma_{cr}^2 - 1)^{-1} - (\gamma^2 - 1)^{-1}} \right], \qquad (2)$$

where $Q = n\lambda_p/\ell$; $\beta = v/c$; $\gamma = (1-\beta^2)^{-1}$, and $\gamma_{cr} = [1 + (Q^2 - \sin^2\theta)^{-1}]^{1/2}$ is the critical energy required for the excitation of resonant radiation. For $Q << 1$, $\lambda \simeq \lambda_p(Q \pm \sqrt{\gamma_{cr}^{-2} - \gamma^{-2}})$. Here $\gamma_{cr} \simeq (Q^2 - \theta^2)^{-1/2} >> 1$, such that only an ultra-relativistic beam can excite radiation. On the other hand, when $Q >> 1$, the critical kinetic energy turns out to be extremely low, $(eU/mc^2)_{cr} = \gamma_{cr} - 1 \simeq 1/2Q^2$, which is less than 10 KeV for all conventional materials if $\ell \sim 100$ Å. For sufficiently higher (but still non-relativistic) energies eU, Eq(2) gives simply

$$\lambda \simeq \frac{\ell}{n}(\frac{1}{\beta} - \cos\theta) \simeq \frac{\ell}{n}(\sqrt{mc^2/2eU} - \cos\theta); \quad eU << mc^2 = 0.51 \text{ MeV}. \quad (3)$$

For instance, if $\ell = 100$ Å, $n = 1$, $eU = 75$ KeV, and $\theta = 45°$, one has $\lambda = 113.6$ Å; for $n = 10$, $\lambda = 11.3$ Å.

The resonant radiation (i.e. spontaneous emission in the system) can provide a narrowband source of radiation. A single electron traversing multilayer structure with N layers radiates energy I in a solid angle $d\Omega$ in the frequency interval between ω and $\omega + d\omega$, given by [1-4]

$$d^2I/d\omega d\Omega = (d^2I_o/d\omega d\Omega) \cdot 4\sin^2(\xi\ell_1/\ell)\sin^2(\xi N/2)/\sin^2\xi \quad (4)$$

where $\xi = (\frac{1}{\beta} - \sqrt{\overline{\epsilon}}\cos\theta)\pi\ell/\lambda$, and $d^2I_o/d\Omega d\omega$ is a radiation produced by a single interface. According to Ginsburg-Frank theory [1-3], for non-relativistic electrons ($\beta^2 << 1$) and small variations of ϵ ($|\Delta\epsilon| = |\epsilon_1 - \epsilon_2| << \overline{\epsilon}$) the distribution of single-interface radiation is given by

$$d^2I_o/d\Omega d\omega = e^2\beta^2[\Delta\epsilon(\omega)]^2\sin^2\theta/4\pi^2c \quad (5)$$

If the number N of layers is sufficiently large, ($N >> |\overline{\epsilon}/\Delta\epsilon|$), Eq(4) provides for very narrow spectral peaks of radiation for each particular angle θ (with central wavelength determined by Eq(3)), which also implies that any frequency is radiated in a very narrow intervals of angles. Noting that $\Delta\epsilon = \lambda^2(\lambda_1^{-2} - \lambda_2^{-2})$, and integrating Eq(4) over ω and Ω (with $d\Omega = 2\pi\sin\theta d\theta$), one gets the total radiation in each order n

$$I \simeq 16e^2L\ell^2(\lambda_1^{-2} - \lambda_2^{-2})^2\sin^2(n\pi\ell_1/\ell)/3\beta n^4\pi, \quad (6)$$

(where $L = N\ell/2$ is the total thickness of the structure) with the wavelengths of radiation being in the range $\frac{\ell}{n}(\frac{1}{\beta} - 1) < \lambda < \frac{\ell}{n}(\frac{1}{\beta} + 1)$. The total energy of radiation increases as speed β decreases. In order to calculate the power of resonant radiation emitted by an electron beam with an electrical current J, one must multiply Eqs(4)-(6) by J/e. If $\ell = 100$ Å, $\ell_1 = \ell_2 = \ell/2$, $L = 1$ μm, $eU = 75$KeV, $J = 1$mA, and $\lambda_1 \simeq 400$ Å (e.g. Zn, Cu, Ag or Au), $\lambda_2 \simeq 800$ Å (e.g. Si or Ge, see [6]), the system can provide a radiation of first harmonic ($n = 1$) with a total power ~ 0.1 mW and a mean wavelength ~ 200 Å.

Stimulated Emission (Amplification)

We will derive now an amplification caused by the stimulated emission. This effect may be viewed in the following way. An EM wave having a wave vector component $k_z = k_o\cos\theta$ along the axis z (which coincides with the electron trajectory) produces the higher order spatial harmonics with $k_{z_n} = k_z \pm 2\pi n/\ell$ which is due to the periodicity of medium. The phase velocities of these harmonics along the axis x are, therefore, $v_n = c/\sqrt{\overline{\epsilon}}(\cos\theta \pm 2\pi n/\ell k_o)$. If the resonant condition (1) for λ is fulfilled, one of these phase velocities coincides with the speed of the electron that results in an exchange of energy between the EM wave and the electron. For some frequencies in the neighborhood of resonance, the electron loses energy to the EM wave; this results in a coherent gain of the wave, or stimulated emission.

Essentially, this resembles a common mechanism of amplification for many kinds of microwave devices based on the interaction of electrons with "slow" EM wave. The important point is to find the intensities of the resonant spatial harmonics of the field. In all the previous work on resonant radiation [2-4] it is assumed that $\lambda << \ell$ (which is always valid in the ultra-relativistic case, see the introductory section). This allows one to use the WKB approximation. This approximation is not valid in our case since λ may be of order or longer that ℓ. Instead we will find a solution of the exact wave equation (with periodic parameters) based on the assumption of smallness of variations of susceptibility (i.e., $|\Delta\epsilon/\bar{\epsilon}| << 1$, which is always true for short wavelengths); no assumption is made regarding the ratio λ/ℓ. Furthermore, the spatial variation of $\epsilon(z)$ is usually approximated by a cosine function [2-4]. In this approach, the relative amplitude ρ_n of nth harmonic of the EM field is $\rho_n \sim (\Delta\epsilon/\bar{\epsilon})^n$, so that for small $\Delta\epsilon$ ρ_n is negligible for all but the smallest n (= 1). In our approach, we can treat any arbitrary function of $\epsilon(z)$, in particular the true rectangular function. We show that ρ_n falls off algebraically like the Fourier coefficients of $\epsilon(z)$. Significant radiation is expected even for large n provided the interfaces are sharp enough. In this paper, we approach the problem using a single-particle picture which provides direct insight into the mechanism of the electron-EM wave interaction. The problem can also be treated using either the Boltzmann equation [4] or a quantum mechanical formalism; we plan to address these aspects in a subsequent publication [7].

We consider the exact Maxwell equation for the EM field [9] with $\epsilon(z)$ being an arbitrary periodic function in z. We assume a plane wave; it can be shown that only the EM wave with its electric field $\vec{E}$ polarized in the plane of incidence (i.e. plane x,z) may be amplified by the beam [8]. By virtue of the Floquet's theorem for wave equations with periodic coefficients [9], any component of the EM field can be written as a sum of spatial harmonics:

$$u = u_o \exp(j\vec{k}\cdot\vec{r} - j\omega t)[1 + \sum_{n \neq 0} \rho_n \exp(2jn\pi z/\ell + j\phi_n)]; \tag{7}$$

where $\vec{k}\cdot\vec{r} = k_o x\sin\theta + k_o z\cos\theta$, and ρ_n is the amplitude of the nth spatial harmonics. We make the conventional assumption that there is no retroreflection, which is valid if $N|\Delta\epsilon/\bar{\epsilon}|^2 << 1$, and $|\Delta\epsilon/\epsilon_o| << \cos\theta$. This assumption is strictly true in the vicinity of $\theta = 45^\circ$ (see [8]). Substituting the EM field in the form (7) into the Maxwell equations, collecting the terms with $\exp(j\vec{k}\cdot\vec{r} + 2jn\pi z/\ell)$ for each particular n and retaining only terms that are first order in $a_n (<< \cos\theta)$, where the a_n's are the Fourie coefficients of $\epsilon(z)$: $\epsilon(z) = \bar{\epsilon} + \sum_{n=1}^{\infty} a_n \cos(2n\pi z/\ell + \psi_n)$, one gets the amplitudes u_o, ρ_n of the spatial harmonics of nonvanishing components of electric and magnetic fields (E_x, H_y, E_z): $E_{x_o} = E_o\cos\theta$; $H_{y_o} = E_o\sqrt{\bar{\epsilon}}$; $E_{z_o} = -E_o\sin\theta$, and

$$\begin{pmatrix} \rho_{x_n} \\ \rho_{y_n} \\ \rho_{z_n} \end{pmatrix} = \frac{a_n/2}{q(2\cos\theta + q)} \begin{pmatrix} 1 + q\sin^2\theta/\cos\theta \\ 1 + q\cos\theta \\ 1 - q\cos\theta - q^2 \end{pmatrix}; \tag{8}$$

where E_o is the amplitude of the principal harmonic of total electric field, and $q = 2\pi n/\ell k_o = \lambda n/\ell\sqrt{\bar{\epsilon}}$. Further calculations are based on the conventional model of energy exchange between the EM field and an electron which is used, e.g. in the theory of free-electron lasers (see e.g. [10]). From the Lorentz equation $mc\, d(\vec{\beta}\gamma)/dt = e(\vec{E}+[\vec{\beta}x\vec{H}])$. one gets the equation for the energy $\mathcal{E} = \gamma mc^2$ of electron

$$d\mathcal{E}/dt = ec(\vec{\beta}\vec{E}) = ec(\tilde{\vec{\beta}}\tilde{\vec{E}}+\tilde{\vec{E}}\Delta\vec{\beta}+\tilde{\vec{\beta}}\Delta\vec{E}); \tag{9}$$

where $\vec{E} = \vec{E}[\vec{r}(t),t]$ is the field at the instantaneous location of the electron, $\tilde{\vec{\beta}}$ and $\tilde{\vec{E}}$ are unperturbed vectors, $\Delta\vec{\beta}$ is a small perturbation of electron velocity due to interaction, and $\Delta\vec{E}$ is a small perturbation of the field seen by the electron due its spatial displacement in respect to the unperturbed trajectory, i.e.

$$\Delta\vec{E} = \frac{\partial\tilde{\vec{E}}}{\partial z}\Delta z(t); \Delta z = c\int_0^t \Delta\beta_z dt \tag{10}$$

For the assumed polarization of the field, it follows from the Lorentz equation that

$$\Delta\beta_x = \frac{e}{\gamma mc}\int_0^t (E_x-\beta H_y)dt;\ \Delta\beta_z = \frac{e}{\gamma^3 mc}\int_0^t E_z dt \tag{11}$$

In Eqs (9)-(11) one has to take into account only that particular n^{th} component of the wave which is "resonant" to the speed of electron, i.e. that one with $|\xi-n\pi| << 1$. After substituting $z = c\beta t$ and the amplitudes (8) of the proper resonant harmonic of E_x,H_y, and E_z, into (9)-(11), integrating over the temporal interval $[0,\tau = L/\beta c]$, where τ is a time for an electron to pass through the superlattice, one has to average the result over all the possible phases ϕ_n of the relevant field harmonic. We denote this operation by angle brackets. Note that the term $<\tilde{\vec{\beta}}\tilde{\vec{E}}>$ in (9) vanishes, i.e. the stimulated emission is only due to changes in the electron motion caused by the field. Finally, one gets the total averaged change of the electron energy per pass:

$$<\Delta\mathcal{E}_n> = \pi\frac{e^2E_o^2L^3}{m\beta^3c^2\gamma\lambda}\cdot\frac{1}{(\nu_n\tau)^3}[\rho_{x_n}\cos\theta(\rho_{x_n}\cos\theta-\beta\sqrt{\epsilon}\rho_{y_n})\frac{\nu_n}{\omega}\times$$

$$(1-\cos\nu_n\tau)+\gamma^{-2}\rho_{z_n}^2\sin^2\theta(-2+2\cos\nu_n\tau+\nu_n\tau\sin\nu_n\tau)]; \tag{12}$$

where $\nu_n = \omega[\beta\sqrt{\epsilon}(\cos\theta+q)-1] = \beta c(n\pi-\xi)/\ell$ is a resonant factor. For some ν_n, the change of energy $<\Delta\mathcal{E}>$ becomes negative which constitutes the gain of EM field, $<\Delta\mathcal{E}_{EM}> = -<\Delta\mathcal{E}>$. In the non-relativistic case, the main contribution to the change of energy is due to z-components of $\Delta\vec{\beta}$ and $\Delta\vec{E}$ i.e., in (12) $\frac{\nu}{\omega}$ $(\rho_x^2, \rho_x\rho_y) << \rho_z^2 \simeq (a_n/2)^2(1-\beta\cos\theta)^2$. Replacing the term $(-2+2\cos\nu\tau+\nu\tau\sin\nu\tau)/\nu^3\tau^3$ by its negative extremum $-4/\pi^3$ $(\nu\tau \simeq \pi)$, one gets the maximal EM-wave gain per electron per pass:

$$<\Delta\mathcal{E}_{EM}> = a_n^2e^2E_o^2L^3\sin^2\theta(1/\beta-\cos\theta)^2/\pi^2mc^2\beta\lambda \tag{13}$$

In order to obtain an amplification Γ per pass in the system bombarded by an electron beam with the density of electric current $i(A/cm^2)$, one has to multiply $<\Delta\mathcal{E}_{EM}>$ by i/e and divide by the energy flux of incident EM wave per unity area of the interface $E_o^2\cos\theta/2R$, where $R = 377\Omega$ is the vacuum impedance. One has also to take into account that for rectangular form of $\epsilon(z)$, $a_n/2 = (\Delta\epsilon/n\pi)\sin(n\pi\ell_1/\ell)$ with $\Delta\epsilon = \lambda^2(\lambda_1^{-2}-\lambda_2^{-2})$. Bearing in mind a resonant condition (3), one finally gets the maximal EM wave amplification per pass:

$$\Gamma = 8\mu ieRL^3\sin^2(\pi n\ell_1/\ell)\sin^2\theta/mc^2\ell\pi^4\cos\theta, \tag{14}$$

where

$$\mu = \beta^{-1}\ell^4(\lambda_1^{-2}-\lambda_2^{-2})^2(1/\beta-\cos\theta)^5 n^{-5} = \lambda^5\ell^{-1}(\lambda_1^{-2}-\lambda_2^{-2})^2(\cos\theta+n\lambda/\ell).$$

If $\lambda = 140$Å, $\ell = 100$Å, $\ell_1 = \ell_2 = 50$Å, $n = 1$, $L = 1\mu$m, $\lambda_1 \simeq 400$ Å, $\lambda_2 \simeq 800$ Å, $\theta = 45^\circ$, and $i = 5\times10^{10}$ A/cm^2 [4] (i.g. beam of 2 μm diameter with a current $\sim 1.5\times10^3$ A), one gets an amplification $\Gamma \simeq 5\%$ per pass. The required speed of electrons is $\beta \simeq 0.474$ which corresponds to energy eU = 69 KeV. For larger λ the amplification increases drastically. With the mirrors situated outside the superlattice to form a Fabry-Perot resonator to provide feedback, the system becomes a short-wave laser which may transform significant portion of energy of electron beam into coherent radiation. It is obvious that the amplifiers and lasers based on the proposed principle should work in the short pulse regime of operation, with the duration of current pulse being determined by the heating, ionization, diffusion of absorbed electrons, etc. As a rough approximation, the per atom heating rate caused by the energy losses of the electron beam [11], is

$$(n^e/N^a)d\mathcal{E}/dt = 4\pi i Z(mv^2)^{-1}\ln(\gamma^2 mv^3/e^2\omega) \quad (15)$$

where $n^e = i/ev$ is the electron beam density, N^a is the atomic density of the material, Z is the atomic number. For the parameters mentioned above, the duration of the current pulse must be shorter than $\sim10^{-13}$ sec in order for the energy transfer per atom to be of order of ~1 eV or less. One may note though that in the case of ultra-relativistic beams [4] with energy ~50 GeV for the same current, the losses (15) are even greater, such that one needs even shorter pulses. It will probably be hard to obtain such short and powerful pulses, but it may prove worthwhile. However, the first step is to attain spontaneous resonant emission as described in the beginning of this letter. The spontaneous emission can be used as a very narrowband source of radiation by selecting narrow range of angles [Eq. (4)]. It should also be noted that the spontaneous radiaton intensity depends on the total current in the electron beam unlike the stimulated emission gain which depends on the current density. Since there is no constraint on the current density, the conditions to obtan spontaneous emission are much more relaxed. According to Eqs. (4-6), the electrical current required to observe spontaneous emission with the same spotsize of the beam is 10^6–10^9 less than that required for stimulated emission.

Conclusion

In conclusion, we have demonstrated the feasibility of generating far-ultraviolet and soft X-ray radiation by electron beams with relatively low, non-relativistic energies, traversing the solid-state superlattice composed of very thin periodic layers. The use of low energies is a desirable feature as compared with ultra-relativistic beams. The proposed system can be used as a very efficient noncoherent source of narrowband radiation, and, under special conditions, as an amplifier and laser.

We thankfully appreciate very useful discussions with G. Ascarelli, R. Gunshor, as well as with P. Kelley, A. Calawa, N. Economou, and all participants at the seminar given by A. E. K. at Lincoln Lab, M.I.T.

This work is supported by the US Air Force Office of Scientific Research.

References

1. V. L. Ginzburg and I. M. Frank, Zh. Eksp. Theor. Fiz., 16, 15 (1946) (in Russian).
2. G. M. Garibyan, Sov. Phys. JETP, 33, 23 (1971).
3. M. L. Ter-Mikaelian, High Energy Electromagnetic Processes in Condensed Media, Wiley Interscience, New York, 1972.
4. M. A. Piestrup and P. F. Finman, IEEE J. of Quan. Elect., QE–19, 357 (1983).
5. See for example, A. Madhukar, J. Vac. Sci. Technol., 20, 149 (1982); A.C. Gossard, Thin Solid Films, 57, 3 (1979).
6. P. M. Platzman and P. A. Wolff, Waves and Interactions in Solid State Plasmas, Academic Press, New York, 1973.
7. S. Datta and A. E. Kaplan, to be published elsewhere.
8. Incidentally, this fact brings up a substantial advantage. For the polarization in the plane of incidence, there is an angle θ at which the retroreflection at the layer interfaces vanishes altogether; it is the so called Brewster angle. In our case ($|\Delta\epsilon| << \overline{\epsilon}$) this angle is $\theta \sim 45^\circ$
9. L. M. Brekhovskikh, Waves in layered media, 2nd edition, Academic Press, N.Y., 1980; S. M. Rytov, Zh. Eksp. Teor. Fiz. 29, 605 (1955) [Sov. Phys. - JETP, 2, 466 (1956)].
10. A. Fruchtman, J. Appl. Phys. 54, 4289 (1983).
11. J. D. Jackson, Classical Electrodynamics, Eq (13.13), John Wiley & Sons, Inc., N.Y., 1962.

Multilayer Optics for the CUV: Synthesis, Experimental Results, and Projected Performance

Troy W. Barbee, Jr.

Materials Sciences Engineering

Stanford University, Stanford, CA

Advanced vapor deposition techniques have yielded a new class of multi-layer optics for soft X-rays. Applications of these processes to fabrication of XUV optics promises significant improvements. Synthesis processes, current experimental results, and projected performance are discussed.

0094-243X/84/1190311-01 $3.00

MIRRORS FOR THE EXTREME ULTRAVIOLET

Eberhard Spiller
IBM T.J. Watson Research Center

ABSTRACT

The possibilities to obtain high reflectivity mirrors for wavelengths below λ = 2000 Å are reviewed. For wavelengths above λ = 1200 Å reflectivities close to R = 100% are theoretically possible and values up to R = 95% have been obtained with either all dielectric multilayers or with metal-dielectric multilayers. Below 1000 Å all materials become absorbing and single-layer reflectivities decrease with decreasing wavelengths to unmeasurably low values below λ = 200 Å. However, multilayer design can compensate for this loss in reflectivity and enhance reflectivities by 3 to 5 orders of magnitude. Multilayer reflectivities above 20% have been measured for the λ = 50-200 Å wavelength range.

INTRODUCTION

The extreme ultraviolet is a difficult region for high quality and low loss optical elements. All dielectric materials become absorbing when the photon energy is increased above their bandgap, and below λ = 1100 Å there are no transparent materials available which can be used in a conventional multilayer coating to enhance the reflectivity of a mirror. A clean aluminum mirror still has a useful normal incidence reflectivity above R = 50% for wavelengths as short as 800 Å, but only if oxidation of its surface can be avoided. Below 800 Å all materials have low normal incidence reflectivity and this reflectivity decreases further with decreasing wavelength to a value around 1% at λ = 200 Å. At still shorter wavelengths nearly all electrons in a material can be considered as free, and the reflectivity drops proportional to λ^4 reaching values of less than 10^{-5} around λ = 50 Å. However, the absorption decreases even stronger with decreasing wavelength and this decreasing absorption makes multilayers possible again at shorter wavelengths. The peak reflectivity of a multilayer at a specific wavelength is limited by the lowest absorption available at this wavelength and is now increasing with decreasing wavelength. At very short wavelengths in the x-ray region very high reflectivities can be obtained in theory. In a practical coating in this region the performance is not limited by available optical constants but by the roughness (or sharpness) of the boundaries in a multilayer and by thickness errors.

We will discuss in this paper the theoretical possibilities and practical difficulties to obtain near normal incidence mirrors for the ultraviolet covering the wavelength region from $\lambda \approx$ 2000 Å down to the x-ray region. We will first give some theoretical background and then discuss the region

0094-243X/84/1190312-12 $3.00

above $\lambda = 1100$ Å, where some dielectric materials are still available and the wavelength region below $\lambda = 1000$ Å where all materials are absorbing.

REFLECTIVITY AND OPTICAL CONSTANTS

The Fresnel equations give the reflectivity at the boundary of two materials with the (complex) refractive indices n_1 and n_2. For normal incidence the reflected amplitude is given by

$$r = \frac{n_1 - n_2}{n_1 + n_2}. \tag{1}$$

In the x-ray region the optical constants are usually expressed in the form

$$n = 1 - \delta - ik = 1 - \frac{r_o \lambda^2}{2\pi} N_{at}(f_1 + if_2), \tag{2}$$

where $r_0 = e^2/mc^2 = 2.82 \times 10^{-13}$ cm is the classical electron radius, N_{at} is the number of atoms per unit volume and $f = f_1 + if_2$ is the atomic scattering factor. The atomic scattering factors and the mass absorption coefficients of all elements have been tabulated for photon energies above E = 100 eV or wavelengths shorter than $\lambda = 124$ Å.[1,2] For these wavelengths the optical constants of each compound can be obtained from Eq. 2 by adding the f-values of each element in the compound weighted with the corresponding atom density. This superposition neglects the modification of the wavefunction of an atom during the formation of molecules and solids.

For the interface between vacuum ($n_1 = 1$) and a mirror material with index $n = 1 - \delta - ik$, we obtain from Eq. 1 for the reflected amplitude

$$r = \frac{1-n}{1+n} = \frac{\delta + ik}{2 - \delta - ik} \tag{3}$$

and for the reflectivity

$$R = rr^* = \frac{\delta^2 + k^2}{(2-\delta)^2 + k^2} \tag{4}$$

For wavelengths $\lambda < 200$ Å δ and k are small and we can simplify Eq. 4 to

$$R = \frac{1}{4}(\delta^2 + k^2), \tag{5}$$

For the general case that the incident medium is not air, Eq. 5 has to be replaced by

$$R = \frac{1}{4}(\Delta\delta^2 + \Delta k^2) \tag{6}$$

where $\Delta\delta$ and Δk are now the differences in the optical constants of the materials on both sides of a boundary.

High normal incidence reflectivity at the boundary between vacuum and a material requires that either $n \rightarrow 0$, $\delta \rightarrow 1$ or $k \rightarrow \infty$ and some metals like aluminum approach the required optical constants and give good reflectivity. It is worthwhile to note that the reflectivity in Eq. 4 does not change if the sign of k is changed; a material with gain gives the same reflectivity as the material with the corresponding absorption.

All dielectric materials ($k=0$, no absorption) have refractive indices larger than 1 (or δ negative). They could give large reflectivities only for the case that $n \rightarrow \infty$. However, all available dielectrics have refractive indices between 1.3 and 3 in the visible and ultraviolet, and high reflectivity cannot be obtained from a single boundary.

MULTILAYER MIRRORS

The reflectivity of a single boundary can be enhanced by constructing a multilayer structure with many boundaries spaced such that the reflections from all the boundaries add in phase to produce a very high total reflectivity. This principle leads to the quarter wave stack, a design where a layer with high (H) and low (L) refractive index alternate and where each layer has an optical thickness $nd = \lambda/4$ for a normal incidence mirror (Fig. 1a). The detailed theory of optical multilayers is treated in many textbooks and will therefore not be repeated here.[3]

The minimum number of periods N_{min} required to approach a reflectivity close to 100% can be roughly estimated by the condition that $2 \cdot N_{min} |r| = 1$, where the factor of 2 is due to the fact that there are 2 boundaries per period. We obtain with Eq. 1

$$N_{min} = \left| \frac{n_1 + n_2}{2 \cdot \Delta n} \right| , \tag{7a}$$

with $\Delta n = n_1 - n_2$. For XUV wavelengths where δ, $k << 1$ we can use Eq. 6 to obtain

$$N_{min} = 1/\sqrt{\Delta\delta^2 + \Delta k^2} \; . \tag{7b}$$

If two absorption- and scatter-free materials of different indeces are available reflectivities approaching 100% become possible and values R > 99.9% have been reported for the visible.

The performance of the quarter wave stack deteriorates very fast if one or both of the materials is absorbing. For the case that there is still one absorption-free material available, i.e., an absorber is alternated with a nonabsorbing spacer material, a reflectivity close to 100% can still be obtained with the design of Fig. 1b. Very thin layers of the absorber are spaced $\lambda/2$ apart such that they are centered in the nodes of the standing waves produced by the superposition of the incident and reflected wave. In the limit of very many very thin films the intensity in the nodes and the absorption losses approach zero and the reflectivity approaches 100%.[4]

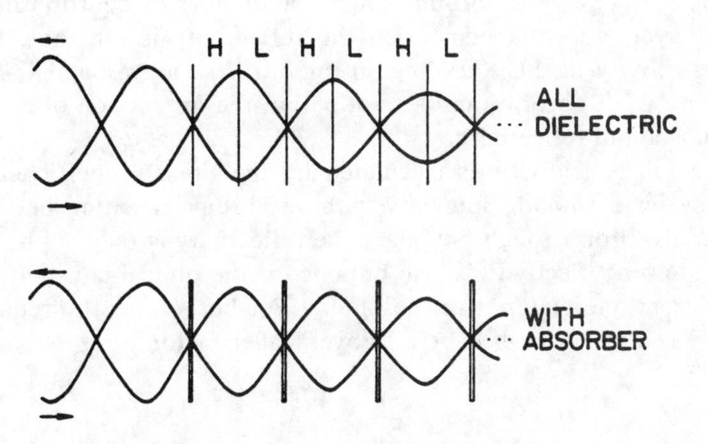

Fig. 1: Two design concepts for high reflectivity multilayer mirrors. In the quarter wave stack (a), absorption-free materials of high (H) and low (L) refractive index alternate such that all boundaries add in phase to the reflected wave. In the design (b), thin films of a strong absorber are separated by an absorption-free spacer material such that each absorber film is located in the node of the standing wave produced by the superposition of incident and reflected wave. Theoretically, both designs can reach 100% reflectivity. Actual designs for the XUV range are between the two designs, and have lower reflectivities due to absorption in the spacer material.

For wavelengths $\lambda < 1100$ Å, no material is absorption-free and the number of layers that can still contribute to the reflectivity is limited by the absorption length in the material. Using the design of Fig. 1b the absorption length is determined by the spacer material and the maximum number of layers (each layer is $\lambda/2$ thick) is given by

$$N_{max} = \frac{1}{2\pi k} . \tag{8}$$

A comparison of N_{max} and N_{min} gives an estimate for the improvements which are possible with a multilayer coating. For $N_{max} >> N_{min}$ there is little limitation by absorption and very high reflectivities close to 100% can be obtained. It is also possible to put coatings for several wavelengths on top of each other to produce broad band mirrors or to design mirrors with a specific reflectivity versus wavelength curve.

For $N_{max} \approx N_{min}$ absorption limits the number of contributing layers in a multilayer, an enhancement in the reflectivity is still obtained and typical reflectivity in the XUV region for this case are around $R \approx 20\%$. For $N_{max} << N_{min}$ absorption does not permit the production of a multilayer with substantial reflectivity.

The full reflectivity as calculated by the Fresnel coefficient is obtained only for a smooth boundary with an abrupt transition between the two materials. For a rough boundary the reflectivity is reduced by scattering, and a lower reflectivity is also obtained if the optical constants change gradually from one material to the other. For both cases the reduction in reflectivity is often described by a Debye-Waller factor

$$R = R_0 \, e^{-\left(\frac{4\pi\sigma}{\lambda}\right)^2} , \tag{9}$$

where R_0 is the reflectivity of perfect boundaries and σ is the rms heights of the roughness or thickness of the transition layer.[5]

Our discussion permits the definition of a strategy for the selection of materials for a multilayer:

1) Select as spacer material the material with the smallest absorption constant available at the design wavelength (for largest N_{max}).

2) Select a second material with the largest possible reflection coefficient at the boundary with the spacer material (for minimum N_{min}, Eq. 7).

3) If several materials give similar values for N_{min} choose the one with the smallest absorption.

4) Make sure the materials can be deposited with sharp, smooth boundaries ($\sigma \leq \lambda/20$).

MIRRORS FOR THE NEAR VACUUM UV (λ = 1100-2000 Å)

For wavelengths above λ = 1100 Å, a clean Al mirror has a reflectivity above 90%. Some dielectric materials with low absorption are still available which permit the fabrication of a dielectric quarter-wave stack especially for the longer wavelengths. Furthermore, it is theoretically possible to approach reflectivities close to 100% for the entire region with the design idea of Fig. 1b using thin films of Al and spacer layers of MgF_2 or LiF. There are, however, some practical difficulties to realize these possibilities.

Clean aluminum mirrors (and probably also Be) have high reflectivities down to about λ = 800 Å. However, aluminum oxide becomes absorbing around λ = 1600 Å and, therefore, the hgih reflectivity at the shorter wavelengths is only maintained if the freshly-deposited mirror is kept in ultra-high vacuum.

The oxidation can be avoided if the aluminum mirror is immediately coated with a protective layer of MgF_2 or LiF. The reflectivity can be peaked at a specific wavelength by adjusting the thickness of the protective layer. With LiF as a protective layer, reflectivities above 80% have been obtained for wavelengths as short as λ = 1026 Å, while the MgF_2 coatings become already absorbing at λ = 1200 Å. LiF is slightly hygroscopic and can also become absorbing by the generation of F-centers by EUV light; therefore, the reflectivity of LiF-coated Al mirrors decreases with aging while MgF_2 coatings are more stable.[6]

The reflectivity of an Al mirror can be further enhanced by overcoating the aluminum film with additional layers of Al and MgF_2. The additional very thin Al films are positioned in the nodes of the standing waves in front of an Al mirror to minimize their absorption and are spaced in such a way that they all add in phase to the reflected wave. Table I shows the thicknesses for several designs with 2 additional Al mirrors (a total of 6 layers in the coatings). The reflectivity of a single Al film can be increased or one can maintain the reflectivity and increase transmission. The decrease in absorption shown in Table I has been experimentally verified for wavelengths around λ = 2000 Å and it should be possible to obtain a similar performance for wavelengths down to λ = 1200 Å.[8]

For wavelengths λ > 1500 Å, a few dielectric materials with different refractive indices (MgF_2, CaF_2, ThF_4) are still available and can be used to fabricate dielectric mirorrs corresponding to the design of Fig. 1a. These mirrors have been reviewed by Stelmack and Flint[9] and reflectivities of 95% have been obtained. It can be expected that these dielectric mirrors have considerably lower damage thresholds than the mirrors containing aluminum.

TABLE I.

Physical thicknesses of the Al and MgF_2 layers in 6-layer structures designed for use as low-loss Fabry-Perot mirrors. Also given are the values of the reflectivity, absorption and transmission at the wavelength specified. Optical constants from Ref. 7.

	THICKNESS (Å)			
	A	B	C	D
1. Al	400	200	180	116.7
2. MgF_2	635	635	600	579.4
3. Al	152	120	71.2	116.7
4. MgF_2	653	653	600	579.4
5. Al	117	100	49.3	116.7
6. MgF_2	667	667	580	579.4
Reflectivity	0.955	0.926	0.879	0.881
Absorption	0.0436	0.0448	0.0503	0.0454
Transmission	0.0017	0.0294	0.0705	0.0739
Wavelength (Å)	2200	2150	1925	1975

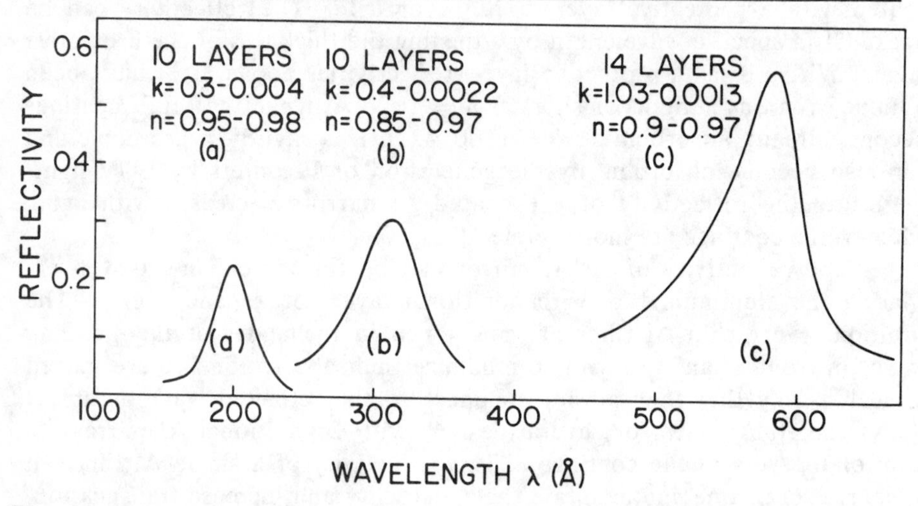

Fig. 2: Reflectivity as a function of wavelength of mirror coatings for the XUV region. The thickness of the layers counted from the substrate are (in Å) (a) 61.4, 40.3, 66.6, 33.4, 71.9, 28.9, 75.2, 26, 77.5, 24.7; (b) 98.9, 51.9, 101.9, 45, 112.5, 38.3, 116.4, 34.9, 119.8, 34; (c) 198.4, 100.3, 237.9, 61, 250.3, 48.7, 257.7, 42.8, 262.8, 36.6, 268, 32.7, 269.4, 32.8. The optical constants used for the substrate are (a) n = 0.95, k = 0.08; (b) n = 0.89, k = 0.902; (c) n = 0.85, k = 0.47. The last layer of each coating (i.e., the first layer toward the incident light) is always a layer of the material with the high k.

For the experimental realization of coatings one has to realize that practically all contaminants are absorbing and small traces of impurities will increase the absorption of a film. Therefore, good vacuum and/or fast deposition rates are required to obtain good performance. Another problem is caused by the fact that a surface plasma wave can be excited along the Al- MgF_2 interface for wavelengths $\lambda > 1500$ Å. Incident energy can be coupled into the surface plasmon by roughness and the roughness of MgF_2 increases with the total MgF_2 thickness within a multilayer. This coupling to the surface plasmon has limited the numbers of layers in the (Al-MgF_2)-multilayer to 6 for wavelengths $\lambda > 1500$ Å. The surface plasmon should be less limiting for wavelengths $\lambda < 1500$ Å, but up to now this has not been confirmed by experiments.[8]

MIRRORS FOR THE XUV ($\lambda < 1000$ Å)

For wavelengths $\lambda = 500 - 1000$ Å, noble heavy metals have been used as mirror coatings for the last 30 years.[6] Maximum normal incidence reflectivities are between 20% and 40% in this region. For wavelengths below $\lambda = 500$ Å, the reflectivity of all materials decreases with decreasing wavelengths to best values around 1% at $\lambda = 200$ Å and $R = 10^{-5}$ for $\lambda = 50$ Å. Around $\lambda = 1000$ Å, the known absorption constants of materials are all between 0.1 and 1. This high absorption does not permit penetration into a multilayer deep enough to enhance the reflectivity (see Eq. 8). Below $\lambda = 500$ Å, δ and k also become small (see Eq. 2) and a multilayer can be used to enhance the reflectivity. Figure 2 shows the calculated performance of three multilayers for this region; the spacer materials are Si for curve (a) and Mg for curves (b) and (c), while the optical constants of the stronger absorber are typical for the heavy metals like Au, Pt, Ir.

Below $\lambda = 200$ Å, the photon energies are higher than most electronic transitions. Each electronic transition contributes a one to the value of f_1 in Eq. 2 such that at any energy f_1 can be considered as the effective number of free electrons at this energy. For higher energies, f_1 approaches the total number of electrons per atom. The value of f_2 is large at a resonance and decreases for larger distances from a resonance. As a consequence, the values for k decrease faster with decreasing wavelengths than the values of δ, and at very short wavelengths materials approach dielectrics with $\delta << k$. The value of N_{max} (Eq. 8) which is limited by the absorption constants, increases faster than the value of N_{min} (Eq. 7) which is determined by the available δ values, and the theoretical peak reflectivity of a multilayer increases with decreasing wavelength. This behavior is demonstrated in Fig. 3 and Fig. 4 for multilayer systems of Re-C and Ni-C. The figures show the general trend of increasing performance towards shorter wavelengths and also the drop in performance at the absorption edges (carbon at $\lambda = 44.8$ Å and Ni at $\lambda = 14.5$ Å).

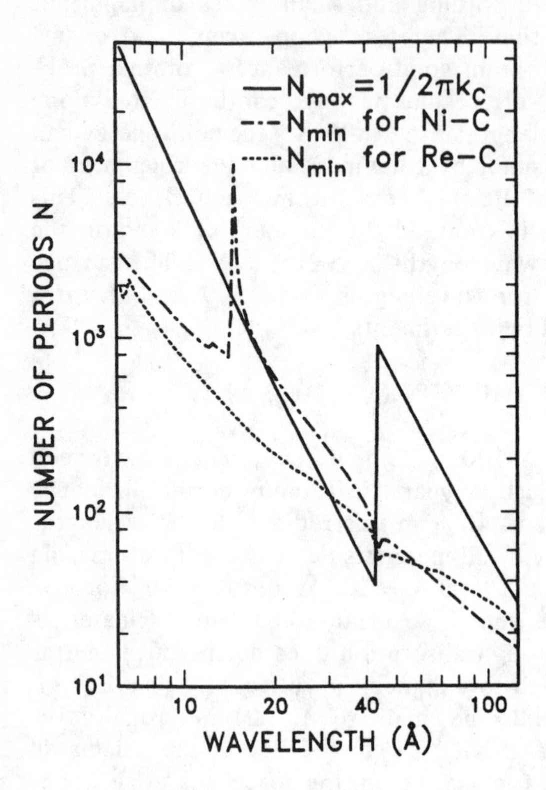

Fig. 3: The maximum number N_{max} of periods possible in a multilayer with a carbon spacer layer due to the absorption of carbon, and the minimum number in a Ni-C and Re-C system required for reflectivities approaching 1. Optical constants from Ref. 1, normal incidence.

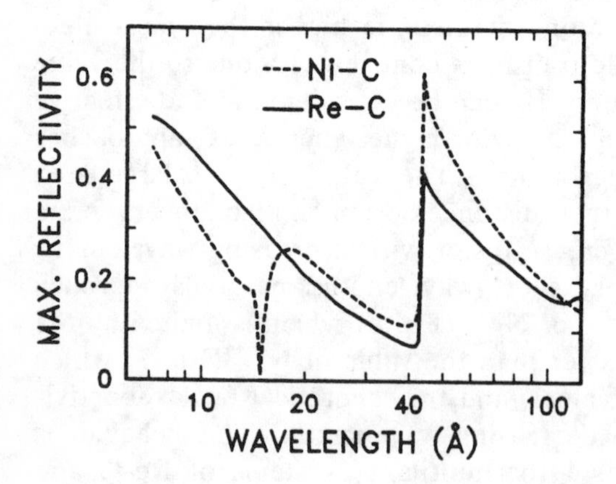

Fig. 4: Maximum reflectivity obtainable with periodic Ni-C and Re-C multilayers for normal incidence, a large number of periods and smooth, sharp boundaries. Optical constants from Ref. 1.

Normal incidence mirror coatings have been experimentally realized mostly for the wavelength region λ = 50 - 250 Å with carbon as a spacer material.[10-15] Measured peak reflectivities in this region are about 25% and Fig. 5 is a summary of the peak reflectivities obtained from various evaporated material combinations in this region. All the measurements are obtained from coatings with periods between 30 and 100 Å and several points have been obtained from each coating by measuring the reflectivities at different angles of incidence. The coatings have also been tested at much shorter wavelengths by using them near grazing incidence and these measurements confirm the higher reflectivities for shorter wavelengths as predicted by theory. Reflectivities up to 80% have been obtained at λ = 1.54 Å and grazing incidence. There is presently a gap in the performance of the mirros for λ = 44 - 30 Å due to the increased absorption of carbon. Carbon is the most successful spacer material found up to now; it forms very smooth and stable boundaries. The best multilayers can be described by effective values of $\sigma \approx$ 2-3 Å (Eq. 9), permitting the fabrication of multilayers with periods as short a 25 Å (λ_{normal} = 50 Å) without too much loss in reflectivity. The deposited materials are amorphous or polycrystalline and one can hope that epitaxial growth of the multilayer might permit one to fabricate still thinner layers. There is no material combination of 2 materials which has suitable optical constants and matching lattices for such an epitaxial growth. It might, however, be possible that for very thin layers the lattice mismatch can be compensated by strain.

Experimental results for wavelengths λ > 250 Å do not exist, probably mostly due to insufficient interest in these wavelengths up to now. The spacer layer of carbon would have to be replaced by other elements like Al and Mg and one would have to find compatible materials which prohibit diffusion of the spacer layer. We do not expect any drastic difficulties to realize a performance close to theoretical values for this wavelength region.

In addition to normal incidence mirrors the multilayers can also be used as spectrometers, beam splitters, polarizers and imagers, and these applications have been reviewed in the literature.[10-13]

For the practical realization of the multilayers, roughness and thickness control are the most important consideration. It turns out that the interface roughness or mixing is mostly a material property. The same materials have been deposited by different deposition methods (sputtering[10] and evaporation[11,14,15]) with similar performance. We obtain sufficient thickness control in our evaporation system by monitoring the x-ray reflectivity in situ during the film deposition. This method prevents the accumulation of thickness errors and permits one to recognize and correct any errors while they occur.

Contamination presents no problem in coatings below λ = 1000 Å. All materials are absorbing in this region and all optical constants are the same within a factor of 100. Therefore, impurities in the order of a few

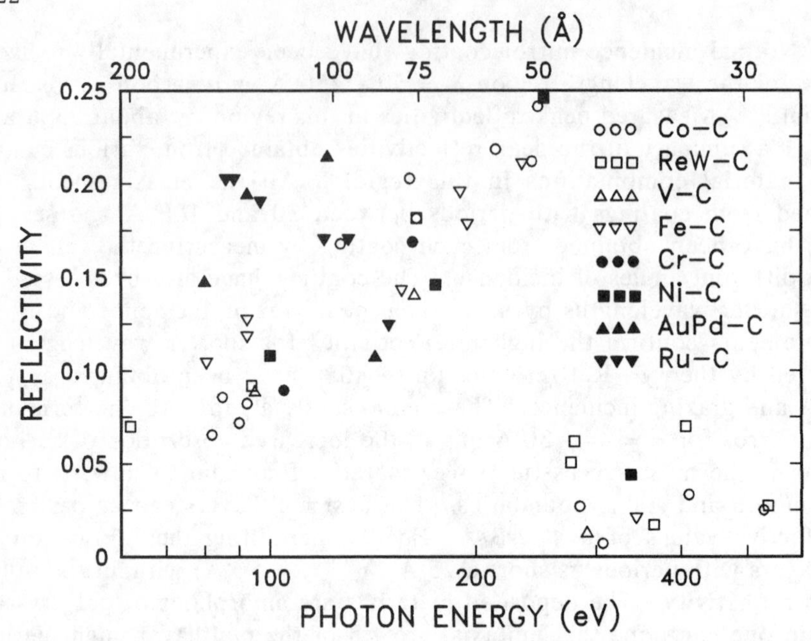

Fig. 5: Measured maximum reflectivity versus photon energy for different multilayer systems. The periods of the multilayers are between 35 Å and 100 Å; the peak reflectivity is shifted to different energies by changing the angle of incidence.

percent will not change the optical properties very much and requirements on the cleanliness of the evaporation system are minimal. Addition of nearly any impurities in order to change the structure of a film (e.g., to make it amorphous or to produce sharper boundaries) is, therefore, allowed.

Data on the high power laser damage of the described mirror coatings are not known to the author. Except for the all dielectric mirrors above λ = 1500 Å all coatings have absorption losses. Many of the successful coatings in the XUV region use high temperature materials as C and W and can be operated at rather high temperatures. However, problems have to be expected when these mirrors are used for high power short pulses.

REFERENCES

1. B.L. Henke, P. Lee, T.J. Tanaka, R.L. Shimabukuro, and B.K. Fujikawa, Low Energy X-Ray Diagnostics-1981, AIP Proc. **75**, eds. D.T. Attwood and B.L. Henke, p.124 (1981).

2. B.L. Bracewell and W.J. Veigele, Developments in Applied Spectroscopy **9**, eds. E.L. Grove and A.J. Perkins, (Plenum, New York 1971) p.375.

3. O.S. Heavens, Optical Properties of Thin Films (Dover, NY 1965); A. Vasicek, Optics of Thin Films (North-Holland, Amsterdam 1960); H.A. Macleod, Thin-Film Optical Filters, (Elsevier, NY 1969); M. Born and E. Wolf, Principles of Optics, 5th ed. (Pergamon Press, 1975); P.H. Berning, Theory and Calculations of Optical Thin Films, in Physics of Thin Films, ed. by G. Hass, 1, 69 (Academic Press, NY 1963).

4. E. Spiller, Appl. Phys. Lett. **20**, (1972); Appl. Opt. 16, 89 (1976); Proc. ICO-IX, Space Optics, Natl. Acad. Science. Washington (1974) p.581.

5. H.E. Bennett and J.M. Bennett, in Physics of Thin Films, Vol.4, eds. G. Hass and R.E. Thun, (Academic Press, NY 1967) p.1

6. G. Hass and W.R. Hunter, Proc. ICO-IX, Space Optics, (Natl. Acad. Sci., Washington 1974) p.525; J.A..R. Samson, Techniques of Vacuum Ultraviolet Spectroscopy, (Wiley, New York 1967).

7. G. Hass, in American Institute of Physics Handbook, 3rd Edition, E.E. Gray, Ed., (McGraw-Hill, New York 1972) Chapter 6g.

8. E. Spiller, Optik **39**, 118 (1973); Appl. Opt. **13**, 1209 (1974).

9. L.A. Stelmack and B.K. Flint, Electro-Optical System Design, Sept. 1980, p.39; ibid., Nov. 1980, p.41.

10. T.W. Barbee in AIP Proc. **75**, 131 (1981).

11. E. Spiller in AIP Proc. **75**, 124 (1981); RAL Symposium on New Techniques in X-Ray and XUV Optics, eds. B.J. Kent and B.E. Patchett, (Rutherford Appleton Lab., 1982) p.50

12. D.T. Attwood, N.M. Ceglio *et al.*., Top. Meeting on Laser Techniques for Extreme Ultraviolet Spectroscopy, Boulder, CO (March 1982).

13. D.J. Nagel, J.V. Gilfrich, T.W. Barbee, Nucl. Instru. and Methods, **195**, 63 (1982); J.V. Gilfrich, D.J. Nagel, T.W. Barbee, Appl. Spectroscopy **36**, 58 (1982).

14. S.V. Gapanov, S.A. Gusev, B.M. Luskin, N.N. Salashchenko, E.S. Gluskin, Opt. Comm. **38**, 7 (1981); Opt. Comm. **48**, 229 (1983).

15. P. Dhez, in X-Ray Microscopy, Symposium at Göttingen, Sept. 1983, (Springer-Verlag, to be published).

VUV LASER PHOTOLITHOGRAPHY

J. C. White, H. G. Craighead, R. E. Howard
L. D. Jackel, and O. R. Wood, II

AT&T Bell Laboratories
Holmdel, New Jersey 07733

ABSTRACT

An F_2 excimer laser at 157 nm has been used for the first time as an exposure source for high resolution photolithography. At this short wavelength, conventional glass and quartz mask substrates are opaque, and therefore alkaline-earth halides and sapphire were used as mask substrates. The masks were patterned by e-beam lithography and mask features as narrow as 150 nm have been replicated and represent the smallest features yet produced by contact photolithography.

INTRODUCTION

The steady miniaturization of components in integrated circuits demands ever higher resolution lithography. Optical lithography is the best developed and most widely used lithographic technique. Because it has higher throughput and lower cost than other techniques, there are strong incentives for extending the process to shorter wavelengths, thereby reducing diffraction effects and improving resolution. Since there are materials suitable for making transparent optical components down to wavelengths of about 120 nm, it is possible to extend optical lithography to this transmission limit.

Excimer lasers are a promising light source for approaching this short wavelength limit because they provide high efficiency, high average power beams in the Vacuum Ultraviolet (VUV). In addition, the narrow linewidth of the laser radiation eliminates the problem of chromatic aberration in projection lithography systems. In several recent experiments, excimer lasers, most notably the ArF excimer laser operating at 193 nm [1-3], have been used to produce features as small

0094-243X/84/1190324-06 $3.00

as 500 nm.

In this paper we describe experiments using a molecular fluorine (i.e., F_2 dimer) laser operating at 157 nm as a Vacuum Ultraviolet (VUV) source for high resolution photolithography. The F_2 excimer laser is now the only commercially available laser with high average power at a wavelength as short at 157 nm. Using this laser as a source for contact lithography, we made resist lines as narrow as 150 nm, the smallest features yet reproduced by photolithography.

EXPERIMENTAL TECHNIQUE

Conventional masks with quartz or glass substrates are opaque in the VUV, so two alternative mask systems were developed using substrates that are transparent at 157 nm. The first mask system used patterned polyimide on CaF_2 substrates and the second used patterned nichrome on sapphire substrates.

The CaF_2 masks were made by coating the substrates with a trilevel [4,5] resist structure consisting of 100 nm of polyimide followed by 25 nm of thermally-evaporated germanium and then 100 nm of polymethyl methacrylate (PMMA), a high-resolution e-beam lithography resist. Masks were formed by e-beam exposing and developing the PMMA and sequentially etching the patterns into the lower layers by anisotropic reactive ion etching. Minimum features on the masks were about 100 nm wide. Measurements showed that 100 nm of polyimide transmitted less than 10% of the 157 nm radiation. Coupled with strong absorption in the germanium, the polyimide on CaF_2 produced high-contrast photo-masks transparent to visible light. The latter feature is important for making accurate mask-to-substrate alignments. These masks were used for the exposures shown in the following figures.

Sapphire masks were made by using e-beam exposure to pattern a 100 nm thick PMMA on the substrates. Before exposure, the PMMA was covered by 20 nm of aluminum to prevent charging. (The germanium film performed this function on the CaF_2 masks.) This layer was removed by etching in a dilute KOH solution before development. A 25 nm thick layer of nichrome was evaporated on the developed resist stencils and liftoff was performed to make the final nichrome-sapphire mask. Transmission measurements of the sapphire substrates were in good agreement with published values [6] and showed that a 0.5 mm thick sample was about 80% transparent.

The 157 nm exposure source used a Lambda Physik model 101 excimer laser pulsed at 10 Hz with 10 nsec duration and ~0.1 mJ/cm^2 energy density. Since the laser was highly multimode with little pulse-to-pulse coherence, the problem of laser speckle was eliminated. The exposure was performed in a simple vacuum apparatus to avoid strong light absorption from atmospheric oxygen [6]. To expose samples, resist-coated silicon substrates were pressed against a mask in a ring holder. Visible Newton's rings were observed around the sample and the entire patterned area was kept within the first fringe, indicating close proximity of mask and sample.

EXPERIMENTAL RESULTS

The sensitivity and resolution of several resists were studied since little or no data is available on resist properties in the VUV. Initial measurements were made on PMMA because it is the polymer resist with the highest known resolution. The attenuation length for unexposed PMMA is several times larger than the resist thickness, but becomes much shorter as the PMMA is exposed; this makes complete exposure of the resist nearly impossible. Initial measurements of the absorption coefficient of PMMA further into the VUV indicate that this problem is likely to get worse as one utilizes deeper VUV wavelengths.

This problem was circumvented by using a thin layer of PMMA in a trilevel resist process similar to that used in the polyimide-on-CaF_2 mask fabrication. A 50 nm thick PMMA layer was coated on top of a 25 nm thick germanium layer deposited on 250 nm of polyimide. This thin upper PMMA layer was readily exposed by 0.5 J/cm^2 and developed in a 3:7 cellosolve-methanol solution for 10 seconds. Reactive ion etching was used to transfer the pattern into the germanium layer and from there into the thicker polymer layer. In some trilevel patterns irregularities in the mask less 100 nm were reproduced in the resist.

The exposure properties of the resists AZ 2400 [7], HPR-204 [8], and two compositions of the copolymer [9] of methyl methacrylate (MMA) and methacrylic acid (MAA) were tested. Of these, the copolymer was the most sensitive. Unlike the PMMA, this copolymer became more transparent (up to 70% transmission) with exposure. This property permitted thicker resist layers to be exposed and developed.

Fig. 1. shows a developed pattern in a 200 nm thick copolymer resist (9% MAA) at an exposure dose of 0.2 J/cm^2. The exposed resist was developed for 4 seconds in 3:7 cellosolve-methanol solution. There was no difficulty developing completely through this resist thickness. This resist was also as a mask for the reactive ion etching of silicon as shown in Fig. 2.

Fig. 1. Developed pattern in copoloymer. The resist thickness is 200 nm.

Contact exposures were also done using the nichrome on sapphire masks. Features as small as 250 nm were reproduced in 200 nm thick copolymer resist. These exposures were not performed in a clean environment, and it possible that dirt particles caused poor contact between the mask and the resist-coated substrate. The CaF_2 masks are much softer than the sapphire masks, and it is likely that better contact was achieved with CaF_2 even in the presence of the dirt.

Fig. 2. Relief pattern etched into the silicon substrate by reactive ion etching with SF_6 using a developed copolymer pattern as an etch mask.

In conclusion, a complete process for vacuum ultraviolet photolithography at 157 nm has been described. Experiments have shown that the high resolution polymer PMMA has an optical attenuation length that decreases with photon exposure, requiring special methods for its use. A copolymer (MMA-MAA) resist was found to have a higher sensitivity and was usable in a single layer at 157 nm. Both the PMMA and the copolymer were successfully patterned with features as small as 150 nm. The copolymer was used as a mask for reactive ion etching of silicon. Some of the techniques explored may be applied to projection lithography or patterning by photon assisted processes.

REFERENCES

[1] K. Jain and C. G. Wilson, Appl. Phys. B 28, 206 (1982).

[2] K. Jain and C. G. Wilson, Digest of the 1982 Symposium of VLSI Technology, Oisi, Japan, September 1982.

[3] Y. Kawamura, K. Toyoda, and S, Namba, J. Appl. Phys. 53, 6489 (1982).

[4] L.D. Jackel, R.E. Howard, E.L. Hu, D.M. Tennant, P.Grabbe, Appl. Phys. Lett. 39, 268 (1981).

[5] P. Grabbe, E.L. Hu and R.E. Howard, J. Vac. Sci. Technol. 21, 33, (1982).

[6] James A. R. Samson, <u>Techniques of Vacuum Ultraviolet Spectroscopy</u>, (Pied Publications, Lincoln, Nebraska, 1980.)

[7] Shipley Company, Newton, MA.

[8] P.A. Hunt Chemical Corp., Palisades Park, NJ.

[9] I. Haller, R. Feder, M. Hatzakis, and E. Spiller, J. Electrochem. Soc. 26, 154 (1979).

Diagnostics for an for an XUV/Soft X-ray Laser

R. L. Kauffman, D. L. Matthews, N. Ceglio, and H. Medecki
Lawrence Livermore National Laboratory, Livermore, CA 94550

ABSTRACT

We have begun investigating the production of an XUV/soft x-ray laser, using our high-powered glass lasers as drivers. A major diagnostic for lasing is the measure of the absolute power produced in the lasing line. I have developed a spectrograph to time-resolve lasing lines in the energy range from 50 eV to greater than 200 eV. The spectrograph combines a transmission grating and x-ray streak camera to produce a flat field instrument. A cylindrical mirror is used in front of the grating to image the source and act as a collecting optic. The efficiency of the components is calibrated so that absolute intensities can be measured. I will compare the performance of this instrument with reflection grating systems. I will also discuss planned improvements to the system which should increase total throughput, image quality, and resolving power.

The soft x-ray region has proven to be a formidable barrier for extending lasing to shorter wavelengths.[1] Techniques used at longer wavelengths, such as resonance cavities, cannot be extended to soft x-ray wavelengths because ordinary reflective and refractive optics are not available for this regime. Usual methods for diagnosing lasing and establishing its characteristics also cannot be extended directly from the optical regime for similar reasons. The lasing medium itself differs greatly for soft x-ray lasers. Most lasing schemes use excited states of multiply-charged ions, requiring the medium to be a high energy density plasma pumped by pulsed high-powered sources. The nature of these sources also makes the task of diagnosing lasing difficult.

At LLNL we have concentrated on measuring the output power directly from a long cylindrical plasma.[2] The diagnostics are absolutely calibrated to measure output efficiencies directly. Our systems are designed to have high gain lengths so that in the exponentiation region small variations in gain translate into large changes in the signal, making the output intensity a sensitive diagnostic. We also can test exponentiation of a particular system by correlating the signal with changes in the plasma length. Saturation levels are predicted to be well above the fluorescence levels of the system and are easily observed by an absolute intensity measurement. Our measurements are also time-resolved, aiding in comparing our observations with predictions, since all systems are pulsed and lasing is predicted at only certain times during the pulse. Time resolution also helps in background suppression since background emission times are much longer than lasing times.

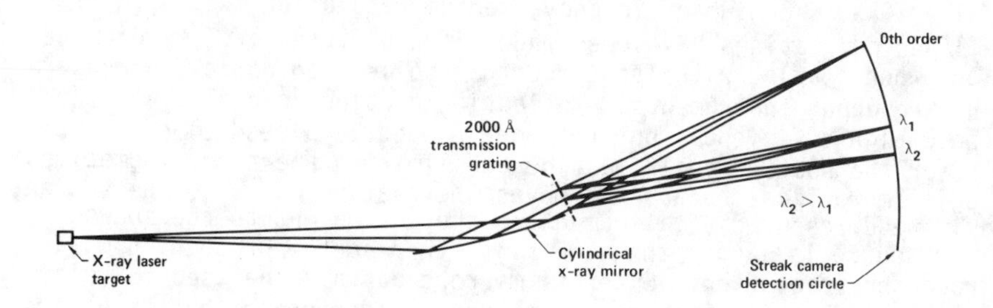

Fig. 1 Schematic of the transmission grating streaked spectrograph.

The principal diagnostic for measuring the lasing output is a transmission grating streaked spectrograph (TGSS). A schematic of the instrument is shown in Fig. 1. Components of the spectrograph are a cylindrically-curved x-ray mirror for light collection, a transmission grating for wavelength dispersion, and a soft x-ray streak camera for x-ray detection. The instrument is designed to have high spectral resolution, E/dE>200, continuous wavelength coverage from 80 Å to 300 Å, good time resolution, ∿20 ps, and high sensitivity, $\sim 10^{20}$ photons/sec-sr. In addition, we have designed an optical alignment system pointing the instrument to the target with an accuracy of better than 0.1 mrad.

The TGSS characteristics can be compared to a grazing incidence grating spectrograph, which is normally used in this energy region.[3] The transmission grating, the dispersing element for the TGSS, is separate from the cylindrical mirror focusing element, compared with the grazing incidence spectrograph where the dispersing and focusing elements are combined into a single element. This allows the TGSS to have a nearly flat field, normal incidence detection plane, in contrast to the grazing incidence spectrograph whose detection plane must lie on a Rowland circle. The TGSS geometry is much more conducive to being used with active detectors (like x-ray streak cameras or silicon diode arrays) where the incident x-rays must pass through a cathode substrate, or detector dead layer, before reaching the active detection region.

A grazing incidence grating has much higher spectral resolution than the TGSS. Resolutions for commercially available instruments can be greater than 10^4 in the soft x-ray range. These resolutions, in practice, are limited by the slit, or source, size and the number of periods of the grating that is sampled, assuming the instrument is well aligned. For a grazing incidence grating, the number of periods sampled is quite large. Slit, or source, sizes are set by the desired resolution and sensitivity. For the TGSS the number of periods sampled and the dispersion is much smaller because the grating is illuminated at normal incidence. For a 2000 Å period grating with a detector plane 1 m away, the grating dispersion is 2 Å/mm. For 10^3 resolution the image size must be less than 50 μm, and at least a 200 μm aperture of the grating must be illuminated to sample a sufficient number of periods. A simple slit imaging system cannot be used to obtain even these moderate resolutions. Instead high resolution x-ray optics must be employed, as well as aperturing of the source.

The TGSS is a high-efficiency, medium resolution instrument. Small period gratings have been made with apertures of 5 mm diameter[4], which, when coupled with large aperture collection optics, produce high throughput instruments. Reflection gratings can also be made to have high efficiency, but the source must be placed close to the slit, or the source itself must be the limiting aperture. Sometimes this is impractical because it moves the grating close to the source, and it requires precise alignment of the instrument to the source. If the source is larger than the slit, then source broadening will degrade the resolution. External mirrors can also be used to relay the source image for a grazing incidence spectrograph, but these are also subject to high alignment tolerances.[5]

Our instrument at LLNL has been designed specifically for the x-ray laser experiment. It optimizes efficiency, while maintaining high resolution required for resolving the x-ray line above the background continuum emission. As illustrated in Fig. 1, x rays from the target are collected and focused at the detection plane by a cylindrically-curved grazing incidence x-ray mirror. The mirror acts like one element of a Kirkpatrick-Baez x-ray microscope producing a line focus perpendicular to the plane of dispersion.[6] The mirror has an 9.75 m radius of curvature and operates at an angle of incidence of 4°. It is 58 cm from the target, nearly midway between the target and detection circle. This produces a magnification near unity. Spherical aberrations for these conditions are estimated to be less than 10 μm, which is much less than target sizes, although off axis, finite source size aberrations may degrade the attainable resolution. The transmission grating is placed directly behind the mirror and disperses the x rays from the zeroth order, established by the reflected direction of the mirror. The grating has a periodicity of 2000 Å and a line-to-space ratio of 1.9. The grating is 62 cm from the detection plane for a dispersion of 3.2 Å/mm. The active area of the grating is about 3 mm diameter, and the dimensions of the mirror overfill the grating area with x rays.

An LLNL-designed soft x-ray streak camera is used as a detector.[7] The camera is designed to pivot around the grating from the D.C. to 9° to allow continuous coverage through the desired wavelength region. The LLNL streak camera has a temporal response of 20 psec, providing good time resolution for the experiment. The entrance slit of the streak camera is aligned with the dispersion plane of the grating. With the 1.2 cm slit, a 38 Å portion of the spectrum is measured at one time. To cover other regions of the spectrum, the streak camera must be moved along the detection circle. The cathode has a 100 μm spatial resolution corresponding to a wavelength resolution of 0.32 Å. Neglecting source size effects, the resolving power, $E/\Delta E$, is 250 for 80 Å and increases for the longer wavelengths.

The instrument has been absolutely calibrated to determine its overall efficiency. The efficiency of the system is the throughput of the system combined with the efficiency of each component of the system. The mirror is coated with Ni and has a reflectivity calculated to be 75% or greater at these angles and energies. Reflectivity measurements at higher energy agree with the calculated efficiency. The grating efficiency has not been measured, but it is calculated to be 10% diffracted into first order. Because of a super structure support grid on the grating, this efficiency is reduced to an efficiency of 2.5% of the incident light.

Calibrating the streak camera is the most difficult part of the instrument. Well characterized x-ray sources with sufficient brightness do not exist in this energy region. We have calibrated the instrument at higher energy and extrapolated the calibrations to energies of interest using measured quantum yields for the streak camera photocathode modified for the streak camera geometry.

For these experiments the photocathode is 900 Å CsI film deposited on a 1000 Å parylene substrate. A 50 Å layer of Al is placed over the parylene to increase electron conduction. CsI has been shown to have high sensitivity in previous studies.[8] For streak cameras photoemission occurs at the rear surface of the cathode after passing through the substrate. Most photoemission measurements are from the front surface. We have estimated the quantum efficiency of the cathode using the measured front surface yields of CsI[10] and the expression by Henke, et. al.,[9] to translate these to rear surface yields needed for x-ray streak camera efficiencies. The calculated rear surface quantum efficiency including substrate absorption is shown in Fig. 2, as well as the back-to-front surface ratio. The large increase in the CsI yield, around 100 Å, observed in the front surface yield is moderated in the rear surface yield. This increased yield is due to increased absorption in the CsI near the surface where electrons can more easily escape. For the streak camera geometry, the photons must traverse the cathode before reaching the electron emitting surface, so that more are absorbed away from the cathode surface and do not contribute to the quantum yield.

The streak camera is calibrated using a laser-produced plasma as an x-ray source. Details of the calibration technique have been described elsewhere.[10] The streak camera response is compared to absolutely calibrated x-ray diode signals. Broad band emission channels for both instruments are defined using K- and L-edge transmission filters in the range from 200 eV to 800 eV. An example of the calibration data for the 200 eV channel is shown in Fig. 3. The

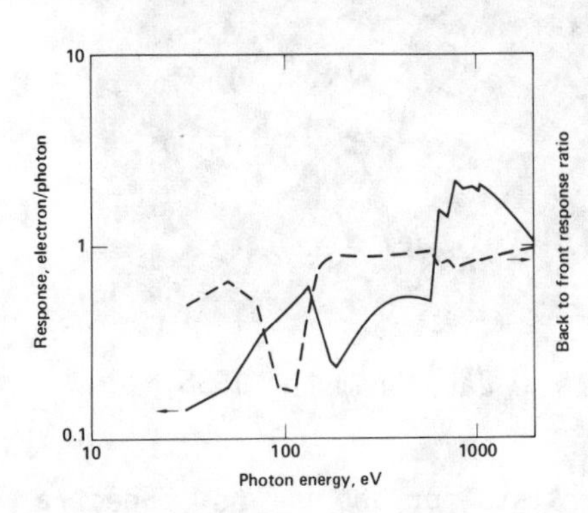

Fig. 2 Energy dependence of the CsI cathode response and the predicted rear surface to front surface secondary electron yield.

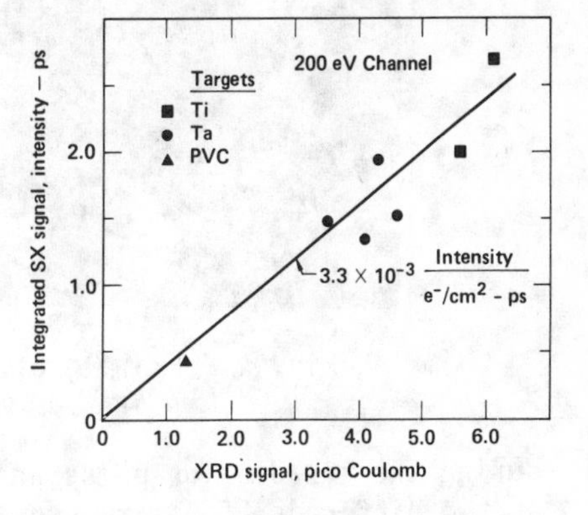

Fig. 3 X-ray streak camera calibration versus an x-ray diode monitor.

data demonstrates the linear response of the streak camera. The slope of the data gives the correlation between intensity, or exposure, on film to current density from the photocathode. This quantity, coupled with the cathode response, determines for the x-ray response of the x-ray streak camera.

An example of the data and demonstration of the instrument performance is shown in Fig. 4. The spectrograph is viewing the edge of a Formvar foil irradiated on both sides by 100 ps pulses of 450 Joules of 0.525 μm light, focused in a cylindrical spot 2 mm by 12 mm. The intense line feature in the middle of the cathode is the 3-2 transition from O VIII at 102.4 Å. The line turns on early

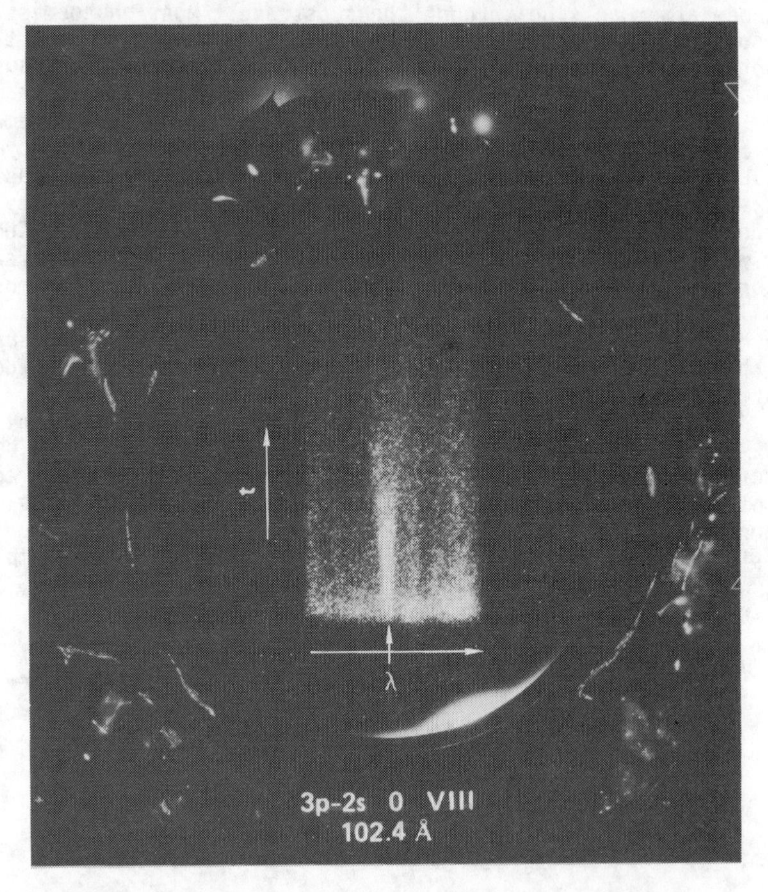

Fig. 4 Example of the raw data from the TGSS.

during the irradiation pulse and persists for about 2 ns. Spectra at two different times are shown in Fig. 5. The upper trace is early in time during the laser pulse. The line is relatively sharp with a high continuum level. Later in time, as shown in the lower trace, the continuum level is reduced, but the line is as intense, although it is slightly broader, due to expansion of the exploding foil.

In summary, we have built a streaked spectrograph for time-resolving laser plasma spectra above 80 Å. It uses a cylindrically-curved mirror for light collection, a transmission grating for spectral dispersion, and a soft x-ray streak camera for time-resolved

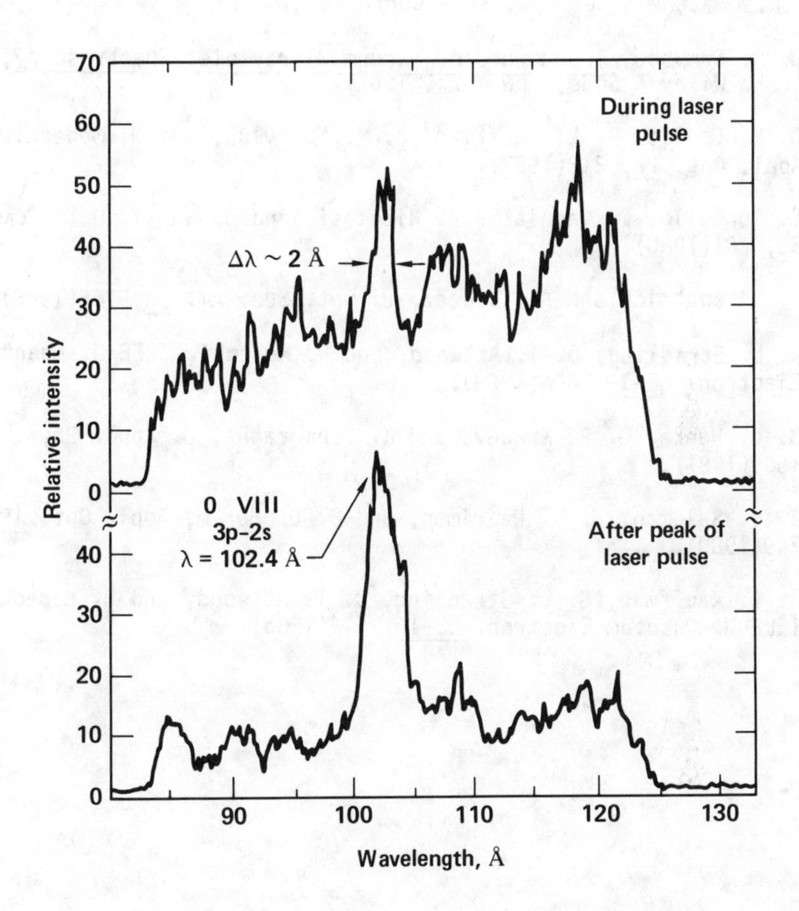

Fig. 5 TGSS spectra from during the laser pulse and after the laser pulse, showing the O VIII 3-2 transition.

detection. This configuration produces a versatile, moderate resolution instrument in this region. We have absolutely calibrated the instrument to make power measurements of x-ray line intensities from possible x-ray laser targets. We are presently modifying the instrument to increase its efficiency. The cylindrical mirror is being replaced by a two-dimensional focusing ellipsoidal mirror to increase collection efficiency. We also are increasing the aperture size of the grating for an effective 10 mrad divergence acceptance, which more nearly matches our present experiment.

References

1. R. W. Waynant and R. C. Elton, Proc. IEEE, Vol 64, 1059(1976).

2. D. L. Matthews, et al., this conference.

3. J. R. Sampson, Techniques of Vacuum Ultraviolet Spectroscopy, (John Wiley & Sons, New York, 1967).

4. N. M. Ceglio, R. L. Kauffman, A. M. Hawryluk, and H. Medecki, Appl. Opt. 22, 318(1983).

5. G. Tondello, E. Jannitti, P. Nicolosi, and D. Santi, Opt. Comm. 32, 281(1980).

6. P. Kirkpatrick and A. V. Baez, J. Opt. Soc. Am. 38, 766(1948).

7. G. L. Stradling, D. T. Attwood, and H. Medecki, IEEE J. Quantum Electron. QE-19, 604(1983).

8. B. L. Henke, J. P. Krauev, and K. Premaratne, J. Appl. Phys. 52, 1509(1981).

9. E. B. Saloman, J. S. Pearlmon, and B. L. Henke, Appl. Opt. 19, 749(1980).

10. R. L. Kauffman, G. L. Stradling, D. T. Attwood, and H. Medecki, IEEE J. Quantum Electron. QE-19, 616(1983).

TUNABLE RESONANT ENHANCEMENT IN NONLINEAR OPTICAL FREQUENCY MIXING*

A. V. Smith
Sandia National Laboratories, Albuquerque, NM 87115

ABSTRACT

The mechanism of four-photon resonant enhancement of third harmonic generation is studied using two input frequencies and shown to lead to tunable resonant enhancements.

Recent reports of resonant enhancement of third order sum frequency conversion efficiencies in atomic gases when the three photon level is coupled by a strong optical field to a bound state has prompted our study of this phenomenon in order to determine the nature of the process and to evaluate its potential as a general means of improving the efficiency of VUV generation by nonlinear optical mixing. Resonant enhancement has been reported for Na where the three photon level lies in the ionization continuum[1] and for Hg where the three photon level lies near the 6^1P level.[2] These processes are illustrated in Fig.1. Because the effect allows

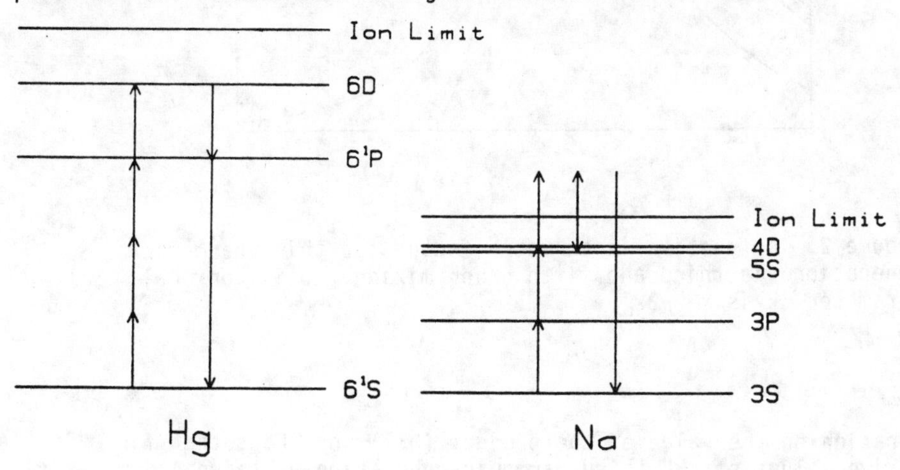

Figure 1.

*This work performed at Sandia National Laboratories supported by the U. S. Department of Energy under Contract Number DE-AC04-76DP00789.

ne to, in effect, create a resonance at any desired three photon level by using a tunable laser this has appeal as a technique to increase efficiency of 4-wave sum frequency mixing in general. There are several processes which might reasonably be expected to cause such resonant enhancement. These are discussed briefly below.

Process A: Perhaps the most straightforward accounting is that a fifth order nonlinear interaction involving $\chi_5(3\omega;\omega,\omega,\omega,\omega,-\omega)$ is more effective in light generation than $\chi_3(3\omega;\omega,\omega,\omega)$. This could occur for example in regions of normal dispersion where propagation effects prevent the third order process from producing light for tight focusing conditions but do not prevent the fifth order process from doing so. Thus, even though χ_5 may be smaller than χ_3 it can result in greater radiated energy at the third harmonic frequency. Fig 2 shows the dependence of the third harmonic

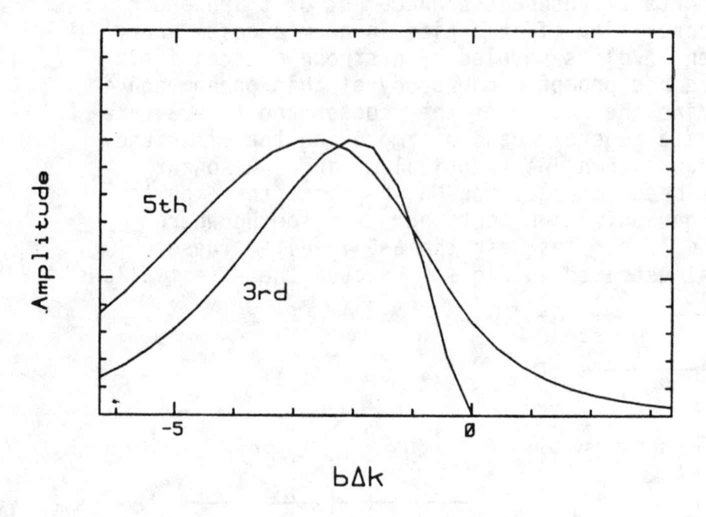

Figure 2. Refractive index match factor for third harmonic generation via third and fifth order mixing. b is confocal parameter Δk is $k_{3\omega}-3k_{\omega}$.

generation on the value of the product ($b\Delta k$) and shows that for positive values of Δk, third harmonic generation by third order process is prohibited while generation by the fifth order process is not.

Process B: Third harmonic generation via χ_3 does not occur for positive Δk because of destructive interference between the third harmonic light radiated before with that radiated after the focal point. It follows that if ionization in the region of the focal point alters the amplitude or phase of light passing through it, the cancellation may be incomplete allowing net harmonic output.

Process C: Third harmonic light could be generated in a cascade process involving first the generation of fourth harmonic followed by mixing with the fundamental to produce third harmonic. Both these processes are forbidden in a centrosymmetric medium. However, ionization can produce local electric fields which destroy this symmetry allowing mixing by terms $\chi(4\omega;\omega,\omega,\omega,\omega,0)$ and $\chi(3\omega;4\omega,-\omega,0)$. Efficient second harmonic generation in atomic vapors has been seen and attributed to this mechanism[3]. Another possibility is that ground state population depletion destroys the symmetry allowing electric quadrupole mixing to express itself.[4]

In order to clarify the nature of four-photon-resonant third harmonic generation we have examined the process in mercury vapor near four-photon resonances with the states 6D, and 7^1S. Typical experimental conditions were: 8ns light pulses of energy ranging from 1 to 60 mJ focused by a 25cm lens into .1-10 torr of Hg vapor with 10 torr He buffer. The output was dispersed by a quartz prism or a small monochromator and detected with a solar blind PMT. Scanning a single laser in the vicinity of the 6D and 7S resonances revealed resonant enhancement of third harmonic in both cases. These scans are shown in figure 3. The shapes of the resonances at 6D may be accounted for by ac stark shifts of the 6^1D and 6^1P levels plus index match considerations for 6^1P allowing harmonic generation only on the high frequency side of 6^1P. The 6^3D level experiences much smaller stark shifts because of its much weaker transition strength to 6^1P.

The value of $b\Delta k$ is positive for the 6D resonances so harmonic generation is allowed by fifth but not third order processes. At the Hg densities used in these scans, $b\Delta k$ is estimated to be about +1.0 at the D lines. The behavior of the third harmonic signal near the 7S resonance is not well understood. The red shift of the major peak is most likely a stark shift due to the 9^1P level which is nearly resonant at the 5-photon level. The red shift is compensated at higher Hg pressures by a blue shift of unknown origin so that at 10 torr Hg pressure the peak third harmonic coincides with the unperturbed 7S position. The origin of the secondary peak is unknown. We have also measured photoionization current near the 7S resonance and find it follows the third harmonic signal quite faithfully except the secondary peak is absent or strongly attenuated. The size of the peak ion signal indicates that only one in 10^5 Hg atoms in the focal volume is ionized. Again, $b\Delta k$ is positive in this region.

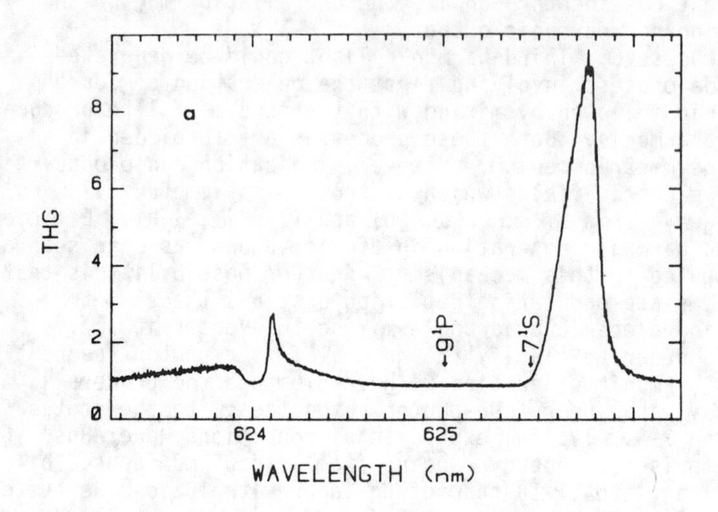

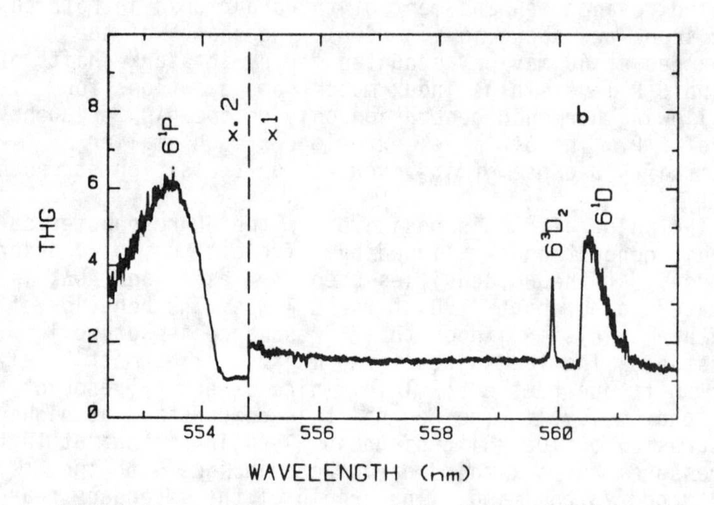

Figure 3. Third harmonic energy as function of fundamental wavelength (a) near 7^1S resonance and (b) 6D resonance. Note vertical scale change in 3b.

To get more detailed information we next did two-color experiments at both resonances. Light of two frequencies was used to reach the four-photon level by absorption of three photons of frequency ω_1 and one photon of frequency ω_2. For the 7S resonance, using lasers tuned to approximately 646 nm and 570nm for sources 1 and 2 respectively, we find that the output occurs only at $(3\omega_1+\omega_2)-\omega_1=2\omega_1+\omega_2$ and $(3\omega_1+\omega_2)-\omega_2=3\omega_1$ with none at $\omega_1+2\omega_2$ nor at $3\omega_2$. We did not look for light at 156nm corresponding to radiation near the 7S-6S transition i.e. at $3\omega_1+\omega_2$.

Similar experiments on the 6D resonance were performed with two wavelengths separated by only a few wavenumbers using a monochromator to analyze the output in the third harmonic range. This case is different from the 7S case in that the 6P resonance level lies only 522 cm^{-1} from the three photon level. When $3\omega_1+\omega_2$ is coincident with 6^3D_2 and ω_1 is off resonance by $-12.8cm^{-1}$, we find output at $2\omega_1+\omega_2$ and $3\omega_1$ as for 7S but we also find comparable but diminishing levels at $\omega_1+2\omega_2$, $4\omega_1-\omega_2$, $5\omega_1-2\omega_2$, $6\omega_1-3\omega_2$.

Based on these observations, we have reached some conclusions about the cause of four-photon resonant mixing. The fact that resonance occurs on 7S means a cascade involving a quadrupole transition in not involved since a 7S-6S transition is forbidden. Also, the fact that for 7S, only $3\omega_1$ and $2\omega_1+\omega_2$ occur without $\omega_1+2\omega_2$ or $3\omega_2$ indicates that process B above is not responsible, and the very weak ionization seems to rule out any mixing involving local fields produced by electron-ion separation. The observations are consistent with fifth-order mixing. The observation of decreasing intensity in progressively higher order terms near 6D resonance further indicates that under conditions encountered here, a perturbation description is still valid.

In conclusion, we have made a start toward understanding the role of higher order processes in third harmonic generation. A number of features are still unexplained, and an assesment of the practical implications for VUV generation awaits further research.

REFERENCES

1. L. I. Pavlov, S. S. Dimov, D. I. Metchkov, G. M. Milev K. V. Stamenov, G. B. Altshuller, Phys. Lett. 89a 441-443, (1982).
2. D. Normand, J. Morollec, J. Reif, J. Phys. B: At. Mol. Phys., 16, L227-L232, (1983).
3. T. Mossberg, A. Flusberg, S. R. Hartmann, Opt. Comm., 25, 121-124, (1978).
4. W. Jamroz, P. E. LaRocque, B. P. Stoicheff, Opt. Lett. 7, 148-150, (1982).
5. R. R. Freeman, J. E. Bjorkholm, R. Panock, W. E. Cooke, Laser Spectroscopy, V. A. R. W. McKellar, T. Oka, B. P. Stoicheff, eds. Springer-Verlag, Berlin, 1981) p, 453-456.

GENERATION OF VUV RADIATION IN THE KILOWATT POWER RANGE DOWN TO 130 nm BY STIMULATED RAMAN SCATTERING IN HYDROGEN

H.F. Döbele, M.Röwekamp and B.Rückle
Fachbereich Physik, Universität GH Essen
D-4300 Essen 1, Federal Republic of Germany

ABSTRACT

Two schemes of VUV-generation by stimulated Raman scattering in H_2 are discussed in the following. In the first case, tuned, narrow-bandwidth dye laser radiation in the green yields 13 AS components down to 138 nm. In this application, continuous tunability is the main feature. In the second version, radiation in the spectral interval around 193 nm is amplified in ArF* and is used for Raman excitation. Tuning is possible over approximately 2 nm. VUV power is especially remarkable in this case.

MOTIVATION

The impulse to generate VUV radiation originates in our case from plasma physics, where the application of 'low-Z-materials' for the wall in contact with the hot plasma is investigated for some years. Reduction of radiation losses is one of the expected advantages. The necessity arises then to diagnose the concentration of the impurities released by the influence of the hot plasma with spatial and temporal resolution and close to the place where they are produced, i.e. the wall. For the analogous situation with metallic walls, laser-induced fluorescence with dye lasers in the visible is already successfully applied /1/. Extension of this method to low-Z-materials requires tuned and narrow-bandwidth VUV radiation. For the intensity, the saturation intensity defined by the equality of spontaneous and induced de-excitation rates of the level populated by optical pumping

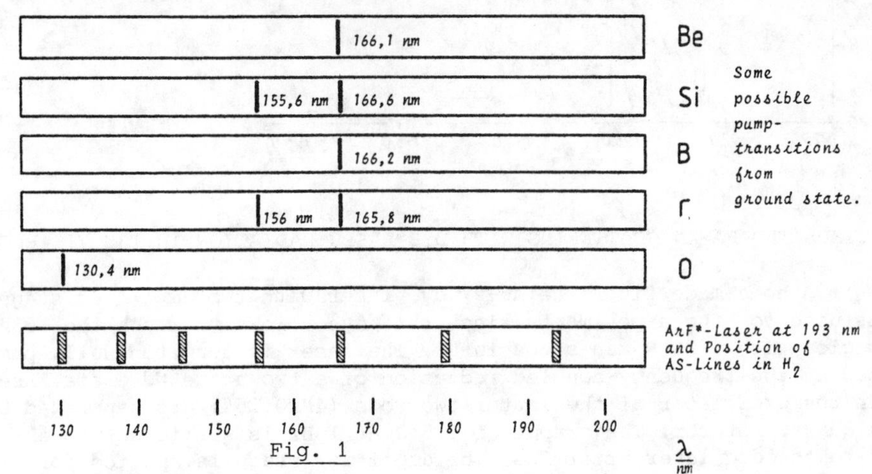

Fig. 1

yields a reasonable measure that should be reached or better surpassed. Typical values for interesting transitions are a few hundred watt for reasonably choosen line-width and imaging conditions.

Fig.1 shows some important VUV-lines. Furthermore, those spectral positions are indicated, that correspond to anti-Stokes lines of hydrogen irradiated with 193 nm-radiation. The figure indicates that SRS may result in a radiation source suited for our intended application provided the conversion process turns out efficient enough and tunability is sufficient.

STATE OF THE ART

Hydrogen has been studied by several authors as an anti-Stokes Raman medium because of its large Raman shift /2/. The shortest wavelength obtained with a visible pump laser is 138 nm as reported by SCHOMBURG et al. /3/, whereas HARGROVE and PAISNER /4/ identified 7 AS-components - the shortest at 123.7 nm - starting with an ArF* laser in the VUV. Although SRS with an ArF* laser results in considerable power /5/, the restricted tunability represents a serious disadvantage. The situation can be improved somehow by using ArF* as an amplifier medium rather than a laser /6/. Fig.2 illustrates the spectral width over which ArF* shows amplification. This interval is approximately twice as wide as in the case of the ArF* laser.

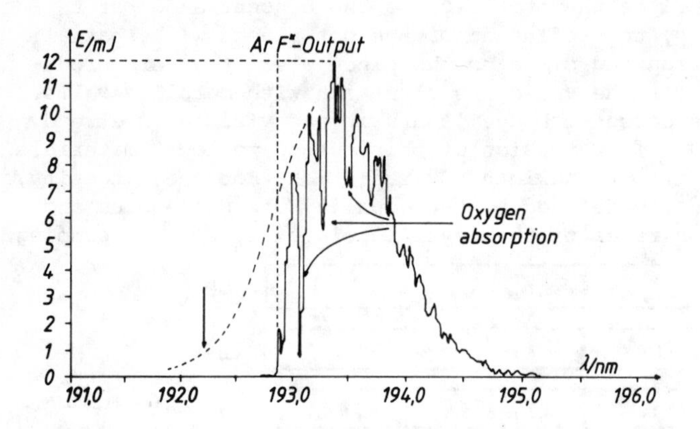

Fig. 2

Spectral width of radiation amplified in ArF* (after /6/)

STIMULATED RAMAN SCATTERING WITH DYE LASER RADIATION IN THE VISIBLE

A home-made fluoresceine-27 dye laser with two amplifier stages is used for the experiments since this dye allows to reach the 193nm region with 8 AS Raman steps in H_2. The laser is longitudinally pumped by the frequency-doubled radiation of a two-stage Nd:Glass laser. In the oscillator of the latter two rods (4"x0.25") are separated by a quartz rotator. The output of 0.5 J at 1 Hz is sufficient to excite the dye laser after SHG. The amplifier stage is applied for single-shot operation and yields 5 J in a 20 ns-pulse. After SHG the energy at 527 nm reaches up to 1.3 J with typical values around 0.7J. When pumped under these conditions, the dye laser produces 15-ns-pulses of approximately 160 mJ at 548 nm. The tuning range is 533 nm to 565 nm. A prism arrangement in front of the holographic grating expands the beam by a factor 36 in one dimension and provides bandwidth reduction to 1 cm^{-1}.

The Raman-cell consists of a 2" diam stainless steel tube. A suprasil lens (f=0.6 m or 0.8 m) acting simultaneously as cell window is used to focus the radiation. The radiation leaves the cell through a CaF_2 window and is then dispersed by a CaF_2 prism. The energy was measured with the aid of a pyroelectric energy meter (Laser Precision RjP 735). In fig.3 the energy of the 8th AS component at 193 nm is shown for different pressures and dye laser energies. Around 10 kW of 193-nm-radiation (to be amplified later) can be generated this way.

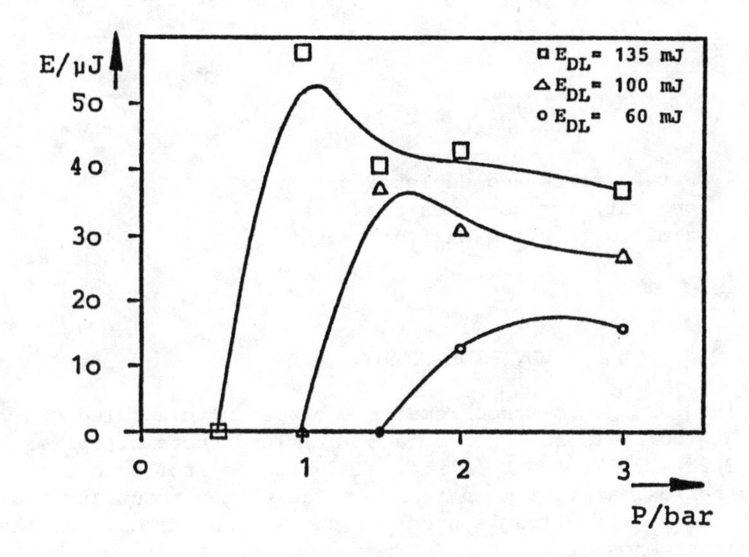

Fig. 3

Fig.4 summarizes the energy in the various Stokes- and anti-Stokes orders for 100 mJ dye laser energy.

Although only fluoresceine-27 was tried, the existence of dyes of good efficiency for neighbouring spectral intervals ensures continuous tunability at least from 138 nm onwards. The VUV intensities have been determined in the following way:

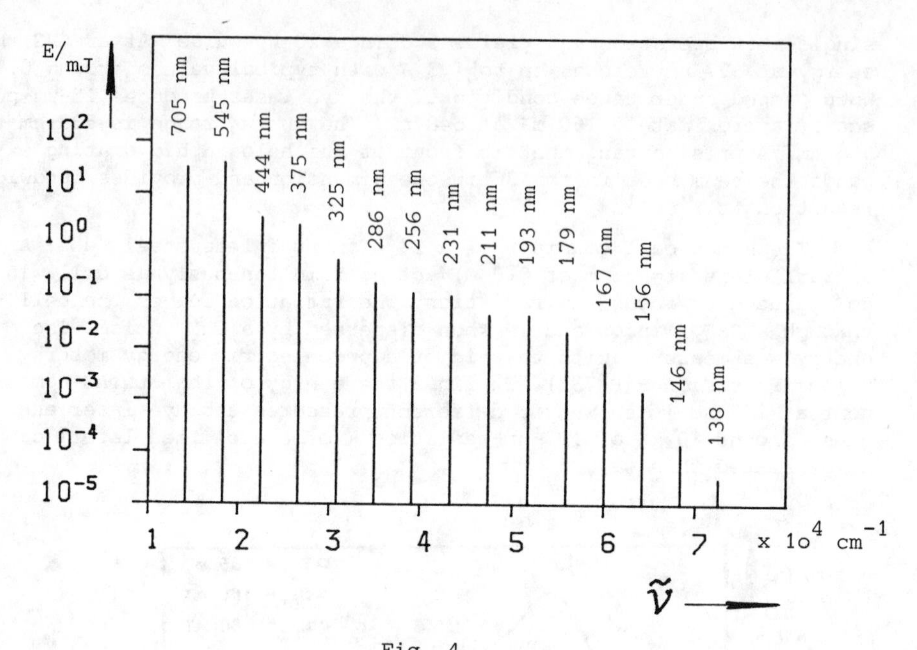

Fig. 4

First Stokes and depleted pump for p=5 bar; for AS-lines pressure is optimized to yield maximum output.

VUV - DETECTION

A SPEX 1500 SP vacuum-monochromator equipped with a ruled grating blazed at 150 nm is followed by a solar-blind photomultiplier (EMI-GENCOM G 26 L 315). A MgF_2 plate, sand-blasted from both sides, is placed at approximately 1 m in front of the slit without any imaging element and acts as light source for the monochromator. For the intensity measurements of VUV Raman lines, the exit window of the cell is just followed by this plate to ensure depolarisation and uniform illumination of the slit regardless of the intensity distribution at the cell exit.

Relative calibration of the monochromator/photomultiplier combination is performed with the aid of a deuterium lamp. An absolute scale can be obtained by irradiating the evacuated Raman cell with 193-nm-radiation of previously measured energy and equating both signals. Interference filters of known transmission characteristics (manufacturers data) were used to avoid interference from grating ghosts.

STIMULATED RAMAN SCATTERING USING 193-nm-RADIATION AMPLIFIED IN ArF*

Amplification of the narrow-bandwidth radiation tuned to 193 nm as described above was obtained in a LUMONICS TE 262 section followed by an EMG 150 system of LAMBDA PHYSIK - see fig.5. At a resulting output energy of 40 mJ air breakdown is regularly observed with a f=1 m CaF_2 lens.

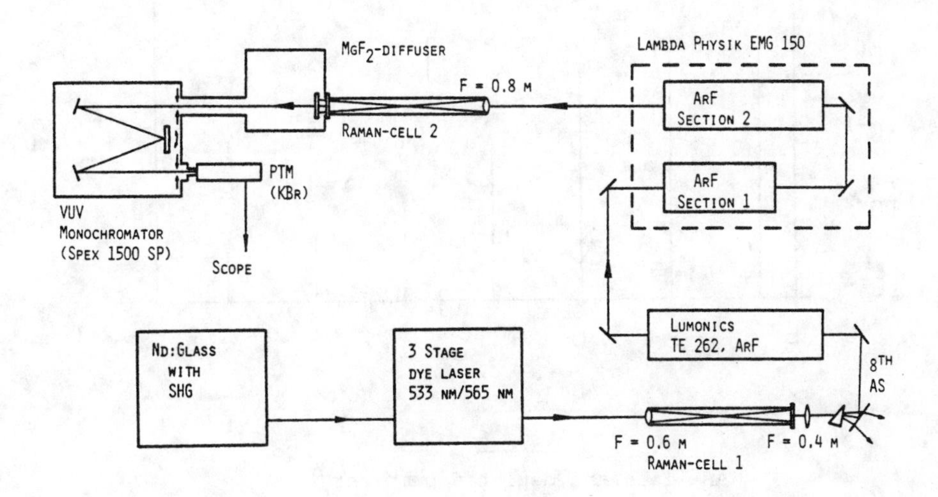

Fig. 5

Optimum conversion to VUV is again obtained at low filling pressure. When passing to higher AS-orders in succession, the optimum pressure for conversion decreases being as low as 1 bar for 130 nm. The power in the various AS-components is listed in fig.6.

The results indicate that SRS at H_2 is indeed capable of generating VUV radiation at high power - high enough to envisage application to plasma impurity studies. For wavelengths longer than 150 nm it may even be sufficient to rely on dye laser system alone and to have full freedom in tuning. Tunability is also sufficient in the case of radiation amplified in ArF* to pump most of the transitions of fig.1.

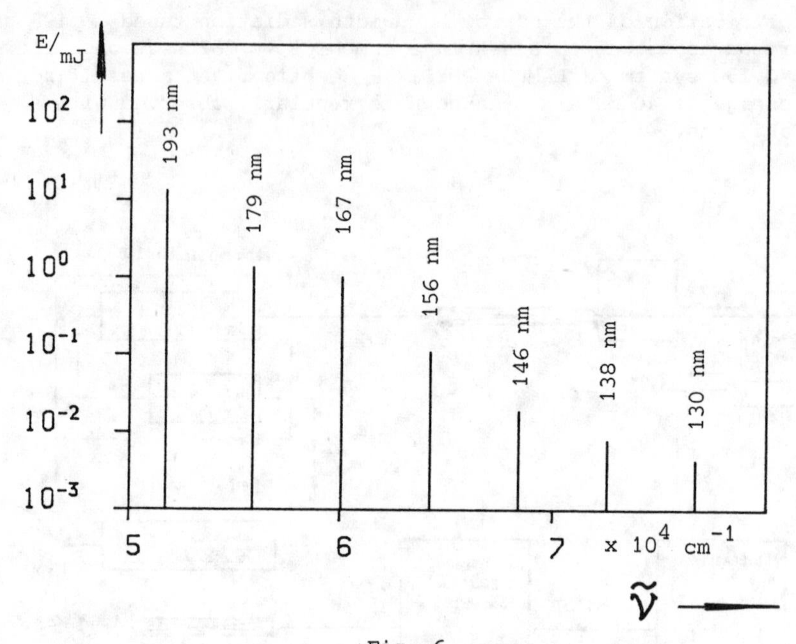

Fig. 6

AS-lines and depleted pump for p=1 bar.

REFERENCES

/1/ P.Bogen and E.Hintz, Comments Plasma Phys. Controlled Fusion 4, 115 (1978)

/2/ H.Mennicke, Ph.D.Thesis, Technische Universität München (1971)
V.Wilke and W.Schmidt, Appl.Phys. 18, 177 (1979)
R.S.Hargrove and J.A.Paisner, in: Technical Digest, Topical Meeting on Excimer Lasers -OSA- (1979)

/3/ H.Schomburg, H.F.Döbele and B.Rückle, Appl.Phys.B30, 131(1983)

/4/ R.S.Hargrove and J.A. Paisner, Technical Digest, Conference on Electro-Optics and Lasers, Anaheim, Oct.1979

/5/ H.F.Döbele and B.Rückle, to appear in Appl.Opt.,april 1984

/6/ H.Schomburg, H.F.Döbele and B.Rückle, Appl.Phys.B28, 201(1982)

EFFICIENT GENERATION OF TUNABLE RADIATION BELOW 100 nm IN KRYPTON

Keith D. Bonin, Mark B. Morris and T. J. McIlrath
Institute for Physical Science and Technology
University of Maryland, College Park, MD 20742

ABSTRACT

Two-photon resonantly-enhanced four-wave mixing in krypton was used to generate tunable coherent VUV radiation below 100 nm. Maximum efficiencies on the order of 10^{-5} were achieved.

INTRODUCTION

These experiments demonstrate efficient generation of tunable coherent radiation near 942.3Å by using the $4p^6\ ^1S_0 - 5p[2\text{-}1/2]_2$ two-photon resonance to enhance the power efficiency. The generation of coherent radiation below the LiF cutoff at 105nm is no longer uncommon[1] and coherent light at wavelengths as short as 35nm has been generated[2]. Some of the more recent work has involved tunable sources[3,4] and a few groups have actually applied the radiation to a particular spectroscopic problem[4]. The inherent difficulty with generation in this region involves both a lack of transmitting material other than thin metal films and a lack of non-absorbing non-linear media. Previous researchers have made use of differential pumping or pulsed supersonic nozzles[2,5] to overcome these difficulties. This experiment used a novel rotating disk to separate the generating cell containing krypton from a second cell in which high-resolution absorption and ionization experiments on xenon were performed. The generated radiation was used to scan the 11s' antoionizing resonance in xenon near 942Å and we determined its position to an accuracy of 2 parts in 10^6 using a Fizeau interferometer-based wavemeter[6]. We also measured the Fano width Γ of the resonance and the Fano asymmetry parameter q. These measurements are described in another paper in these proceedings.[7]

EXPERIMENTAL

The driver laser for this experiment was a Q-switched Nd:YAG laser at 1.064 μm producing 700mJ, 10ns pulses at 10 Hz. The Nd:YAG laser was doubled and used to pump a Hansch type dye laser which produced 35 mJ pulses at 544 nm. The Nd:YAG and dye laser had linewidths which were $< .2\text{cm}^{-1}$ and $< .05\text{cm}^{-1}$ respectively. The dye laser was doubled to 272 nm by a KDP type I crystal. The 272 nm beam and the remaining 1.064 μm radiation were summed in a KDP type I crystal to produce radiation around 216.67nm, the two-photon resonance in krypton. Typical energies for the two-photon resonant beam were .5mJ. A second dye laser, pumped with the remaining 544 nm radiation, produced radiation between 620-730 nm with a linewidth $\simeq .08\ \text{cm}^{-1}$. This second dye laser at ω_2 was pressure tuned with freon CF_2CL_2, which has a large index of refraction, while the two-photon resonant laser at ω_1, was kept fixed. The laser outputs at ω_1 and ω_2 were focused into a cell containing krypton gas, generating sum frequency radiation at $\omega_4 = 2\omega_1 + \omega_2$ near 940 nm (see Fig. 1).

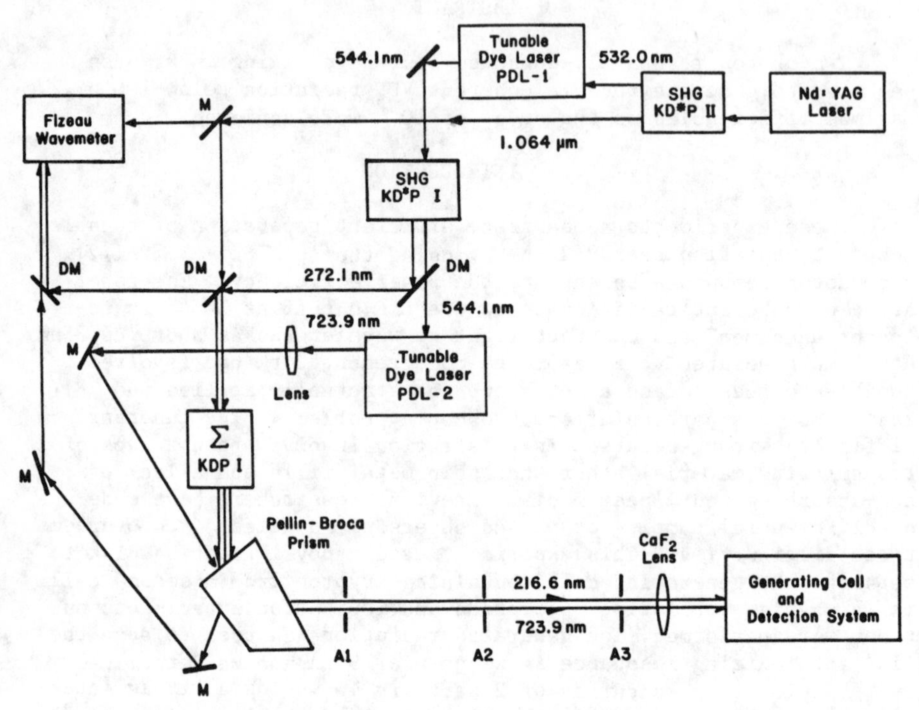

Fig. 1. Laser driving system for VUV generation.

The two beams ω_1, ω_2 were focused with a 250 mm focal length CaF_2 achromatic lens into a glass krypton cell separated from a similar xenon cell by the rotating disk aperture. The rotating disk had a 2 mm hole on a 4 inch radius and the laser was fired when the aperture was opened. The rotating disk effected a savings in gas of over a factor of ten when the pressure differential was 10 torr and it provided a large, uniform, easily monitored region of gas for the non-linear mixing to take place in. A buffer gas could be used in the Xe cell to further reduce the gas flow from the generating region to the experimental region and to reduce the tendency of the disk to bind. The rotating disk could only withstand pressure differentials up to $\sim$50 torr before it would start binding and become difficult to rotate. The duty cycle of the aperture when the disk is rotating is 6.27×10^{-3}. Thus, most of the leakage comes from gas escaping between the disks when the aperture is closed.

The Kr gas is transparent at 94 nm and a positive gas flow was always maintained between the Kr and Xe cells so that there was no self absorption in the source. The 94 nm radiation is above the first ionization limit in Xe and absorption and associated ionization was monitored by collection of ions on a pair of plates placed directly inside the Xe cell and maintained at +10V and ground potential. Absorption was also monitored by observing the VUV radiation transmitted by the Xe gas. The VUV beam was isolated from the pump beams by a .5 m McPherson monochromator with a 600g/mm grating combined with a 150 nm thick indium filter to block scattered radiation. The 94 nm radiation was finally measured using fluorescence from a sodium salicylate coated window with a photomultiplier tube (Fig. 2). Photodiodes to detect the 216.67 nm radiation and the near IR beam were also placed inside the monochromator to allow normalization of the VUV signal.

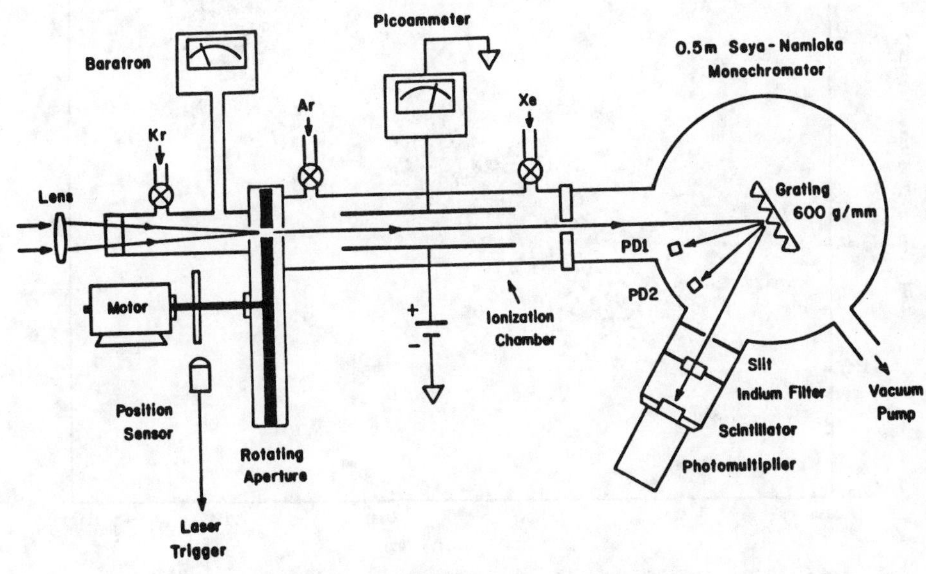

Fig. 2. VUV generating cell and detection apparatus.

The ionization signal that resulted when xenon was put in the second chamber was amplified with a picoammeter. All four signals, the VUV, the UV, the IR, and the ionization were then put into sample and hold circuits and subsequently read by analog to digital converters on a (DEC) minicomputer.

RESULTS AND DISCUSSION

The generated radiation was chosen to overlap the $^1S_0 \rightarrow 5p^5(^2P_{3/2})$ 11s' transition in Xe at 94.23 nm. In order to record the profile of the resonance, the output was tuned over

30 cm^{-1}, corresponding to ~0.03 nm at 94.2 nm, by tuning the input dye laser at ω_2. With input pump beams in the energy range of .5 mJ we obtained power conversion efficiencies $P_{VUV}/P_{UV} \lesssim 10^{-5}$. This corresponds to about 10^9-10^{10} photons/pulse providing ample signals for both the ionization and absorption scans.

Mesurements were made of the dependence of the VUV intensity on the tunable near IR beam input ($\omega_2 \simeq 724$ nm) and on the UV input beam ($\omega_1 = 216.67$ nm). The dependence of the VUV on the UV (two-photon resonant beam) was not quadratic as can be seen in the log-log plot of Fig. 3. This particular plot indicates a slope

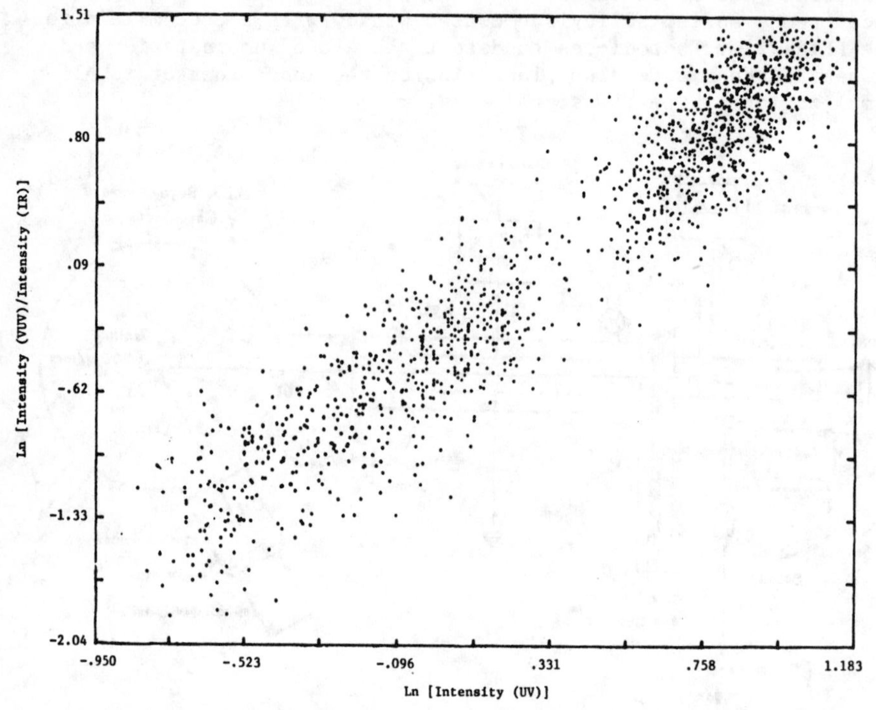

Fig. 3. VUV intensity vs. UV (two-photon resonant) intensity:slope = 1.54, $I(\omega_4)$ vs. $I(\omega_1)$.

near 1.5. Other plots, at slightly different detunings off resonance indicate slopes of 1.0 to 1.5. The deviation from a quadratic behavior is due to saturation effects and is not surprising at our power densities (10^{11} W/cm^2). The tunable beam ω_2 also saturated, providing a less than limited dependence of the 94.2 nm output on the ω_2 input as shown in Fig. 4.

In a rather large bandwidth region (~4 cm^{-1}) around the two-photon resonance, breakdown could easily be seen with the unaided eye. For a collimated Gaussian beam of our size (~6 mm dia.) an intensity in the focal spot of 10^{11} W/cm^2 is predicted. The Kr gas was chosen as the non-linear medium because it was expected to have

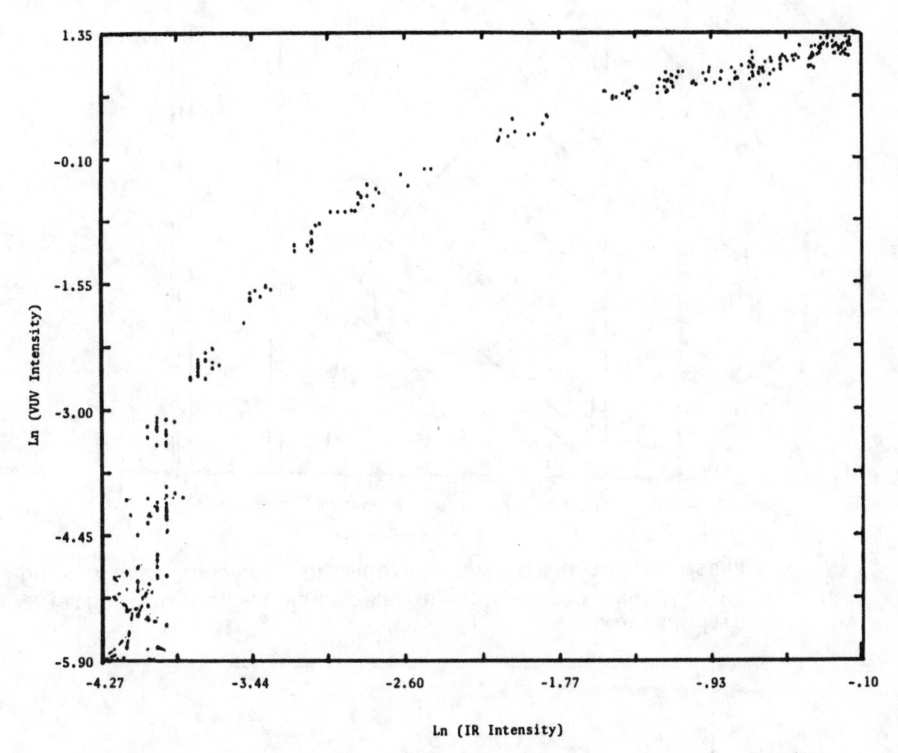

Fig. 4. VUV intensity vs. IR intensity: $I(\omega_4)$ vs. $I(\omega_2)$ with $I(\omega_1)$ fixed.

a negative dispersion for generation at 94.2 nm. A plot of the expected phase mismatch per atom, calculated from known oscillator strengths, is shown in Fig. 5. Using the values from Fig. 5 for 94.2 nm and using an estimated b value of 1 mm for the 216.6 nm input beam at focus, the generation of 94.2 nm radiation is expected to maximize at a pressure of 11.7 torr for our configuration. A plot of the VUV output versus Kr pressure for generation of 94.228 nm is shown in Fig. 6 where it is seen that the actual maximum occurs at ∿5 torr pressure.

It is clear from Fig. 5 that Kr gas provides negative dispersion for a significant portion of the spectral region between 95 nm and 90 nm. We have used the region around 92.8 nm to generate tunable VUV radiation with which to scan the 17s' resonance in Xe and it is clear that other resonances can be scanned using additional available regions.

CONCLUSIONS

These experiments have demonstrated that an efficient, high repetition rate, source of tunable coherent VUV below 100 nm can be generated by use of a two-photon resonance in krypton. Power

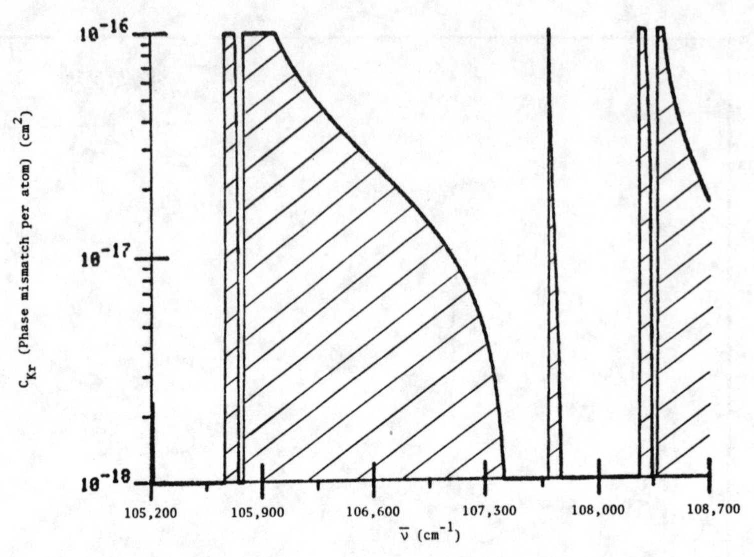

Fig. 5. Phase mismatch/atom vs. wavenumber for four-wave mixing in krypton (cross-hatched areas are regions of negative dispersion).

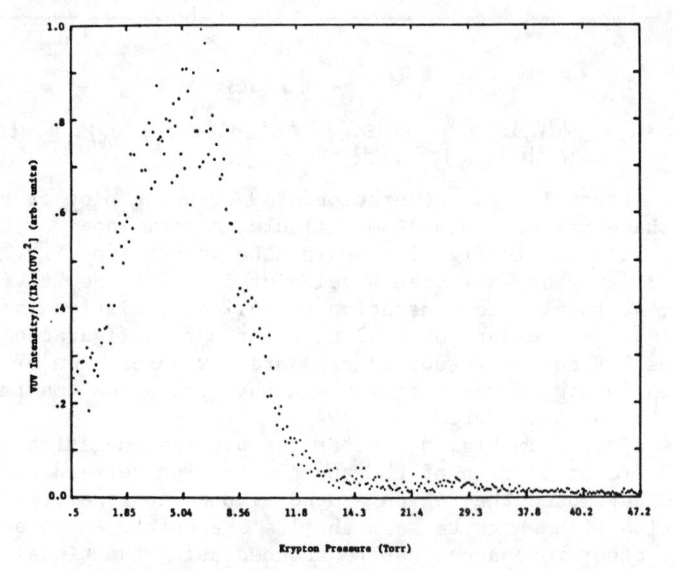

Fig. 6. VUV intensity vs. krypton pressure: $I(\omega_4)/[I(\omega_1)^2 I(\omega_2)]$ vs. P_{Kr}.

efficiencies on the order of 10^{-5} were achieved. The generating system used a novel rotating disk mechanism to keep gas losses from differential pumping to a minimum. This work was supported by NSF Grant CPE 81-19250 and by the Dissertation Research Committee of the Graduate School at the University of Maryland.

REFERENCES

1. J. Reintjes, C. Y. She, and R. C. Eckhardt, IEEE J. Quantum Electron., QE-14, 581 (1978). Triveni Srinivasan, Hans Egger, Herbert Pummer and Charles K. Rhodes, J. Quantum Electron QE-19, 1270 (1983). R. R. Freeman, R. M. Jopson, and J. Bokor in Laser Techniques for Extreme Ultraviolet Spectroscopy ed. T. J. McIlrath and R. R. Freeman (AIP, New York, 1982) pp. 422-430.
2. J. Bokor, P. H. Bucksbaum and R. R. Freeman, Opt. Lett., 8, 217 (1983).
3. R. Hilbig and R. Wallenstein, Opt. Comm., 44, 283 (1983).
4. E. E. Marinero, C. T. Rettner, R. N. Zare and A. H. Kung, Chem. Phys. Lett., 95, 486 (1983) M. Rothschild, H. Egger, R. T. Hawkins, J. Bokor, H. Pummer and C. K. Rhodes, Phys. Rev. A, 23, 206 (1981).
5. A. H. Kung, Opt. Lett., 8, 24-26 (1983).
6. J. J. Snyder, Laser Spectroscopy III, ed. J. L. Hall and J. L. Carlsten (Springer Verlag, New York, 1977), p. 419-420; J. J. Snyder, Laser Focus, May 1982, pp. 55-61. Mark B. Morris, T. J. McIlrath and J. J. Snyder submitted to Appl Opt.
7. Keith D. Bonin, Mark B. Morris and T. J. McIlrath, "High-Resolution Laser Spectroscopy of Xe($5p^6$ - $5p^5$ 11s') at 94.2 nm", Proc. Conf. on Laser Techniques in the Vacuum, ed. S. Harris and T. B. Lucatorto (AIP, New York, 1984).

CONTINUOUSLY TUNABLE SUM-FREQUENCY GENERATION INVOLVING RYDBERG STATES

R.W. Boyd, D.J. Gauthier, J. Krasinski and M.S. Malcuit
Institute of Optics, University of Rochester, Rochester, N.Y. 14627

ABSTRACT

Extremely large values of the second-order nonlinear susceptibility have been demonstrated in a system involving Rydberg states. The applicability of this technique to the extreme ultraviolet is discussed.

INTRODUCTION

It has recently been shown that extremely large values of the second-order nonlinear susceptibility can be obtained through use of Rydberg atomic states perturbed by an external dc electric field.[1, 2] This mixing process utilizes two tunable lasers and is resonantly enhanced at each intermediate step. The initial experimental studies of this interaction were performed using sodium vapor, whose energy level diagram is shown in Fig. 1.

An incident laser field at frequency ω_2 connects the sodium $3^2S_{1/2}$ ground state to the $3^2P_{3/2}$ excited state while a second laser field at frequency ω_2 completes a two-photon resonance with a sodium Rydberg level. A dc electric field is applied to the atomic system; this field breaks the inversion symmetry of the sodium atom and thus allows the existence of an electric dipole moment oscillating at the sum frequency $\omega_3 = \omega_1 + \omega_2$.[1] The resulting nonlinear polarization is related to the applied field strengths $E(\omega_1)$ and $E(\omega_2)$ through the relation

$$P(\omega_3) = 2\chi^{(2)}(\omega_3 = \omega_1 + \omega_2)\, E(\omega_1)\, E(\omega_2) \tag{1}$$

where the nonlinear susceptibility describing the mixing process is given by[1]

$$\chi^{(2)}(\omega_3 = \omega_1 + \omega_2) = \sum_{jk} \frac{N\, \mu_{ik}\mu_{kj}\mu_{ji}}{2h^2(\omega_{ki}-\omega_2-\omega_1)(\omega_{ji}-\omega_1)} \tag{2}$$

Here N denotes the number density of atoms, and $\mu_{\ell m}$ denotes the electric dipole matrix element connecting levels ℓ and m.

EXPERIMENTAL SETUP

The setup used in our experimental study is shown in Fig. 2. An excimer laser (Lumonics Model 861S-2) operating at 308 nm with XeCl was used to pump two dye lasers. Each laser consisted of a two-grating Littman-style oscillator followed by a power amplifier. One laser (hereafter called the yellow laser) used rhodamine 6G as the laser dye and was tuned to the vicinity of the D_2 ($3^2S_{1/2} - 3^2P_{3/2}$) resonance line. The other laser (the blue laser) used stilbene 1, stilbene 3, POPOP or

DPS as the laser dye and could be tuned over the range 3900 to 4300 Å to access all Rydberg levels with $n > 11$. These lasers produced output energies of typically 0.5mJ with linewidths of ~5 GHz and durations of ~4ns. The output beams from these lasers were combined using a dichroic beam splitter and were loosely focused to obtain a beam diameter of several millimeters inside the sodium vapor cell. Typically, the cell was operated with no buffer gas and with the cell body at a temperature of 410°C and the sidearm at a temperature of 330°C thus producing a sodium number density of 4×10^{14} cm^{-3}.

High voltage pulses were applied to the Stark plates simultaneously with the arrival of the laser pulses. The risetime of the electrical pulse was ∿1ns, and a short pulse duration of ∿10ns was used to avoid establishing a breakdown in the sodium vapor.

The beam exiting the sodium cell was focused onto the entrance slit of a quartz-prism monochromator. Additional spectral discrimination was provided by UV-transmitting interference filters. The ultraviolet signal was detected using a solar blind photomultiplier tube and was processed using a boxcar averger.

EXPERIMENTAL RESULTS

We observe resonantly enhanced sum-frequency generation whenever the yellow laser is tuned near the sodium D lines and the blue laser is tuned to complete a two-photon resonance with a Rydberg state. The UV signal is emitted in the forward direction in a beam whose divergence angle is comparable to that of the incident lasers.

The tuning characteristics of the sum-frequency generation process have been investigated and are shown for a typical case in Fig. 3. Here the intensity of the UV output signal is plotted as a function of the wavelength of the blue laser, with the yellow laser detuned 1.1 Å to the short wavelength side of the D_2 resonance line. Resonance enhancement of the output signal is observed whenever the sum frequency is coincident with a Stark-split resonance of the sodium atom. The $\ell \geq 2$ levels are seen to mix and spread out in a fan as predicted theoretically.[3]

Scans covering a much broader spectral interval but with lower spectral resolution are shown in Fig. 4 for several different values of the detuning of the yellow laser and with the dc electric field fixed at its maximum value of 2000 V/cm. We see that the positions of the resonances excited by the blue laser depend on the detuning of the yellow laser, as expected for a two-photon resonantly enhanced parametric mixing process. These curves also show that at the largest field strengths the various resonances have broadened into one another, leading to broad regions of continuous tunability.

We have found that the maximum output signal is obtained with the yellow laser detuned 1 cm^{-1} to high-frequency side of D_2 and with the blue laser tuned to complete a two-photon resonance with the 13 s level, with a dc field strength of 2000 V/cm. Under these conditions, the measured conversion efficiency (defined as the ratio of the UV output energy to the total energy of the two input pulses) was found to be 3×10^{-4}. The nonlinear susceptibility is found to have the value $\chi^{(2)}=1.2 \times 10^{-8}$ esu. Due to uncertainties regarding our beam profile within the interaction region, this value is probably accurate only to within a factor of three. This measured value is in good qualitative agreement with the predicted value given above.

CONCLUSIONS

We have demonstrated that extremely large values of the second-order nonlinear susceptibility can be obtained through use of Rydberg states perturbed by an external electric field. Values of $\chi^{(2)}$ as 1.2×10^{-8} esu have been obtained in a sodium vapor of density 4×10^{14} cm^{-3}. The splitting of the highly excited levels due to the external electric field creates a broad region of overlapping levels, leading to continuous tunability. This technique can be applied to any system displaying a Rydberg series. When applied to a system with a large ionization potential such as hydrogen or a noble gas it should be capable of producing coherent, tunable radiation in the extreme ultraviolet.

ACKNOWLEDGEMENTS

This work was supported by the U.S. Army Research Office and the Air Force Office of Scientific Research.

REFERENCES

1. R.W. Boyd and L.Q. Xiang, IEEE J. Quantum Electron. QE-18, 1242, 1982.
2. D.J. Gauthier, J. Krasinski, and R.W. Boyd, Optics Lett. 8, 211 (1983), and to be published in Coherence and Quantum Optics V, L. Mandel and E. Wolf, eds., Plenum, New York 1984.
3. M.G. Littman, M.L. Zimmerman, T.W. Ducas, R.R. Freeman, and K. Kleppner, Phys. Rev. Lett. 36, 788 (1976).

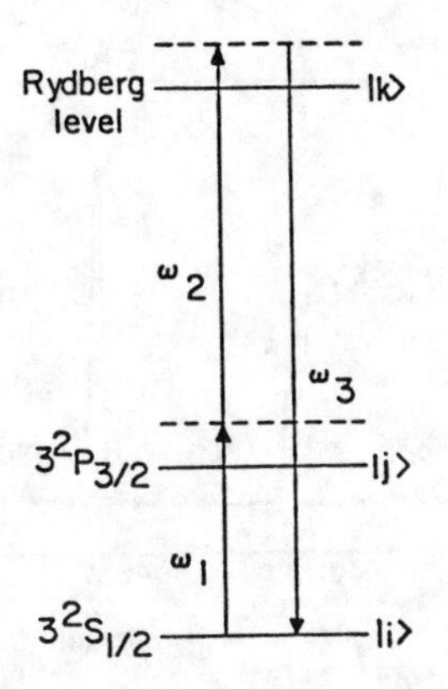

Fig. 1. Optical waves at frequencies ω_1, ω_2, and $\omega_3 = \omega_1 + \omega_2$ interact by means of the nonlinear response of a Rydberg atom.

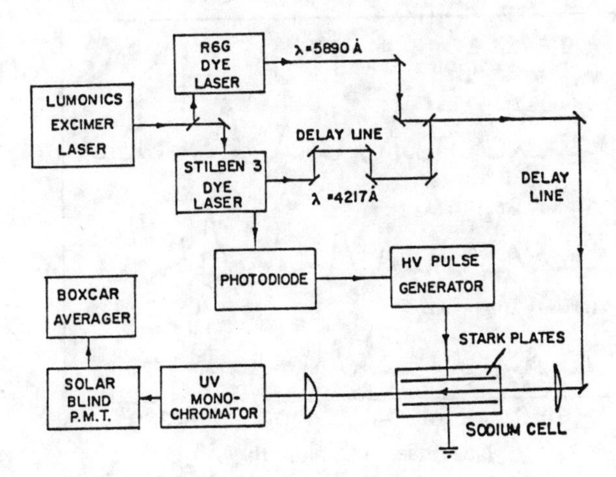

Fig. 2. Experimental setup.

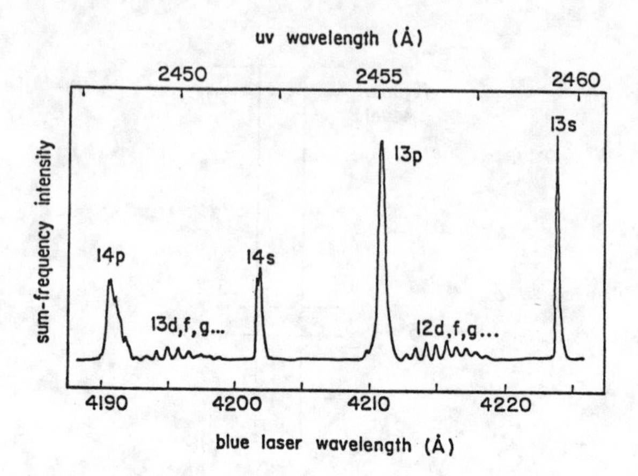

Fig. 3. Measured intensity of the sum-frequency signal as a function of the wavelength of the blue laser.

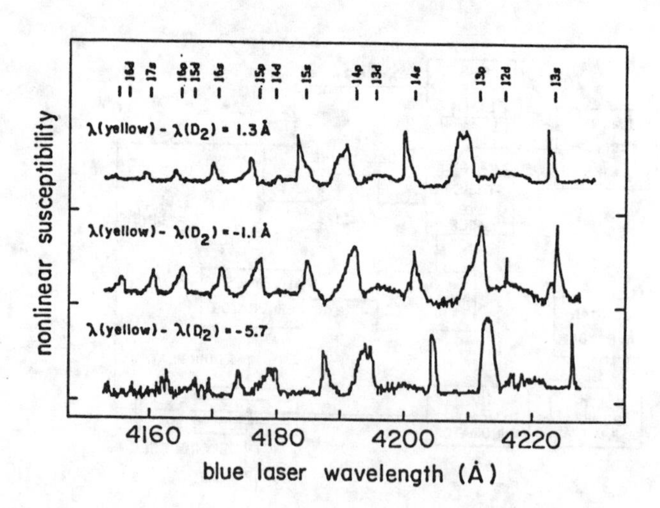

Fig. 4. Tuning characteristics of the sum-frequency generation process for several different values of the yellow laser detuning. The dc field strength is 2000 V/cm.

Sum Frequency Generation Spectroscopy of Rydberg States

A. Schnitzer, K.Hollenberg, W.Behmenburg; Physikalisches
Institut I, Universität Düsseldorf, 4ooo Düsseldorf 1, BRD

Summary

Consider sum frequency generation (SFG) $\omega_3 = 2\omega_1 + \omega_2$ in a phase matched metal-noble gas mixture, using two lasers ω_1 and ω_2, where $2\omega_1$ is tuned to a two photon transition Ω_{g2} and ω_2 is shifted across Rydberg transitions Ω_{2n} (g groundstate, 2 excited state, n Rydberg state of the metal atom). Then the intensity of the SFG spectrum $I_3(\omega_3) \sim |\chi^{(3)}(\omega_3)|^2$ displays resonances and related antiresonances, separated by

$$\Delta\omega_n = \frac{\mu_{2n}\mu_{ng}}{|\chi_{NR}|}; \qquad \Delta\omega_n \gg \Gamma_{ng} \qquad (1)$$

which are caused respectively by constructive and destructive interference of resonant and nonresonant parts of the 3. order nonlinear susceptibility $\chi^{(3)}(\omega_3 \simeq \Omega_{ng}) = \chi^R(\omega_3) + \chi^{NR}$

$$\chi^R(\omega_3) \sim \frac{\mu_{2n}\mu_{ng}}{\Omega_{ng} - \omega_3 - i\Gamma_{ng}} \quad (2) \qquad \chi^{NR} \sim \sum_{n' \neq n} \frac{\mu_{2n'}\mu_{n'g}}{\Omega_{n'g} - \Omega_{ng}} \quad (3)$$

The products $\mu_{2n}\mu_{ng} := x_n$ of the dipole transition matrix elements may then be obtained from a simple measurement of the frequency spacings $\Delta\omega_n$ at a large number of transitions Ω_{gn} and solving iteratively the coupled equation system (1),(3); neither absolute radiation intensity nor vapour density determination is needed. We have applied this method to Cadmium I

0094-243X/84/1190361-02 $3.00

and measured $I_3(\omega_3)$ in the range of the principal series $5s\,^1S_o \rightarrow np\,^1P_1$, n=12-28 (see figure). For evaluation of the corresponding x_n we included theoretical values for n=5-11 [1)2)] contributing to χ^{NR}. The x_n turn out to obey the n^{*-3} law for n=12-28. The matching to the x_n for $n < 12$ is achieved only by using the correct sign of x_5, which is predicted to be negative from theory [2)]. Additional measurement of $|\mu_{ng}|^2$ using the modification of $I_3(\omega_3)$ due to mismatch yields the individual μ_{2n}. This, together with the information on its sign relative to μ_{ng}, opens the possibility to test recent pseudo-potential approaches in the theory of heavy atoms with several valence electrons [3)4)].

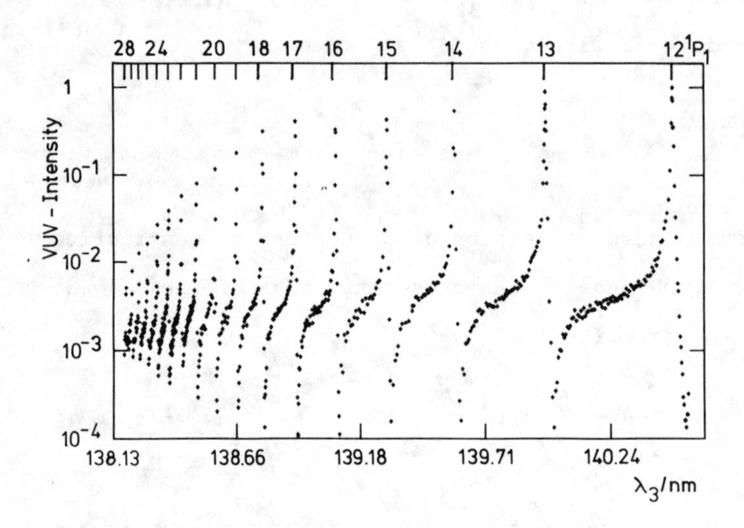

references

1) V.A. Zilitis, Optics and Spectr.(USA), 31, 86(1971)

2) W.H.E. Schwarz, private communication

3) T.C. Chang, P.Habitz, W.H.E. Schwarz, Theor.Chim.Acta 44, 61(1977)

4) L.R.Kahn, P.Baybutt, D.G.Truhlar, J.Chem.Phys.65,3826(1976)

5) A. Schnitzer, K. Hollenberg, W. Behmenburg; appeared in: Opt. Com. 48, 116 (1984)

DIRECT MEASUREMENT OF A VUV TRANSITION OSCILLATOR STRENGTH IN XENON

Steven D. Kramer, Chung H. Chen, and Marvin G. Payne
Oak Ridge National Laboratory, Oak Ridge, Tennessee 37831

ABSTRACT

A new method for accurately measuring the oscillator strengths of gases is presented. The technique is based on determining the phase-matching conditions for a laser four-wave mixing process in a gas mixture. By using high gas pressures, the measurement is independent of the detailed experimental geometry or spatial mode properties of the laser. Using the refractive index of argon as a reference, the oscillator strength of the xenon ground state to $7s[3/2]_1$ transition at 117.04 nm was found to be 0.098 ± 0.012. This value is more precise than previous results obtained from more complicated low-angle electron scattering experiments.

INTRODUCTION

Although optical oscillator strengths (f values) are important atomic quantities, they are generally not known to accuracies of better than 10%.[1] This is true since the standard classical measurement techniques tend, in most cases, to be complicated and indirect.[2] In particular, f values for short wavelength vacuum ultraviolet (VUV) transitions are often not known very precisely.

This paper describes a method for measuring f values by determining the phase-matching condition for four-wave mixing in a gas mixture. By using known values for the refractive index of argon, an absolute value for f can be determined. Although a number of techniques based on phase matching have been developed previously, most depend on having either well-defined plane[3,4] or Gaussian[5,6] incident laser beams. This means that the mode quality and focusing properties of the laser and associated optics have to be carefully controlled, or erroneous results will occur.[7] In one case, where a technique is described that gives results independent of the detailed properties of the laser mode, it is necessary to measure a very weak third-harmonic generated signal.[8] This limits its accuracy. In contrast, our method is based on easily measured parameters such as pressure and visible wavelength that optimize a signal in a way that is independent of the detailed properties of the experimental system.

EXPERIMENTAL METHOD

The oscillator strength, f, of the xenon $7s[3/2]_1$ state was measured by determining the phase-matching condition for four-wave sum mixing in xenon-argon gas mixtures. The VUV mixing scheme for xenon is shown in Fig. 1.[9] The 0.2-mJ/pulse input beams at 252.5 nm and near 1.5 μm were generated in the following manner. The 200-mJ/pulse, second-harmonic output of a 10-Hz, Q-switched Nd:YAG laser (Quanta-Ray DCR-1A) was split into two nearly equal intensity beams and used to pump a pair of dye lasers. The wavelength at 252.5 nm, which was tuned to excite the $5p^56p[3/2]_2$ two-photon allowed transition in xenon, was generated by mixing the doubled output of one visible dye laser (Quanta-Ray PDL) with the residual 1.06-μm pump laser output. While not required, the two-photon resonance increases the VUV generation efficiency. This required that the dye laser be tuned to 661.9 nm. The second visible dye laser (Quanta-Ray PDL) pumped a high pressure hydrogen Raman cell whose second Stokes-shifted output produced tunable radiation from below 1.33 μm to 1.62 μm. In order to cover this wide range, the second dye laser wavelength was varied from about 630 to 690 nm. The linearly polarized beams at 252.5 nm and near 1.5 μm were separated from their respective generating wavelengths by using Pellin-Broca prisms. These beams were then focused with separate 40-cm focal length lenses and made coaxial by use of a dichroic beam splitter before entering the 15-cm long xenon-argon cell in which the VUV light generation occurred. The transverse mode structure of the input laser beams was a somewhat distorted "donut" mode. The generated VUV light pulse, which contained up to 1 μJ of energy, passed through a LiF window and into a nitric oxide ionization chamber, where it was detected by measuring the amount of ionization that was produced.

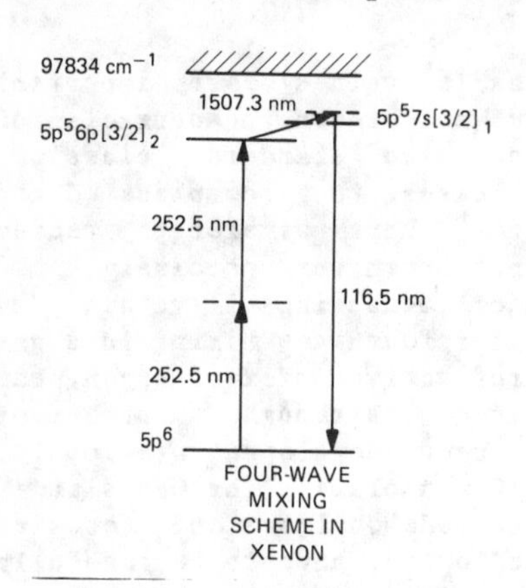

Fig. 1. Xenon four-wave mixing process.

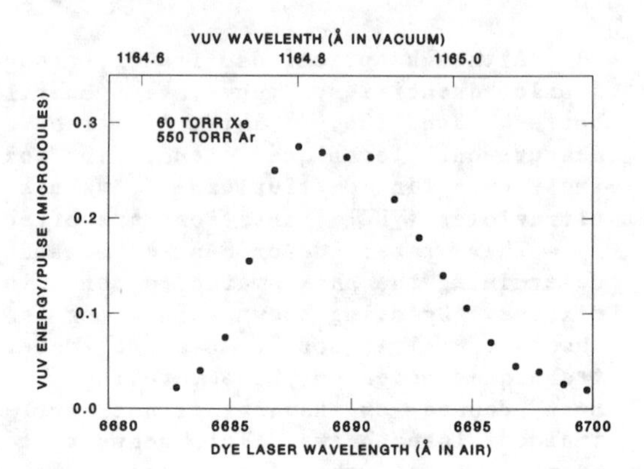

Fig. 2. VUV tuning curve.

In the VUV region probed, which was just above the xenon 7s state from about 115.7 to 116.9 nm, xenon is negatively dispersive.[10] Therefore, the generated VUV output can be increased by mixing the xenon with a positively dispersive gas such as argon. Argon was chosen because its refractive index is well known and varies slowly in this wavelength region.[11,12] The experimental procedure was to fill the generation cell with known partial pressures of high purity argon and xenon (Spectra Gases) which were measured with a capacitance manometer (MKS Baratron Corporation). Xenon pressures used ranged from about 10 to 150 Torr, while argon pressures ranged from 100 to 1800 Torr. The argon to xenon pressure ratio, R, covered the range from about 4 to 180 and could be measured to 0.1% accuracy. Then, for any given R the wavelength for maximum VUV output was determined. The VUV wavelength was varied by tuning the dye laser which pumped the

Raman cell. A typical tuning curve is shown in Fig. 2. Note that only a relative measurement of the VUV intensity over a very small wavelength range was necessary. Since the energy of the intermediate xenon 6p state and the Raman vibrational shift of hydrogen is well known, the accuracy of the VUV wavelength measurement is determined by how well the wavelength of the dye laser is known and how closely the peak in the tuning curve can be determined. The wavelength was measured by using a 0.5-meter monochromator, which had been calibrated with an inert gas calibration lamp. This gave an absolute wavelength accuracy for the dye laser of ± 0.03 nm. This corresponds to an accuracy of ± 0.7 cm^{-1} (± 0.001 nm) in the VUV which is somewhat smaller than the calculated VUV linewidth of 1.5 cm^{-1}. As shown in Fig. 2, the phase-matching curve was often much broader than this width, and in these cases the peak position could be measured to an accuracy in the VUV of about ± 7 cm^{-1} (± 0.01 nm).

The experimental results for ΔE as a function of R are shown in Fig. 3. ΔE is the difference in energy between the VUV photon and the xenon 7s state which gives the maximum VUV signal intensity for a given value of the argon/xenon pressure ratio, R.

It will be shown that if the xenon pressure is high enough, the peak in the phase-matching curve will depend only on R and should be independent of the laser mode structure. To test this, measurements were made, keeping R fixed as the partial pressure of xenon was varied. For xenon pressures above 10 Torr, it was found, within experimental accuracy, that the phase-matching peak was independent of R. Therefore, only experimental results taken above this pressure are used in the analysis. This constancy in R also indicated that effects due to ionization or dimers could be neglected. By using apertures both centered or offset from the laser beam axes, opaque cards to block portions of the beam, filters, deliberately misaligning the lasers, changing the position of the focal spot in the generation cell, and using slightly different focal length focusing lenses, it was possible to drastically alter the laser mode profiles. For xenon pressures above 10 Torr, none of these procedures affected the measured wavelength for maximum VUV generation. The analyzed data were taken using an approximately cylindrically symmetric optical configuration where the focal spot positions and confocal parameters for the two input beams were about equal. This gave a large VUV signal and approximated the conditions of a previous calculation.[7]

The total output of the four-wave mixing process is[5,7]

$$I(\omega_1 + \omega_2 + \omega_3) = N^2 X^2 \; F I(\omega_1) I(\omega_2) I(\omega_3),$$

where $I(\omega_i)$ is the intensity at frequency ω_i; N is the density of the generating gas; X is the third-order susceptibility of the gas; and F is a factor which includes the effects of mode structure, phase matching, and geometry on the conversion efficiency. Absorption and depletion of the beams have been ignored. In general, F is a complicated function of the detailed experimental configuration, but it has been shown under rather general conditions to have a maximum when $\Delta kb = -c$ where Δk is the phase mismatch, b is the confocal parameter, and c is a small positive number.[7] For example, $c = 2$ if Gaussian input beams are used.[5] So, tuning through a dispersion curve would give a maximum VUV signal when $\Delta kb = -c$ if X is independent of frequency. However, if the generated VUV light is close to a one-photon-allowed resonance, X will have a term with a resonant denominator. In this case, F also will have terms with a similar resonant denominator. Therefore, the basic condition phase matching for maximum VUV generation will still hold near resonance, although the exact value for c may be changed from the case when X has no dispersion.

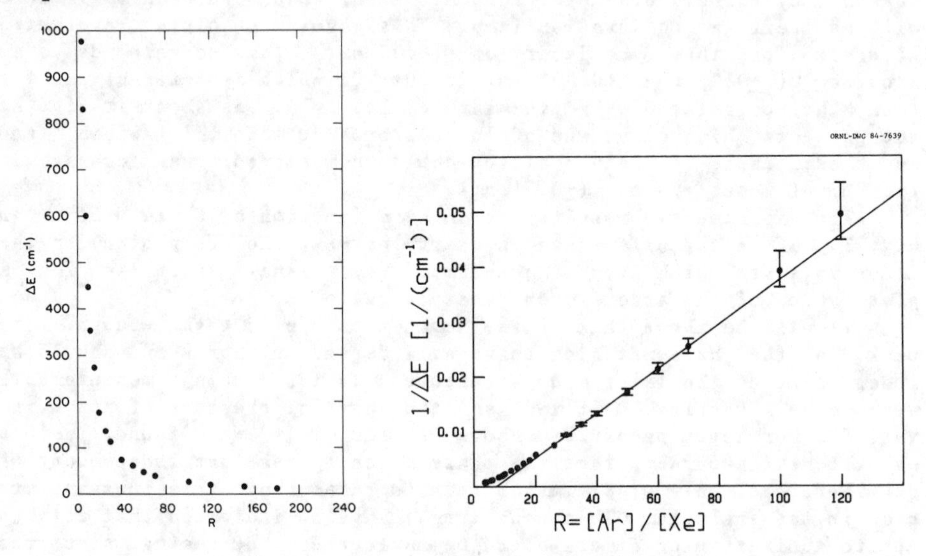

Fig. 3 Conditions for phase matching.

Fig. 4. Phase-matching condition.

The phase mismatch has the form

$$\Delta k = 2k(\lambda_1) + k(\lambda_2) - k(\lambda_3),$$

where λ_1 is the ultraviolet input wavelength, λ_2 is the infrared input wavelength, λ_3 is the generated VUV wavelength, and $k(\lambda_i) = 2\pi n_i/\lambda_i$. Here n_i is the refractive index of the gas mixture at λ_i. The term $n_i = 1 + N(Ar)n_i(Ar) + N(Xe)n_i(Xe)$, where $N(A)$ is the density of the gas A and $n_i(A)$ is the refractive index per atom of gas A at λ_i. Using this expression for Δk in the phase-matching equation, $\Delta kb = -c$, shows that the maximum VUV signal will occur when

$$\alpha R + \beta - n_3(Xe)/\lambda_3 = -c/[2\pi b\ N(Xe)] \quad , \qquad (1)$$

where

$$R = N(Ar)/N(Xe) \ ,$$

$$\alpha = 2[n_1(Ar) - n_2(Ar)]/\lambda_1 + [n_2(Ar) - n_3(Ar)]\lambda_3 \ ,$$

$$\beta = 2[n_1(Xe) - n_2(Xe)]/\lambda_1 + n_2(Xe)/\lambda_3 .$$

Over the small wavelength range covered, α and β are approximately constant. The near infrared and ultraviolet refractive indexes that appear in the expressions for α and β have all been accurately measured.[11,13] They are

$$n_1(Ar) = 1.128\ (\pm\ 0.002) \times 10^{-23}\ cm^3,$$
$$n_2(Ar) = 1.035\ (\pm\ 0.002) \times 10^{-23}\ cm^3,$$
$$n_1(Xe) = 3.04\ (\pm\ 0.01) \times 10^{-23}\ cm^3, \text{ and}$$
$$n_2(Xe) = 2.53\ (\pm\ 0.07) \times 10^{-23}\ cm^3.$$

The indicated error limits include the effects of wavelength dispersion.

Since $n_3(Ar)$ is in the short wavelength VUV region, it is not known as precisely as the other refractive indexes. Chashchina has measured the dispersion curve for the refractive index of argon down to 110.0 nm.[11,12] Calculations and single wavelength measurements near the region of interest in this paper agree with Chaschchina to within about 5%.[5,10,11,13,14] So, using Chashchina's value and the scatter in the other measurements as an estimate of the error gives a reasonable value of $n_3(Ar) = 2.1\ (\pm\ 0.1) \times 10^{-23}\ cm^3$. The estimated error is larger than the dispersion over the small VUV region covered. With these values, $\alpha = -9.0\ (\pm\ 0.9) \times 10^{-19}\ cm^2$ and $\beta = 2.6\ (\pm\ 0.1) \times 10^{-18}\ cm^2$.

The term $-c/2\pi bN(Xe)$ in Eq. (1) contains all the information about the laser mode structure and the geometry of the measurement. If N(Xe) is large, $-\alpha R$, β, and $-n_3(Xe)/\lambda_3 \gg [c/2\pi bN(Xe)]$, and the phase-matching condition reduces to

$$\alpha R + \beta - n_3(Xe)/\lambda_3 = 0 , \tag{2}$$

which is geometry independent.[15] In this work, $c \approx 2$, $b \approx 3$ cm, the smallest value of R is 4, and $n_3(Xe) \sim 10^{-22}\ cm^2$.[10] So, this condition becomes the requirement that the xenon pressure has to be much greater than about 1 Torr.[10] All the data analyzed in this paper was taken at xenon pressures above 10 Torr. Thus, the maximum in the generated VUV light intensity will occur when Eq. (2) holds. This condition is independent of the detailed geometry of the four-wave mixing process.

In this work, the generated VUV photon energy was close to resonance with the xenon 7s transition. For a wavelength, λ_3, close to this resonance yet still far enough away so that absorption is not important, the dispersion in $n_3(Xe)$ is dominated by the presence of the 7s transition, and so $n_3(Xe) = d - (r\lambda_3/4\pi)[f/(1/\lambda_3 - 1/\lambda)]$.[10] The term d is a constant that includes the effects of all other transitions, $r = 2.818 \times 10^{-13}$ cm is the classical electron radius, f is the 7s oscillator strength, and $1/\lambda = 85440.5\ cm^{-1}$ is the energy of the xenon 7s level at 117.04 nm.[16] Putting this expression for $n_3(Xe)$ in Eq. (2) and slightly rearranging gives

$$\frac{1}{\Delta E} = (-4\pi\alpha/rf)R + (4\pi/rf\lambda)(d - \lambda\beta) , \tag{3}$$

where $\Delta E = (1/\lambda_3) - (1/\lambda)$ is the energy difference in wave numbers between the generated VUV photon and the 7s state. Therefore, a straight line should result if the experimentally determined value of $1/\Delta E$, which maximizes the generated VUV signal, is plotted as a function of the corresponding R value. This form of plot is shown in Fig. 4 where the values used are taken from Fig. 3. For R greater than about 15, the points do lie close to a straight line as predicted by this theory. The deviation from linearity for small R is due to the influence of a higher lying state which was not included in the analysis which led to Eq. (3). The straight line in Fig. 4 is a weighed, linear least squares fit to the points with $R > 20$. Its slope is $4.1\ (\pm\ 0.2) \times 10^{-4}\ (1/\text{cm}^{-1})$ which from Eq. (3) is equal to $-4\pi\alpha/rf$). Since α depends only on the known refractive index of argon, the measured slope immediately gives $f = 0.098 \pm 0.012$ for the ground state to $7s[3/2]_1$ transition in xenon.

CONCLUSIONS

The only other oscillator strength measurements of this transition have been made using low-angle electron scattering. This is a complicated method which requires the use of high resolution, electron energy analyzers. Also, it determines only the generalized oscillator strength. The optical oscillator strength is then found by extrapolation to zero electron momentum transfer. This procedure was used by three different groups, and the f values found were 0.11 ± 0.4,[17] $0.0968 \pm .02$,[18,19] and 0.09 ± 0.02.[20,21] The result from this study, 0.098 ± 0.012, is consistent with these measurements and is more precise than any of them.

It should be stressed that to obtain our f value it was necessary to measure only pressure ratios and visible wavelengths. This can be done easily and accurately. Other than precisely known fundamental constants, the only other information needed was the argon refractive index. Most of the imprecision in our f value is due to the uncertainty in the measured VUV refractive index of argon.

ACKNOWLEDGEMENTS

The authors thank S. L. Allman and R. C. Phillips for their assistance. This work was sponsored by the Office of Health and Environmental Research, U.S. Department of Energy under contract W-7405-eng-26 with the Union Carbide Corporation.

REFERENCES

1. W. L. Wiese and G. A. Martin, "Atomic Transition Probabilities," in CRC HANDBOOK OF CHEMISTRY ANY PHYSICS, edited by R. C. Weast (CRC Press, Inc., Boca Raton, Florida, 1983), pp. 319-354.

2. P. G. Wilkinson, J. Quant. Spectrosc. Radiat. Transfer 6, 823-831 (1966).

3. J. J. Wynne and R. Beigang, Phys. Rev. A 23, 2736-2739 (1981).

4. H. Puell and C. R. Vidal, Optics Comm. 19, 279-283 (1976).

5. G. C. Bjorklund, IEEE J. Quant. Elec. QE-11, 287-296 (1975).

6. R. Mahon, T. J. McIlrath, and D. W. Koopman, Appl. Phys. Lett. 33, 305-307 (1978).

7. Y. M. Yiu, T. J. McIlrath, and R. Mahon, Phys. Rev. A 20, 2470-2485 (1979).

8. V. P. Gladushchak, S. A. Moshkalev, G. I. Chaschina, and E. Y. Shreider, Opt. Spectrosc. 51, 608 (1981).

9. S. D. Kramer, C. H. Chen, M. G. Payne, G. S. Hurst, and B. E. Lehmann, Appl. Optics 22, 3271-3275 (1983).

10. R. Mahon, T. J. McIlrath, V. P. Myerscough, and D. W. Koopman, IEEE J. Quant. Elec. QE-15, 444-451 (1979).

11. P. J. Leonard, Atomic Data and Nuclear Data Tables 14, 22 (1974).

12. G. I. Chashchina, V. I. Gladushchak, and E. Y. Shreider, Opt. Spectrosc. 24, 542-543 (1968).

13. P. D. Chopra and D.W.O. Heddle, J. Phys. B: Atom. Molec. Phys. 7, 2421-2427 (1974).

14. A. Bideau, Y. Guern, R. Abjean, and A. Johannin-Gilles, J. Quant. Spectrosc. Radiat. Transfer 25, 395-402 (1981).

15. R. Hilbig and R. Wallenstein, IEEE J. Quant. Elect. QE-17, 1566-1573 (1981).

16. C. M. Moore, ATOMIC ENERGY LEVELS, Vol. 3 (National Bureau of Standards, Washington, D.C., 1971).

17. A. Delage and J. D. Carette, Phys. Rev. A 14, 1345-1353 (1976).

18. J. Geiger, Z. Phys. A 282, 129-141 (1977).

19. J. Geiger, Phys. Lett. 33A, 351-352 (1970).

20. R. J. Celotta, National Bureau of Standards, Washington, D.C. (personal communication).

21. K. T. Lu, Phys. Rev. A 4, 579-596 (1971).

High Energy VUV Pulse Generation by Frequency Conversion

M. L. Dlabal, J. Reintjes and L. L. Tankersly
Laser Physics Branch
Optical Sciences Division
Naval Research Laboratory
Washington, DC 20375

The efficiency with which VUV radiation can be generated by frequency conversion has been limited by various competing processes to values that are typically of the order of 0.1 - 0.3%. As a result, increases in available VUV pulse energy will have to be made through increases in the pump laser energy. However, the use of higher pump laser power can be accompanied by additional limitations associated with various experimental parameters, especially in the VUV where tight focusing and short interaction lengths are usually used. For example the larger focal spots that must be used with stronger pump lasers to avoid breakdown are necessarily accompanied by longer interaction paths that can make absorption in the nonlinear medium and damage to cell windows more important. We have conducted a series of experiments in third harmonic conversion of XeF laser radiation in Xe to determine the extent to which the additional problems associated with high pump power can be overcome. Our initial experiments, done at a pump energy of 20 to 30 mJ (2 MW peak power) have indicated that the conversion efficiency, which was measured to be of the order of 5×10^{-4}, was limited by a combination of breakdown in the focus and absorption of the VUV radiation in the Xe downstream from the nonlinear interaction. The corresponding pulse energy was measured to be of the order of 10 μJ without phase matching. Further increases in VUV pulse energy can be accomplished with a combination of phase matching and increased

0094-243X/84/1190370-02 $3.00 American Institute of Physics

pump energy. Our initial measurements with phase matched mixtures of Xe and Ar indicate that the conversion efficiency increases by a factor of about 2 when a 5:1 Xe:Ar mixture is used. Further improvements in phase matching, the use of various gas combinations and their effects on breakdown and absorption limitations will be discussed. Scaling of our current results indicate that the conversion efficiency can be raised from 2×10^{-4} to 10^{-3} with a corresponding increase in VUV energy to 1 mJ if the pump energy is raised to 250 mJ and the nonlinear medium can be confined to the region of the nonlinear interaction. The usefulness of various gas confinement techniques for the conditions required by our high pump laser power (large apertures, long interaction paths) will be described and the results of conversion efficiency measurements at high pump energy will be reported.

SPATIAL PROFILES AND TIME EVOLUTION OF PLASMAS WHICH ARE CANDIDATES FOR A SOFT X-RAY LASER

C.H. Skinner, C. Keane, H. Milchberg,
S. Suckewer and D. Voorhees
Princeton University, Plasma Physics Laboratory
Princeton, New Jersey 08544

ABSTRACT

We present spatially resolved measurements of EUV line intensities and excited state populations in plasmas produced by the interaction of a CO_2 laser with carbon and aluminum targets of differing configurations. The results are assessed in terms of conditions necessary for a soft X-ray laser.

I. INTRODUCTION

The quest for laser action in the soft X-ray region poses a severe technical challenge, not only in achieving a high population inversion at a single point in space, but also in creating an extended medium of appropriate length, width and spatial uniformity to generate sufficient gain for a practical laser. In the previous paper[1] our experiment on soft X-ray laser development was introduced. This paper presents data from the experiment focussing on spatially resolved CVI line emission and populations in the light of the following requirements for a practical recombination laser. First, the effective length of the plasma should be sufficient for a gain-length product of 10 or more. Second, the plasma width must be low enough to avoid optical trapping on the radiative decay from the lower level of the potential lasing transition. Third, the upper level population of the potential lasing trasition (n = 3 for CVI 182Å) should be high enough for significant gain. A unique feature of our experiment is the use of a magnetic field to maintain a high electron density conducive to fast recombination. We present data showing the effect of the magnetic field on the spatial distribution and time evolution of the plasma. Other studies of CO_2 laser produced plasmas in magnetic fields have been done by Crawford and Hoffman,[2] and Loter _et al_.[3]

II. EXPERIMENTAL SET UP

A 1 kJ CO_2 laser was used to create a plasma from a variety of targets (see Fig. 1 and also the previous paper).[1] A EUV grazing incidence duochromator, equipped with two channel electron multipliers, observed plasma emission in the radial direction. A scanning slit assembly limited the field of view of the duochromator to a region 4 mm x 0.16 mm, the long dimension being parallel to the laser beam. By scanning the slit the transverse distributions of the line intensities were built up on a shot to shot basis. Between runs the target and laser focus were moved to different axial locations and hence a complete spatial picture of the plasma was constructed. A grazing incidence monochromator equipped with a 16 stage electron multiplier (rise time 20 nS) also

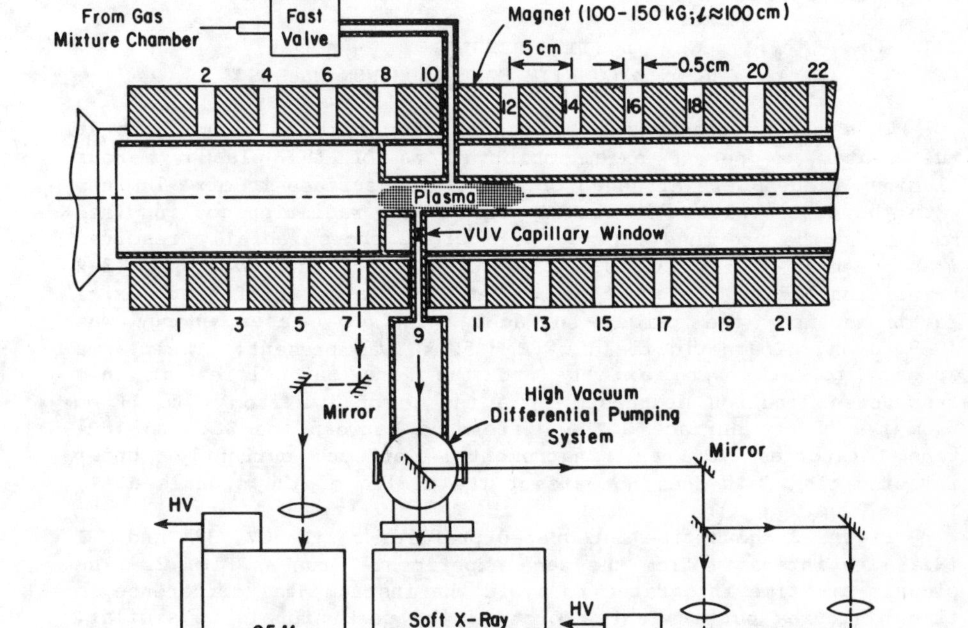

Fig. 1. Experimental arrangement of transverse spectroscopic instruments showing arrangement of beam splitters used in absolute intensity calibration.

observed the plasma emission in the axial direction. A grazing incidence mirror[4] was used to image the plasma on the monochromator entrance slit. The emission in the axial and transverse directions was simultaneously observed by two absolute intensity calibrated air monochromators using beam splitters so that each pair of EUV and air instruments observed practically the same region of the plasma. In this way it was possible to use the branching ratio method[5,6] to provide an in-situ absolute intensity calibration for the VUV instruments. The line pairs used were OVI 150Å/3811Å and NeVIII 88Å/2820Å.

III. RESULTS

A. CARBON DISC WITH FOUR ALUMINUM BLADES

This target consists of a carbon disc with a 1.5 mm hole at the center to permit axial observations of the plasma. Four aluminum blades were arranged on the target surface to mix aluminum with the carbon plasma to provide additional radiation cooling (see Fig. 2 in the previous paper).[1] Figure 2 shows radially resolved measurements of the CVI 33Å (1-2 transition) and CVI 182Å (3-2 transition) line intensities taken at three different axial locations from the target surface. The CO_2 laser energy was 70-90 J and a magnetic field of B = 50 kG was present. The plasma appears to be moving axially and the line intensities at this transverse location decreased by a factor of two from 1 to 11 mm from the target surface. The difference between the 33Å and 182Å time histories is an instrumental effect currently under investigation. It does not affect the timing of the signals at 1, 11 and 22mm.

Figure 3 shows time integrated profiles of the CVI 33Å and CVI 182Å line intensity from the same experimental run as Fig. 2. The signals are time integrated to avoid the instrumental difference in time histories but in fact a plot of the peak signals is similar. In general the CVI emission extends over a diameter of 1 mm or more. Data taken with a higher laser energy shows a larger diameter. A width of 1 mm or more may put a severe limit of the CVI ground state population in order to avoid optical trapping of the CVI 33.7Å radiation, unless the population of the lower level of the potential lasing transition (3-2) can be reduced (e.g., by fast transport).

The data indicates there are large "wings" in the spatial profile of the CO_2 laser focal spot which are responsible for the large plasma width and the next section will describe a carbon blade target designed to overcome this problem. The maximum CVI n=3 level population for this low energy of the CO_2 laser is 7 x 10^{13} cm^{-3} close to the target, based on an overall plasma width of 1 mm. For a laser energy of ~ 600J the column density ($N_3 \cdot \ell$) is significantly higher, but still insufficient for laser action. However, as with all absolute intensity calibrations even though the greatest care was taken with the measurements it was never possible to completely rule out unidentified systematic effects.

The actual population could be higher. A second intensity calibration based on a vacuum spark source is currently underway. Figure 4 shows the effect of a 90 kG magnetic field on the CVI 33Å and 182Å line emissions. For the 90 kG case the CVI emission increased by up to a factor of five and shows a faster decay consistent with fast radiation cooling in the confined plasma column.

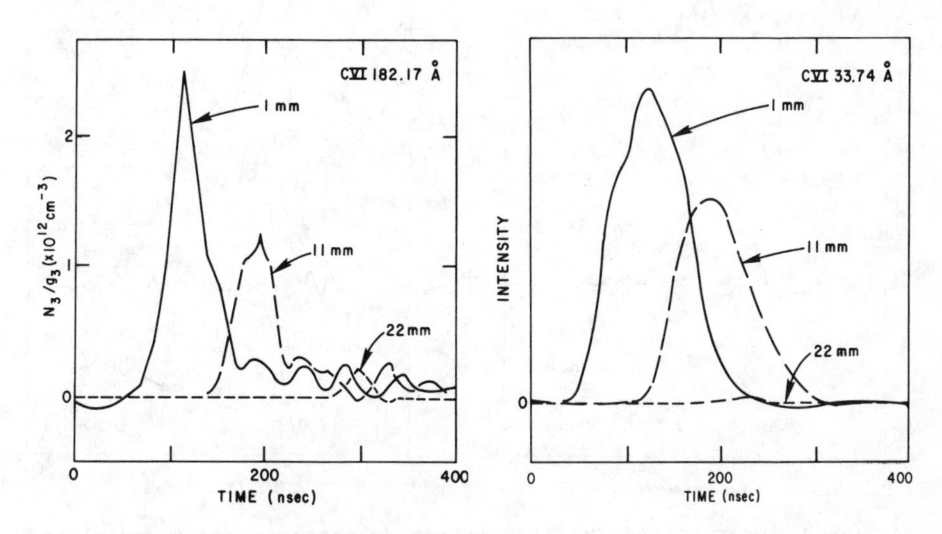

Fig. 2. CVI 182Å and 33Å emission at three axial locations: 1, 11 and 22 mm from the target surface. The radial location corresponds to + 0.7 mm in Fig. 3.

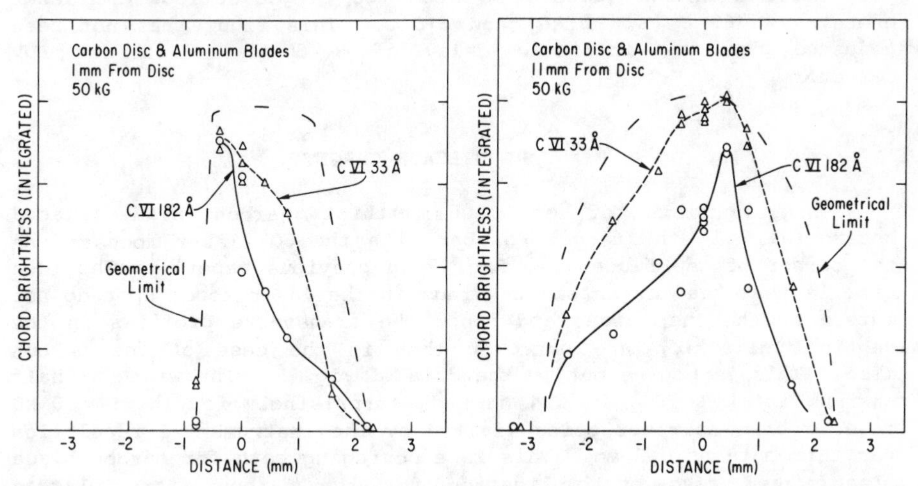

Fig. 3. Transverse scan of CVI 182Å and 33Å 1 and 11 mm from the disc. The outer dashed line represents, at 1 mm, the area not obstructed by the aluminum blades and at 11 mm, is a semi-elipse representing the geometrical limit of the circular feedthrough. The line through the data points are intended as a visual aid.

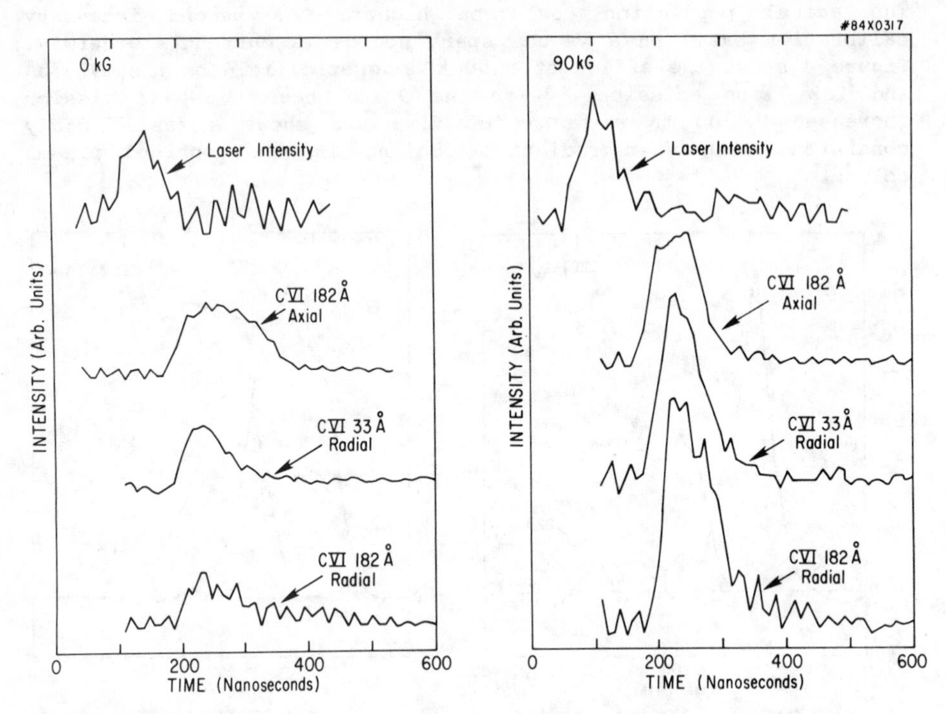

Fig. 4. Time resolved CVI line intensity measurements in axial and transverse directions taken with and without a magnetic field of B = 90 kG. The target was a carbon disc with 4 aluminum blades. The radial data was taken 1 mm from the target surface and laser energy was 460 J (B = 0 kG) and 610 J. This figure has not been adjusted for the intrinsic delay of ~ 60-80 nS in our EUV detectors.

B. CARBON BLADE TARGET

This consists of a 0.3 mm thick carbon blade placed horizontally in the target chamber with the CO_2 laser focussed on the corner of the blade (see Fig. 3 in previous paper).[1] The idea here is that the low intensity wings in the laser focal spot do not interact with the target and hence the transverse profiles in the vertical direction are narrower than in the case of the carbon disc. This is bourne out in the data of Fig. 5. The width at half maximum for CVI 182Å is 0.4 mm and, surprisingly for the B = 0 kG case is even narrower being limited by the instrumental resolution for this data of 0.3 mm. This is encouraging both for carbon blade plasmas and also by implication for the carbon fibre plasmas discussed in the next paper.[7] The spatial profile of the CVI 33Å emission appears to be broader than the CVI 182Å emission. In the horizontal direction the plasma thickness is estimated to be in the range 0.2 - 0.8 mm. In Fig. 5 the maximum population of level n=3 of CVI is 5×10^{13} cm^{-3} for the B = 50 kG case and 2×10^{14} cm^{-3}

for the B = 0 kG case, based on a plasma thickness of 0.2 mm, for a relatively low laser energy of ~ 120J. Initial results with 600J laser energy show an order of magnitude higher column density at B = 50 kG. These numbers are encouraging and are approaching that required for lasing action. A spatial scan at 11 mm axial distance from the target showed N_3 populations reduced by more than an order of magnitude so that at present the carbon blade plasma is clearly of very limited axial length. Further target development is necessary to combine the high CVI N_3 population and narrow width of the carbon blade plasma with the longer axial extent of the carbon disc plasma.

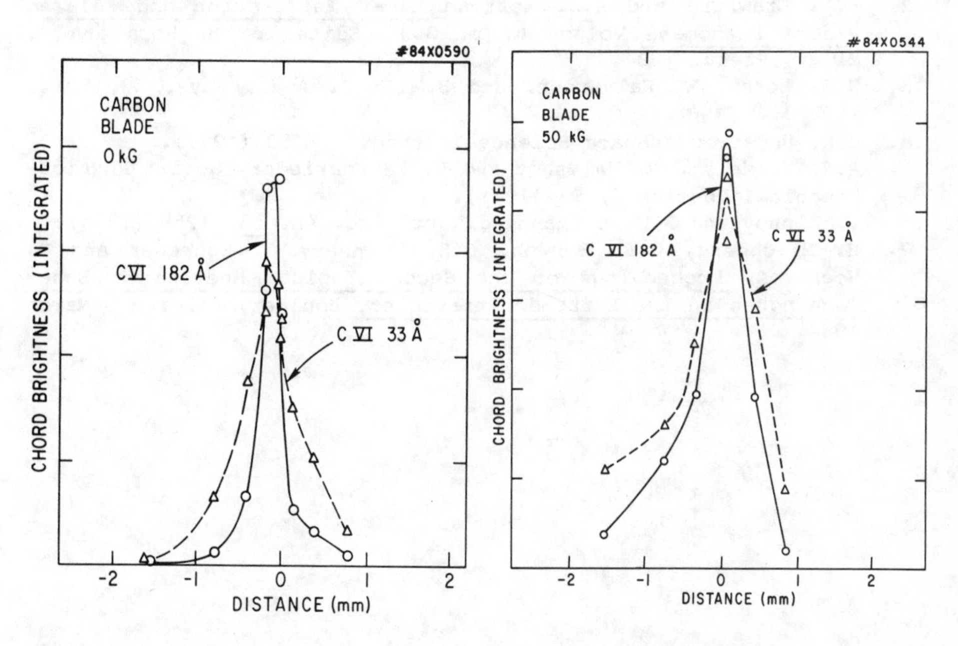

Fig. 5. Vertical distribution of CVI 182Å and 33Å line emissions with B = 0 and 50 kG showing very narrow profiles at B = 0 kG. The laser energy used here was 120 J.

CONCLUSIONS

We have investigated the spatial characteristics and excited state populations of plasmas produced by the interaction of a CO_2 laser with carbon and aluminum targets. The populations and geometry show promise but further target development is necessary to generate a long, narrow plasma suitable for a soft X-ray laser.

ACKNOWLEDGEMENTS

The authors would like to thank J.L. Schwob for productive discussions and help in the operation of the VUV duochromator.

They also thank D. DiCicco and V. Vasilotas for helpful technical assistance in the experiment. This work is supported by the United States Air Force Office of Research, Contract No. AFOSR-84-0025.

REFERENCES

1. S. Suckewer, C. Keane, H. Milchberg, C.H. Skinner and D. Voorhees Proceedings of the Second Topical Meeting on Laser Techniques in the Extreme Ultraviolet, Boulder, Colorado, March 1984.
2. E.A. Crawford and A.L. Hoffman Laser Interaction and Related Plasma Phenomena Volume 6. p. 353 Edited by H. Hora and G. Miley, Plenum 1984.
3. N.G. Loter, W. Halverson, and B. Lax J. Appl. Phys. 52, 5014, (1981).
4. J.H. Underwood, Space Science Instrum. 3, 259 (1977).
5. A.N. Zaidel, G.M. Malyshev and E. Ya Schrieder, Soviet Physics-Technical Physics 6, 93 (1961).
6. E. Hinnov and F.W. Hofmann, J. Opt. Soc. Am. 53, 1259 (1963).
7. H. Milchberg, J.L. Schwob, C.H. Skinner, S. Suckewer and D. Voorhees, Proceedings of the Second Topical Meeting on Laser Techinques in the Extreme Ultraviolet, Boulder, Colorado, March 1984.

SOFT X-RAY SPECTRA, POPULATION INVERSIONS AND GAINS IN A RECOMBINING PLASMA COLUMN

H. Milchberg, J.L. Schwob, C.H. Skinner,
S. Suckewer and D. Voorhees
Princeton University, Plasma Physics Laboratory
Princeton, New Jersey 08544

ABSTRACT

Time integrated and time resolved soft X-ray spectra have been measured from recombining CO_2 laser produced plasmas. Determinations of population inversions and gain will be discussed.

I. INTRODUCTION

In this paper, which complements the previous two, (Ref. 1, 2) we will present data showing gain and population inversions in hydrogenic carbon, CVI, and the Li-like ions CIV, OVI, FVII and NeVIII.

Prior to converting our EUV instruments (described in Ref. 1, 2) for time resolved measurements of the emission from various targets, spectra were recorded on photographic plates (Kodak #101). For proper exposure of these plates, between 5 and 15 shots were required.

These spectra were studied to (a) estimate the relative concentrations of the ion species present in the plasmas for different CO_2 laser powers and target conditions, (b) obtain time averaged estimates of plasma temperatures and densities via recombination continuua, dielectronic satellites and Stark broadening, and (c) identify line ratios indicative of population inversions and gain in selected species.

Conditions were found such that emissions from the potential lasant species (CVI, CIV, OVI, FVII, NeVIII) dominated the plasma emission. Time averaged electron temperatures during the recombination phase (Te $<$ 40 eV, from recombination continuua) and densities ($n_e \gtrsim 10^{18}$ cm^{-3}, from Stark broadening) were consistent with conditions required for population inversions and gain[3] (Te $\lesssim$ 20 eV, $n_e \sim 10^{18} - 10^{19}$ cm^{-3}).

II. RESULTS

A. H-LIKE CVI ION

Spectra of carbon plasmas produced from carbon (C)-discs, with and without aluminum blades (Fig. 2 in Ref. 1) were analyzed using two methods to determine the presence of enhanced axial emission of the CVI 182.17Å line (3-2 transition). First, it was found that by comparing, in the axial and transverse spectra, the 182.17Å line intensity to intensities of other CVI lines, especially 33.74Å (2-1 transition), gain-length (k•ℓ) products of up to 4.0 were

0094-243X/84/1190379-08 $3.00

indicated[4]. Secondly, in the axial spectra, variations in the 182.17 to 28.47Å (3-1 transition) branching ratio were correlated with plasma conditions. Since the CVI 182Å and 28Å lines share the same upper level the ratio of the line intensities from spontaneous emission is constant under optically thin conditions. With gain on 182Å, the ratio of 182Å to 28Å line intensities should increase. The measured variation in this ratio was approximately 0.8:1 to 3.1:1. Note that the possibility of optical trapping of 28Å radiation was ruled out because the ratio of its intensity to that of CVI 26.99Å line (4-1 transition) and higher members of the CVI Lyman series remained constant to within 30% as the 3-2 to 3-1 ratio increased.

Using these methods, the highest gain was indicated in the case of carbon discs in the presence of B = 50 to 90 kG magnetic fields, with and without Al blades. The increase in enhancement due to the Al blades was less certain as it fell within the scatter of the enhancements estimated with the above methods.

In order to measure enhanced axial emission for the case of time resolved EUV measurements, a relative calibration of the axial and transverse EUV instruments was performed (Fig. 1). The CO_2 laser was focused on vertical graphite fibers of diameter 75-300 so as to produce localized plasmas seen completely by both instruments. This method of calibration has been used in the Hull University experiments[5] with much finer fibers (< 6μ dia.). By focusing the CO_2 laser at the fiber tips and noting that the same relative sensitivity of axial and transverse instruments was obtained within the above range of fiber diameters, it was concluded that the fibers were not blocking the view of the instruments. Measured relative sensitivities were similar at 33.74Å, 40.27Å and 182.17Å. In addition, the EUV instruments were absolute intensity calibrated using the branching ratio method (see Ref. 2).

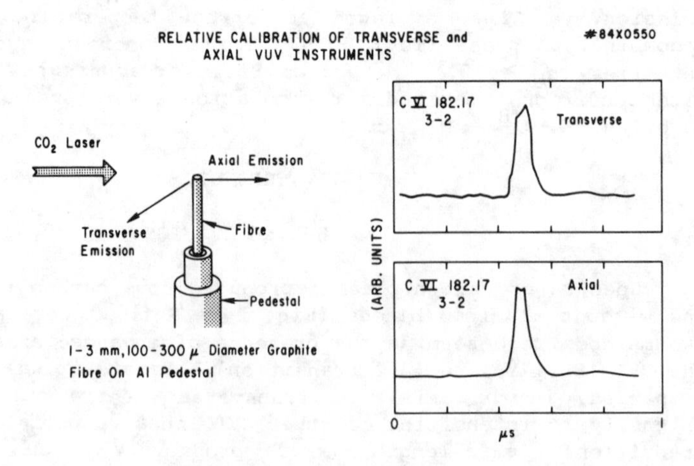

Fig. 1 Relative calibration at 182.17Å. Fiber diameter was 200μ.

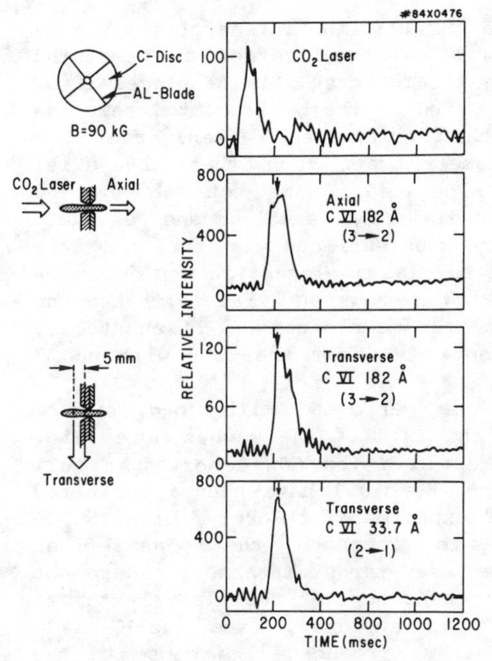

Fig. 2.
Time resolved CVI signals for carbon disc with Al blades and B=90 kG magnetic field. Laser energy 600J.

Fig. 3.
Axial and transverse CVI 182.17Å (3-2) line intensities. Vertical scales adjusted with relative calibration data.

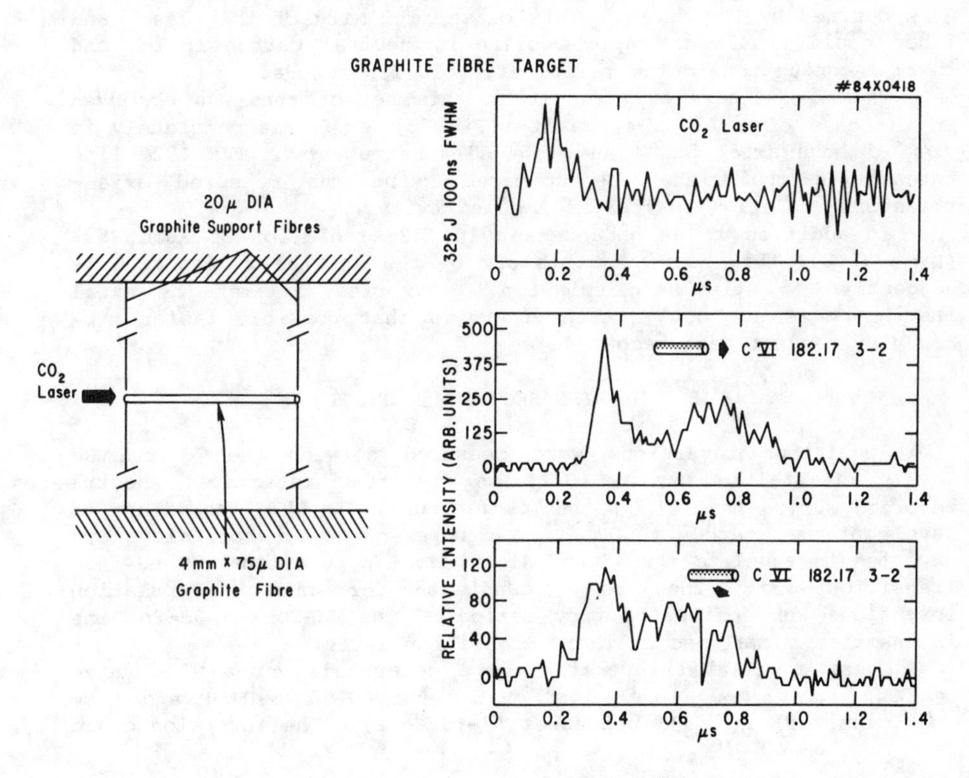

Time resolved CVI 33.74 and 182.17 line intensities for a C-disc (without Al blades) in a B = 90 kG magnetic field is presented in Ref. 1. Here, we present data for the case with Al blades (Fig. 2). The vertical scales were adjusted on the basis of the relative sensitivity determined for the axial and transverse EUV instruments. However, the geometry was such that the axial instrument could possibly view elongated plasma not seen by the transverse instrument. From radial and axial scans of 182 emission (Fig. 2, 3, Ref. 2) one can estimate for this geometry that an apparent enhancement due to plasma elongation could not be larger than approximately 2.0. Fig. 2 shows an axial to transverse enhancement (including that due to plasma elongation) of about 5.5, which is less than the enhancement of 10 for the case with no Al blades.[1]

Some initial experiments were performed with long, narrow graphite fibers suspended along the axis of the magnet bore (Fig. 3, Ref. 1) and illuminated end-on with the CO_2 laser. At Hull University much narrower fibers (6 dia.) have been illuminated by a line focused laser[5]. In our experiment, thicker fibers (60 - 200.) were chosen in order to increase the plasma-fiber interaction. The dynamics of the laser-target interaction were not studied, and it is not known at present how far the plasma extended along the fiber length. There is evidence, (see the C-blade experiment[2]) however, that the plasma produced lies close to the axis defined by the fiber. This close proximity of the plasma and fiber could result in rapid cooling by neutral carbon influx and thermal conduction to the relatively cold fiber core.

The most interesting result in this set of runs was observed in the case of a 75 x 4 mm fiber(Fig. 3) which was completely in view of both axial and transverse EUV instruments. The 182 line intensities could then be compared using the measured axial-transverse relative instrumental sensitivity.

In addition to an enhancement in 182 radiation of about 5.3 (K = 2.8), the sharp time history of the axial 182 emission is suggestive of stimulated emission. The rise time of the axial instrument is not better than 20 ns so that possible faster time behavior has not been recorded.

B. LI-LIKE SEQUENCE RESULTS

Population inversions were measured between the 4d and 3d levels of Li-like oxygen (OVI) on the time integrated spectra recorded during initial CO_2 gas target runs. In OVI, the 4d and 4f levels are separated by .02 eV so that electron-ion collisions are expected to equalize the 4d and 4f sublevel populations. The 4f-3d transition would then be a candidate for gain. Population inversions were estimated from ratios of the 4d-2p and 3d-2p line intensities normalized by their transition rates.

These population inversions were generated over a wide range of CO_2 laser power (10^{11} - 10^{12} W/cm^2 for 60-80 ns FWHM and 1 mm spot diameter) and fill pressures (3-10 torr). The inversion could

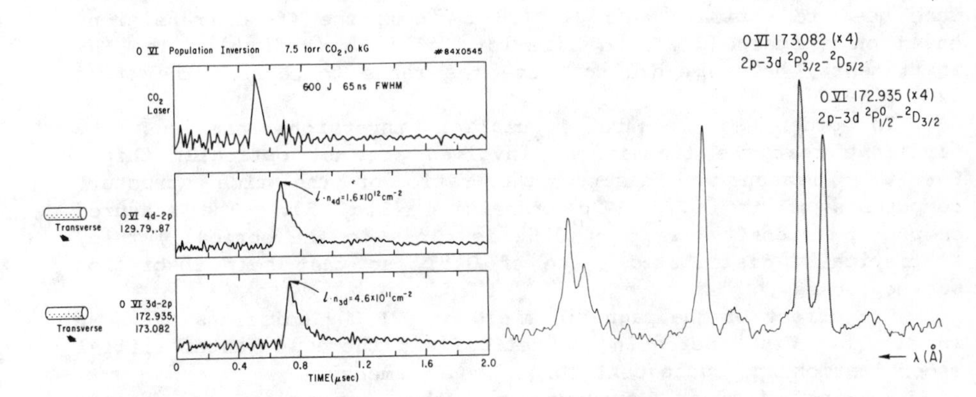

Fig. 4.
Intensities for OVI 129Å and OVI 173Å lines for 7.5 torr CO_2 target pressure, B=0 kG. Column sublevel population densities are $\ell \cdot n_{4d} \approx 1.6 \times 10^{12}\ cm^{-2}$ and $\ell \cdot n_{3d} \approx 4.6 \times 10^{11}\ cm^{-2}$ ($n_k = N_k/g_k$).

Fig. 5.
Typical OVI 3d-2p fine structure splitting in time integrated spectrum [component intensity ratio = 0.62; optically thin (theoretical) ratio = 0.55]

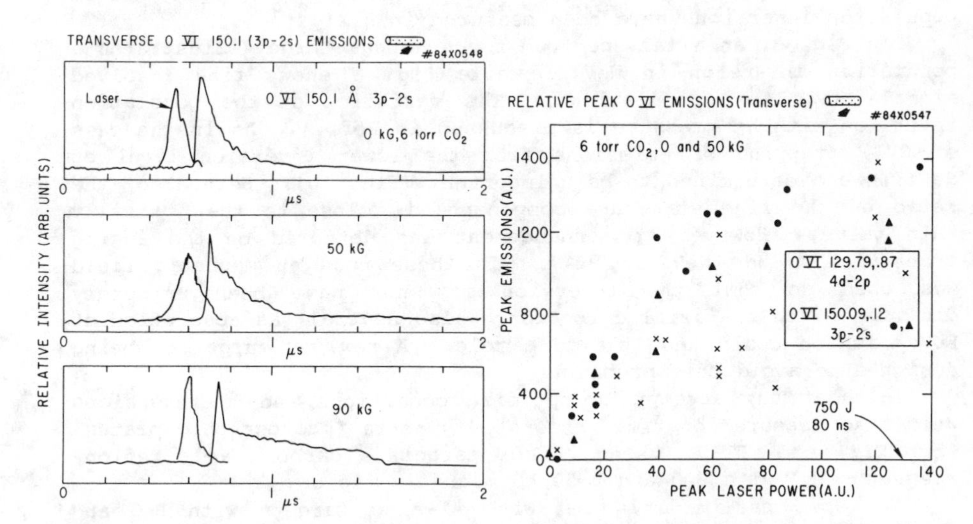

Fig. 6.
Increasing contribution of recombination peak to total signal shape as magnetic field is increased.

Fig. 7.
Saturation of OVI emissions at moderate laser power.

be eliminated by adding 5% Xe to the gas target mix[4]. Time resolved transverse data (Fig. 4) shows population levels which were used to estimate gain of 1.8 cm^{-1} on the 4f-3d transition, based on a Stark broadened linewidth of 0.1Å (Ref. 6). The EUV instruments, however, did not have the range to measure the 4f-3d 520Å line.

In order to measure population inversions one must be confident that the transitions involved are not optically thick. For this reason we measured the ratio of the time structure components of the OVI 3d-2p transition (Fig. 5). The measured component intensity ratio of 0.62 is close to the optically thin, statistically distributed value of 0.55, so that self absorption was neglected.

The effect of the magnetic field on OVI EUV emissions is shown in Fig. 6. With increasing magnetic field there is faster initial recombination and subsequent longer confinement.

Lending further support to the existence of strong recombination and inversions is the transverse data of Fig. 7, showing an apparent saturation of OVI emissions at moderate laser power. This is consistent with the ionization of OVI to the He-like species and subsequent recombination. The wider scatter of the OVI 129Å (4d-2p) points on this graph demonstrates the variation of the 4d-3d population inversion.

The encouraging results in OVI prompted an investigation of other members of the Li-like series. It has been suggested, and demonstrated for LiI and BeII, that the Li-like sequence is well suited for population inversions and gain[7]; for example, 4d-3d population inversions have been measured in AlXI.[8,9]

In Fig. 8, an axial spectrum for a Ne gas target illustrates a population inversion in NeVIII while Fig. 9 shows time resolved signals and level populations. The variation of the population inversion with Ne pressure is discussed in Ref. 1. As in the case of OVI, trapping of emissions from the lower inversion level of NeVIII were determined to be unimportant (Fig. 10). Here also, the ratio of the fine structure components is close to the optically thin value. However, no enhancement was detected on the lasing transition 4f-3d (NeVIII 292Å). In these runs no magnetic field was used so that the inversion may not have been uniformly distributed in the axial direction. Also neutral gas absorption of EUV emission could have played a role. A new gas target is being designed to avoid this problem.

Without any attempt to optimize conditions, 4d-3d inversions were also measured on time integrated spectra (photographic plates) for CIV and FVII using disc targets (carbon and teflon, respectively) with B=0 and B=50 kG magnetic fields.

In some preliminary runs with Aℓ-disc targets with B=0 and B=50 kG, no population inversions were detected in Li-like AlXI. Plasma radiation in this case may have prevented the temperature from rising high enough to appreciably ionize AlXI to He-like AlXII.

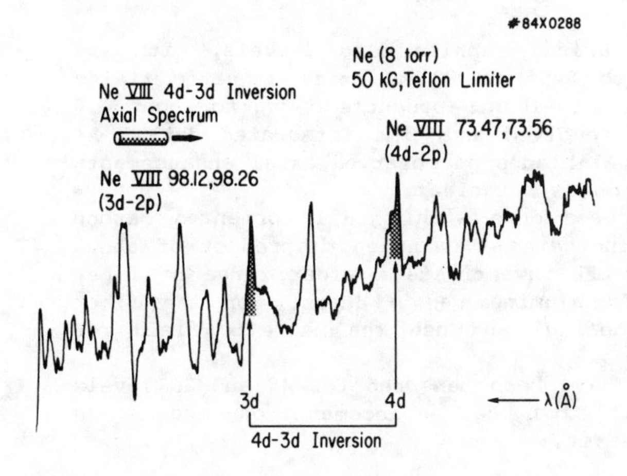

Fig. 8
NeVIII 4d-3d population inversion 3.7:1. Time integrated spectrum.

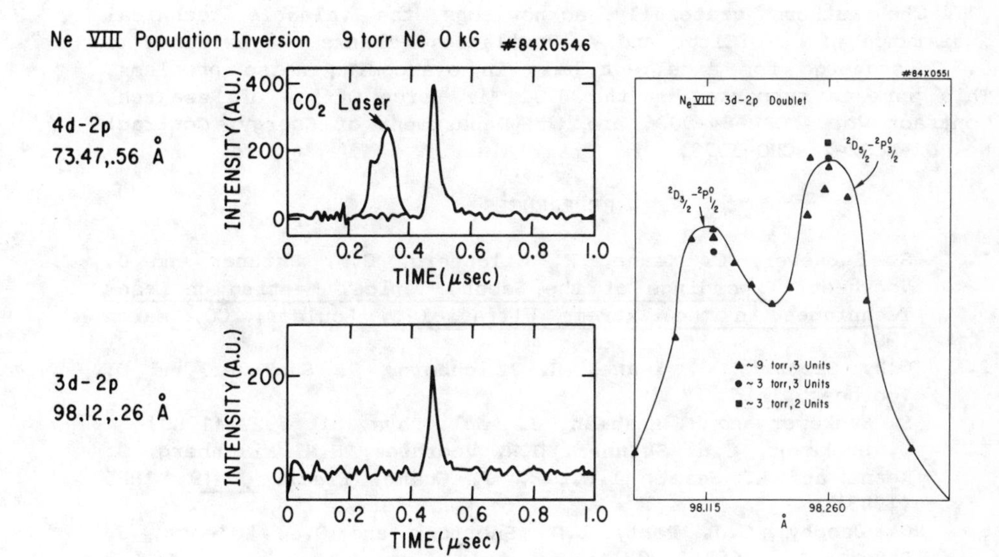

Fig. 9.
NeVIII signals demonstrating 4d-3d population inversion $\approx 5.3:1$. Column sublevel population densities are $\ell . n_{4d} \approx 1.6 \times 10^{12}$ cm^{-2} and and $\ell . n_{3d} \approx 3 \times 10^{11}$ cm^{-2}.

Fig. 10.
Fine structure of NeVIII 3d-2p transitions. Peak values of time resolved signals in third order normalized to the first order.

III. CONCLUSIONS

The plasmas produced using graphite disc targets, with and without Al blades, and with B=50 to 90 kG axial magnetic fields have been shown to produce gain-length products of up to about 4.0 as indicated by both time resolved and time integrated data. At present, the effect of the Al blades on observed axial enhancements of CVI 182.17 line radiation is not clear.

Preliminary work has been done with axially oriented carbon fibers. Our best result indicates a gain-length product of about 3.0. In the future, we will investigate a wider range of fiber lengths and diameters, add aluminum as a deposit on the fiber surfaces (for radiative cooling) and use the magnetic field for confinement.

Population inversions have been measured for 4d and 3d levels in selected Li-like ions, but no enhancement on the 4f-3d transition has been seen as yet.

ACKNOWLEDGMENTS

The authors gratefully acknowledge the valuable technical assistance of D. DiCicco and V. Vasilatos. Thanks are extended to G. Cutsogeorge for excellent help in overcoming noise problems. This work is supported by the U.S. Air Force Office of Research, Contract No. AFOSR-84-0025 and the Department of Energy, Contract No. DE-AC02-76-CHO-3073.

REFERENCES

1. S. Suckewer, C. Keane, H. Milchberg, C.H. Skinner and D. Voorhees, Procedings of the Second Topical Meeting on Laser Techniques in the Extreme Ultraviolet, Boulder, CO, March 1984.
2. C.H. Skinner, C. Keane, H. Milchberg, S. Suckewer and D. Voorhees, ibid.
3. S. Suckewer and H. Fishman, J. Appl. Phys. 51, 1922 (1980)
4. S. Suckewer, C.H. Skinner, D.R. Voorhees, H.M. Milchberg, C. Keane and A. Semet, I.E.E.E. J. Quant. Elect. QE-19, 1855 (1983)
5. D. Jacoby, G.J. Pert, L.D. Shorrock and G.J. Talents, J. Phys, B 15, 3557 (1982)
6. P.C. Kepple and H.R. Griem, Phys. Rev. A 26, 484 (1982)
7. W.T. Silfvast and O.R. Wood II, in Laser Techniques for Extreme Ultraviolet Spectroscopy, T.J. McIlrath and R.R. Freeman, Eds. (A.I.P., N.Y., 1982) p. 128
8. E. Ya. Kononov, K.N. Koshelev, Yu. A. Levykin, Yu. V. Sidel'nikov and S.S. Chirilov, Sov. J. Quant. Electron. 6, 308 (1976)
9. G. Jamelot, P. Jaegle, A. Carillon, A. Bideau, C. Moller, H. Guennon, and A. Sureau, in Proceedings of the International Conference on Lasers 81, New Orleans, LA, 1981, p. 178.

UV FLUORESCENCE BY OPTICAL PUMPING WITH LINE RADIATION

James Trebes and Mahadevan Krishnan
Yale University, New Haven, Connecticut 06520

ABSTRACT

Optical pumping of CII ions in a vacuum arc discharge using AℓIII ions in a laser produced plasma is described. The CII, 2p-5d, 560.437 Å transition was selectively pumped by line radiation from the AℓIII, 3p-5s transition at 560.433 Å. The wavelength mismatch is less than the Doppler width of the AℓIII line. Four transitions in CII, from the 5d, 5f, 4s, and $2p^2$ levels were studied simultaneously to examine the collisional-radiative redistribution of the pumped, 5d population. Electron density and temperature were measured in the C plasma. The Aℓ plasma was characterized by measurements and numerical modeling in order to estimate the intensity of the AℓIII pump line. A collisional-radiative model of the CII level populations was constructed with the measured density and temperature as inputs. Comparison of this model with the measurements allows discussion of the feasibility of building a UV laser with such a pumping scheme.

INTRODUCTION

Among the many approaches to the production of short wavelength population inversions is that of optical pumping with line radiation. In this approach, intense line radiation in one ion species is used to pump a nearly coincident transition from the ground state to a highly excited state in another ion species. The pumped, upper level may then be inverted with respect to lower lying levels. A survey of prior research in this field is given in a companion paper in these proceedings.[1] Recently, the work of Hagelstein[2] has motivated an experimental program[3] to test the feasibility of pumping soft X-ray lasers with such a scheme. Trebes and Krishnan[4,5] have demonstrated UV fluorescence by the combined effects of optical pumping and collisional transfer. This paper presents experimental results of the simultaneous measurement of enhanced fluorescence on four different transitions in CII, corresponding to four distinct upper states, when only one of these states was optically pumped with AℓIII line radiation. Also presented are measurements and estimates of electron density and temperature in the C plasma as well as the Aℓ pump plasma. These measurements enabled the development of a multi-level, collisional-radiative model for CII. The experimental observations are discussed in light of the model. Prospects for building a UV laser are discussed.

0094-243X/84/1190387-22 $3.00

FLUORESCENCE MEASUREMENTS

This section begins by describing the experimental apparatus and characterizing the carbon discharge plasma. Then the measurements of fluorescence are described.

Figure 1 is a schematic diagram of the experimental apparatus. The carbon plasma is produced in a laser initiated vacuum arc, between the negative carbon cathode and a grounded carbon anode as shown. Two Co_2 TEA lasers are used. Laser I is focused on the cathode and triggers a vacuum arc discharge. The power supply for this discharge is a pulse-forming network with an external, impedance matching resistor. A typical oscillogram of discharge current vs time is shown in Fig. 2a. The flat-topped current duration is about 60 μs. Figures 2b, 2c, 2d, and 2e show typical line radiation vs time from lines of CI, CII, CIII, and CIV at wavelengths indicated on the figures. CI radiation is present only during the rising portion of the current pulse and after decay of the pulse. The CII and CIII radiation exhibits quasi-steady behavior, but the CIV intensity is seen to decrease during the latter portion of the discharge, although the current is constant. Such a decrease may be due to a decrease in electron temperature in the constant current arc, which in turn may be caused by dynamic effects. Based on these observations, it was decided to attempt optical pumping of the C plasma at a time of 40-45 μs after arc initiation, when quasi-steady conditions were observed.

The aluminum pump plasma was produced by focusing Laser II after the selected delay of typically 43 μs, onto an Aℓ rod target, shown in Fig. 1. To ensure reproducibility, the Aℓ rod was replaced after every ten laser shots. With each new target, five shots were fired to clean the target surface and then data were obtained with the next 5 shots. Earlier experiments[4,5] had shown that the 5d level in CII was pumped by AℓIII, 560.433 Å line radiation from the adjacent Aℓ laser produced plasma. Furthermore, the optical excitation was shown to be collisionally transferred to the 5f, higher angular momentum level. Enhanced fluorescence was measured on the 5d-3p and 5f-3d transitions at 2138 and 2993 Å, respectively. The primary motivation for the experiments described in this paper was to unravel the collisional-radiative kinetics in CII, following selective optical pumping. Toward this end, four different wavelengths in CII were monitored simultaneously. The wavelengths selected are shown in Fig. 3. The 2138 Å and 2993 Å lines were expected to show enhanced fluorescence due to optical pumping as before. The 3920 Å line would show fluorescence only if the 4s upper state were strongly coupled by collisions and radiative transitions to the 5d level. The 1335 Å line stems from an n = 2 level which is much lower in energy than the 5d and 4s levels. Furthermore, the $2p^2$ upper state of this line is not accessible by dipole, single electron transitions from the n = 5 or n = 4 shells. Therefore no fluorescence was expected at the

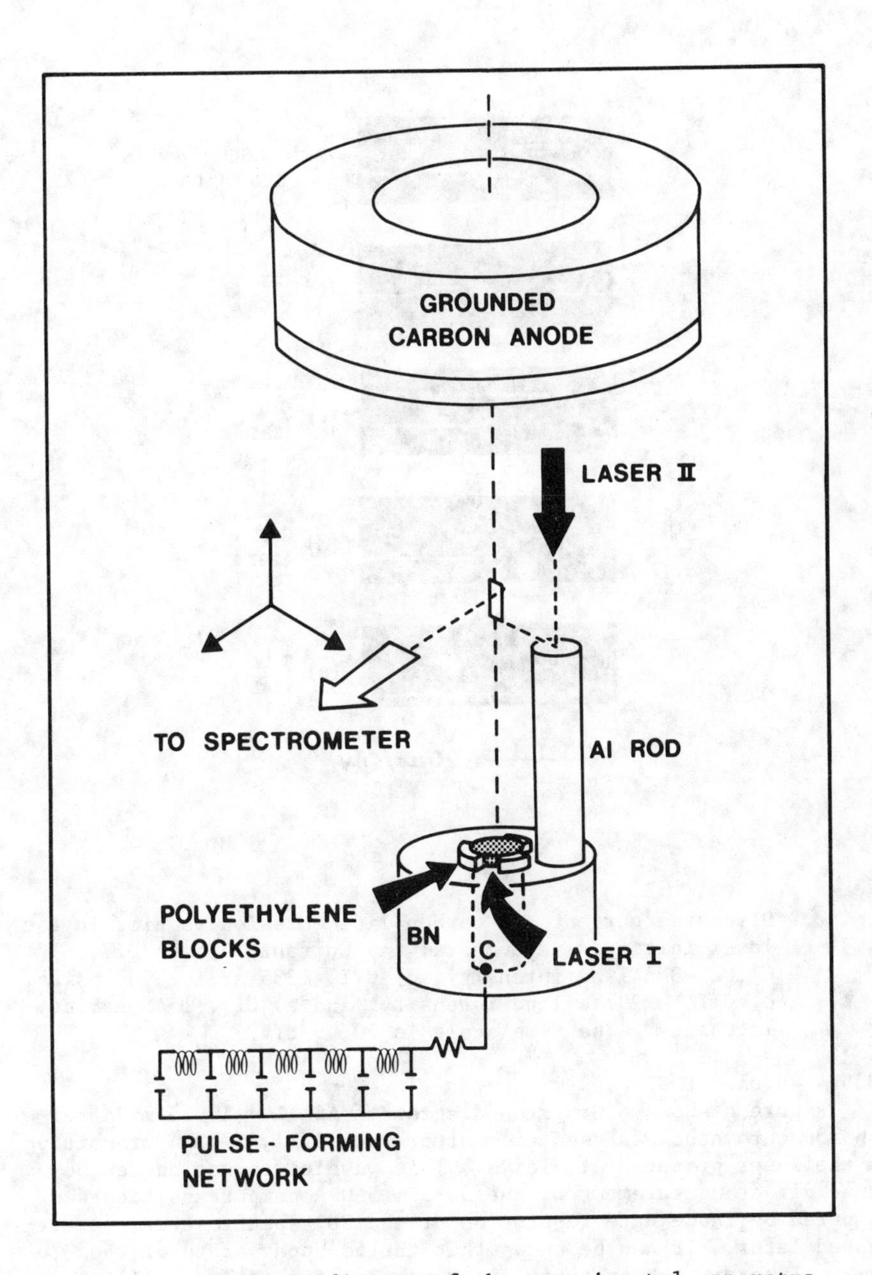

Fig. 1. Schematic diagram of the experimental apparatus.

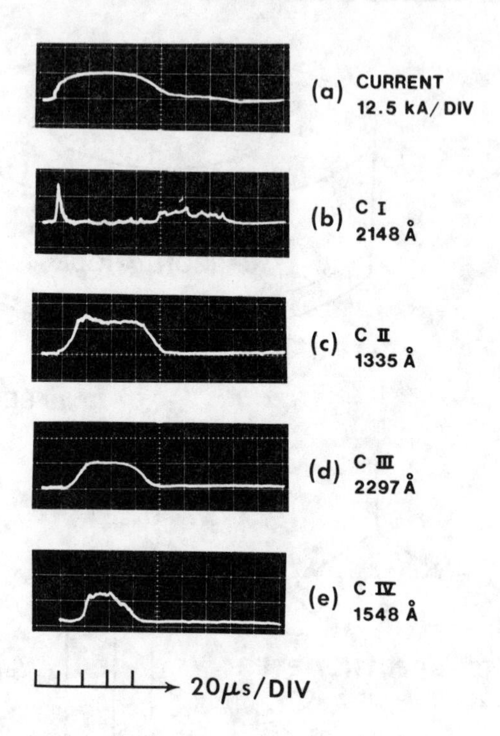

Fig. 2. Discharge current and carbon line emission vs time in the laser-initiated vacuum arc: a) Current 12.5 kA/DIV, b) CI, 2148 Å line intensity, c) CII, 1335 Å line intensity, d) CIII, 2297 Å line intensity, and e) CIV, 1548 Å line intensity. The time scale is 20 μs/DIV.

1335 Å wavelength.

Figure 4 shows a Grotrian diagram of AℓIII. A .25 m Jarrel-Ash monochromator was used to monitor the AℓIII, 3713 Å transition in the laser produced Aℓ plasma. This wavelength was chosen because direct measurement of the 5p-3s, 560 Å pump transition was hampered by inadequate resolution of the XUV spectrometer, as discussed later. It can be shown that the 5s upper state of the 3713 Å line and the 5p state are strongly coupled by collisions in an expanding, laser produced plasma.[6] Therefore the intensity of the 3713 Å line does provide a measure of the duration and relative intensity of the 560 Å pump line.

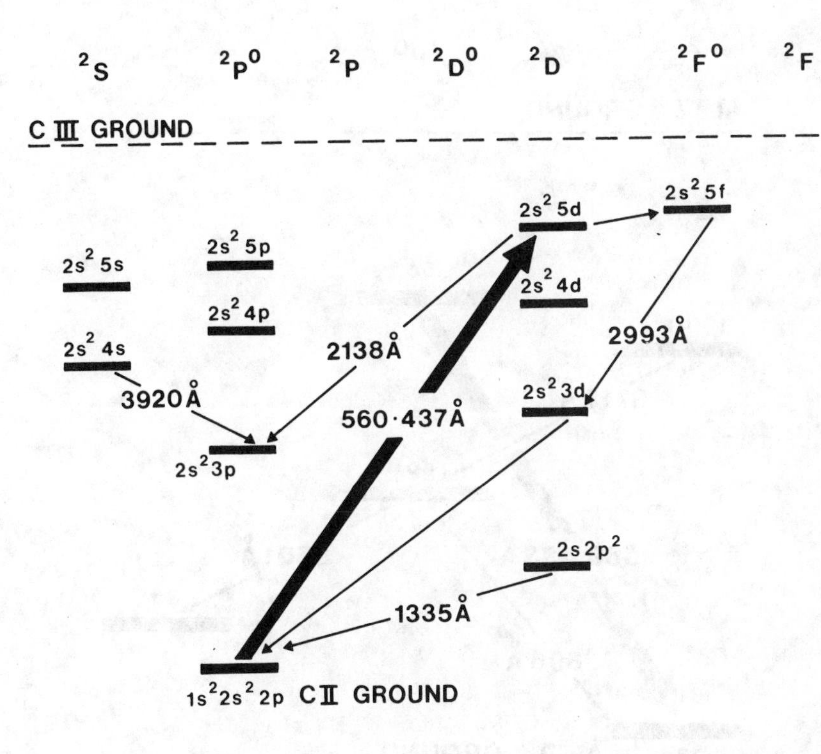

Fig. 3. Partial Grotrian diagram of CII, showing the wavelengths studied.

Figure 5 shows the arrangement of the spectrometers used to monitor the five wavelengths discussed above. All spectrometers, with the exception of the Jarrel-Ash, were focused onto the same local plasma region, 15 mm downstream of the cathode, along the discharge axis. The Jarrel-Ash was focused to a region 2 mm downstream of the Aℓ disc along the laser plasma axis.

In the first series of experiments, a fixed discharge current of 3.4 kA was arbitrarily chosen and the Aℓ plasma was produced 43 μs after discharge initiation. Ideally, the electron temperature in the carbon plasma should be such that the ground state population of CII is high, while the excited state population of the 5d level is low. Since it was difficult to directly measure the CII ground state population, the optimum discharge configuration was not established.

The observed fluorescence is shown in Fig. 6. Figure 6a shows three traces. The upper trace shows the CII, 2138 Å intensity, the middle trace the CII, 2993 Å intensity, and the lower trace the AℓIII, 3713 Å intensity vs time. Significant enhanced

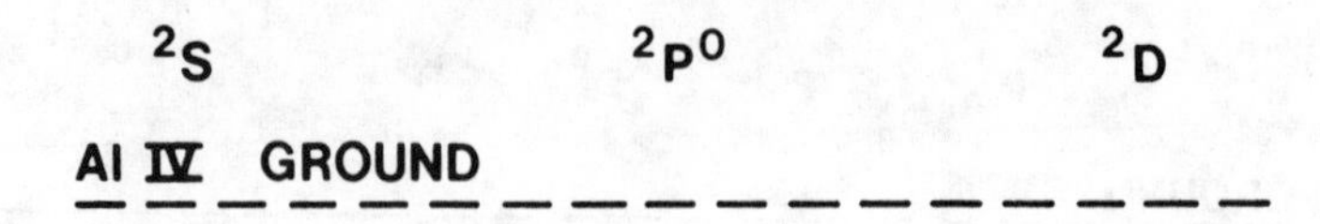

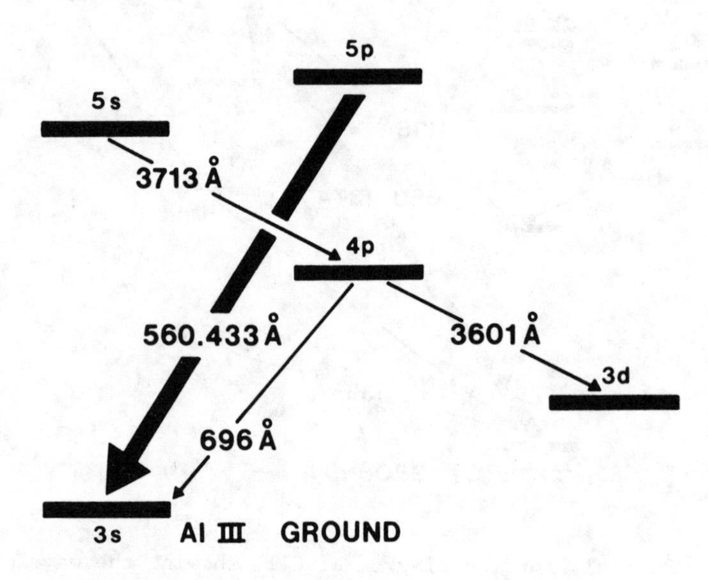

Fig. 4. Partial Grotrian diagram of AℓIII.

fluorescence is observed on both CII lines, coincident with the Aℓ plasma. Figure 6b shows the CII, 3920 Å intensity (upper trace) and the CII, 1335 Å intensity (middle trace) vs time. The lower trace is again the AℓIII, 3713 Å intensity. It is observed that both of these CII wavelengths also show enhanced fluorescence due to the Aℓ plasma. To ensure that the observed fluorescence was not due to continuum or to the wings of neighboring Aℓ lines from the Aℓ plasma leaking into the spectrometers, the experiments were repeated with no carbon discharge and only the Aℓ laser produced plasma. The results are shown in Figs. 6c and 6d. Some spurious background does appear, but the observed fluorescence in Figs. 6a and 6b is larger than this background for all four wavelengths.

The observed fluorescence at 1335 Å was puzzling, since the $2p^2$ upper state is not directly coupled to the upper levels. However, this state is strongly coupled to the CII ground state by a dipole allowed transition. It was possible that electrons from the denser and hotter Aℓ plasma collisionally excited this transition. This conjecture was supported by the different time scale

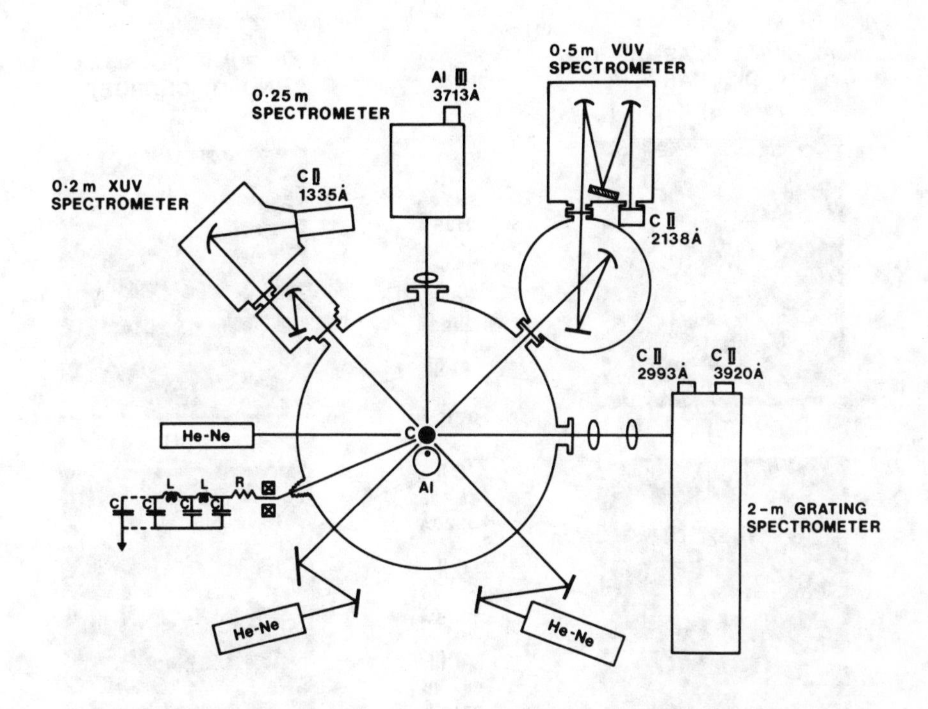

Fig. 5. Schematic diagram of the spectrometer arrangement.

for the 1335 Å fluorescence as compared with that for each of the other three wavelengths. Figure 7 shows the data of Fig. 6 on two timescales, 20 μs/DIV and 2 μs/DIV. The 1335 Å fluorescence is seen to decay more rapidly than the other lines. To test this conjecture, the Aℓ rod target was replaced with a Mg rod and the experiments were repeated. Since no lines of Mg are coincident with any CII transitions, no optical pumping was expected. Figure 8 shows the results. Figure 8a shows the time evolution of the CII, 2138, 2993, and AℓIII, 3713 Å line radiation. Figure 8b shows the CII, 3920, 1335, and AℓIII, 3713 Å lines. As with Fig. 6, the background measurements are shown in Figs. 8c and 8d. It is seen from the figures that although there appears to be some enhanced fluorescence coincident with a Mg pump plasma, the apparent enhancements on all but the 1335 Å line are in fact due to spurious background. The rather large background at 2138 Å is probably due to the wings of a bright, MgIII line at 2135 Å. The 1335 Å line shows enhanced fluorescence clearly above the background, suggesting that for both the Aℓ and Mg plasmas, this transition was probably excited by electrons penetrating the C discharge from the laser produced plasma. Further work is needed to confirm this. Nevertheless, these results serve to underscore the point made in earlier papers[4,5] that optical pumping with line radiation is best achieved by using two separate plasmas which are well isolated

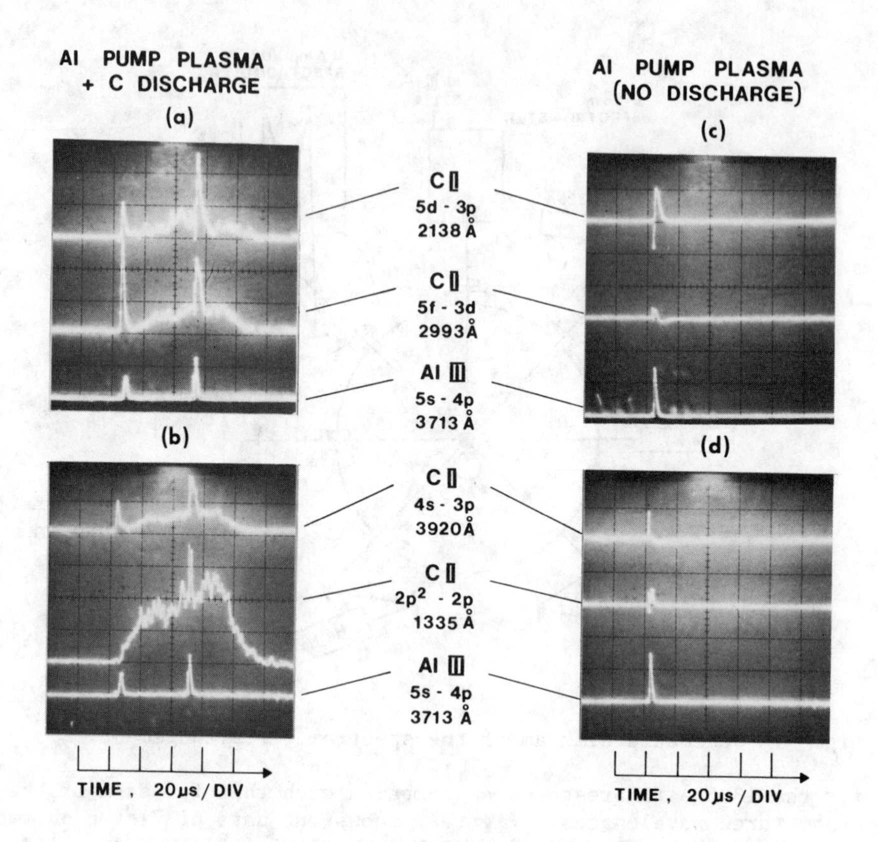

Fig. 6. Optically pumped fluorescence in CII: a) intensity vs time of CII, 2138 and 2993 Å lines, and AℓIII, 3713 Å line; b) intensity vs time of CII, 3920 and 1335 Å lines, and AℓIII, 3713 Å line. 6c) and d) same lines as in 6a and b, but with no C discharge to show spurious background signals due to Aℓ laser produced plasma alone.

from each other.

The preceding discussion has shown that in the carbon discharge, selective optical pumping to the 5d level is accompanied by strong coupling of the pumped level to other levels in the n = 5 and n = 4 shells. It is possible that the n = 3 levels are also coupled to higher levels. Such coupling is not conducive to producing population inversions at the 2138 and 2993 Å UV wavelengths. A detailed, collisional-radiative model must be constructed to examine such coupling. Necessary inputs to such a model are the electron density n_e and electron temperature T_e in the carbon plasma. Experiments were performed to measure these parameters. The results are described in the next two sections.

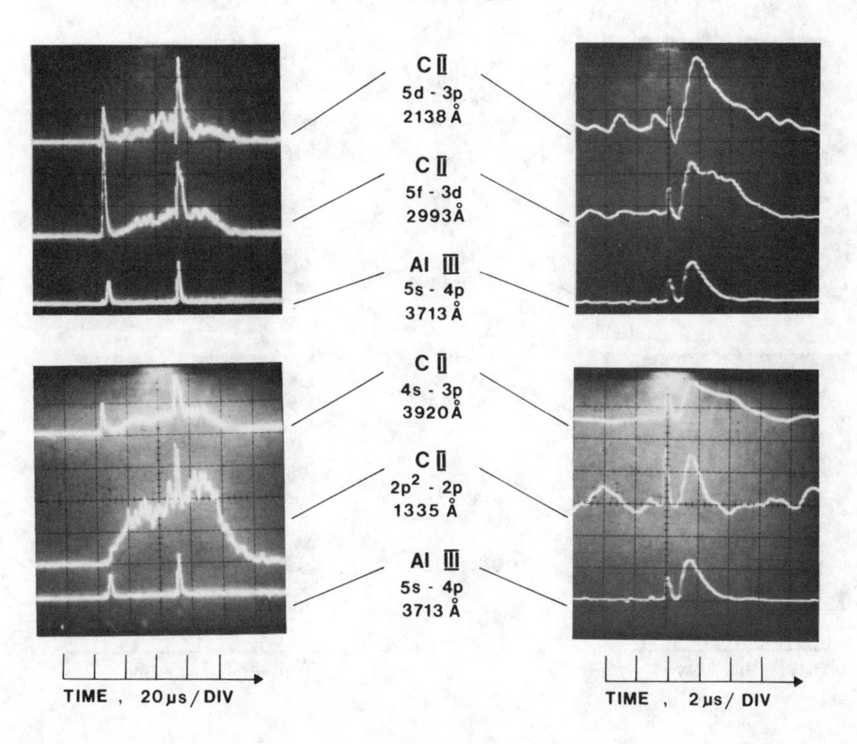

Fig. 7. Data of Fig. 6a and 6b displayed on two timescales of 20 μs/DIV and 2 μs/DIV, respectively.

ELECTRON DENSITY MEASUREMENT

Electron density in a laser-initiated carbon vacuum arc similar to that used in these experiments was measured by Keren and Hirshfield,[7] by measuring the refraction of a far-infrared laser beam after propagation through the plasma. For a cathode identical to that used here, but with the vacuum vessel walls serving as anode, n_e was measured for discharge currents I from 2 to 8 kA. The measurements were fit to a power law dependence on current, viz.:

$$n_e = 3 \times 10^{14}\ I^{1.9}\ \cos\phi / r^{2.4} \qquad (1)$$

where: r is the radius in cm and ϕ is the angle from the discharge axis. For a radius of 1.5 cm in these experiments, the above formula gives a density on axis of 1.2×10^{15} cm^{-3}, for a current of

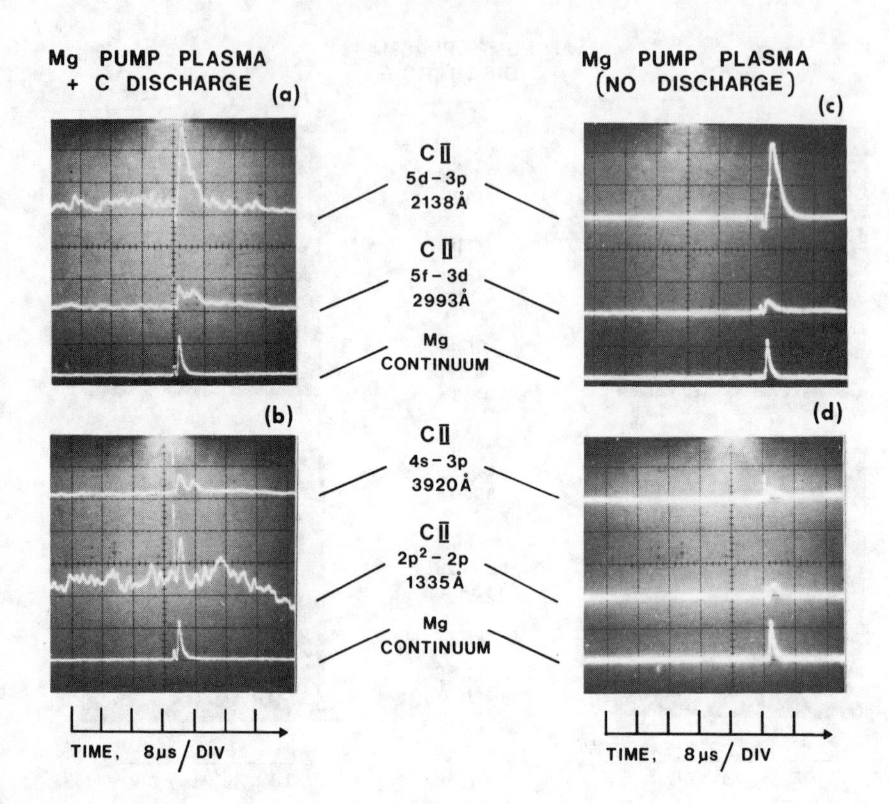

Fig. 8. C discharge with an adjacent Mg laser produced plasma: a) intensity vs time of CII, 2138 and 2993 Å lines, and AℓIII, 3713 Å line; b) intensity vs time of CII, 3920 and 1335 Å lines, and AℓIII, 3713 Å line. 8c) and d) same lines as in a and b, but with no C discharge, to show spurious background signals due to Mg laser produced plasma alone.

3.4 kA. To corroborate this estimate, hydrogen atoms were introduced into the arc and the Stark width of the H_β line was measured. The hydrogen was introduced by arranging three segments of polyethylene on the surface of the boron nitride insulator (Fig. 1). The 2 m Ebert spectrometer was used for these measurements, with a resolution of 0.2 Å. The Stark profile was measured by scanning a photomultiplier across the focal plane at 0.2 Å intervals. Ten shots were fired at each wavelength and the average line intensity was recorded. To ensure reproducibility of the data, the Ly_α, 1216 Å line intensity and the CII, 1335 Å line intensity were monitored simultaneously with the H_β signal. Repro-

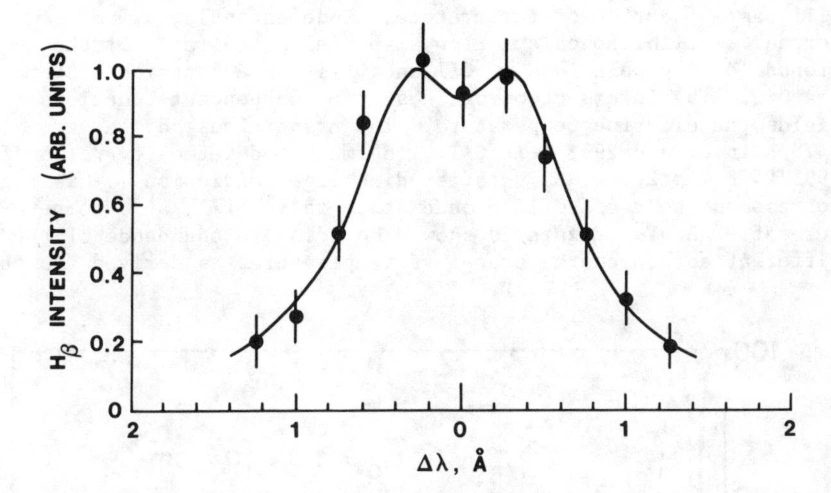

Fig. 9. H_β line intensity vs wavelength in the C discharge. The curve through the measured points is a convolution of a Stark profile with a Doppler profile.

ducible data were obtained by cleaning the entire cathode assembly after every 30 shots. The measured H_β line intensities at 3.4 kA, at 50 μs after arc initiation are shown in Fig. 9. The curve through these points is a best fit of a Stark profile,[8] convoluted with a Doppler profile. The density obtained is 6×10^{14} cm^{-3}, for a best fit temperature of 0.4 eV. The dip at the center of the Stark profile is a sensitive measure of temperature, because of the convolution of the Doppler profile. For example, a higher assumed temperature of 1 eV would completely wash out the dip. In the next section it will be shown that the temperature in the C arc on the axis is about 3 eV. Modified coronal calculations show that the H_β intensity is very sensitive to temperature.[6] It would appear therefore that the measured H_β signals originated predominantly from outer regions of the arc, where the temperature is lower. The measured density of 6×10^{14} cm^{-3} is thus a lower bound and consistent with the earlier estimate of 1.2×10^{15} cm^{-3}, from the scaling law of Keren and Hirshfield.[7]

ELECTRON TEMPERATURE MEASUREMENT

At densities of $\sim 1 \times 10^{15}$ cm^{-3} and temperatures of ~ 3 eV, it can be shown[6] that high lying levels of CI and CII are in thermodynamic equilibrium with the ground levels of CII and CIII, respectively. Measurement of the relative intensities of transitions from such higher levels in CI and CII thus leads to a determination of the ratio of the ground level populations of CII and CIII, as a function of temperature. Independently, a modified coronal equilibrium calculation also yields the ratio of the ground level populations of CII and CIII, as a function of temperature. The intersection of these two independent functions then yields the electron temperature. The transitions chosen were at 2478 Å in CI and 2993 Å in CII. The measured intensity ratio of 2993/2478 was 27, at 50 μs after discharge initiation. This ratio corresponds to a CII/CIII ground state ratio of 1, at a temperature of 3.25 eV. Figure 10 shows the relative abundance of the different carbon charge states vs temperature, as derived from the

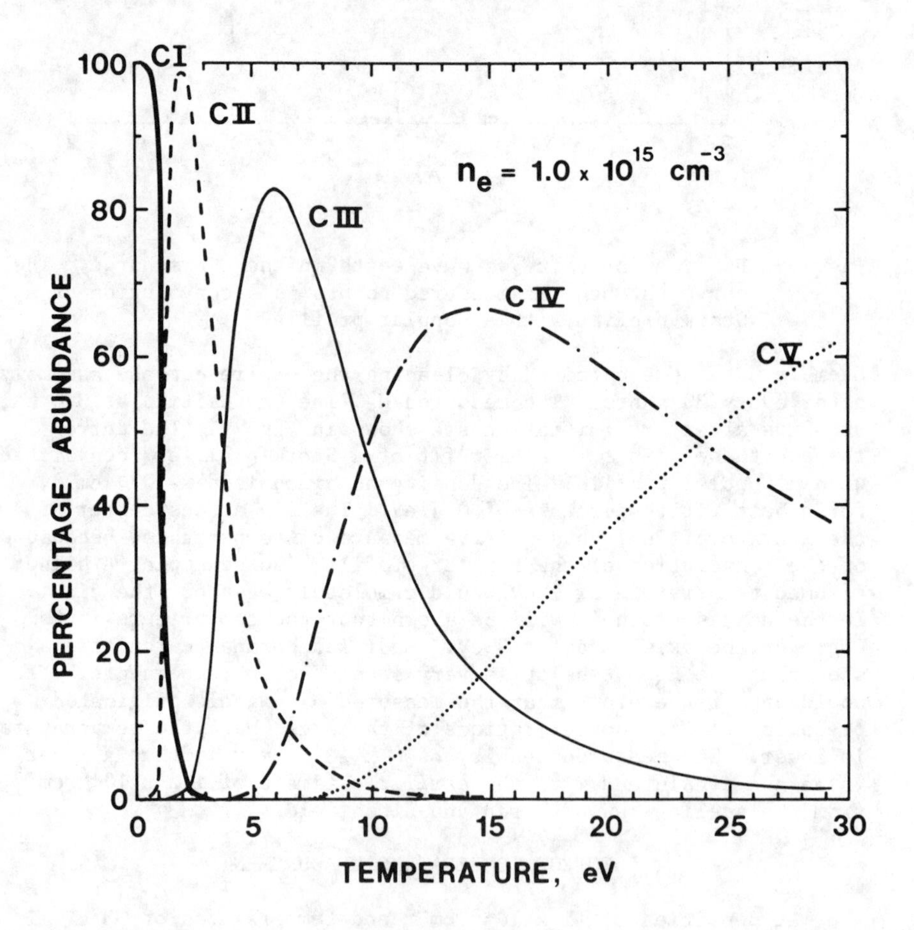

Fig. 10. Relative abundance of C charge states vs temperature, obtained from a modified coronal equilibrium calculation. $n_e = 1 \times 10^{15}$ cm^{-3}.

modified coronal equilibrium model.[1] The coronal equilibrium also gives a CII/CIII ground state ratio of 1 at the same temperature of 3.25 eV. The uncertainty in the measurement translates to a temperature uncertainty of ±0.25 eV. At the measured temperature of 3.25 eV, Fig. 10 shows that there is a negligible fraction of CIV in the plasma. At 3.4 kA, the resonance line of CIV at 1548 Å could not be detected.

To recapitulate, the carbon discharge plasma was found to have an electron density of 1×10^{15} cm^{-3} and an electron temperature of 3.25 ± 0.25 eV. Under these conditions, the n = 5 levels of CII are in thermodynamic equilibrium with the CIII ground state, thus causing the optical excitation of the 5d level to be rapidly distributed by collisions over a large number of higher levels. Population inversions and lasing on the 5d-3p and 5f-3d transitions are therefore difficult to achieve. In the next section, the Aℓ pump plasma is examined with a view to estimating the 560 Å pump line intensity.

PUMP LINE INTENSITY

In these experiments, it was not possible to measure directly the density and temperature of the laser produced plasma or to measure the absolute intensity of the pump transition. However, some measurements were made which together with numerical modeling allowed a reasonable estimate to be made of the 560 Å pump radiation. The initial temperature of the Aℓ plasma was determined by using an empirical relation[9] based on the laser focal spot intensity on the Aℓ target. The temperature was found to be ∿10 eV. The initial density at a distance of 0.1 mm downstream of the target surface was estimated as 5×10^{18} cm^{-3}, based on measurements of Tonon and Rabeau[10] in a similar plasma. At this density and temperature, the modified coronal model[1] predicts that most of the ions will be in charge states higher than AℓIII. The AℓIII must therefore be formed in the expansion phase of the laser produced plasma. This expansion leads to a rapid decrease in density and temperature, such that "freezing" of the populations occurs, with the resultant charge state distribution characterized by an effective temperature which is much higher than the local electron temperature. Such a non-equilibrium expansion makes it hard to interpret spectroscopic measurements in the expansion phase of the plasma.

One approach to determining the temperature was by measuring the relative intensity of the 3713 Å, 5s-4p and 3601 Å, 4p-3d transitions in AℓIII (see Fig. 4). If the collisional coupling rates between the 5s and 4p levels greatly exceed the radiative rate, then the two levels are in collisional equilibrium and the 3713/3601 line ratio yields an excitation temperature for AℓIII. It is shown[6] that this is the case for $n_e > 1 \times 10^{16}$ cm^{-3}. The excitation temperatures obtained as a function of time at various axial positions downstream of the target surface are plotted in Fig. 11. Curiously, the temperature appears to increase well

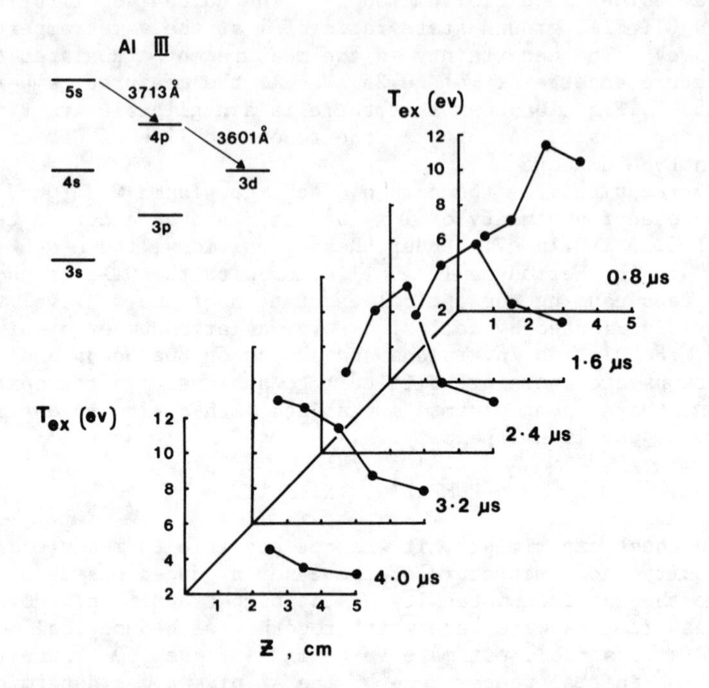

Fig. 11. Excitation temperature derived from the AℓIII 3713/3601 Å line ratio plotted as a function of time and of distance downstream from the Aℓ target.

after the laser pulse, reaches a maximum and then decreases. This anomalous behavior persists for up to 3 μs at distances up to 3 cm downstream. The cause of this anomaly is optical trapping of the 3601 Å radiation, which causes an increase in intensity of the 3601 Å line and results in an anomalously low temperature derived from the line ratio. At greater distances downstream and for later times in the expansion, the plasma density is sufficiently low that the 3601 Å line is optically thin and the derived temperature decreases with time as expected. The density of the 3d level required to give an optical depth of unity at 3601 Å is found to be 1×10^{13} cm^{-3}, for an ion temperature of 1 eV. With this lower bound, and assuming that the excitation temperature is 11 eV (based on the line ratios), the lower bound on the 5s and 5p densities is found to be 1×10^{12} cm^{-3}. For distances less than 3 cm and times shorter than 3 μs, the densities are clearly higher. The persistence of intense, AℓIII radiation for such long

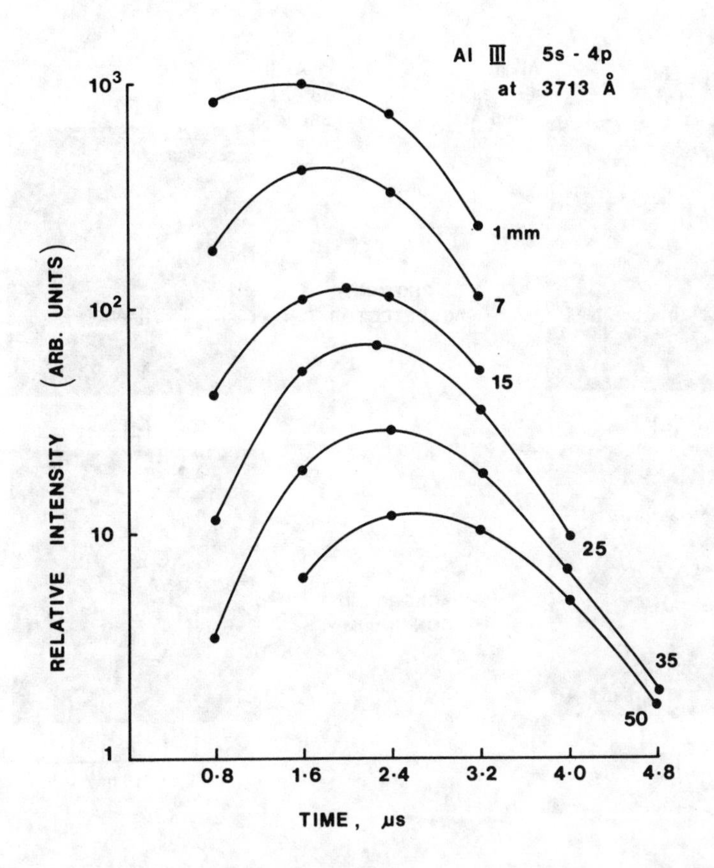

Fig. 12. AℓIII, 3713 Å line intensity vs time, for several axial positions downstream of the Aℓ target, along the target normal.

times in the plasma expansion is further verified in Fig. 12, which shows the AℓIII, 3713 Å intensity vs time, for several positions downstream. The intensity is highest for distances less than 3 cm at times less than 3 μs. The above detailed description of the AℓIII, 3713 Å line intensity indicates the presence of excited states of AℓIII in the rapidly expanding laser produced plasma. Direct measurement of the AℓIII, 560 Å line shape was not possible because the 1.5 Å resolution of the XUV spectrometer was about 150 times larger than the estimated Doppler width of the 560 Å line. Even to detect this line, therefore, the line intensity must be 150 times higher than that of the continuum background. Time-resolved intensity measurements of the 560 Å as well as the AℓIII, 4p-3s, 696 Å line are shown in Fig. 13. Figure 13a shows the 696 Å intensity vs time, while Fig. 13b shows the back-

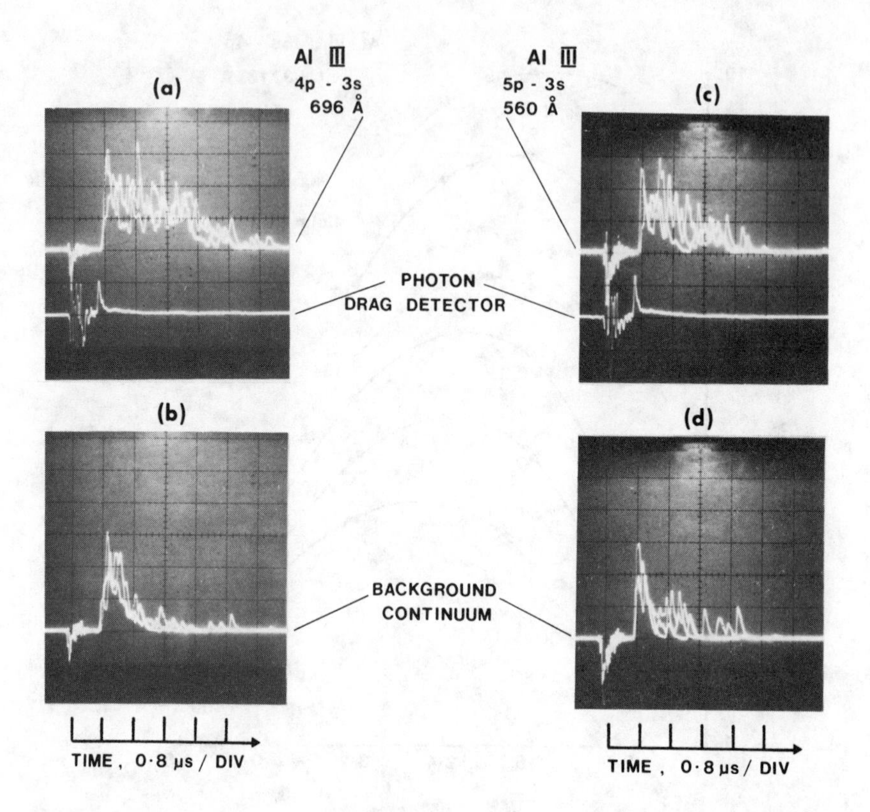

Fig. 13. Time-resolved AℓIII line intensities in laser produced plasma: a) AℓIII, 4p–3s, 696 Å intensity vs time; b) continuum background; c) AℓIII, 5p–3s, 560 Å intensity vs time; d) continuum background. The lower trace in a and c is the signal from the photon drag detector which samples the Co_2 laser.

ground continuum, measured 5 Å from line center. Figure 13c shows the 560 Å intensity vs time, and Fig. 13d shows the background continuum, 5 Å from line center. In each case, the figure shows a superposition of three laser shots on the photograph. The lower trace on Figs. 13a and 13c is the output of a photon drag detector which samples a portion of the Co_2 laser beam. Both 696 Å and 560 Å line radiation signals emerge above the continuum background for about 3 µs. This time duration is consistent with the observation of enhanced fluorescence also for 3 µs. In conclusion, the lack of an absolutely calibrated XUV spectrometer as well as the non-equilibrium expansion of the laser plasma made it impossible to accurately estimate the pump line intensity.

To summarize the experimental measurements, optical pumping of CII with AℓIII line radiation was studied by examining simultaneously four different wavelengths in CII. Enhanced fluorescence was observed from the pumped 5d level as well as from the neighboring 5f level. In addition, fluorescence was also observed from the 4s and $2p^2$ levels. The 4s level was probably fed by collisions from the n = 5 shell whereas the $2p^2$ level was probably collisionally excited from the CII ground state by electrons from the laser plasma. The electron density in the C arc was measured to be 1.5×10^{15} cm^{-3}. The electron temperature was estimated to be 3.3 eV. The AℓIII, 560 Å pump line radiation was found to persist for up to 3 μs in the laser produced plasma. In the next section, the collisional-radiative kinetics of the optically pumped CII ions are examined. The experimental observations are compared with the model and the feasibility of pumping a laser using such a scheme is discussed.

COLLISIONAL-RADIATIVE KINETICS IN CII

Using the measured density and temperature as inputs, a collisional-radiative model was developed for CII. This model, described in detail in Trebes,[6] included for a given level: collisional excitation and de-excitation, collisional ionization, radiative and three-body recombination from the CIII ground state, and dipole allowed radiative transitions. The energy levels and some oscillator strengths were obtained from standard references.[11,12] Other oscillator strengths and various rate coefficients were obtained from Morgan.[13] The model developed allows calculation of the distribution of the populations in many levels, subsequent to the selective optical pumping of the 5d level. Some results are shown in Fig. 14. Figure 14a shows the energy levels, some allowed transitions and their radiative lifetimes in ns. From the radiative lifetimes indicated in Fig. 14a, it appears that the 5f-3d transition at 2993 Å is a potential candidate for a quasi-cw laser, since the 5f radiative lifetime is much longer than the 3d radiative lifetime. However, one major drawback is the strong coupling of the 3d lower level with the CII ground state, which can lead to optical trapping of the 3d-2p, 800 Å radiation. From the modified coronal model, the CII ground state density was estimated to be 6.7×10^{14} cm^{-3}, for the measured n_e and T_e. For this ground state density, the optical depth at line center of the transition is 27.2, for an assumed transverse plasma dimension of 1 cm. Using the Holstein escape factor,[14] the modified lifetime of the 3d level is 16 ns. Since this modified lifetime is still shorter than the radiative lifetime of the 5f level, quasi-cw lasing is possible, but if the transverse plasma dimension is higher or if the CII ground state density is slightly higher, then optical trapping increases the lifetime of the 3d level to a value higher than the 5f lifetime, thus destroying the possibility of a quasi-cw laser at 2993 Å. A second deleterious consequence of the rather

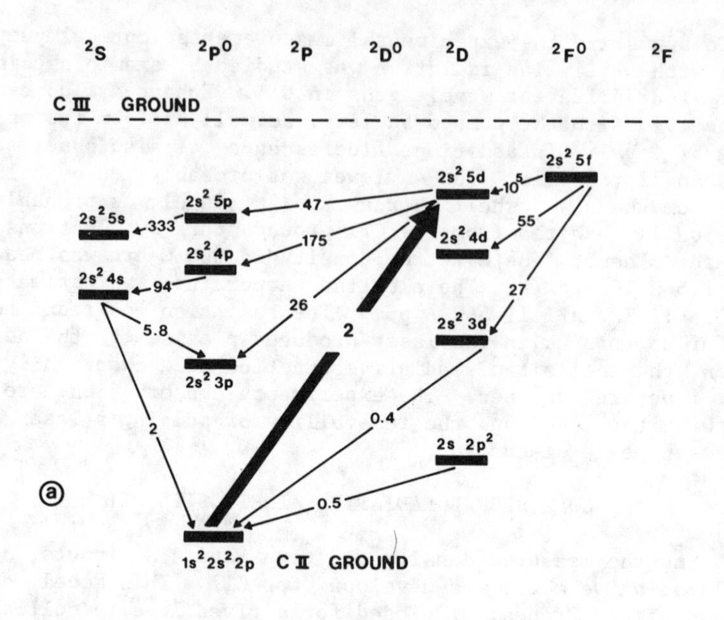

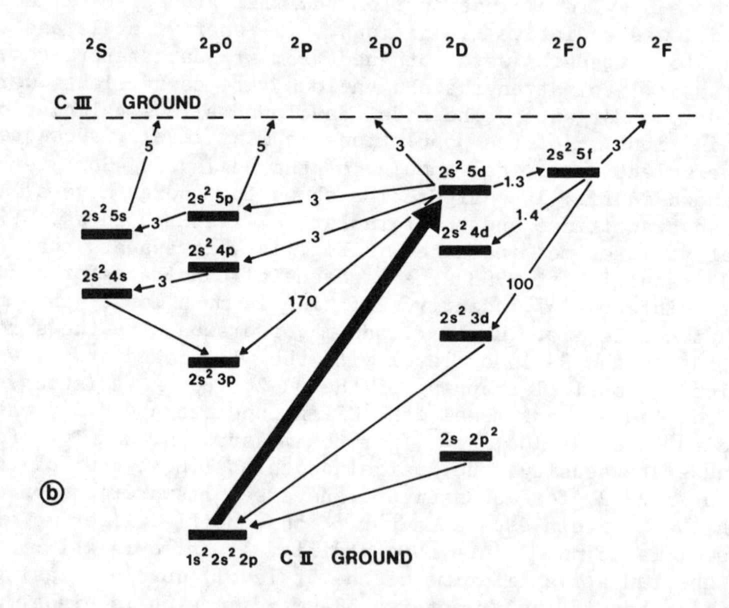

Fig. 14. Radiative and collisional times for transitions in CII. a) Radiative lifetimes (in ns) and b) collisional ionization and de-excitation times (in ns), for selected transitions in CII.

high n_e and T_e in the carbon plasma is that collisional rates are high enough to tend to thermalize the n = 3, 4, and 5 shells in CII. Figure 14b shows the collisional ionization rates for the n = 5 levels, the collisional transfer rates between these levels, and some de-excitation rates for the transitions considered in Fig. 14a. From the figure, it is observed that the collisional transfer time from 5d to 5f is short compared to the 5d radiative lifetime. This is desirable in order to transfer the pumped electrons in the 5d to the 5f, potential upper laser level. However, the figure also shows that collisional ionization from the 5d and 5f levels proceeds as rapidly as the collisional transfer. Furthermore, the collisional de-excitation times for the 5d and the 5f levels are an order of magnitude or more shorter than the radiative lifetimes. This means that the optically pumped 5d population is rapidly distributed by collisions to the other levels in the n = 4 and 5 shells, as well as to higher lying levels and the CIII ground state. It is clear that conditions in the carbon plasma are not optimal for a 5f-3d laser, since selective optical pumping of the 5d level is not accompanied by collisional transfer exclusively to the 5f level. Another conclusion that may be drawn from the above analysis of collisional rates is that for a given strength of the optical pumping, enhanced fluorescence should be seen simultaneously on several transitions from the high lying levels of CII, since the collisional coupling times between levels are orders of magnitued shorter than the 3 μs duration of optical pumping. This conclusion is verified by the observed fluorescence at 3920 Å from the 4s level, discussed earlier.

The above discussion has revealed much of the kinetics of the CII ions without regard to the actual strength of the optical pump. Using the estimate of the lower bound of the AℓIII, 5p population of 1 x 10^{12} cm^{-3}, it is possible to estimate the degree of enhanced fluorescence expected from the pumped 5d level. With the CII ground state density of 6.7 x 10^{14} cm^{-3}, convolution of the AℓIII and CII Doppler line shapes and inclusion of the solid angle subtended by the observed volume to the Aℓ plasma yields[6] an expression for the lower bound of the optical pump rate:

$$\text{Pump Rate} = 9.7 \times 10^{11} \frac{1}{\sqrt{T_{A\ell}\, T_C}} L_{A\ell} \times \Delta\Omega \times n_{5p} \times A_{5p-3s} \quad I' \qquad (2)$$

where $T_{A\ell}$ and T_C are the ion temperatures of Aℓ and C, respectively, $L_{A\ell}$ is the effective Aℓ plasma dimension, $\Delta\Omega$ the solid angle subtended by the pumped carbon volume to the Aℓ plasma, n_{5p} is the lower bound estimate of the AℓIII, 5p population, A_{5p-3s} is the 560 Å transition probability and I' is a numerical estimate of the convolution integral of the two Doppler line shapes. For the conditions discussed throughout this paper, the lower bound of the pump rate is determined to be 10^{18} electrons/cm^3 s. For the 5d level, the sum of all possible collisional and radiative de-excitation rates as well as the collisional ionization rate is 10^{10} s^{-1}. Thus the steady state population enhancement of the 5d level by optical pumping is 10^8 cm^{-3} or higher. In the absence of optical pumping, the 5d population is estimated from the collisional-radiative model to also be about 10^8 cm^{-3}. These estimates imply that the enhanced fluorescence should be comparable to the spontaneous emission from the 5d level. If the pump line intensity is up to a factor of ten higher than the estimated lower bound, the enhanced fluorescence should then be a factor of ten above the spontaneous emission. In the earlier measurements[4,5] and in this work, the observed fluorescence was always between one and ten times the spontaneous emission, in agreement with the predictions of the model. In addition, the model predicts that the degree of enhancement at the 2993 Å wavelength should be lower than that at 2138 Å, since only a fraction of the enhanced 5d population is transferred to the 5f level. This is also in agreement with the experimental observations.

To summarize, a detailed examination of the collisional-radiative kinetics in CII has revealed that the density and temperature in the carbon plasma are far from ideal for creating a population inversion. Collisions dominate the kinetics and cause strong coupling between the n = 3, 4, and 5 levels, as well as higher levels. Thus, although the optical pumping is selective, the pumped population is dispersed over many channels.

CONCLUSION

A detailed experimental and theoretical study has been made of optical pumping in CII ions using AℓIII line radiation. This study has shown that although the optical pumping itself is selective, the pumped 5d population in CII is dispersed into many competing channels by collisional and radiative processes. Also, the strong collisional coupling between levels in the n = 3, 4, and 5 shells renders it difficult to sustain an inversion between 5f and 3d, a potential laser transition. Optical trapping also increases the 3d lifetime and further contributes to spoiling the chance for an inversion. These deleterious effects may be alleviated somewhat by a better choice of density and temperature for the carbon plasma. For example, as n_e is reduced, the collisional rates for various decay channels out of the 5d level are all proportionally reduced. For $n_e \simeq 10^{13}$ cm^{-3}, the 5d-5f electron collisional transfer time is 300 ns, which is still short compared to the optical pumping duration of 3 μs. For transitions within a given shell, particularly when the energy gap is very small compared to the temperature, the ion collisional transfer rate can sometimes exceed the electron collisional rate. For the 5d-5f transition, ion collisions are more important than electron collisions.[6] Thus the collisional transfer time is even less than 300 ns. At this n_e, 5d-3p radiative decay is ten times as rapid as collisional ionization or collisional de-excitation. If the collisional transfer time from 5d-5f is comparable with the 5d-3p radiative decay, up to 50% of the optically pumped 5d population may be transferred to 5f. Also, at this lower n_e, the CII ground state density is lower, and optical trapping of the 3d level is avoided. Finally, the 5f-3d radiative decay is then five times as rapid as collisional de-excitation via 5f-4d. Therefore a population inversion may be produced between 5f and 3d. It is important to point out that although the CII ground state density is only 6.7×10^{12} cm^{-3}, if only 10% of the ground state is optically pumped to the 5d level, then the 5f population would be about 3.0×10^{11} cm^{-3}. At 2993 Å, this corresponds to a small signal gain of .1 cm^{-1}. If the gain medium is 10 cm in length, the resultant net gain is quite sufficient to sustain oscillation in a cavity.

In retrospect, the CII-AℓIII combination was chosen because of the good coincidence between the pump and absorption line wavelengths. In these experiments, the plasma conditions were not optimized for lasing. The CII plasma was too dense and hot, while the AℓIII pump species were produced in the non-equilibrium expansion of a laser produced plasma under conditions far from ideal for maximizing the pump line intensity. Nevertheless, some enhanced fluorescence was observed, with only 1×10^{-6} of the CII ground state being pumped to the 5d level. Under optimized conditions, 10% or more of the ground state population may be pumped to the 5d level and a very high gain laser is possible. These arguments are tempered by the observation that the major stumbling block of this particular ion combination is the unfavorable energy level structure of the CII ion. Firstly, the pumped 5d level is too close to the CIII ground state and thus readily ionized. Secondly, the $n = 4$ levels are close to the $n = 5$ levels and strongly coupled to them by superelastic collisions. Finally, since the 3d lower level of the potential laser transition is directly coupled to the ground state, optical trapping is a serious concern. A better approach to producing a laser using optical pumping with line radiation would consider both the line coincidences as well as the atomic level structure of the pumped ion. Just such considerations have led to the proposal of a new class of optically pumped lasers in Be-like ions. These schemes are described in a companion paper in these proceedings.[1]

ACKNOWLEDGMENTS

We are grateful to Dr. W.L. Morgan and to Dr. R.D. Cowan for providing atomic data on CII and AℓIII. This research is supported by the Air Force Office of Scientific Research (Grant # 81-0077).

REFERENCES

1. M. Krishnan and J. Trebes, these proceedings.
2. P. Hagelstein, Ph.D Thesis, "Physics of short wavelength laser

design," Lawrence Livermore Laboratory Report URCL-53100 (1981).
3. D.L. Matthews, P. Hagelstein, E.M. Campbell, A. Toor, R.L. Kauffman, L. Koppel, W. Halsey, D. Phillion, and R. Price, IEEE J. Quant. Elect. QE-19, 1786 (1983).
4. J. Trebes and M. Krishnan, Phys. Rev. Lett. 50, 679 (1983).
5. J. Trebes and M. Krishnan, IEEE J. Quant. Elect. QE-19, 1870 (1983).
6. J. Trebes, Ph.D Thesis, Yale University, unpublished.
7. H. Keren and J.L. Hirshfield, Appl. Phys. Lett. 36, 128 (1980).
8. H. Greim, Plasma Spectroscopy (McGraw Hill, New York, 1964), p. 270.
9. A. Montes, M. Hubbard, C. Kler, and I.J. Spalding, Appl. Phys. Lett. 36, 652 (1980).
10. Tonon and Rabeau, Plasma Physics 15, 871 (1973).
11. S. Bashkin and J.O. Stoner, Atomic Energy Levels and Grotrian Diagrams, Vols. I and II (North Holland Publishing Co., New York, 1975, 1978).
12. J. Reader, et al., "Wavelengths and Transition Probabilities for Atoms and Ions," NSRDS - NBS, 68 (1980).
13. Dr. W.L. Morgan, personal communication.
14. T. Holstein, Phys. Rev. 72, 1212 (1947).

A SOFT X-RAY SOURCE BASED ON A LOW DIVERGENCE, HIGH REPETITION RATE ULTRAVIOLET LASER

E.A. Crawford, A.L. Hoffman, R.D. Milroy, D.C. Quimby, G.F. Albrecht
Mathematical Sciences Northwest, Inc.
Bellevue, Washington 98004

ABSTRACT

We find, using numerical modeling, that small size plasmas produced by a short-pulse, low-energy uv laser are potentially efficient soft x-ray sources.

INTRODUCTION

The soft x-ray emission of dense plasmas produced by irradiation of solid targets with high intensity lasers has been studied extensively in recent years. Experiments motivated by requirements of the Inertial Confinement Fusion (ICF) program have dominated these studies.[1] In addition, other experiments have been directed toward the efficient production of soft x-rays for microlithographic[2] and other industrial application. These experiments have primarily been done using relatively large (E>1 J) lasers operated at low repetition rates, although recently, experiments[3] have been done at moderate intensity and repetition rates.

We have recently concluded a study, funded in part by the National Science Foundataion, on the use of sub-joule size, sub-nanosecond, low divergence ultraviolet lasers for the efficient production of soft x-rays for microlithography applications. We find that such small lasers, focused to about 30μ spot sizes and 10^{14} w/cm^2 intensities, are potentially very useful sources of 1-2 keV radiation. These conclusions are based on numerical computations performed with a one dimensional (z-t) code, CORK, which models the atomic physics and hydrodynamics of the plasma as it ablates from the solid target.

We have found that for 1 GW laser power, the x-ray conversion efficiency is a strong function of spot size, peaking at about 20μ diameters. The conversion efficiency is only weakly dependent on pulse length for the time scales exceeding 100 psec. Better conversion efficiencies are also obtained at shorter wavelengths and for energy matched L and M shell radiators.

Axial Hydrodynamic Code With One Zone Radial Expansion

The CORK[4] code employs an axial Lagrangian coordinate system. The radial motion is approximated with a one zone sharp boundary model, which encompasses radial inertial effects. Separate electron and ion temperatures and electron and ion flux-limited, thermal conduction are included. A flux-limiting value of $f_f = 0.3$ is used. Electrons are heated only through classical inverse bremsstrahlung absorption of a laser beam which propagates axially through the plasma plume to the critical surface. The fraction of laser energy reflected at the critical surface is calculated and is assumed to propagate backwards (anti-parallel) to the incident beam.

Atomic Physics and Radiation Model

The CORK code includes a time-dependent collisional-radiative atomic physics package which calculates the ionization level of the plasma and radiation rates. Line radiation, continuum radiation, and ionization energy, are included in the plasma energy balance. In addition, CORK is structured to tabulate the radiated energy as a function of wavelength. The atomic physics procedure requires rate coefficients for ionization, recombination, and radiation in three-dimensional tables which are dependent on temperature, density, and ionization state. This data is provided by an atomic physics code, CRAFTY. CRAFTY accounts for electron impact excitation and de-excitation, spontaneous emission, electron impact ionization, radiative recombination and collisional (three-body) recombination. Bremsstrahlung, recombination radiation and line radiation are calculated.

The model generates a synthetic spectrum of this radiation, sorted into bins of 0.2 keV width each. An optically thin model is used and only radiation into one hemisphere is included in the energy balance. Radiation in the hemisphere toward the target, which actually will tend to raise the temperature of the blowoff plasma, is ignored.

Temperature Scaling in Laser Produced Aluminum Plasma Plumes

A series of computations using the CORK code was made to determine the effect of laser intensity, spot size, and wavelengths on the plasma electron temperature. The results are summarized in Fig. 1 in terms of peak electron temperature at t = 100 psec.

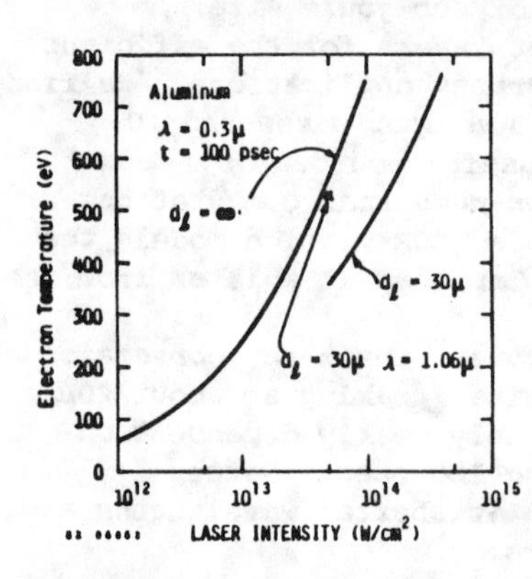

Figure 1. Peak Electron Temperature Dependence on Laser Intensity. The $d_\ell = \infty$ Curve Implys no Radial Expansion.

An axial profile of the electron and ion temperatures, electron density, and laser power is shown on Fig. 2 for a 10^{14}w/cm^2 intensity. Most of the plasma energy resides on the low density side of the 10^{22} cm^{-3} critical density surface, as we would anticipate from simple scaling for $I\lambda^2 << 10^{14}\mu^2$w/cm^2. Each circled point on the density curve represents one Lagrangian cell containing the same number of ions.

When radial expansion is insignificant, the scale length increases with time because the ablation process is essentially a one dimensional, time dependent, hydrodynamic expansion. Since the laser continues to heat the expanding plasma, the temperature at the front increases continually with time and the laser heating region eventually moves away from the neighborhood of the critical surface.

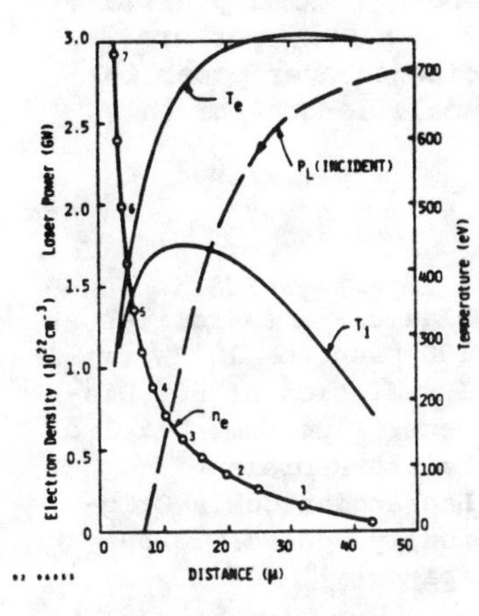

Figure 2. Plasma Plume Profiles at 100 psec for 10^{14}W/cm^2 UV Irradiation on Aluminum with no Radial Expansion. Circled points Represent Lagrangian Cells Originally Spaced 0.05μ apart.

A more realistic calculation includes the effects of radial expansion. A temperature-versus-intensity curve is shown on Fig. 1 for a 30μ spot size. In this case, most of the heating occurs near the critical surface, and the peak electron temperature moves toward the higher electron density region. A detailed axial plot at 100 psec is shown on Fig. 3 for a 60μ spot size and 10^{14}w/cm^2 intensity.

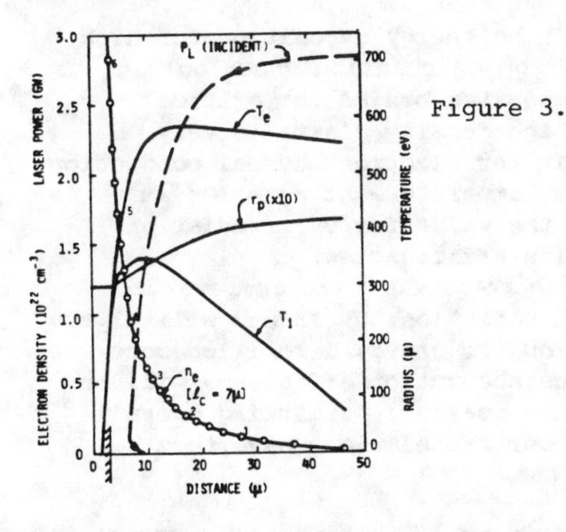

Figure 3. Plasma Plume Profiles at 100 psec for 10^{14}W/cm^2 UV Irradiation of 60μ Diameter Aluminum. Circled Points Represent Lagrangian Cells Originally Spaced 0.025μ apart.

Since radial expansion is included, this profile does not change with time, except at the very low density leading edge. The typical length scale for the density profile is about one-quarter of the laser radius, which is characteristic of a three-dimensional expansion. For the Z=11 plasma, the incident UV laser radiation is calculated to be fully absorbed.

One point is shown on Fig. 1 corresponding to irradiation by a 5×10^{13} w/cm^2, 1.06μ infrared laser. The peak temperature is higher than that produced by UV irradiation at the same intensity, but not as high as would be produced by a UV laser operating at an equal value of $I\lambda^2$. The principal reason for the lower temperature is incomplete absorption at the longer wavelength. Axial profiles for 10^{14}w/cm^2 IR laser irradiation intensity on a 60μ spot are shown on Fig. 4. About 35 percent of the incident laser power is reflected due to an insufficient absorption scale length for this longer laser wavelength.

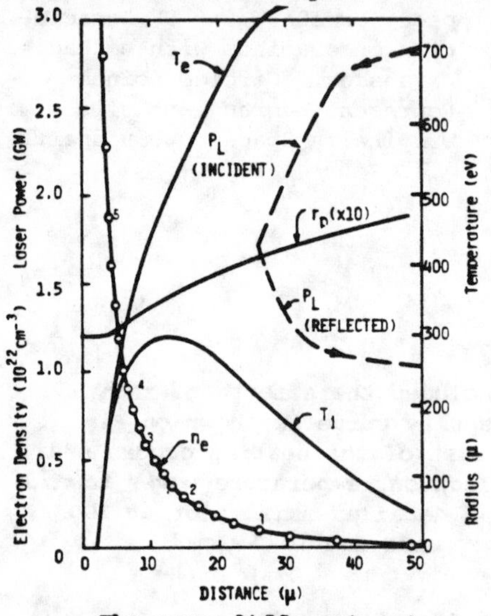

Figure 4. Plasma Plume Profiles at 100 psec for 10^{14}W/cm^2 IR Irradiation of 60μ Diameter Aluminum. Circled Points Represent Lagrangian Cells Originally Spaced 0.025μ apart.

The very different nature of the energy deposition for high $I\lambda^2$ irradiation is illustrated by the calculation shown on Fig. 4. Most of the laser energy is now deposited behind the critical surface. The temperature of the high density plasma is very dependent on the assumed flux limit for electron thermal conduction. For the $f_f = 0.3$ value chosen, the temperature at $n_e = 10^{22}$cm^{-3} is within a factor of two-thirds of the value for UV irradiation, allowing reasonable x-ray conversion efficiencies.

As long as $I\lambda^2$ is high enough to produce the temperatures needed for efficient K- or L-shell radiation, UV lasers will always be more efficient than IR lasers, due to energy deposition occurring at a higher density. Moreover, the absorption efficiency will be higher and the spot size can be made smaller, permitting overall lower laser energies to be used. Our calculations have thus concentrated on UV irradiation.

X-Ray Conversion Efficiencies

For aluminum plasmas, we define an x-ray conversion efficiency based on radiation in the 1.5–2.0 keV energy interval emitted in all directions. Curves of conversion efficiency as a function of laser pulse length, are shown on Fig. 5 for the 5 x 10^{13}w/cm^2 intensity level. Very high conversion efficiencies arise only for large spot sizes and long pulse times. The 5 x 10^{13}w/cm^2 intensity level is insufficient to produce high temperatures near the critical surface.

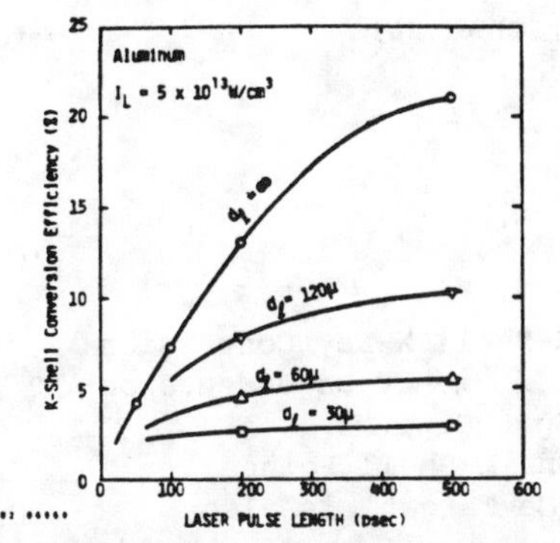

Figure 5. K-Shell X-Ray Conversion Efficiency Dependence on Laser Pulse Time.

For small spot sizes, quasi-steady conditions are reached after about 50-100 psec, and the x-ray conversion efficiency increases very little for longer pulse lengths. The low conversion efficiency is a consequence of the electron temperature reaching only 300-400 eV. Simple considerations show that temperatures exceeding 500 eV are required to efficiently excite 1.5 keV radiation. For 30μ spot sizes, this requires intensities exceeding 10^{14} w/cm².

The x-ray conversion efficiencies corresponding to the examples on Fig. 1 are shown on Fig. 6.

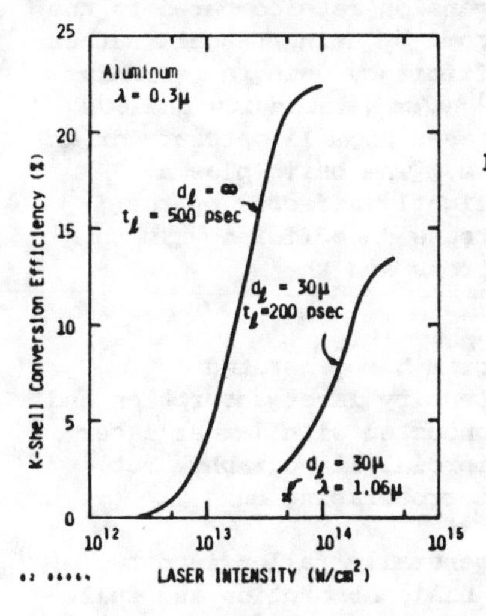

Figure 6. K-Shell Conversion Efficiency Dependence on Laser Intensity for Infinite and Small Spot Sizes.

A very strong dependence on intensity and spot size is evident due to the exponential dependence of the K-shell excitation rate on temperature. High efficiencies can be realized for small spot sizes, but only at very high intensities. The tradeoff between the spot size or the intensity is shown on Fig. 7. The x-ray conversion efficiency is plotted as a function of spot size for various laser intensities. The dashed curves are the loci of constant laser energy. For the range of conditions plotted, it is always more efficient to decrease the spot size and raise the intensity on target. This will probably hold true as long as $I\lambda^2$ is below

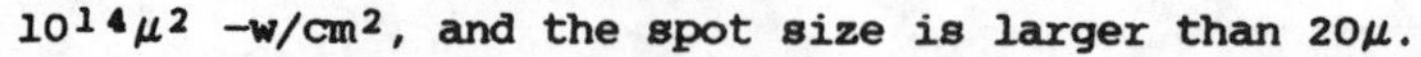
$10^{14}\mu^2$ -w/cm^2, and the spot size is larger than 20μ.

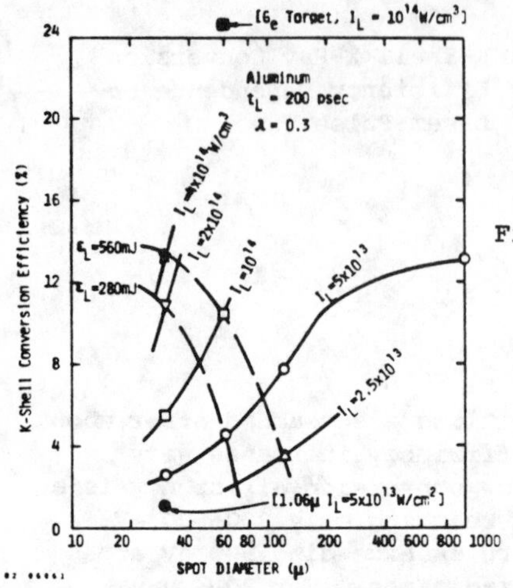

Figure 7. K-Shell X-Ray Conversion Efficiency Dependence on Spot Size. One L-Shell Point (Ge Target) is also shown.

X-ray conversion efficiencies of 10 percent are predicted for 200 psec pulses of several hundred millijoules focused onto $30-60\mu$ spot sizes. The principal cause of inefficiency for these high intensity irradiations is an excessive expansion rate compared to the radiative cooling rate. This can be overcome by using L-shell rather than K-shell radiators. The calculated efficiency using a germanium target is also shown on Fig. 7. For a 10^{14}w/cm^2 intensity and 60μ spot size, the conversion efficiency increases from 11 percent for an aluminum target to 26 percent for germanium. The basic plasma profiles and peak temperatures are only slightly affected by the target material, and this 2-1/2 times increase in efficiency is thus due solely to the additional L-shell electrons and the correspondingly higher radiation rates.

Summary and Comparison with Experiments

The inertial confinement fusion program has generated considerable data on short-pulse, high-intensity laser absorption and x-ray production. The experiments were conducted with larger laser energies than would be reasonable for commercial lithography, but, some of the observations are relevant in corroborating our small-pulse energy calculations.

Flat plate target experiments have generally fallen into two categories. The first involves measuring basic absorption and full spectrum x-ray production. The principal conclusion that can be drawn from these experiments is that at intensities of 10^{14}-10^{15}w/cm^2, UV laser light is almost completely absorbed. For high Z materials, the absorbed energy is radiated in the 50-1000 eV spectral range into the hemisphere toward the laser (2π steradians) with about 60 percent efficiency.[5] Most experiments have been conducted with 1.06μ IR laser light for which the absorption and x-ray conversion efficiency are less by about 1/3 to 1/2.[6]

The second category of experiments is directed at producing intense single-line emission for backlighting diagnostics. Experiments have primarily involved exciting He-like resonance lines at energies from 1.5-8 keV, and efficiencies of 1 to 2 percent into 4π steradians have been measured.[1,2] Higher intensities are required for the higher energy lines. UV lasers give about a factor of two higher efficiency for the lower energy range.

Several experiments have also been conducted aimed at producing keV x-rays for lithographic purposes. Conversion efficiencies of 5 percent into 4π steradians have been measured at the Naval Research Laboratory with 10^{14}w/cm^2, 1.06μ laser pulses.[7] This is consistent with the 1.06μ data points shown on Fig. 7. The NRL researchers estimate that the x-ray conversion efficiency could be doubled by using UV lasers. Experiments with copper L-shell radiators have been carried out at Battelle Columbus Laboratories, again using 1.06μ radiation at 10^{14}w/cm^2. Using only 300-400 mJ, 200 psec pulses focused to 50μ diameters, they claim 10-20 percent conversion efficiencies of 1-1.5 keV x-rays into 2π steradians.[8] The energy range is slightly lower than that produced in our Germanium calculations, but, the results corroborate the high computed efficiencies obtainable with L-shell radiators.

All experiments thus show that laser radiation at 10^{14}w/cm^2 can be efficiently absorbed, with the total absorption approaching 100 percent for UV light. The high conversion efficiencies measured with 1.06μ light are extremely encouraging, since the physical assumptions used in our numerical calculations (classical inverse bremsstrahlung absorption and a high flux limit on electron thermal conduction), are far less certain for high values of $I\lambda^2$. We are much more confident of the UV calculations. The factor of two higher efficiency for predicted x-ray production using UV light, can be accounted for almost solely by an increase in the absorption coefficient. This increased absorption at shorter wavelengths has been well documented experimentally. It thus appears reasonably certain that 200 psec UV laser pulses of energies as low as a few hundred millijoules, can be converted to 1.5 keV L-shell x-rays into 2π steradians with efficiencies exceeding 10 percent.

References

1. D.L. Matthews, et al., "Characterization of Laser Produced Plasma X-Ray Sources For Use in X-Ray Radiography," J. Appl. Phys. 54, 4260 (1983).
2. B. Yaakobi, et al., "High X-Ray Conversion Efficiency with Target Irradiation by a Frequency Tripled Nd: Glass Laser," Optics Communications 38, 197 (1981).
3. D.R. Nagel, et al., "Repetitively Pulsed Soft X-Ray Plasma Source," to be published in Applied Optics (1984).
4. R.D. Milroy and L.C. Steinhauer, Phys. Fluids 24, 339 (1981).
5. K.R. Manes, "Review of Lawrence Livermore National Laboratory Plasma Interaction Experiments Using Upconverted Nd-Glass Lasers," Bull. Am. Phys. Soc._26, 972 (1981).
6. H. Nishimura, et al., Phys. Rev. A 23, 2011 (1981).
7. D.J. Nagel, private communication.
8. H.M. Epstein and B.E. Cabell, "Microlithography of Integrated Circuits With Laser Plasma X-Ray Sources," CLEO '82 Paper FW3 (Opt. Soc. of America, Wash., D.C., 1982).

X-Rays Generated by Laser Irradiated High-Z Targets

X. Fortin, D. Babonneau, D. Billon, J.L. Bocher, G. Di Bona
and G. Thiell

Commissariat à l'Energie Atomique
Centre d'Etudes de Limeil - Valenton
BP. 27 - 94190 Villeneuve Saint Georges (France)

The X-ray emission from laser irradiated high-Z targets has received much attention in the recent years. The two main centers of interest were : spectral studies for the determination of plasma parameters and conversion efficiencies for understanding the plasma energetics. We have performed these kinds of experiments at Limeil during the three past years using spherical and planar targets (Glass, Aluminum, Gold, Copper) and modifying irradiation conditions (laser wavelength and flux).

In this paper, we present a description of the diagnostics and the methods used to unfold the experimental results. To analyse these data we have three threoretical tools :

- The Limeil one-dimensional FCI 1 code.
- Corine : a model for emission and transfer of the radiation in laser plasma, using stationary temperature and density profiles and an atomic physics coupling local thermodynamical and coronal equilibria.
- Analytical expressions to explain qualitatively the experimental results about X-ray conversion efficiencies.

The set of experimental and theoretical results are coherent with other laboratories' ones and shows the growth of X-ray efficiency when target Z and pulse duration increase and when the laser wavelength λ decreases. The maximum of the efficiency appears for a laser flux ϕ about $\phi\lambda^2 \sim 5.x\ 10^{13}\ W/cm^2\ \mu m^2$.

THE EXCITATION OF METASTABLE EXTREME ULTRAVIOLET LEVELS

R. G. Caro, J. C. Wang,* J. F. Young, and S. E. Harris
Edward L. Ginzton Laboratory
Stanford University, Stanford, California 94305

ABSTRACT

The excitation of large densities of high energy metastable atoms and ions is described. Ionic metastable levels are produced as a result of photoionization by x-rays emitted from a laser-produced plasma. For the case of metastable levels in neutral atomic species, excitation is caused by a "photoionization electron source." The electrons involved in this excitation are created when an "absorber" gas is photoionized by a burst of laser-produced x-rays.

INTRODUCTION

The characteristic radiative lifetimes of strong transitions in the extreme ultraviolet (XUV) are between 10 and 100 ps. In order to minimize the necessary peak pumping power required to realize laser action on such XUV transitions, a number of three- and four-level laser systems have been proposed which rely on the storage of large populations in energetic metastable levels of atoms and ions.[1-3] In this work we describe an investigation of the excitation of metastable levels in several different ionic and atomic species which are of interest for such XUV laser systems.

Very large populations of atoms and ions in excited metastable levels have been produced using x-rays emitted from a laser-produced plasma as the primary excitation source. Ionic metastable states have been created in large densities as a result of photoionization by these x-rays.[4] Clearly, however, an alternative mechanism is required for the excitation of metastable levels in neutral atoms. In this work, such levels have been created by electron excitation following the conversion of laser-produced x-rays to energetic electrons by photoionization of an "absorbing" gas.[5] Both of these excitation techniques have been shown to produce populations of energetic metastable levels more than two orders of magnitude larger than have been achieved by alternative excitation methods.[6-7] These population densities now approach those necessary to be able to test a number of proposals for XUV laser systems.[1-3]

SOFT X-RAY PUMPING OF METASTABLE LEVELS OF IONS

The simplest example of energetic metastable levels that are of relevance to this work is the isoelectronic series corresponding to the He(1s2s) level. In Fig. 1(a) is shown an energy level diagram

*Currently at Stanford Research Systems, 460 California Avenue, Palo Alto, California 94306.

0094-243X/84/1190417-10 $3.00

of Li^+, while in Fig. 1(b) is depicted the photoionization cross section of Li and the emission spectrum of a 30 eV blackbody. It can be seen from this data that photoionization of Li atoms by x-rays from a 30 eV blackbody will result in efficient production of $Li^+(1s2s)$ metastable ions.

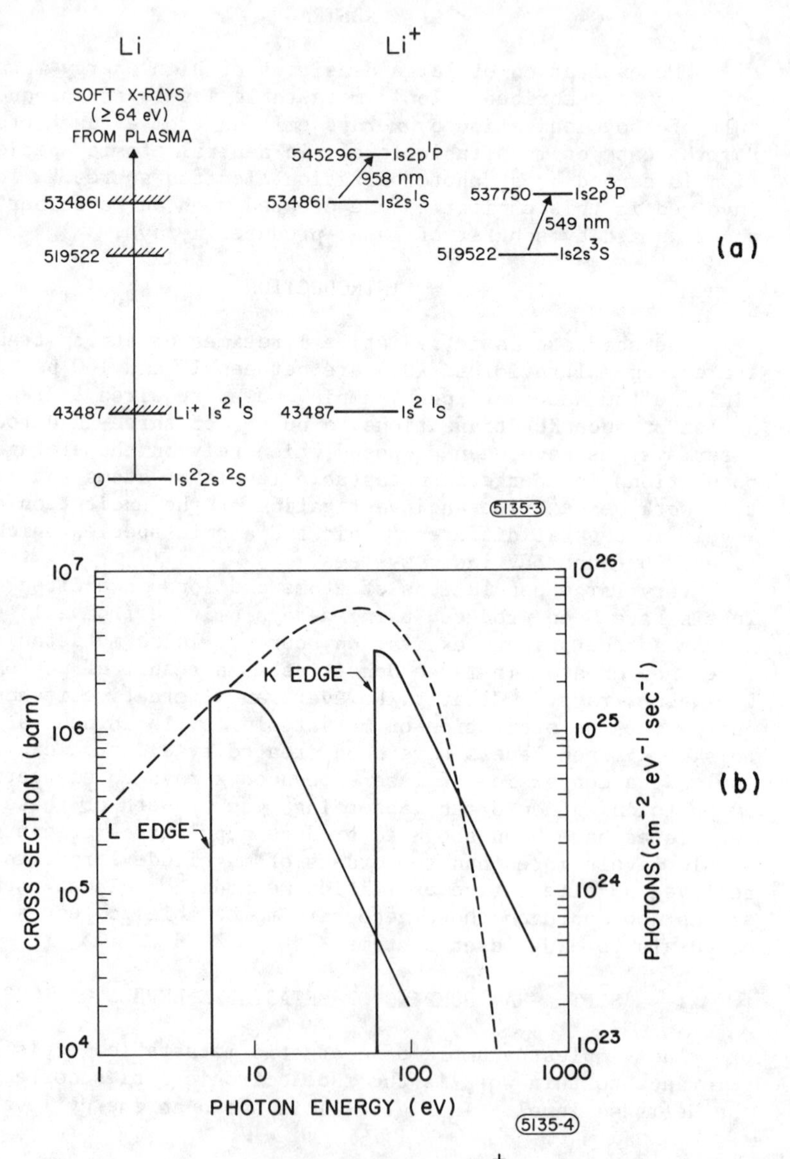

Fig. 1. (a) Energy level diagram of Li and Li^+. The energies are in cm^{-1}. (b) Photoionization cross section of Li (solid curve); emission spectrum of a 30 eV blackbody (dashed curve).

In this work, a laser-produced plasma acts as the source of blackbody radiation for this excitation. The configuration of the experiment described here is shown in Fig. 2. A massive plane target was placed inside a heat pipe containing Li vapor at a density

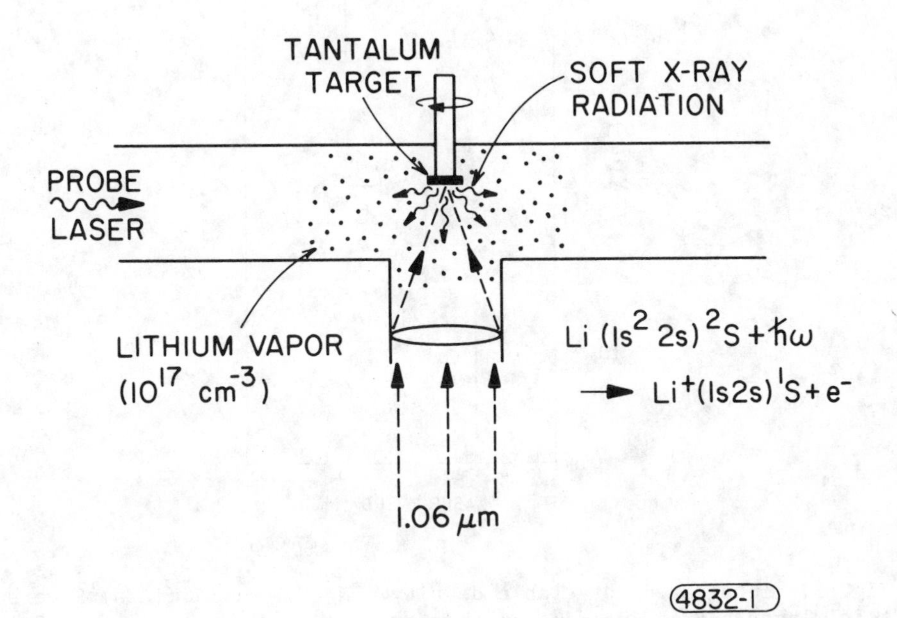

Fig. 2. Schematic of experimental configuration.

of 10^{16} to 10^{17} atoms cm^{-3}. The 1.06 μm laser beam was focused onto the Ta target by a lens with an effective f number of $f/8$. The emitted soft x-ray radiation propagated into the surrounding Li vapor causing inner-shell photoionization and therefore production of Li$^+$ (1s2s)^{1}S and ^{3}S ions. The metastable Li$^+$(1s2s)^{1}S and Li$^+$(1s2s)^{3}S population densities were determined by measuring the absorption of laser probe beams as a function of wavelength at the Li$^+$[(1s2s)^{1}S – (1s2p)^{1}P] and Li$^+$[(1s2s)^{3}S – (1s2p)^{3}P] transitions at 958.1 nm and 548.5 nm [Fig. 1(a)]. The absorption traces were fitted to numerically generated Voigt profiles.

Figure 3 shows the measured value of integrated Li$^+$(1s2s)^{1}S population, N^*L, as a function of the total laser energy incident on the target. The laser pulse width was 600 ps and the maximum energy of Fig. 3. corresponds to an intensity on target of 10^{13} W cm^{-2}. The laser probe beam at 958.1 nm [(1s2s)^{1}S – (1s2p)^{1}P] was obtained by Raman downshifting the pulsed 685.2 nm output of a Quanta-Ray dye laser. The Li vapor density for this experiment was $\sim 10^{17}$ atoms cm^{-3}. The laser probe beam had a diameter of 2 mm and was centered about 1 mm from the Ta target (Fig. 2). Assuming a 45° conical soft x-ray radiation pattern, the maximum measured value of N^*L from Fig. 3 implies a singlet metastable population of 3×10^{14} ions cm^{-3}.

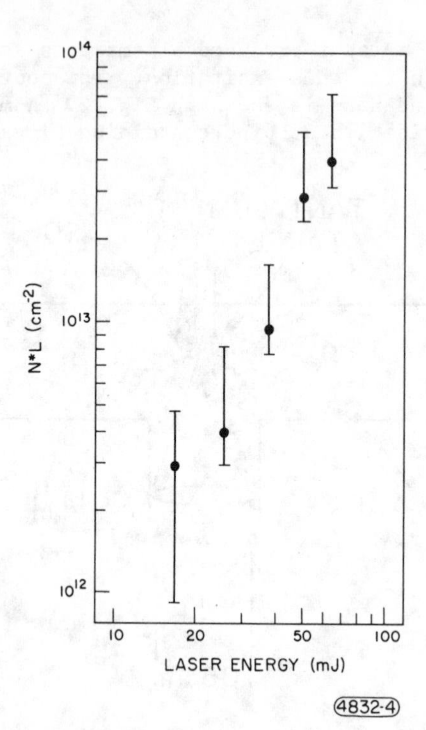

Fig. 3. Integrated metastable density, N^*L, for $Li^+(1s2s)\,^1S$ versus total laser energy incident on the target. Li vapor density = 10^{17} cm^{-3}. The perpendicular probe-target distance is R = 1 mm. The time delay between the peaks of the probe and the 1.06 μm lasers is ∿ 4 ns.

Figure 4 shows the results of a measurement of N^*L for metastable Li^+ singlets as a function of the time delay, t, between the peaks of the 958.1 nm probe pulse and the 1.06 μm laser pulse incident on the Ta target. The time origin in Fig. 4 has an experimental uncertainty of ± 2 ns. The slope of the graph of Fig. 4 corresponds to a singlet metastable (1/e) decay time of 5.3 ns. Such a decay time is consistent with calculated rates for electron collisional de-excitation of the Li^+ metastable ions.

From geometrical considerations[8] it is possible to conclude that the efficiency of conversion from laser energy to soft x-rays is in excess of 10%. The same argument[8] leads us to deduce that the effective temperature of the soft x-ray emitting blackbody is 10-100 eV. This relatively high conversion efficiency is believed to be due to the high atomic weight of the target (Z = 73) and the relatively low laser intensities (10^{13} W cm^{-2}) involved.

A similar experiment has been performed using Na as the species to be photoionized. In this case populations of 10^{16} cm^{-3} were measured in the $Na^+(2p^5 3s)\,^3P_2$ level at 33 eV. This represents 5% of the available Na population and is a measure of the effectiveness of this photoionization excitation technique.

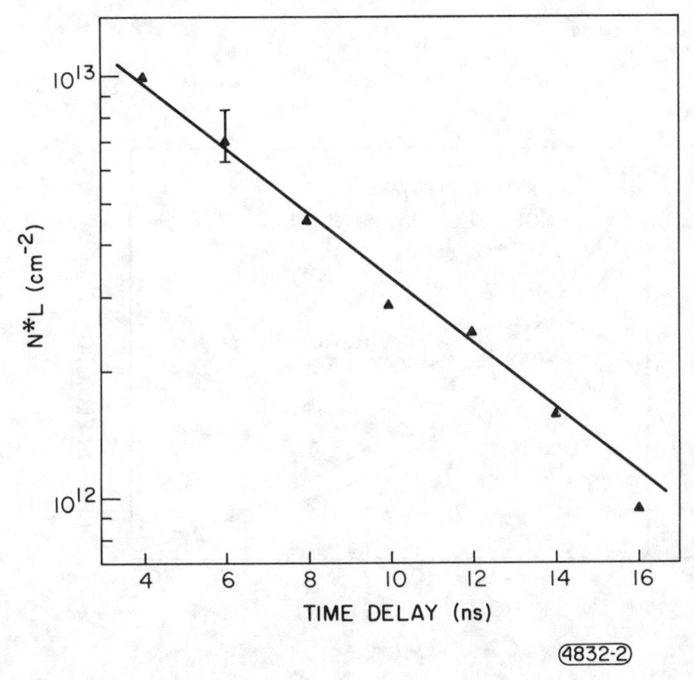

Fig. 4. N^*L for $Li^+(1s2s)\,^1S$ versus the time delay between the peaks of the probe laser and the 1.06 μm laser. Li vapor density = 10^{17} cm^{-3}; total incident laser energy ∿ 50 mJ; R ∿ 4.5 mm .

EXCITATION OF METASTABLE LEVELS IN ATOMS

In addition to the isoelectronic series of He(1s2s), there is another important class of metastable levels of high energies: the metastable quartet levels in the alkali metals.[9] In neutral species, the photoionization excitation technique described earlier is no longer appropriate. Instead, we have produced large densities of neutral metastable species by electron excitation using a "photoionization electron source."

This photoionization electron source[5] (PES) produces a sub-nanosecond pulse of hot electrons ideal for the production of highly excited atoms. In order to produce this burst of electrons, the soft x-rays emitted from a laser-produced plasma have been used to photoionize an "absorber" rare gas with the consequent production of large densities of hot electrons. The high density and temperature of the electron distribution produced by the PES allow efficient excitation of both dipole-allowed and dipole-forbidden transitions to states with energies as high as 100 eV. In addition, the short duration of the pumping pulse — determined by the pulse length of the plasma-generating laser — makes this source well suited to the study of dynamic interactions involving such states.

Figure 5(a) shows the good spectral overlap of the emission of a 30 eV blackbody and the photoionization cross section of Ne. The

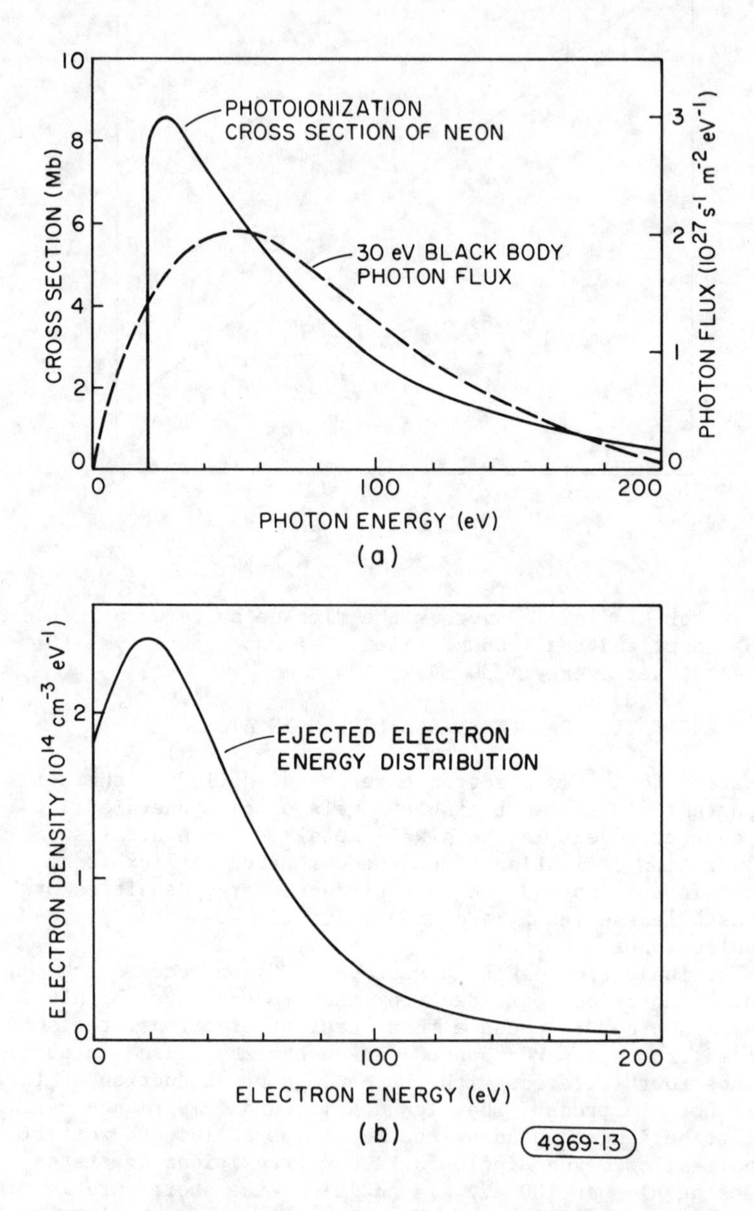

Fig. 5. (a) Photoionization cross section of Ne and the x-ray spectrum 1 mm from the target. (b) Predicted ejected electron distribution for a Ne density of 3×10^{17} cm^{-3}.

predicted electron distribution for a Ne absorber gas density of 3×10^{17} cm^{-3} is plotted in Fig. 5(b). This electron distribution has a maximum at 20 eV and a mean electron energy of 45 eV. The total ejected electron density is 2×10^{16} cm^{-3}. The photoionization electron source described here can be considered to produce an effective current density of 3×10^5 A cm^{-2} for 600 ps.

In order to investigate the excitation of the $Li(1s2s2p)^4P^o$ level by the PES, the plasma cell of Fig. 2 was heated to provide a target Li atom density of 10^{17} cm^{-3}. Varying amounts of Ne absorber gas were then added. The appropriate energy level diagrams are shown in Fig. 6. The number density-length product of $Li(^4P^o)$ atoms, N^*L, was determined by measuring the absorption of a probe dye laser beam, as a function of wavelength at the $Li[(1s2s2p)^4P^o - (1s2p^2)^4P]$ transition at 371 nm. All measurements were taken at a distance of 1 mm from the Ta target where, due to geometrical considerations,[8] the number density, N^*, is given by $N^* = 6.4\ N^*L$. The dye laser was pumped by a harmonic of the same Nd:YAG laser beam which produced the laser plasma and thus had a pulse length of 600 ps.

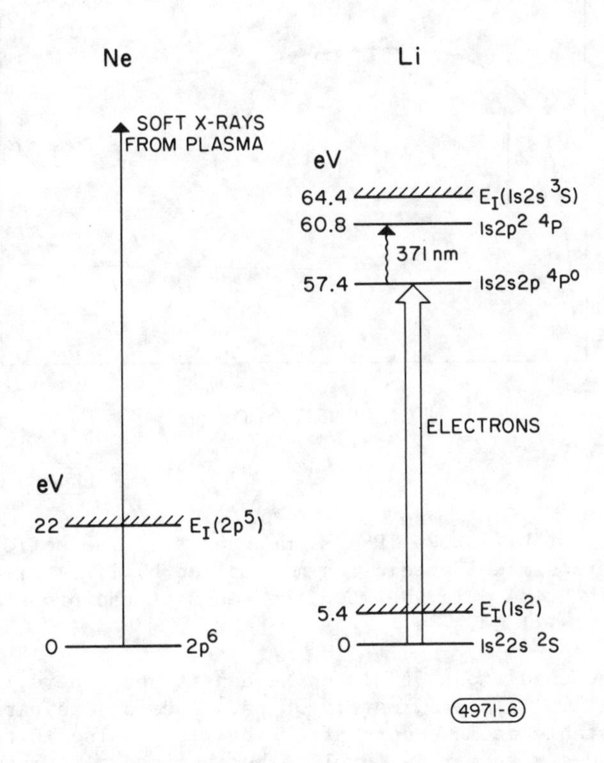

Fig. 6. Energy level diagram showing photoionization of the absorber atom and the subsequent electron excitation of the target atom. The laser probe transition is also indicated.

Figure 7 shows the measured dependence of N^*L on the Ne density. The prediction of a rate equation model is plotted as the solid curve. This model predicts both the magnitude and shape of the experimental curve very well. The saturation of N^*L at high Ne densities can be explained by noting that the density of $Li(^4P^o)$ atoms depends not only on the number of hot electrons, but also on their cooling time. At low Ne densities this cooling time is determined by inelastic collisions with Li atoms. At high Ne densities, however, electron cooling is dominated by collisions with Ne atoms. In this regime, as the Ne density increases the electron density increases but the cooling time decreases. Thus, the magnitude of the $Li(^4P^o)$ population approaches a constant value.

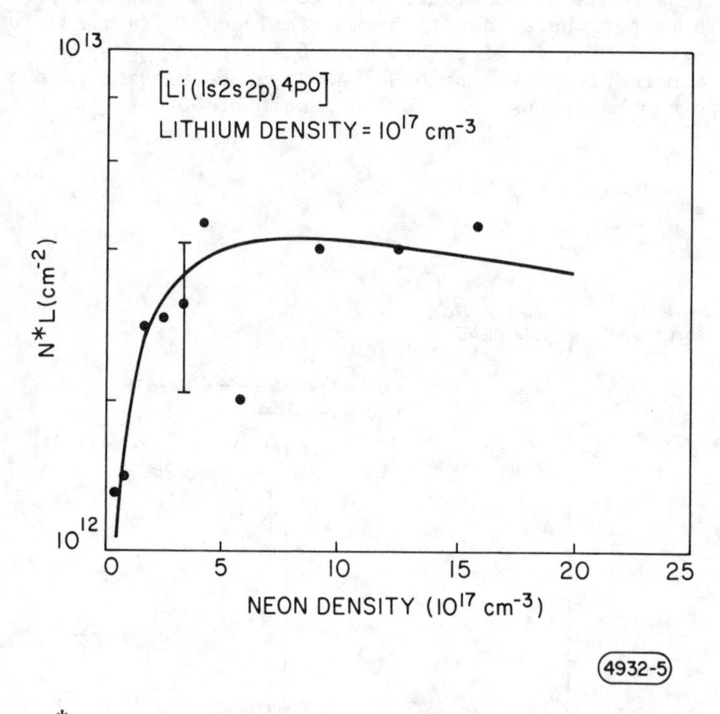

Fig. 7. N^*L for $Li(1s2s2p)\,^4P^o$ vs. Ne density. The solid curve is the rate equation model prediction multiplied by 1.2. Li density = 10^{17} cm^{-3}. The time delay between the peaks of the probe and 1.06 μm laser pulses is ∿ 1 ns.

Under the conditions of Li and Ne density used here, the electron cooling time is approximately 50 ps. Thus no excitation of $Li(^4P^o)$ metastable atoms occurs after the x-ray pulse is terminated, allowing the decay from this level to be observed. To measure the rate of this decay, a variable delay was inserted in the path of the probe beam. Figure 8 shows the dependence of the $Li(^4P^o)$ population on the delay between the probe laser and the plasma pulses. The observed decay time of 2.5 ± 0.5 ns is comparable to that obtained from Fig. 4 for the decay of the $Li^+(1s2s)^1S$ ion.

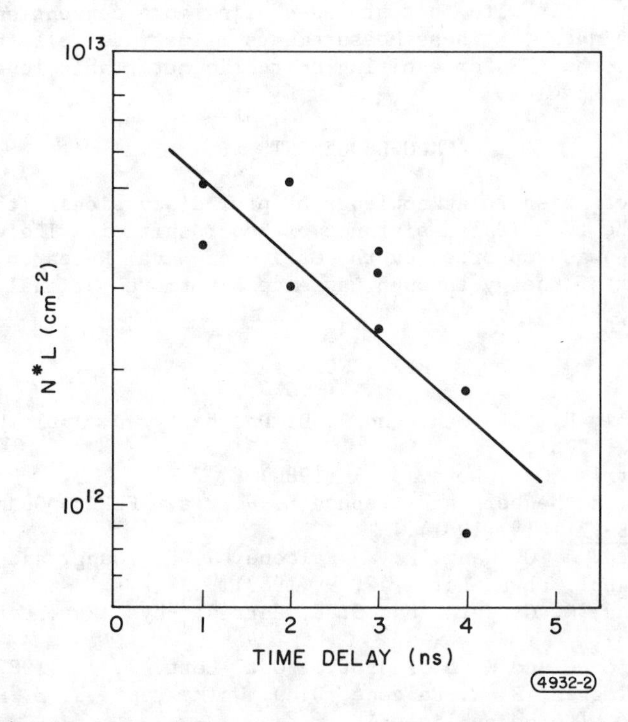

Fig. 8. N^*L for $Li(1s2s2p)^4P^o$ vs. the time delay between the peaks of the probe and 1.06 µm laser pulses. Ne density = 3×10^{17} cm^{-3}; Li density = 10^{17} cm^{-3}. The solid line is a fit to the data yielding τ_{decay} = 2.5 ns .

The electron density, at a time 1 ns after the peak of the x-ray pulse, has been measured to be $(1.3 \pm 0.8) \times 10^{15}$ cm^{-3} [from the Stark broadening of the $Li(1s^2 4d)$ level]. Assuming that electrons are responsible for the de-excitation, these measurements imply a de-excitation rate constant for the $Li(1s2s2p)^4P^o$ level of 3(+ 7 - 1) × 10^{-7} cm^3s^{-1}. This is typical of a dipole-allowed electron excitation of several eV. It can thus be concluded that no anomalously rapid de-excitation of the $Li(^4P^o)$ level occurs at electron densities of $\sim 10^{15}$ cm^{-3}.

A similar series of experiments has been performed for the Na $(2p^5 3s3p)^4D_{7/2}$ level. This system has the advantage that Na can be used both as "absorber" and "target" species, eliminating the need to mix vapors in the cell. Populations of 2×10^{13} atoms cm^{-3} have been measured in the Na $^4D_{7/2}$ state after excitation by the PES.

As a useful comparison with other excitation techniques, the PES was used to excite the $He(1s2s)^1S$ level. For 2×10^{18} atoms cm^{-3} of

He acting as both "absorber" and "target" species, a He(1s2s) population of 3×10^{14} cm^{-3} was measured. This corresponds to a fractional excitation to the He(1s2s)^{1}S level of 2×10^{-4} and thus compares very favorably with that produced using more conventional excitation techniques.[10] These measurements illustrate well the effectiveness of the PES for exciting energetic metastable levels in neutral atomic species.

ACKNOWLEDGEMENTS

The authors wish to acknowledge helpful discussions with R. W. Falcone and the technical assistance of Ben Yoshizumi. The work described here was supported by the Office of Naval Research and the Department of Energy through Lawrence Livermore National Laboratories.

REFERENCES

1. S. A. Mani, H. A. Hyman, and J. D. Daugherty, J. Appl. Phys. 47, 3099 (1976).
2. S. E. Harris, Opt. Lett. 5, 1 (1980).
3. Joshua E. Rothenberg and Stephen E. Harris, IEEE J. Quant. Elect. QE-17, 418 (1981).
4. R. G. Caro, J. C. Wang, R. W. Falcone, J. F. Young, and S. E. Harris, Appl. Phys. Lett. 42, 9 (1983).
5. J. C. Wang, R. G. Caro, and S. E. Harris, Phys. Rev. Lett. 51, 767 (1983).
6. R. W. Falcone and K. D. Pedrotti, Opt. Lett. 7, 74 (1982).
7. D. E. Holmgren, R. W. Falcone, D. J. Walker, and S. E. Harris, Opt. Lett. (to be published).
8. R. G. Caro, J. C. Wang, J. F. Young, and S. E. Harris, Phys. Rev. A (to be published).
9. P. Feldman and R. Novick, Phys. Rev. 160, 143 (1967).
10. H. Ninomiya, S. Horiguchi, and H. Osumi, J. Phys. D 14, 35 (1981).

PHOTOIONIZATION LASERS PUMPED BY BROADBAND SOFT-X-RAY RADIATION FROM LASER-PRODUCED PLASMAS

W. T. Silfvast, O. R. Wood, II,
J. J. Macklin and H. Lundberg
AT&T Bell Laboratories, Holmdel, New Jersey 07733

ABSTRACT

Large population inversions have been obtained in Cd and Zn ions using broadband soft-x-ray radiation from laser-produced plasmas to photoionize and thereby remove inner-shell d-electrons from the neutral atoms. This technique has presently produced lasers in the ultraviolet, visible and near infrared. The use of optical pumping to transfer the population produced by photoionization to higher lying levels may produce VUV and XUV lasers in the same and other species. Gains as high as 40 cm^{-1} and inversion densities as high as 10^{15} cm^{-3} have already been observed. These large inversions suggest that extractable energies approaching 1 mJ/cm^3 may be possible for directly-pumped lasers operating in the visible and for transfer-pumped lasers operating at shorter wavelengths.

INTRODUCTION

In 1967 Duguay and Rentzepis[1] suggested the possibility of producing population inversions at short wavelengths by x-ray photoionization of inner-shell electrons in atomic species. They pointed out that in many species, the photoionization cross section for removal of an inner p-electron is significantly larger than for the removal of an outer s-electron over a broad wavelength range in the x-ray spectral region. Consequently a pumping source with sufficient brightness in that wavelength region might be used to produce an inversion between the core-excited inner-shell ionic state and the outer-electron-removed ion ground state. The first such soft-x-ray pumped photoionization lasers were recently produced in Cd vapor[2] at 441.6 nm and 325.0 nm and in Zn vapor[3] at 747.8 nm. These new lasers are characterized by: (1) the removal of an inner-shell d-electron rather than an s- or p-electron (d-electron removal has a higher photoionization cross section[4]); (2) the use of species with closed outer shells (to provide more selective pumping) and with lower-laser-levels that can not be directly excited from the pumped species (e.g., the neutral ground state); and (3) the use of a laser-produced plasma soft-x-ray pumping source produced within the active medium[5] (to provide more efficient coupling of the pump source to the active medium).

A partial energy level diagram of singly-ionized and neutral Cd is shown in Figure 1. Since the neutral

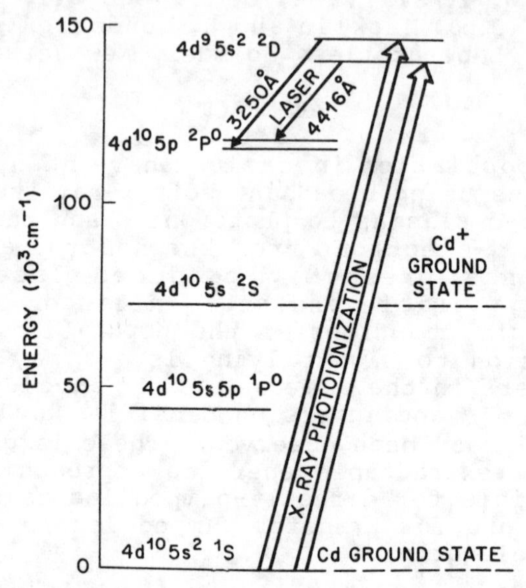

Fig. 1. Energy level diagram for directly-pumped Cd photoionization laser.

ground state is a $4d^{10}5s^2\ ^1S_0$ state, if an outer s-electron is removed by photoionization the species will be left in the $4d^{10}5s\ ^2S$ ion ground state. However, if an inner-shell d-electron is removed by photoionization the ion will be left in the $4d^9 5s^2\ ^2D_{5/2,3/2}$ states. The closed shell formed by the outer s-electrons prevents their interaction with the d-electron removed by photoionization. Such an interaction would produce states other than the ^2D states and, thus, would dilute the effect of the broadband photoionization pumping.

The wavelength dependence of the photoionization cross sections for Cd[6] is shown in Fig. 2. The cross section for s-electron removal is highest (1 Mb) at the threshold for the process (120 nm) and decreases rapidly with decreasing wavelength. The cross section for d-electron removal reaches threshold at 70 nm, increases to its maximum value (15 Mb) near 30 nm and becomes negligible near 10 nm. Due to the low probability of simultaneously removing one s-electron and exciting the other s-electron to a p state, the cross section for producing the $4d^{10}5p\ ^2P$ states (the lower laser levels) is more than a factor of 100 lower[7] than that for removing the d-electron. The Zn atom has a similar

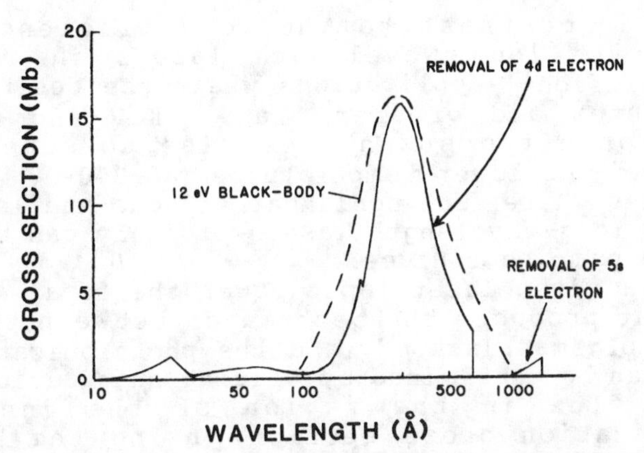

Fig. 2. Photoionization cross section versus wavelength for Cd vapor (solid curve) where the region between 100 and 600 A depicts d-electron removal together with calculated emission spectrum of 12 eV blackbody (dashed curve).

energy level arrangement and photoionization cross section. In Hg, although a similar shell structure and favorable d-electron cross section exists, a directly-pumped photoionization laser is not possible because the 6p states (which would normally be the lower-laser-levels) lie higher in energy than the 5d states populated by the inner-shell photoionization process.

The good overlap between the cross section for d-electron removal shown in Fig. 2 and the spectral emission from a 12 eV blackbody (also shown in Fig. 2) suggests that blackbody emission from a laser-produced plasma could be used as an effective photoionization source. A high-power laser focused onto a solid target of high-Z material has been previously shown[8] to produce blackbody emission with efficiencies as high as 20%. By equating the rate of energy absorption from a focused high-intensity laser at the target surface to the rate of heat conduction away from the interaction region, it is possible to show that the temperature, T, at the target surface is related to the laser intensity I and the laser wavelength λ in the following way:[9]

$$T \sim (I \lambda^2)^{2/3} \quad (1)$$

Hence, the temperature of the blackbody source can be adjusted by varying I or by choosing the appropriate laser wavelength. Some examples of the temperatures recorded over a wide range of intensities and for a variety of high power lasers are shown in Ref. 10. Eqn. 1 also indicates that a given temperature can be produced more easily with a longer wavelength laser.

This is in contrast to the low efficiencies that characterize long wavelength lasers in inertially confined fusion[11] applications where the desired plasma temperatures are greater than 1 KeV and nonlinear absorption effects can give rise to superthermal electrons.[9] At lower temperatures (40-200 eV) and lower intensities where such nonlinear effects are less likely to occur, long-wavelength lasers probably can be used as effective pumping sources.

For a given laser input power the focal spot size, r_{opt}, that produces the best match between the laser-produced plasma blackbody and the photoionization cross section can be determined by writing an equation for the pumping flux in the region of the inner-shell photoionization cross section, using the blackbody spectral distribution equation, substituting a relationship for temperature similar to Eqn. 1 and differentiating with respect to the radius of the blackbody surface area. This gives:

$$r_{opt} = 5.26 \times 10^{-3} \lambda_1^{3/4} \lambda_L P_L^{1/2} \quad (2)$$

where λ_1 is the wavelength at which the photoionization cross section maximizes, λ_L is the laser wavelength and P_L is the laser power. A plot of the soft-x-ray flux from a blackbody peaked at 30 nm produced by a 1.06 μm laser as a function of focal radius is shown in Fig. 3.

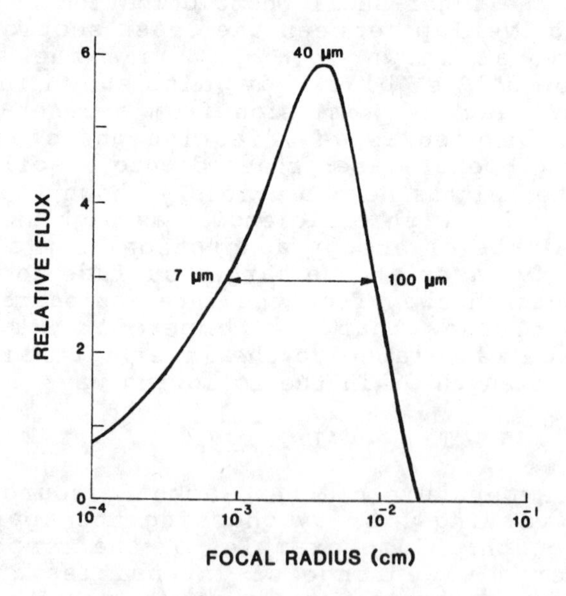

Fig. 3. Calculated dependence of soft-x-ray flux as a function of focal radius.

The pumping flux does not appear to be a strong function of spot radius. At radii smaller than r_{opt}, the flux decreases with decreasing spot radius, even though the blackbody temperature is increasing with increasing laser intensity, because of the reduced radiating area. At radii larger than r_{opt}, the flux decreases with increasing spot radius because the blackbody temperature and thus the pumping flux in the bandwidth of the photoionization cross section is dropping faster than the area of emission is increasing. The results shown in Fig. 3 are in qualitative agreement with experimental results in Cd and Zn.

EXPERIMENTAL ARRANGEMENT

The experimental arrangement[2,3] for producing photoionization lasers in Cd and Zn is shown in Fig. 4.

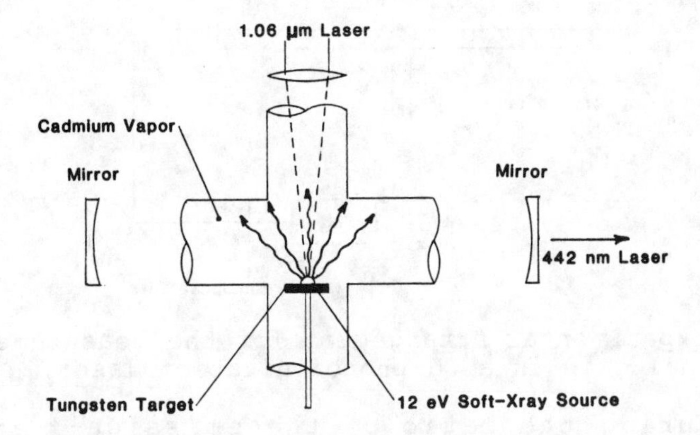

Fig. 4. Experimental arrangement for soft-x-ray pumped photoionization laser in cadmium.

A 2.5 cm dia. heat pipe in the shape of a cross was used to provide Cd or Zn vapors at pressures in the range from 1 to 10 Torr. One axis of the cross was used for the observation of gain and for the extraction of laser energy and the other axis was used for input of a 1.06 μm beam from a Nd:YAG laser or a 10.6 μm beam from a CO_2 TEA laser. When focused with a 25 cm focal length lens onto a tungsten target located in the central region of the heat pipe, the laser intensity, approximately 10^{11} W/cm^2 for the 1.06 μm laser and 10^9 W/cm^2 for the 10.6 μm laser, produced a soft-x-ray source with an approximately 12 - 13 eV blackbody distribution for the duration of the laser pulse. In the directly-pumped Cd photoionization laser, the output from a CW He-Cd probe laser, operating at the photoionization laser wavelengths, was used to determine the pumping conditions that maximized the population in the core-excited inner d-electron state. In the experiments where the output from a dye laser was used to transfer the population from this d-electron state to higher lying states, radiation from the dye laser was brought into the heat pipe through one end and stimulated emission was observed from the same end in order to minimize scattered light. The presence of gain due to stimulated emission was determined by (1) probing the gain directly using a He-Cd laser (as shown in Fig. 5),

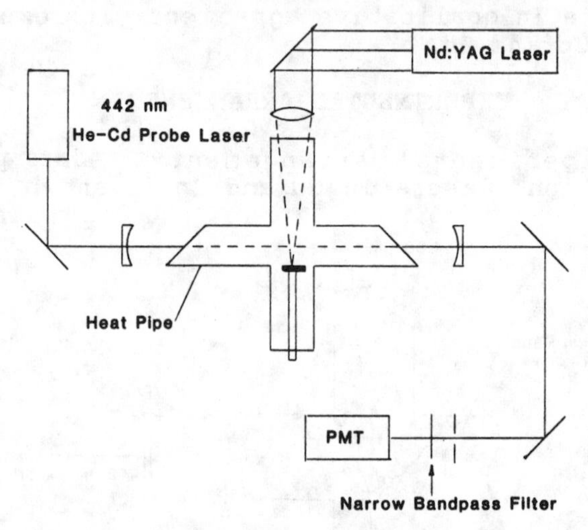

Fig. 5. Experimental arrangement for the measurement of small-signal gain in a Cd photoionization laser.

(2) measuring the ratio of the emission[12] in the presence or absence of a mirror located behind the heat pipe , and (3) measuring the enhancement obtained when an optical resonator was placed around the heat pipe.

RESULTS FOR DIRECTLY-PUMPED PHOTOIONIZATION LASERS

The directly-pumped photoionization lasers in Cd at 441.6 nm and 325.0 nm were studied with both 1.06 µm and 10.6 µm laser-produced plasma soft-x-ray sources. The results of these experiments (partially described in Ref. 2) are summarized in Table I. Gain coefficients as high as 6 cm^{-1}, energy outputs up to 0.1 mJ and power outputs of up to 10 kW for durations of 10 nsec were observed for 300 mJ, 10 nsec duration input pulses of 1.06 µm radiation. In this case the high density of electrons produced by the large pumping flux near the target was found[2] to diminish the gain near the target. Because of this electron quenching effect, an attempt was made to increase the pumping rate by using 60 mJ, 70 psec duration input pulses from a 1.06 μm laser. Gain coefficients as high as 40 cm^{-1} (a population greater

Table I. Summary of Results in Cadmium.

1.06 μm Source	
10 nsec, 300 mJ Pulse Gain - 6 cm^{-1} Laser Output - 10 kW, 0.1 mJ	70 psec, 60 mJ Pulse Gain - 40 cm^{-1}
10.6 μm Source	
50 nsec, 300 mJ Pulse Gain - 2 cm^{-1} Laser Output - 2 kW, 0.05 mJ	

than 10^{15} cm^{-3}) in a volume with dimensions of the order to 1 - 2 mm were observed under these conditions. Gain coefficients of over 2 cm^{-1}, energy outputs up to 0.05 mJ and power outputs of 2 kW for durations of the order of 25 nsec were observed for 300 mJ, 40 nsec duration input pulses of 10.6 μm radiation. Because of risetime and pulselength differences between the 1.06 and 10.6 μm source lasers, it was not possible to make an exact comparison of the performance of the two sources, however, the soft-x-ray flux produced by the 10.6 μm laser seemed to be about as effective as the soft-x-ray flux produced by the 1.06 μm laser at least in the spectral region of interest for Cd and Zn and the 10.6 μm laser was able to pump the Cd photoionization laser with an input intensity 2 orders of magnitude lower than that required by the 1.06 μm laser.

RESULTS FOR TRANSFER-PUMPED PHOTOIONIZATION LASERS

Stimulated emission on several UV transitions in Cd^{+} was observed for the first time by transferring population from inner-shell d-electron states populated by photoionization to outer-shell p-electron states using narrow frequency dye lasers tuned to relatively weak core-linking transitions. The population inversions in the UV at 231, 257 and 275 nm resulted from a 2-step pumping process. The first step involved a core-linking inner-shell transfer from the photoionization pumped $4d^{9}5s^{2}$ core-excited states to the higher lying, core-filled $4d^{10}6p$ states in Cd^{+} using dye laser wavelengths of 382.7 and 504.4 nm. The second step involved stimulated emission from the 6p states to the lower-lying 6s or 5d states at 806.6, 853.0 and 2000 nm. The energy levels and transitions relevant to the transfer-pumped photoionization lasers in Cd are shown

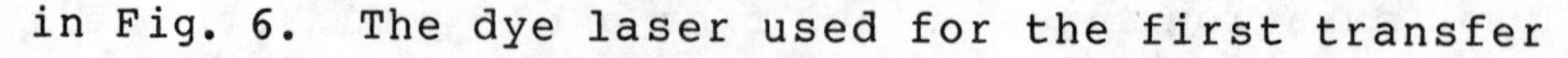
in Fig. 6. The dye laser used for the first transfer

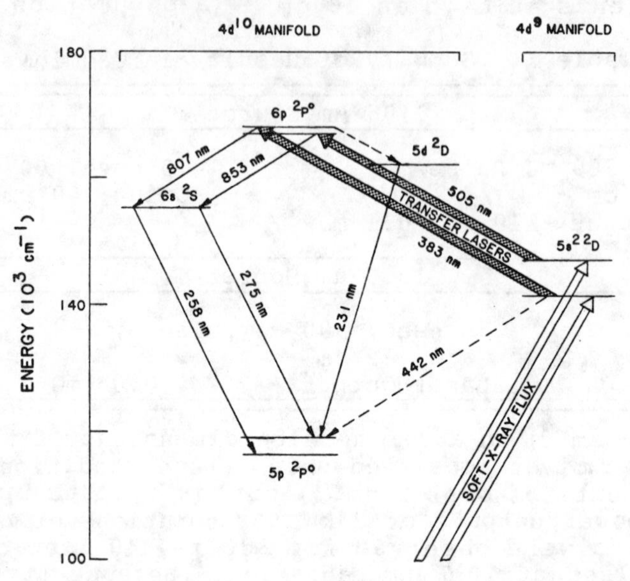

Fig. 6. Energy level diagram for transfer-pumped Cd photoionization lasers.

step provided pulse energies in the range 1 - 10 mJ in a bandwidth of 0.3 cm^{-1}. This resulted in intensities well above the saturation intensity for the core-linking transitions even though the 2-electron transitions from the core-excited d-electron manifold are thought to be relatively weak (oscillator strengths approximately = 0.01).

Stimulated emission both on the infrared and on the UV transitions was observed only during the first few nanoseconds of the transfer pulse. When a fast detector was used, some pulses at 806.6 nm were found to be as short as 400 psec in duration. The short pulse duration may have been due to the limiting effect of the high stimulated emission cross section (10^{-11} cm^2) on the 6p - 6s transition.

The presence of stimulated emission on the UV transitions was confirmed by the observation of a non-linear increase in the UV emission with increasing intensity at the 6p - 6s and 6p - 5d transfer wavelengths. In addition, the UV emission exhibited the large intensity fluctuations one would expect from small fluctuations in the gain. Gain coefficients as high as 7 cm^{-1} at 806.9 nm and 2 - 3 cm^{-1} at 274.8 nm were observed.

The observation of laser action in Cd in the UV demonstrates that the large population densities, produced by broadband soft-x-ray pumping with laser-produced plasmas, can be transferred to higher-lying levels. The success of the two-step transfer technique suggests the possibility of obtaining a number of similar inversions with respect to the lower-lying 5p level. These could result in lasers with wavelengths as short as 117 nm. This same technique could also be applied in Hg and Zn as well as in other elements which have a similar inner-shell electronic configuration. These could eventually result in a number of lasers in the UV and VUV.

POSSIBLE TRANSFER-PUMPED LASERS IN THE VUV

The generation of new VUV lasers may be possible using the double transfer technique described above. Since the 5p level is the lowest lying level above the Cd^+ ground state, transitions from high-lying levels in the Cd^+ spectrum to this level will produce the shortest wavelengths. Inversions to the Cd^+ 5s ground state might also be possible via a single transfer from the inner shell states. This could result in a laser with a wavelength shorter than 80 nm, however, inversions to the ground state are much less likely than inversions to the resonance level due in part to autoionization and in part to rapid filling of the ion ground state by electron excitation from the neutral ground state. Two transitions that appear to be particularly good candidates for VUV lasers are shown in Fig. 7. The

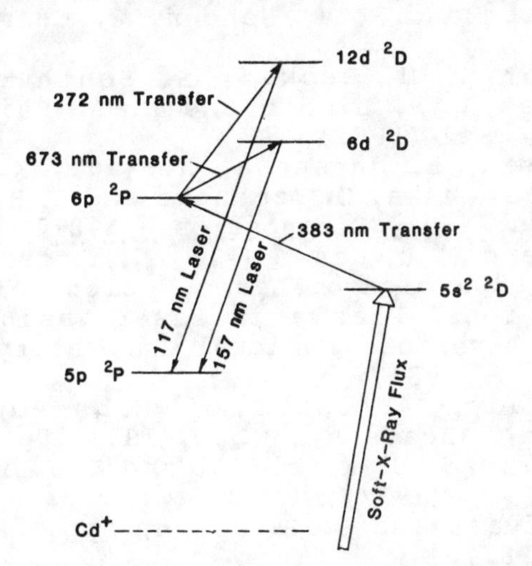

Fig. 7. Energy level diagram for two possible transfer-pumped Cd photoionization lasers in the VUV.

first candidate involves a 2-step transfer, first on a

transition from the $4d^95s^2$ level to the $4d^{10}6p$ level at 382.7 nm (already demonstrated) and then on a transition from the $4d^{10}6p$ level to the $4d^{10}6d$ level at 673 nm. The resulting VUV transition from the $4d^{10}6d$ level to the $4d^{10}5p$ level at 157.2 nm is expected to be a strong one because of its high oscillator strength (f = 0.1). The other candidate is one of the shorter wavelength laser possibilities (117.0 nm), but places more stringent requirements on the transfer laser because of the lower oscillator strengths of the transfer and potential laser transitions.

REFERENCES

1. M. A. Duguay and P. M. Rentzepis, Appl. Phys. Lett. 10, 350 (1967).
2. W. T. Silfvast, J. J. Macklin and O. R. Wood, II, Opt. Lett. 8, 551 (1983).
3. H. Lundberg, J. J. Macklin, W. T. Silfvast, and O. R. Wood, II, unpublished.
4. J. Berkowitz, "Photoabsorption, Photoionization and Photoelectron Spectroscopy," (Academic, N. Y., 1979), p. 48.
5. R. G. Caro, J. C. Wang, R. W. Falcone, J. R. Young and S. E. Harris, Appl. Phys. Lett. 42, 9 (1983).
6. Theoretical dependence of cross section by E. J. McGuire, Technical Memorandum SC-TM-68-70, Sandia Laboratories, Albuquerque, New Mexico, January 1968 normalized to experimental results of R. B. Cairns, H. Harrison and R. I. Schoen, J. Chem. Phys. 51, 5440 (1969).
7. P. H. Kobrin, U. Becker, S. Southworth, C. M. Truesdale, D. W. Lindle and D. A. Shirley, Phys. Rev. A 26, 842 (1982).
8. H. Nishimura, F. Matsuoka, M. Yagi, K. Yamada, H. Niki, T. Yamanaka, C. Yamanaka and G. H. McCall, in "Low Energy X-Ray Diagnostics - 1981," edited by D. Attwood and B. L. Henke (AIP, N. Y. 1981), p. 261.
9. C. E. Max, "Physics of Laser Fusion Vol. I. Theory of the Coronal Plasma in Laser Fusion Targets," Lawrence Livermore National Laboratory Report No. UCRL-53107, (1981).
10. J. F. Ready, "Effects of High-Power Laser Radiation," (Academic, N. Y., 1971), p. 194.
11. D. W. Forslund, J. M. Kindel and K. Lee, Phys. Rev. Lett. 39, 284 (1977).
12. W. T. Silfvast and J. S. Deech, Appl. Phys. Letts. 11, 97 (1967).

DYNAMICAL ASPECTS OF THE PICOSECOND X-RAY GENERATION FROM LASER-PRODUCED PLASMAS

Hiroto Kuroda[*,†] and Noboru Nakano[*]
* The Institute for Solid State Physics, The University of Tokyo, Roppongi, Minato-ku, Tokyo 106 Japan
† National Laboratory for High Energy Physics, Tsukuba, Ibaraki 305 Japan

ABSTRACT

Systematic studies of the picosecond X-ray generation from picosecond laser produced plasmas are performed both experimentally and theoretically. Transient characteristics of continuum x-rays are analyzed computationally by the aid of the newly extended transient collisional radiative model applied to multiply-ionized high density ions. Dynamical aspects of x-ray line emissions are studied with special interests on dynamical atomic processes which are found out to play important roles. Advanced understanding is obtained in the physics underlying the ultrafast x-ray generation from laser-produced plasmas.

INTRODUCTION

A number of experimental and theoretical studies have been performed as to the generation of x-rays from laser-produced plasmas [1~16]. A keen interest has been paid on such an x-ray generation as a tool of a diagnosis of electron densities and temperatures from points of view of nuclear fusion research [1~4]. However, in spite of the keen interest, theoretical analyses ever carried out were mainly limitted to the treatment of steady state plasmas. In addition, no experiments have been performed which show clearly transient characteristics of generated x-rays and their relevance to important atomic processes. As a result, neither transient characteristics of generated x-rays nor dynamical aspects of the x-ray generation from laser-produced plasmas become already clarified. The purpose of the present paper is to make clear dynamical aspects of the x-ray generation from picosecond laser-produced plasmas which are transient and do not stay in steady state equilibrium. It is widely recognized that x-ray lasers, if developed in the future, will be closely related to the ultrafast x-ray spectroscopy without saying of the x-ray pumping, the energy transfer process, their temporal gains and emission processes. Further, apart from x-ray lasers or laser-induced x-ray driven fusion, flush x-ray pulses with ultra-short durations are now developing their applications to such new attractive fields as the x-ray lithography and x-ray diffraction studies of biomedical materials. Therefore, to make systematic studies of the dynamical x-ray generation along the line which we emphasize here, is of keen interest and urgent.

In the following section, we describe our success in making clear transient characteristics of the continuum x-ray generation from both experimental and theoretical points of view. The clear dependence of intensities of generated x-rays on atomic number was observed and successfully analyzed computationally by the aid of the tansient collisional radiative model which were extended by us [17]. In the next section, dynamical characteristics of x-ray line emissions and dynamical atomic processes associated with them are described with emphases on the ionizing phase and the recombining phase[18]. Discussions are given on how much electron densities and temperatures which are estimated by various methods are affected by the fact that picosecond laser produced plasmas do not stay in steady state equilibrium. Next, one of examples of experimental advances in the picosecond x-ray spectroscopy is briefly described. We have developed a grazing incidence XUV spectrograph with a flat-field focal patterns by using newly designed gratings with varied grooves. This spectrograph will be proved as very useful in the coupling with picosecond x-ray streak cameras. Through the present paper, dynamical aspects of the picosecond x-ray generation are described with the emphasis on various dynamical atomic processes which have been so far remained obscure.

CONTINUUM X-RAY RADIATIONS

Various target materials are used in order to investigate the dependence of the intensity of emitted continuum X-rays on the atomic number of target materials. An advantage using a picosecond laser is to make theoretical analysis easier, because effects brought about by the expansion of laser plasmas can be neglected. However, rate equations based upon the collisional radiative model must be solved, since within the duration of laser pulse the ionizing process cannot reach equilibrium.

The dependence of the emitted X-ray intensity on the atomic number of target materials was investigated in the past using nanosecond lasers[9~16]. In those investigations, experimental results were analyzed based on the coronal equilibrium model . The positions of peaks and valleys appearing on the intensity curve as a function of the atomic number were explained[10,12]. However, the intensity variation with the atomic number could not be understood. This may be mainly due to the oversimplification of the model, and partly due to neglect of the effect caused by ion motion. In our experiments using a 30 picosecond laser, this effect can be safely ignored. Transient rate equations describing the ionizing process in target plasmas must be solved, because at least about 1×10^{-10} sec is necessary for the ionizing process to reach equilibrium, as shown later.

In our analysis, we adopt the transient collisional radiative model (the CR model) instead of the steady state coronal model. The CR model includes the collisional ionization and recombination processes and the radiative recombination process. Equations employed in the model are expressed as follows:

$$dn_{j,0}/dt = n_e \cdot (n_{j,0} \cdot \alpha_j - n_{j,0} \cdot (S_j + \alpha_{j-1}) + n_{j-1,0} \cdot S_{j-1}); 0 \leq j \leq Z \quad (1)$$

Here $n_{j,0}$ is population density of ion species with the atomic number Z in charge state j, S_j is collisional excitation rate (cm^3/sec), α_j is recombination rate(cm^6/sec). The power densities of emitted X-rays due to bremsstrahlung(P_{ff}) and recombination radiation (P_{fb}) were calculated using expressions given in ref.16. The assumptions adopted in solving these equations are as follows:

(1) T_e in these equations is assumed to have a constant value through

ionization processes as an approximation. In reality, T_e is changed spatially and temporally, but this assumption is taken in order to make the analysis easier and to understand specific properties of X-rays emitted from laser plasmas.

(2) Within 30 ps, expansion of a plasma does not take place, then the total population density $n_p (=\sum n_j)$ has a constant value . We adopt $1-10^{-6}$, 10^{-6} and 0 as initial values of f_0, f_1, and f_j $(=n_j/n_p, j\neq 0,1)$, respectively.

(3) Line radiation is neglected.

(4) The value of n_p is determined by the condition that n_e ($=\sum j.f_j.n_p$) becomes 10^{21} cm^{-3} at the end of the laser pulse(30 picosecond duration). The temporal dependence of the popoulation density(f_j) is shown in Fig.1 for aluminum(Z=13) and tungsten(Z=74) ,assuming 300 eV for T_e. Here the dotted line indicates the simplified solution obtained by assuming a constant value of n_e (10^{21} cm^{-3}) from the first stage of the ionization and neglecting the three body collisional recombination.

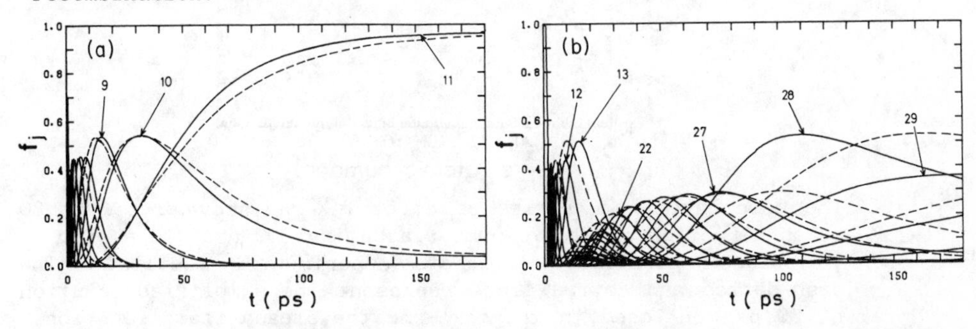

Fig.1 Temporal dependence of the calculated population density f_j in charge state j for plasmas assuming 300 eV as electron temperatures. The solid lines indicate the solutions obtained by assuming that electron densities become 10^{21} cm^{-3} at 30 ps. The dotted lines indicate the solutions obtained for constant electron densities of 10^{21} cm^{-3}. (a) Z=13,(b) Z=74.

As shown in Fig.1, results indicated by solid and dotted lines do not show a marked difference for 13 of Z. However, for 74 of Z, some disagreement is observed.

In Fig.2, the intensities of X-rays above 2 keV at 30 ps are shown as a function of the atomic number (Z), assuming 300 and 400 eV for T_e. The dash-dotted line and dotted line represent the simplified solution at 30 ps as mentioned above and the steady state solution, respectively.

A picosecond high power YAG laser with output energy of about 100 mJ is focused normally with a cone angle 10° onto a solid target suspended in a high vacuum chamber($\sim 10^{-6}$ Torr), using a lens of 10 cm focal length. The focal spot diameter is about 100 μm. X-ray signals detected by photomultiplier tubes through a Be-filter(of 100 μm thickness) were indicated by closed circles in Fig.2 for various kinds of targets. The cut-off energy(E_c) for this filter is about 2 keV. These experimental results could not be explained by the conventional steady state model, but could be fully understood by carring out new transient analysis.

It should be emphasized that the measurement of the dependence of X-ray intensity on the atomic number Z has been carried out for the first time by using a picosecond laser, and also that the results of theoretical calculations performed by adopting the transient CR model show good agreement with experimental observations. These results are surely useful and important to obtain further understanding of the characteristics of emitted X-rays from picosecond laser-produced plasmas.

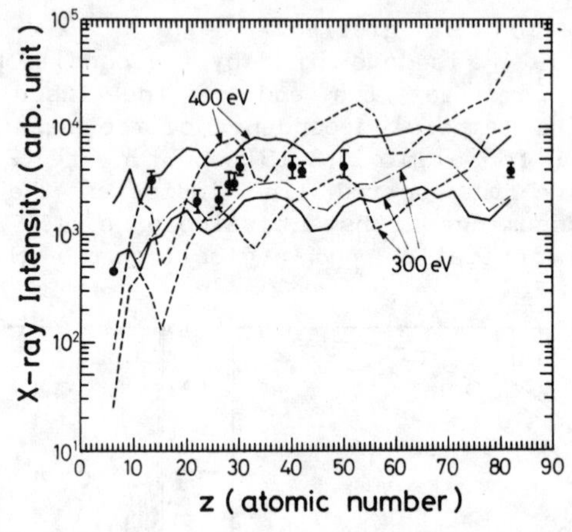

Fig.2 The calculated intensities of emitted X-ray above 2 keV at 30 ps as a fuction of atomic number, assuming 300 and 400 eV for T_e. Closed circles (●) denote experimental results. The dash-dotted and dotted lines represent the simplified solution at 30 ps mentioned in the text and the steady state solution, respectively.

X-RAY LINE RADIATIONS

For the purpose of diagnosis of electron temperatures and electron densities in plasmas, the measurement of the intensity ratio of two kinds of x-ray line spectra, such as satellite, intercombination and resonance lines, has been regarded as a useful and feasible tool. Until now, much experimental and theoretical work has been carried out along this line. However, argument based just on the intensity ratio is apt to allow ambiguity, and few systematic studies have been done concerning the whole understanding of the intensity of various x-ray lines and its dependence on electron temperatures and densities. The purpose of this section is to clarify the different characteristics of various kinds of x-ray lines from multiply ionized ions and their dependence on electron temperatures and densities in plasmas. For this purpose we have calculated intensities of x-ray lines originating from lithium, helium and hydrogen-like ions extensively, taking aluminum plasma as an example, in the framework of a model which includes all atomic processes taking place in high density plasmas.

To estimate intensities of line emissions, it is necessary to solve the eq.(2) which describe the dynamics of number densities of the excited state $n_{j,1}$, in addition to solving the eq.(1). An excited state $n_{j,1}$ distributed in each charge state ($0 \leq j \leq Z$) can be calculated by solving following equations.

$$
\begin{aligned}
dn_{j,l}/dt = {} & n_{j,l}\sum_m\Big\{-n_e\big(D_{e,j}(m,l) + C_{e,j}(l,m)\big)+R_{d,j}(l,m) \\
& -n_e C_{i,j}(l) - A_{i,j}(l)\Big\} + \sum_m\Big\{n_e\Big[C_{e,j}(m,l) \\
& + D_{e,j}(l,m)\Big]+R_{d,j}(m,l)\Big\}\, n_{j,m} + n_e\Big\{n_e D_{i,j}(l) \\
& + E_{c,j}(l) + R_{r,j}(l)\Big\}\, n_{j+1,0} \\
& = 0, \qquad (2)
\end{aligned}
$$

where $R_{d,j}(l,m)$ is a radiative decay rate, $C_{e,j}(l,m)(D_{e,j}(m,l))$ is a collisional excitation (deexcitation) rate, $A_{i,j}(l)$ is an auto-ionization rate, $E_{c,j}(l)$ is an electron capture rate, $R_{r,j}(l)$ is a radiative recombination rate, and $C_{i,j}(l)$ $(D_{i,j}(l))$ is a collisional ionization (three-body recombination) rate, respectively.

Aluminum ions	Levels
Hydrogen-like ion	1s,2p,3p,4p,5p
Helium-like ion	$1s^2\,{}^1S;1s2p\,{}^1P,{}^3P;1s3p\,{}^1P,{}^3P;1s4p\,{}^1P,{}^3P;1s5p\,{}^1P,{}^3P;1s2s\,{}^1S,{}^3S;2s2p\,{}^1P,{}^3P;2p^2\,{}^3P,{}^1D,{}^1S$
Lithium-like ion	$1s^2 2s\,{}^2S;1s^2 2p\,{}^2P;1s(2s2p\,{}^1P)\,{}^2P,1s(2s2p\,{}^3P)\,{}^2P;1s2p^2\,{}^2P,{}^2D$

Table.1 Energy levels

In Table.1,energy levels in which we are mostly interested are shown. To make the situation more simple and applicable to resonable computations, we make disscussions by characterizing the plasmas by two seperated temporal phases, in which one is the ionizing phase and the other is the recombining phase.

IONIZING PHASE

X-ray spectra located between 5.5 Å and 8 Å regions which are simulated at t=30 ps and at the steady state are shown in Fig.3(a) and (b),respectively. It is easily distinguished that these two spectra are entirely different from each other.

Figure 4(a) shows our calculated results representing how much x-ray line intensities are changed as functions of time in picosecond regions. Intensities of a resonance line of helium-like ions I_r and its satellite line I_s represented by curves named as 3 and 4 change monotonously and become saturated and reach to a quasi-steady state after about 70 ps. On the other hand,intensities of a resonance line of a hydrogen-like ion and its satellite line which are labelled by 1 and 2 continue to increase and ionization proceses remain in non-stationary state even at 180 ps after the initiation of ionization processes.

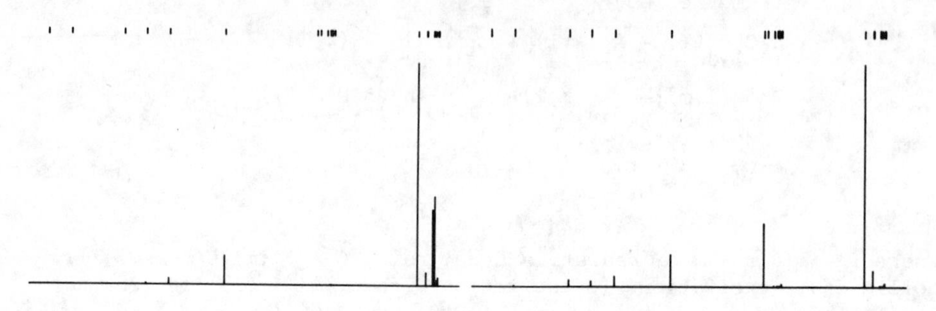

(a) (b)

Fig.3 X-ray spectra located between 5.5 Å and 8 Å. (a) t=30 ps, (b) steady state

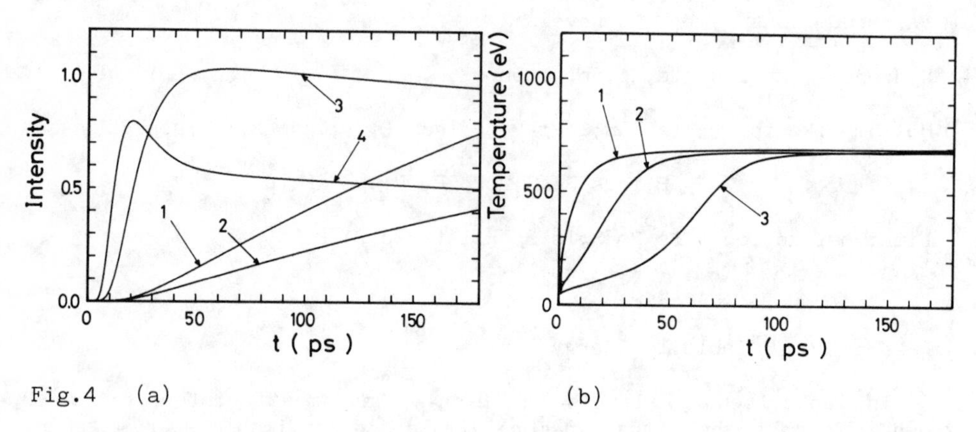

Fig.4 (a) (b)

(a) Temporal dependence of intensities of x-ray line radiation from plasmas with 700 eV of electron temperatures. Curves named as 1,2,3 and 4 indicate lines brought about by transitions written as follows;
1: $2p — 1s$ of a hydrogen-like ion
2: $2p^2\ ^1D_2 — 1s2p\ ^1P_1$ of a helium-like ion
3: $1s2p — 1s^2$ of a helium-like ion
4: $1s2p^2\ ^2D_{5/2} — 1s^2 2p\ ^2P_{3/2}$ of a lithium-like ion.
These curves named as 1,2,3 and 4 are normalized by factors of $2.5x10^{13}$, $1x10^{12}$, $2.5x10^{14}$ and $5x10^{12}$, respectively.

(b) Estimated temperatures from various intensity ratios of resonance lines to their satellite line.
1: a resonance line of a hydrogen-like ion($2p — 1s$) and a satellite line of a helium-like ion($2p^2\ ^1D_2 — 1s2p\ ^1P_1$)
2,3: a resonance line of a helium-like ion($1s2p — 1s^2$) and its satellite line of a lithium-like ion. Curves named as 2 and 3 are obtained from satellite lines brought about by the transition between $1s2p^2\ ^2D_{5/2}$ and $1s^2 2p\ ^2P_{3/2}$ and that between $1s2p^2\ ^2P_{3/2}$ and $1s^2 2p\ ^2P_{3/2}$, respectively.

We will show for the first time that an undesirable error is brought about in the evaluation of an electron temperature for plasmas in a transient state if comparisons of x-ray intensity ratios are made with theoretical results obtained for plasmas in a steady state. Figure 4(b) shows electron temperatures calculated by using various intensity ratios of x-ray lines. It is easily noticed that these values show different temporal changes. The origin of these different rise times τ revealed in the figure is attributed to the temporal variation of the ratio R of the degree of contribution due to the collisional excitation process to that caused by the electron capture process. From eq.(2), it can be understood that τ depends on electron densities n_e and $n_e \cdot \tau$ takes approximately a constant value when an electron temperature is kept at a constant value. Our results will also be applicable to low density plasmas by using this relation.

RECOMBINING PHASE

In recombining plasmas, hydrodynamic motion of plasmas cannot be neglected. Hydrodynamic motion of plasmas can be well discribed by the self-similar solution[19] assuming one dimensional planar expansion of plasmas. As to the validity of adopting the self-similar solution, it was already discussed and concluded that hydrodynamic motion of transient laser plasmas can be characterized well by this formula. The essential points related to transient characteristics of x-rays in a recombing phase are shown to hold true,even if any other hydrodynamic model is used. By using $n_{e,max}$ and T_e obtained from this solution,population densities of ions for ground states $n_{j,0}$ and those for excited states $n_{j,1}$ in charge state for ions with an atomic number Z ($0 \leq j \leq Z$) can be calculated by solving the coupled equations of (1) and (2).

Computations are performed for aluminum plasmas. Temporal changes of $n_{e,max}(t)$ and $T_e(t)$ are shown in Fig.5(a). Initial values of $n_{e,max}(0)$ and $T_e(0)$ are assumed as typical values of 7×10^{22} cm^{-3} and 700 eV,respectively. These values of n_e and T_e and other values for the self-similar solution are chosen by the condition that the temporal change of n_e is similar to that shown in ref.20. In this figure,a schematic block diagram of atomic levels related to these x-ray line emissions is inserted. Temporal changes of intensities of emitted x-ray lines are shown in Fig.5(b) in the case of a resonance line, an intercombination line and satellite lines. It can be easily distiguished that satellite lines decay very fast within about 5 ps, whereas a resonance line and an intercombination line continue to emit for a long time until after 50 ps. These different temporal tendencies can be explained as follows. For plasmas in steady states, emissions of a resonance line and an intercombination line are brought about mainly by the collisional excitation process from the ground state and those of satellite lines are brought about by the electron capture process,respectively. By using levels indicated in Fig.5(a), it can be shown that the rates for these excitation processes C_e (0,1) and E_c (3) strongly depend on electron temperature[21,22]. For instance, at 5 ps,electron temperatures take 140 eV and these rates are decreased as

$$C_{e,j}(t=5\ ps)/C_{e,j}(t=0) \cong 2.4 \times 10^{-4}, E_{c,j}(t=5\ ps)/E_{c,j}(t=0) \cong 1.7 \times 10^{-2}.$$

Therefore, satelllite lines decay within 5 ps. On the other hand, the fact that a resonance line and an intercombination line last for longer time can be explained by the contribution of the radiative recombination process from the next higher charge state of ions. Three-body recombination processes become less important than radiative recombination processes as electron densities are decreased.

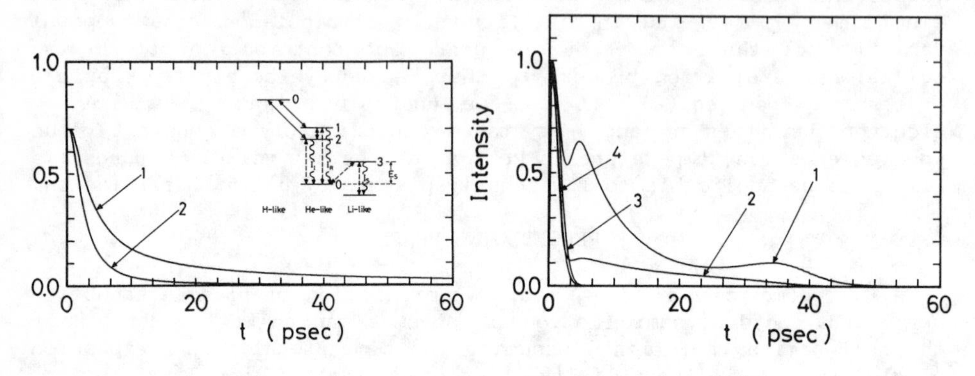

Fig.5(a) (b)

(a) Temporal changes of electron densities n_e indicated as 1 and electron temperature T_e named as 2 obtained by computations based on the self-similar solution. A schematic energy level diagram originating a resonance line of a helium-like ion is also shown. A level named as 0 denotes ground state of each charge state.

(b) Computed results of temporal changes of normalized x-ray intensities of a resonance line, an intercombination line and two satellite lines.
1: $2p1s\,^1P - 1s^2\,S$ (a resonance line of a helium-like ion)
2: $2p1s\,^3P - 1s^2\,S$ (an intercombination line of a helium-like ion)
3: $1s2p^2\,^2D_{5/2} - 1s^2 2p\,^2P_{3/2}$ (a satellite line of a lithium-like ion)
4: $1s2p^2\,^2P_{3/2} - 1s^2 2p\,^2P_{3/2}$ (a satellite line of a lithium-like ion)

Next, we comment on the accuracy of the electron densities and temperatures estimated from the intensity ratios of I_r/I_t and I_s/I_r. Estimated electron densities and temperatures are shown in Figs.6(a) and (b), respectively, together with true values which are shown in Fig.5(a). Disagreement between values estimated from intensity ratios and true values are enhanced as a function of time, already before satellite lines decay as shown in Fig.5(b). This is due to the fact that dominant atomic processes responsible for originating x-ray line emissions are changed from one to another as electron temperatures are decreased, as is dicussed above. Therefore, it is concluded that an effort to make an estimation of electron densities and temperatures becomes of no use after the time when satellite lines are diminished to a negligibly low level.

Next, comments on the comparison of our results with experimental ones obtained by M.H.Key et al.[20] will be made. Our results concerning temporal change of intensities of emitted x-ray lines, as shown in Fig.5(b), agree well qualitatively with their experimental results. However, an interpretation given by them was rather incorrect. At first, they attributed the origin of the fast decay of satellite lines observed to the fact that the intensity ratio between a satellite line and a resonance line I_s/I_r is decreased as temperature is decreased when a steady state model is used. However, contrary to their statement, it can be easily shown that this intensity ratio is increased as electron temperature is decreased[23]. Secondly, they estimated electron densities at the time long after satellite lines are already diminished, by comparing experimental results of I_r/I_t with theoretical n_e dependence obtained by assuming that

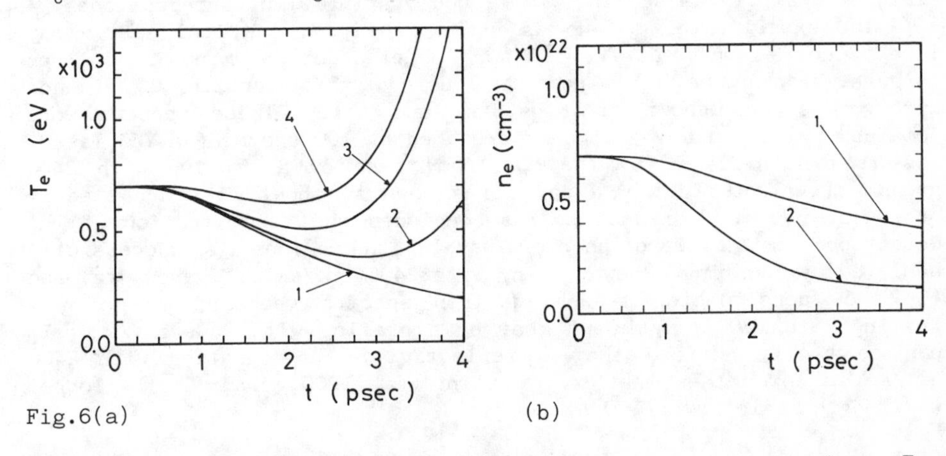

Fig.6(a) (b)

(a) Temporal changes of estimated values for electron temperatures T_e. A curve indicated as 1 represents T_e computed by the self-similar solution. The other curves labeled as 2,3 and 4 show temperatures determined from x-ray intensity ratios as follows.
2:a resonance line of a hydrogen-like ion(2p 1s) and its satellite line($2p^2\ ^1D_2$ $1s2p\ ^1P_1$).
3:a resonance line of a helium-like ion(1s2p $1s^2$) and its satellite line($1s2p^2\ ^2D_{5/2}$ $1s^22p\ ^2P_{3/2}$).
4:a resonance line of a helium-like ion(1s2p $1s^2$) and its satellite line($1s2p^2\ ^2P_{3/2}$ $1s^22p\ ^2P_{3/2}$).

(b) Temporal changes of electron densities n_e. Curves represented as 1 and 2 denote n_e calculated by the self-similar solution and that evaluated from an x-ray intensity ratio between a resonance line and an intercombination line of a helium-like ion, respectively.

collisional excitations from the ground state to emitting levels play an important role as excitation processes. However, these obtained values of n_e are far from accurate, as was described above, related to Fig.6(b). An incorrect interpretation based on the similar argument is also found in ref.4. They obtained spatially resolved spectrogram for a resonance, an intercombination and satellite lines. They estimated electron densities from the ratio I_r/I_t over the distance of 0.3 mm where no satellite lines were observed. In the expanding plasmas, spatial distributions of densities are closely connected to temporal changes, therefore, similar arguments described above can be applied also to their experimental situation. In any case, it can be said that the estimation of electron densities n_e becomes invalid both temporally and spatially when and where satellite lines are already diminished to a negligibly low level.

DEVELOPMENT OF A NEW TYPE OF XUV SPECTROGRAPH

Measurements of time-resolved spectra in XUV and soft x-ray regions with the aid of picosecond streak camaras are strongly desired to clarify dynamics of high density plasmas. For such purposes, we have deyeloped a new type of spectrograph, by which spectra in the 15 to 250 Å wavelength range can be sharply focused on a flat detection field by utilizing a specially designed grating with varied pitches.

The experimental set-up is shown in Fig.7. Typical soft x-ray line spectra emitted from laser plasmas of aluminum and iron targets,which were obtained by using this grazing incidence spectrometer,are shown in Figs.8(a) and (b). These spectra were obtained by accumulating signals produced by 100 shots of a YAG laser. As is distinguished by the spectrum taken by a grating with 2400 gr./mm shown in Fig.8(b),x-ray lines down to about 15 A are clearly observed. From this fact,it was concluded that an extention of the detection limit toward shorter wavelength regions is successfully achieved by adopting a new grating with 2400 gr./mm. Figures 9(a) and (b) show densitometer traces of iron spectra measured by using a grating with 1200 gr./mm and that by a grating with 2400 gr./mm. It can be easily noticed that a reflectivity of a grating with 2400 gr./mm is lower than that of a grating with 1200 gr./mm in the longer wavelength region over 100 A.

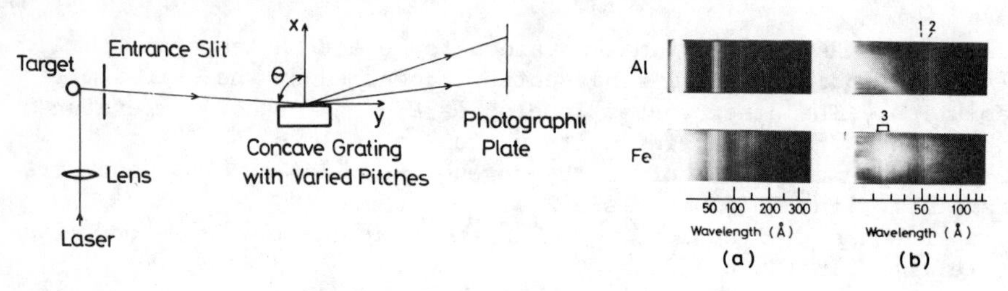

Fig.7 Fig.8

Fig.7 Schematic and design specifications of the flat-field spectrometer.

Fig.8 Typical soft x-ray line spectra obtained on x-ray films. (a) and (b) show the spectra from laser produced plasmas by using gratings with 1200 gr./mm and 2400 gr./mm,respectively for an aluminum target and an iron target.

Experimental results concerning x-ray spectra are extremely different from those in steady states as shown in Fig.9 (a) and (b). However, the functional dependence of observed spectra can be qualitatively explained by those in a transient state. In addition,we present that transient solutions are essential for giving a correct explanation for experimental results. Since,x-ray line radiations from iron plasmas in charge states of 16 (Fe XVII) and 17 (Fe XVIII) are observed as shown in Fig.8, it can be concluded that both charge states are generated to some extents. This fact agrees well with charge states computed, assuming a transient state.

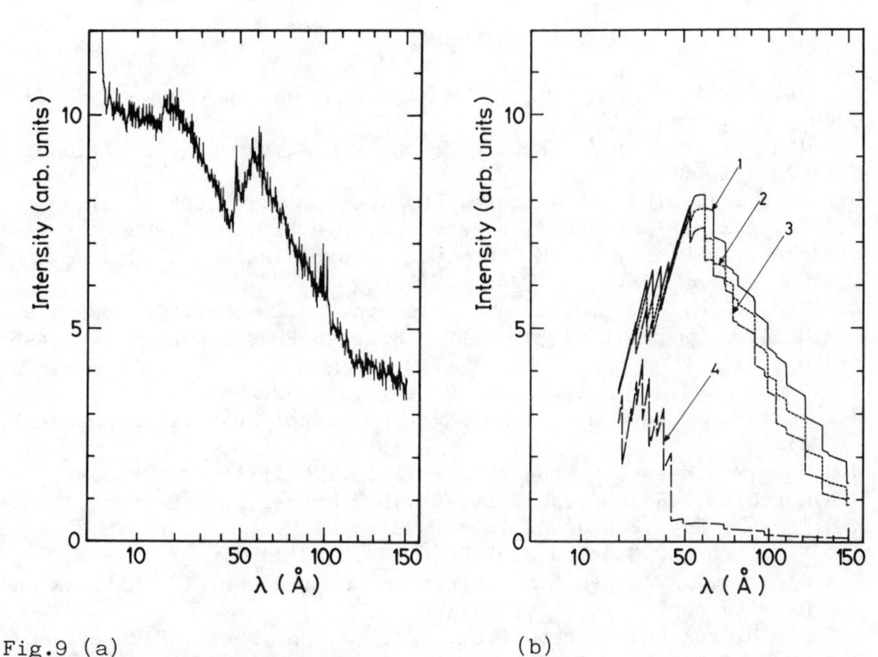

Fig.9 (a) (b)

(a) Micro-densitometer traces of photographic plates. (a) and (b) indicate results for a grating with 1200 gr./mm and that with 2400 gr./mm, respectively.

(b) X-ray spectra emitted from iron plasmas obtained by summing up from first to fourth order reflected ones under the assumption that higher order components of x-ray spectra have the similar reflectivities by each other. 1:t=10 ps,2:t=20 ps,3:t=30 ps,4:in steady state. (a) T_e=300 eV,(b) T_e=700 eV.

CONCLUSION

We have perfomed, experimentally and theoretically, extensive and systematic studies of the x-ray generation from laser-produced plasmas. As to continuum x-rays, the strong dependence of x-ray intensities on the atomic number was found out. The computational analysis based on the transient collisional radiative model which we extended clarified the physics associated with such a dependence and related transient characteristics. Various characteristics of x-ray line emissions are also analyzed successfully with taking into account such dynamical atomic processes as electron captures and collisional excitation processes. As a result, it was also shown that special attention should be paied in the evaluation of electron temperatures and densities when transient non-steady state plsamas are treated. In a word, we could make advances in the understanding of physics underlying the ultrashort x-ray generation from laser-produced plasmas. More elaborated efforts will be necessary to extend our studies along the line aiming x-ray lasers. To clarify physics underlying those effects such as due to the opacity and the deviation of electron energies from Maxwell distributions will be necessary as the first step.

REFERENCES

1. E.V.Aglitskii, V.A.Boiko, A.V.Vinogradov and E.A.Yukov, Sov.J. Quant.Electron.,4,322(1974)
2. A.V.Vinogradov, I.Yu.Skobelev and E.A.Yukov, Sov.J.Quant. Electron.,5,630(1975)
3. B.Yaakobi, I.Pelah and J.Hoose, Phys.Rev.Lett.,37,836(1976)
4. V.A.Boiko,S.A.Pikuz and A.Ya.Faenov,J.Phys.B12,1889(1979)
5. B.Yaakobi,S.Skupsky,R.L.McCrory,C.F.Hooper,H.Deckman,P.Bourke and J.M.Soures,Phys.Rev.Lett.,44,1072(1980)
6. A.Hauer,K.B.Mitchell,D.B.van Hulsteyn,T.H.Tan,E.J.Linnebur and M.M.Mueller and P.C.Kepple and H.R.Griem,Phys.Rev.Lett.,45,1495 (1980)
7. J.G.Lunney and J.F.Sely,Phys.Rev.Lett.,46,342(1981)
8. H.Kuroda,H.Masuko and S.Maekawa,Jpn.J.Appl.Phys.,17,Supplement (1978)
9. H.D.Shay,R.A.Haas,W.L.Kruer,M.J.Boyle,D.W.Phillion,V.C.Rupert, H.N.Kornblum,F.Rainer,V.W.Slivinsky,L.N.Koppel,L.Richards and K.G.Tirsell,Phys.Fluids,21,1634(1978)
10. R.D.Bleach and D.Nagel,J.Appl.Phys.,49,3832(1978)
11. K.M.Glibert,J.P.Anthes,M.A.Gusinow and N.A.Palmer,R.R.Whitlock and D.J.Nagel,J.Appl.Phys.,51,1449(1980)
12. H.Pepin,B.Grek and F.Rheault,D.J.Nagel,J.Appl.Phys.,48,3312(1980)
13. B.Yaakobi,P.Bourke,Y.Conturie,J.Delettrez,J.M.Forsyth,R.D.Frankel ,L.M.Goldman,R.L.McCrory,W.Seka,J.M.Soures,A.J.Burek, and R.E.Deslattes,Opt.Commun.,38,196(1981)
14. W.C.Mead,E.M.Campbell,K.G.Estabrook,R.E.Turner,W.L.Kruer, P.H.Y.Lee,B.Pruett,V.C.Rupert,K.G.Tirsell,G.L.Stradling, F.Ze,C.E.Max and M.D.Rosen,Phys.Rev.Lett.,47,1289(1981)
15. D.Colombant and G.F.tonon,J.Appl.Phys.,44,3524(1973)
16. D.Mosher,Phys.Rev.A10,2330(1974)
17. N.Nakano and H.Kuroda,Phys.Rev.A27,2168(1983)
18. N.Nakano and H.Kuroda,Phys.Rev.A(to be published),Appl.Phys.Lett. (to be published)
19. J.M.Dawson,Phys.Fluids,7,981(1964)
20. M.H.Key,C.L.S.Lewis,J.G.Lunney,A.Moore,J.M.Ward and R.K.Thareja, Phys.Rev.Lett.,44,1669(1980)
21. D.Duston and J.J.Duderstadt,J.Appl.Phys.,49,4388(1978)
22. D.Duston and J.Davis,Phys.Rev.,A21,932(1980)
23. C.R.Bhalla,A.H.Gabriel,L.P.Prenyakov,Mon.Not.R.Astro.Soc.,172, 359(1975)

EXPERIMENTAL AND SIMULATION STUDIES ON SOFT-X-RAY EMISSION FROM 0.53 μm-LASER IRRADIATED SOLID TARGETS

T. Mochizuki, T. Yabe, K. Okada and C. Yamanaka
Institute of Laser Engineering, Osaka University
2-6 Yamada-oka, Suita, Osaka, 565 Japan

ABSTRACT

The spectrum-resolved radiant energy in 0.1∿1.6 keV range from various plane targets irradiated by a 0.53μm laser at 0.1∿1.0 nsec pulse through a f/1.6 lens with a nominal incidence angle of 54° were obtained. Atomic number dependences of the spectrum and X-ray conversion efficiency are described. The conversion efficiency increased as laser pulse duration.

These X-ray emission spectra are investigated by the 1-D hydrodynamic Lagrangian code HIMICO and 2-D particle-in-cell code IZANAMI, both of which are coupled with non local thermodynamic equilibrium (non-LTE) average ion model and multi-group radiation transport. The physical meaning of the experimentally obtained spectrum is clarified and the total emitted power density proves to have maximum at certain combination of density and temperature.

INTRODUCTION

At shorter-wavelength laser irradiations on solid targets, energy transport into dense matter occurs more efficiently. In this situation the emission of soft X-rays by dense target plasma becomes also more efficient especially at high Z material. X-ray conversion efficiency has been measured at 0.5μm at average intensities from 3×10^{13} to 5×10^{15} W/cm^2 [1]. Similar experiments have been conducted at 10μm [2], 1.05μm, 0.53μm, 0.35μm [3], and produced conversion efficiencies by assuming blackbody spectrum and the laser intensity dependence of effective radiation temperature. Extensive investigation on overall characteristics of 0.53μm laser-produced Au plasmas has been reported by Mead et al [4], but the detailed dynamical behavior of emission and the optical thickness of the plasma were not discussed. Z-dependence of the X-ray conversion efficiency has been studied experimetally to some extent by Glibert et al [5] but their photon energy range was limited only in the range over 0.7 keV and the absolute spectral intensity were determined indirectly by iteratively comparing the diode signals with a trial spectrum. Recently the interaction between intense soft X-ray radiation and solid target has been investigated by the authors [6]. They found that the interaction is extremely impotant to inertial confinement fusion and thus parametric information including accurate spectrum, Z-dependence etc. is strongly required.

In this paper we present the spectrum-resolved absolute energy measurement of the sub-keV radiation emitted from various Z targets which were irradiated by 0.53μm laser with the intensity of about 10^{14} W/cm^2 at the incidence angle of 54°. The radiant energy spectrum from Au plasmas is compared with the hydrodynamic computer code results. The detailed structure of the emission coupled with the ablation process is described.

EXPERIMENTAL CONDITION AND DIAGNOSTICS

A 0.53μm light from a frequency-doubled GEKKO IV laser with an energy of 4-20 Joule in 0.1ns, 0.15, 0.25 (short pulse) or 1.0 nsec (long pulse) in full width at half maximum was focused onto the target through an aspheric lens of effective f number of 1.6 with a nominal incidence angle of 54°. A spot diameter was varied from 90μm to 330μm to obtain a wide laser intensity range $2\times10^{13}\sim2\times10^{15}$ W/cm^2. Typically it was fixed at 150∿180μm at $I_L\sim2\times10^{14}$ W/cm^2. The irradiations were done near S-polarization. Intensity profile was not directly measured, but a Be filtered X-ray pinhole camera seeing hν>1.5 keV provided no significant nonuniformity in its X-ray image. For comparison, we made intentionally nonuniform irradiation on targets by putting concentric structured mask in the laser beam path in front of the final focusing lens. In this case very rough estimation of scale length of nonuniformity was 30μm.

We used a soft X-ray transmission type grating spectrometer to obtain an X-ray spectrum in 0.1∿2.0 keV range [7]. The dispersive element was a linear array of 0.3μm thick gold bars with an 1μm spatial period in a free standing structure. The spectral resolution was estimated to be about 5 A in case of a 100μm X-ray source size. The dispersed X-ray was usually recorded on a KODAK Type 101 film and also on KODAK NS-5T film when hν>1 keV range was required with higher sensitivity. This spectrometer was mounted at 22.5° off the target normal axis. Filtered 10 channel biplanar X-ray diodes (XRD's) and mini-X-ray-calorimeters were used to measure an X-ray radiant energy from the target. The XRD's had Al photocathodes whose quantum efficiency was measured by R.H.Day et al. [8]. The filter sets used in the present experiment are listed in Table 1. The diode itself had about 80 psec rise time. The signals of the XRD were monitored with the combination of a Tektronix 7104 and 485 oscilloscopes. The temporal response of 485 was not enough, therefore the oscilloscope response was calibrated by using 7104.

1. $C_2HCl_{1.8}$ 1.33μm + C_8H_8 0.1μm
2. Parylene (C_8H_8) 2μm
3. V 0.37μm + C_8H_8 0.5μm
4. C_8H_8 7.4μm
5. Fe 0.5μm
6. Fe 0.5μm + C_8H_8 0.75μm
7. Ni 0.37μm + C_8H_8 0.5μm
8. Zn 0.42μm + C_8H_8 0.5μm
9. Al 1.0μm + C_8H_8 0.5μm
10. Be 45μm

Table 1 Filter sets for XRD.

To monitor the emission image of the X-ray above 1.5 keV, two X-ray pinhole cameras with a 7∿8μm pinhole diameter and a 55μm thick beryllium filter were mounted tangential to the target surface and at 45° titled from the target normal, respectively. A

KODAK XRP-5 film was used to record the X-ray image. The transmission grating spectrometer with KODAK 101 film was also used to see the spectral resolved size of emission region by placing a pinhole of 150μm diameter in front of the grating. In the 0.15 nsec pulse experiments, underdense corona plasmas were observed from the tangential direction by an optical shadowgraph method using a synchronized probe beam (633 nm) of 60 psec pulse duration. A flash X-ray backlighting method was also separately applied to observe tangentially the high density region of the target with 1.4 keV X-ray photons in the short pulse experiments. Expanding ions were always monitored using charge collectors, which were placed typically at 41.6° to the target normal.

Typical X-ray emission spectrum obtained by using the transmission grating spectro-meter is shown in Fig. 1. The Au planar target was irradiated in a 150μm focal spot diameter by a 1.0 nsec pulse laser. The film sensitivities are not calibrated. There can be obviously recognized three band-structures consisting of hν=0.15 keV, 0.2∿0.3 keV and 0.6∿1.2 keV. At the 0.12 nsec pulse with higher intensity $\sim 10^{15}$ W/cm^2, another hump was observed in the spectral range ∿2.0 keV on the KODAK NS-5T film. One-dimensional spatial image of the emission region by the spectrometer with the input pinhole of 150μm diameter showed slightly larger size for hν≤500 eV region than that by the Be filtered X-ray pinhole image. No significant difference in the emission sizes was found for the sub-keV range investigated and they roughly coincided with the nominal laser spot size.

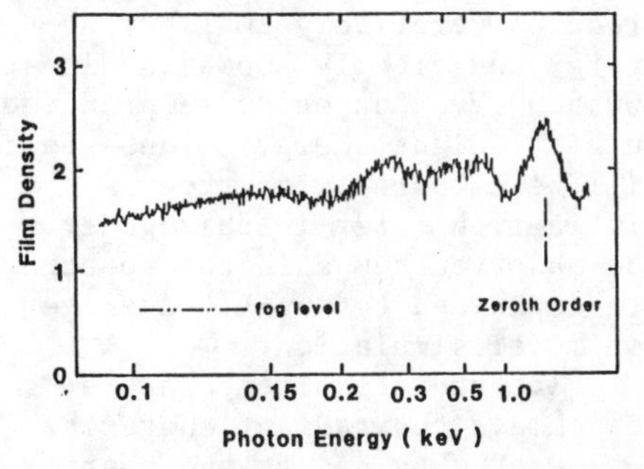

Fig.1 Typical X-ray emission spectra in transmission grating spectrometer at a viewing angle 22.5. Au target was irradiated at $I_L=7.3\times10^{13}$ W/cm^2 in 1.0nsec.

XRD's signal showed the X-ray emission time was nearly equal to the laser pulse duration for 1 nsec and slightly shorter at higher hν region, but it was only ascertained to be less than 500 psec for short laser pulse because of overall temporal response of the detection. XRD's signal was time-integrated to obtain the total electrical charge produced at the diode cathode taking a calibrated value of a cable attenuation and oscilloscope response into account. The uncertainty in the X-ray radiant energy evaluation by XRD's signal data could be considerably reduced by comparing with the spectral shape in Fig. 1, because the non-flatness in the diode sensitivity otherwise produces a large evaluation error.

XRD's signals of the parylene channel at two different angular positions indicated approximate cosine distribution at $I_L=5\times10^{13}\sim 4\times10^{14}$ W/cm^2.

RADIANT ENERGY SPECTRUM FROM Au PLASMAS

Figure 2 shows typical radiant energy spectra obtained at 100 psec and 1.0 nsec laser pulse. The horizontal bars of data points mean the observed energy band. It is found in Fig. 2 that the obtained spectral shape is not of ideal blackbody spectrum and most of the radiant energy concentrated in below 0.7 keV. Shorter pulse produced relatively smaller radiant energy and clearly showed a dip at 300∿500 eV. Longer pulse produced larger radiant energy. These facts will be compared with the simula-tion later in this paper. The observed humps in the spectrum are explained by non-LTE average ion model simulation.

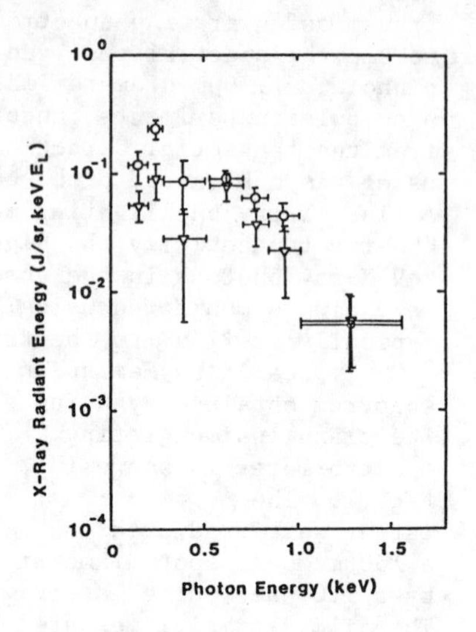

Fig.2 Radiant energy spectra from Au plasma at a viewing angle 22.5 at ∇:I_L=4.0×10^{14}W/cm^2 in 100psec and o:1.0×10^{14}W/cm^2 in 1.0nsec.

The radiation temperature T_R to achieve the radiant energy at hν=250 eV obtained at the short pulse is estimated from the blackbody relation to be 224 eV, on the other hand the nominal T_R to explain the data point at hν=600 eV is 170 eV. These values are 175 eV and 127 eV for 1 nsec pulse, respectively. The spectrum deviates from the blackbody spectrum especially at the laser pulse duration of less than 0.25 nsec, and approaching to blackbody radiation spectrum as the pulse increasing to 1 nsec, but still deviation remains. One can think the reason for two different radiation temperatures as follows; Lower energy photon is produced at high dense plasma which may be optically thick enough to provide cosine angular distribution. Rosseland mean free path for bound-free transition provides 10μm at Te∿T_R=200 eV, Ni=1×10^{22}/cm^2, <z>=20. This value is comparable to or larger than plasma scale length. However for bound-bound transitions which are dominant at high Z target, the plasma may be optically thick. Such situation may not hold on the emission of hν>500 eV. If the plasma is optically thin, then the emissivity decreases and the angular distribution approaches an isotropic one. These situation seems to be responsible for the present results.

We found that the structured laser beam profile enhanced the radiant energy at 200∿300 eV by a factor 2 at a shorter laser pulse compared to the smooth laser beam profile under the same average laser intensity. This suggests that the plasma region where the quasi-blackbody condition is attained becomes hotter at the peak positions of the laser intensity modulation so that the emissivity is enhanced there.

One can conclude that the emission observed can not be described by the blackbody radiation with single radiation temperature. This was not caused by 2D effect because the emission size is almost the same for 100∿800 eV as the laser beam spot sizes used therefore large lower temperature region did not exist around laser beam spot on the target.

At normal incidence, the radiant energy increased compared to 54° incidence angle. This is inferred to be due to the increase in the plasma density by a factor 3 at which the laser energy is deposited, resulting in the increase of laser absorption. In this situation plasma density and temperature may have a different profile from that by 54° incidence angle so that energy transport to the X-ray emitting region becomes more effective.

LASER INTENSITY DEPENDENCE OF X-RAY CONVERSION EFFICIENCY IN Au PLASMAS

The conversion efficiencies from the incident laser energy to the total emitted X-ray radiant energy of 0.1∿1.6 keV have been obtained as shown in Fig.3 by integrating spectral radiant energy such as Fig.2. Here we assumed that the emission angular distribution is in cosine law for the whole hν region investigated at the laser intensity of more than 5×10^{13} W/cm^2. This does not hold around hν∿1 keV, but the error induced in the total conversion efficiency by the ambiguity in the angular distribution is small because the radiant energy at hν>1 keV is much smaller than that at hν=100∿300 eV. In the long pulse irradiations, the conversion efficiency is about 40% and this value is two times as large as that in the 0.25 nsec pulse at the intensity of $\sim10^{14}$ W/cm^2. This means that the increase in laser absorption and the increase in plasma scalelength may have contributed to the increase in the X-ray conversion efficiency. Figure 3 shows that the conversion efficiency is only weakly dependent on I_L. From the beryllium filtered X-ray calorimeter's signal, the total energy above 1.6 keV was estimated to be less than 1% of the incident laser energy.

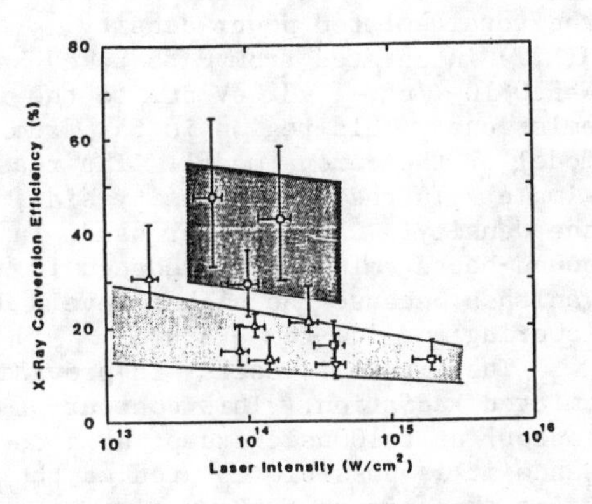

Fig.3 Laser intensity dependence of X-ray conversion efficiency for laser pulse duration, o:1.0nsec, Δ:0.23∿0.25nsec, □:0.15nsec, ∇:0.1nsec.

SIMULATION

These X-ray emission spectra have been investigated by the 1 D by hydrodynamic Lagrangian code HIMICO [9][10] and 2-D particle-in-cell code IZANAMI [11], both of which are coupled with non local thermodynamic equilibrium (non-LTE) average ion model and multi-group radiation transport.

Before we started the calculation to simulate the experimental results, we calculated the emitted power density as a function of plasma density and temperature to obtain a general feature of X-ray generation.

Figure 4 shows the contour of total emitted power density J(N,T) at t=500psec. The solid lines indicate J(N,T)=const. These lines of J(N,T)=const. have a convex shape downwardly at higher temperature region as seen. Therefore, when the parameters N and T_e changed according to NT_e=const, which is the reasonable assumption in a subsonic ablation region, it is found that J(N,T) has the maximum value because the line of NT_e=const. can be tangential to the line of J(N,T)=const. When the parameters N an T_e were changed according to $NT_e=1.6\times10^{23}$eV/cc (or 0.256Mb) it was found that the total emitted power density J(N,T) integrated from 0 to 2 keV has maximum value at $N=5.0\times10^{20}$/cc, T_e=320eV due to the intense and numerous line emissions. This region is far from the validity domain of the LTE model or the corona model. The reason for this maximum is very simple. In the lower density side, the emission is proportional to the density. On the other hand, in the higher density side, bound-bound emission is reduced in number of lines and finally vanishes because the higher levels disappear due to large potential lowering and low temperature.

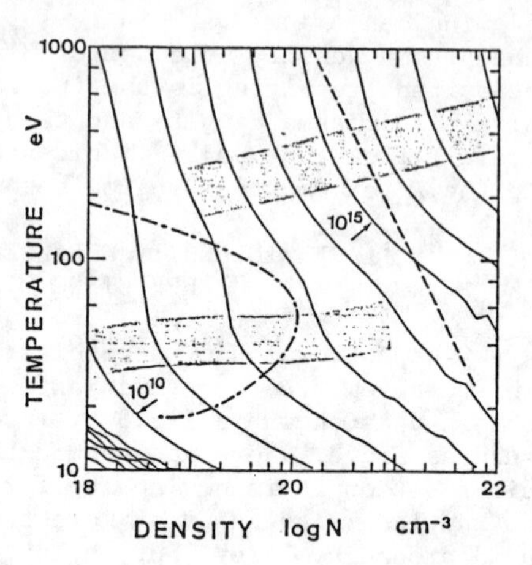

Fig.4 Contour of total emitted power density J(N,T) (W/cm^3) at t=500psec. The maximum of total emision is shown by the shaded region for various choices of the product NT_e.

The figure is useful in providing an idea of the region of emitted radiation. This contour of J at t=500psec agrees with the contour at t=100psec except at a density lower than N $\sim10^{19}$/cc, and hence it is possible to predict the maximum value of total emission using this figure for any time longer than 100psec.

The dielectronic recombination is also one of the important processes in laser produced plasmas. Since this ionization state is reduced due to this process at a lower density region, the line spectra shift towards a lower energy side and some of their intensities are larger than that in the non-LTE model without the dielectronic recombination process.

The calculation to simulate the experimental results assumed that a spherical target having a radius the same as a focal spot radius is irradiated by 150psec and 18.8×2J green laser whose incident angle is set to 54° in order to replicate the experimental incident angle. Furthermore, the flux limitation factor f is set to either 0.01 or 0.03 and the fraction of resonance absorption to 0.03.

Figure 5 shows electron temperature and ion density profiles for f=0.01 at laser peak. A steep profile is due to inhibited thermal electon transport, while a plateau is formed by radiation transport. Figure 6 shows spectral power density of emitted radiation at the sampled points denoted by the arrows in Fig. 5. Bound-free radiation is more sensitive to the temperature and density changes than bound-bound radiation. This trend is

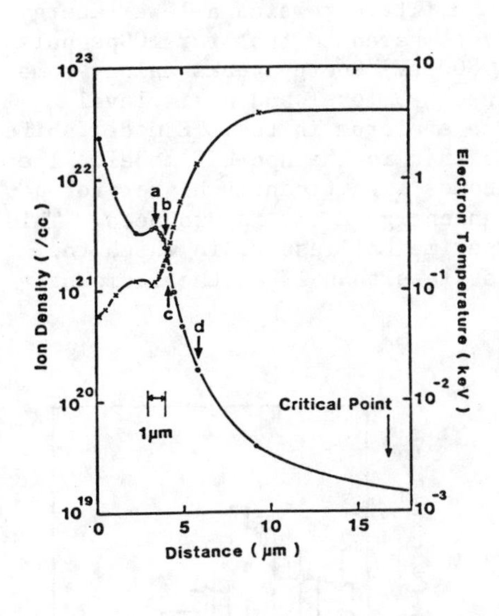

Fig.5 Electron temperature and ion density profiles in ablated Au plasma at laser peak. The laser pulse is 150psec in FWHM and E_L= 18.8×2 J.

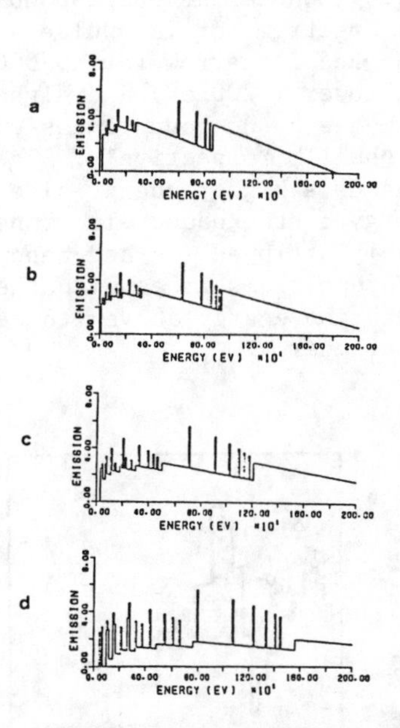

Fig.6 Spectral power density of emitted radiation at the sampled points denoted by the arrows in Fig.5. Vertical axis is in logarithmic scale and in arbitrary units.

apparently seen in the lowest energy line (∿ 600eV) emission from N shell : the line strength remains constant from N_i=4×10^{21}/cc to 4×10^{19}/cc. The bound-free emission less than 1keV is large at the plateau region, while the line emission is large at the steep profile region. Since the line emission determines the total emitted power, the energy transport is one of the most important processes to determine the X-ray conversion efficiency. This is reflected on the conversion efficiencies at 1ns and 100ps laser experiments. The simulations give the efficiencies of 46.7% and 6.7% from laser to X-ray, and the efficiencies of 54.3% and 20.6% from absorbed energy to X-ray energy, for 1ns and 100ps pulses, respectively. In the former case, the laser parameters are 17.9J/1ns and 150μm focal spot, and in the latter case those are 10.6J/100ps and 184μmϕ. It should be noted that the region of steep density gradient is optically thin for $h\nu \gtrsim 500$ eV because the line centers in the neighboring two regions have an energy difference larger than the line width and hence the opacity is mainly determined by bound-free opacity. On the other hand, the optical thickness at the plateau region is relatively large for $h\nu \sim 200$ eV. Accordingly, for a longer pulse, the emission becomes larger because the density scale length is longer : the total energy of the emitted X-ray is mainly given by the line emission and is determined by the volume emission process in the region having steep density gradient.

The total emitted spectra for 1ns and 100ps cases are shown in Fig. 7, where the corresponding experimental data are also shown. The spectrum for 1ns pulse shifts a little towards a lower energy side and is narrow for $h\nu \gtrsim 500$ eV, compared to that for 100ps pulse. The lower ($\sim$200 eV) and higher ($\sim$800 eV) energy peaks mainly come from the bound-bound emissions into $n \gtrsim 5$ level and n = 4 level (N-shell), respectively. The line spectrum in the LTE model shifts towards a higher energy side than that in the non-LTE model. The energy of the bound electrons becomes significantly higher for a highly stripped ion and hence high energy X-ray is produced. This LTE result disagrees with the experimental result, in which the radiation energy above 1.6 keV was less than 1% of the incident laser energy.

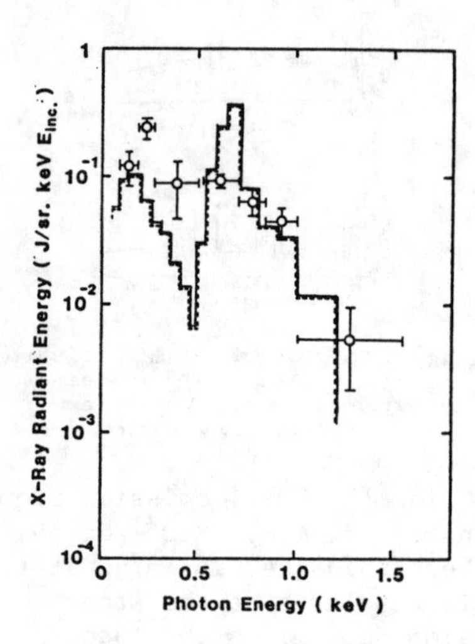

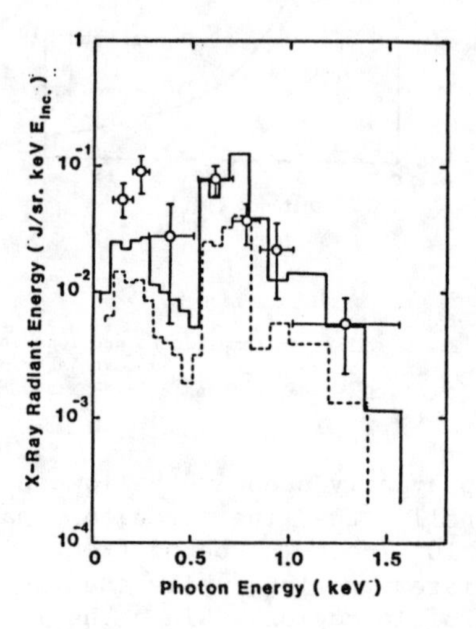

Fig.7 Calculated emitted spectra of Au plasma for a) 1 nsec and b) 100 psec cases. Open circles show experimental values. In the latter case, the solid and dashed lines denote the results for f=0.03 and 0.01, respectively.

The calculated spectra are in excellent agreement with the experimental results at f=0.03 except for lower energy ($\sim$200 eV) component. These results indicate that the detailed X-ray spectrum observed is closely related with the value of energy flux limitation factor in an ablating plasma. There are many possibilities which may improve the discrepancy. Two of these are (1) the laser intensity distribution in the laser spot, e.g., low temperature plasma as a source of low energy X-ray, (2) the contribution from sub-level transitions which are neglected in the average ion model. The simulation can not satisfactorily replicate a difference between 1ns and 100ps pulses concerned with the spectrum dip at around 500 eV. The Stark broadening at $N_i \sim 10^{21}$/cc is about 20eV and may not play an important role here. Because the energy window of X-ray diode for each channel is wide, the direct comparison of the spectrum dip between the simulation and experiments requires a finer resolution of the spectrum.

The 2-D code simulation shows that the laterally directed, radiation from an expanding plasma, especially at a longer laser pulse, illuminates the edge parts of the laser spot and heats them. Such lateral radiative energy transport becomes significant at least in high Z targets.

Z-DEPENDENCE OF X-RAY SPECTRUM AND CONVERSION EFFICIENCY

Similar X-ray generation experiments have been made with other solid target materials at $I_L \sim 1\times10^{14}$ W/cm^2 in 1.0 nsec, a laser spot of 150μmϕ and the incidence angle of 54°. Figure 8 shows X-ray conversion efficiency as a function of Z. The total efficiency is broken down into fractions corresponding to respective photon energy range as indicated by several lines. It should be noted that the conversion efficiency increases gradually with some undulation as Z increasing. In particular the fractional efficiencies in Fig. 8 have undulation structure. These undulations are due to the humps in the spectrum whose position and intensity depend on the material Z. The humps originate in electronic transitions in the n-th shell of ions. The simulation results by our hydrodynamic code described before also predicts undulation structure in fractional efficiency and reproduced well total efficiency change qualitatively. From the shell number identified, the achieved maximum ionization state can be inferred and it reflects the plasma temperature at the emission region.

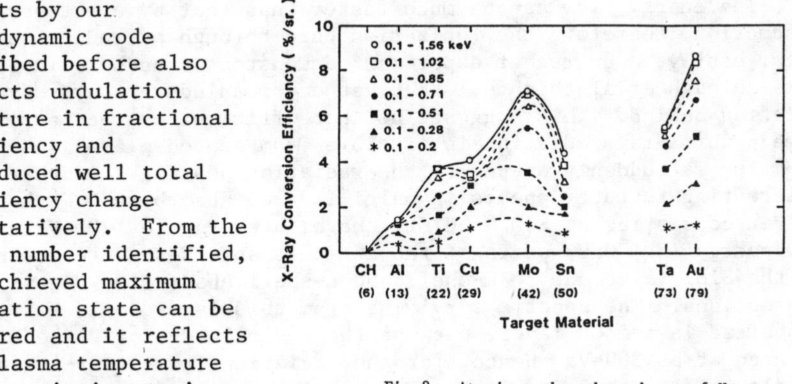

Fig.8 Atomic number dependence of X-ray conversion efficiency.

If we determine the energy range of the photons induced by the transitions of the bounded electrons, then we can rewrite the conversion efficiency in such a way that the contribution of the electrons in respective shell to X-ray generation is clarified. Such contribution is shown in Fig. 9. It is clearly seen that mainly contributed shell becomes outer as Z increasing. Therefore undulation structure in X-ray efficiency can be well explained by the number of shell electrons left at ionization equilibrium.

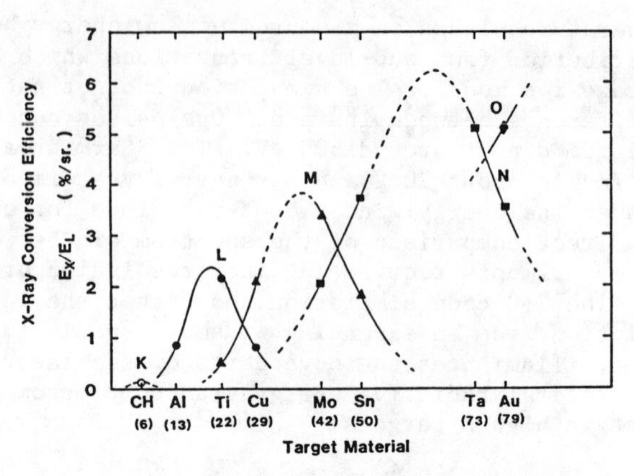

Fig.9 Atomic number dependence of X-ray generation from various shells.

INTERACTION OF INTENSE SOFT X-RAY RADIATION WITH SOLID MATTER

As described in the previous sections, the high X-ray conversion efficiency in laser-produced plasma can provide intense soft X-ray radiation whose intensity reaches 10^{13}-10^{14}W/cm^2. Such intense radiation may be a good candidate as a pumping light to lase in the extreme ultraviolet region.

Let us imagine that the solid matter is irradiated by an externally generated, intense soft X-ray and a plasma is produced. In contrast to the heating by laser-produced thermal electrons, the interaction has various different aspects due to this complexity in opacity manifesting shell structure of bounded electrons in the matter.

The energy flow can be much faster than that by electron conduction, therefore the ionization burn-through by radiation will occur easily. Our recent experiment demonstrated such ionization burn-through of Al thin foil that was accompanied with hydrodynamic ablation behind. This suggests us that efficient extraction of high power X-ray radiation is possible from heated plasma when the end plug is suddenly opened by the radiation heating. Another interesting feature appears in multi-layered absorbers. With the target consisting of SiO_2 and CH, the high energy ($h\nu \sim 700$eV) and low energy ($h\nu \sim 50$eV) parts of the incident soft X-ray are absorbed at the SiO_2 layer due to K-shell and L-shell absorption edges of oxygen ions. The spectra surviving from this layer could be deposited in the CH layer, because the K-shell edge of carbon is located at $h\nu \sim 300$eV. Hence, for the radiation from laser-produced Au plasma discussed previously or the blackbody radiation whose radiation temperature is about 100eV, the bulk energy will be absorbed at the interface between the two layer[6].

Figure 10 shows the high energy X-ray backlighted image of a 0.1μm Au layered SiO_2 foil which was irradiated by the intense soft X-ray from an externally laser-heated Au plasma. It should be noticed that the outer Au layer was peeled off from the SiO_2 layer. This phenomena may be related with the effect described above. The detailed mechanism is now under discussion.

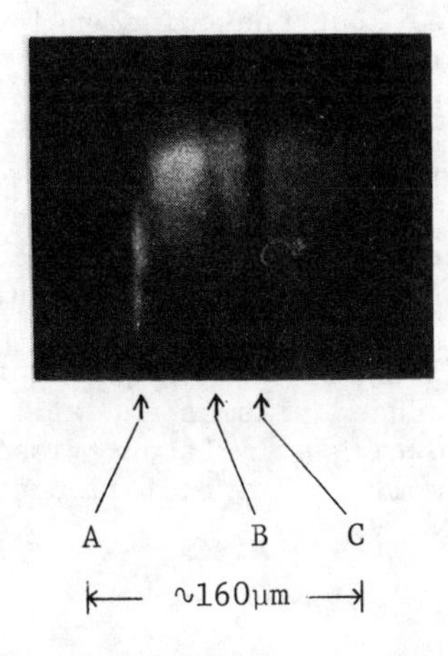

Fig.10 Mo X ray backlighted image of an intense soft X ray irradiated Au (1000Å) coated SiO_2 plane target at 500psec after the irradiation.
A: a soft X ray emitter of plane geometry produced by 0.53μm laser irradiation.
B: Au layer peeled off from the SiO_2 layer.
C: SiO_2 layer.

SUMMARY

Radiant energy spectrum of soft X-ray from 0.53μm laser-produced Au plasma has been found to deviate from the simple blackbody radiation spectrum. The spectrum can be fitted by T_R of about 200 eV for 100∿200 eV photon but the spectral intensity is lower at the $h\nu \geq 500$ eV than the blackbody radiation intensity. The shorter pulse irradiation produced more clearly a dip in the spectrum between 0.3 keV and 0.6 keV. The conversion efficiency increased as laser pulse duration.

Hydrodynamic code coupled with non-LTE average ion model was used to investigate the dynamics of X-ray generation in laser produced plasmas and to explain the X-ray spectra at short and long pulses. Simulation clarified the plasma density and temperature region which are most responsible for X-ray generation in the laser-driven ablation structure of Au target. The broad band-structures in the spectrum have been inferred to be due to the bound-bound and free-bound transitions of the electrons in the M.N.O shells. However the discrepancy between experimental data and simulation result at $h\nu$=100∿300 eV still remains. It will be improved by incorporating more accurate processes in highly excited state of many electrons by which such low energy photons can be generated efficiently. The angular distribution of emission is not yet fully understood. A surrounding colder plasma might be also responsible for it.

The spectral dip around 400 eV was found to diminish at longer pulse. This tendency could not be satisfactorily reproduced by the simulation although the origin of this dip was clarified. Its detailed physical meaning is now under consideration.

Detailed information on soft X-ray emission reflects the transport of absorbed laser energy into the target plasma. It may become good diagnostic tool to investigate the energy transport into high-dense plasma region. The experimental results showed the lateral energy transport is negligible. But the emission level is sensitive to the axial energy transport inhibition factor f especially at a shorter pulse which produces steeper density gradient.

Z dependence of detailed specta in soft X-ray region has been experimentally obtained. It has been concluded that the undulation character of X-ray conversion originates in the ionization state. The simulation results explained well qualitatively the undulation obtained experimentally.

These results described above will give physical insight in X-ray generation dynamics in laser-produced plasmas and will provide suggestive information to producing stimulated emission and X-ray source for microlithography as well as for radiation drive scheme in inertial confinement fusion.

REFERENCES

1) W.C. Mead et al, Phys.Rev.Lett., 47, 1289 (1981)
2) P.D. Rockett, W.C. Priedhosky, D.V. Giovanielli, Phys.Fluid, 25, 1286 (1982)
3) H. Nishimura et al, Phys.Fluid, 26, 1688 (1983)
4) S.C. Mead et al, Phys.Fluid, 26, 2316 (1983)
5) K.M. Glibert, J.P. Anthes, M.A. Gusinow, M.A. Plamer, J.Appl.Phys., 51, 1449 (1980)
6) T. Mochizuki et al, Jpn. J.Appl.Phys., 22, L133 (1983), T.Yabe et al, ibid 22, L88 (1983)
7) K. Okada et al, Jpn.J.Appl.Phys., 22, L671 (1983)
8) R.H. Day, P. Lee, E.B. Saloman and D.J. Nagel, J.Appl.Phys., 52, 6965 (1981)
9) T. Yabe et al, Nucl.Fusion 21, 803 (1981)
10) S. Kiyokawa, T. Yabe and T. Mochizuki, Jpn.J.Appl. Phys., 22, L772
11) A. Nishiguchi and T. Yabe, J.Comput.Phys., 52, 390 (1983)

MEASUREMENTS OF PHOTOIONIZATION CROSS SECTIONS OF IONS IN THE EXTREME ULTRAVIOLET

E. Jannitti
Istituto Gas Ionizzati, Via Gradenigo 6/A, 35131 Padova, Italy

P. Nicolosi and G. Tondello
Istituto di Elettrotecnica ed Elettronica, Via Gradenigo 6/A, 35131 Padova, Italy

Wang Yongchang
Northwest Normal College, Lanzhou, The People's Republic of China

ABSTRACT

The photoionization cross section of Be^+ at 18.2 eV has been measured. The absorption spectrum has been recorded using two laser produced plasmas. The result is in good agreement with theoretical predictions. The absorption spectra of B^+ corresponding to transitions of the type $2s2p\ ^3P^\circ$-$2snd\ ^3D$ and $2s^2\ ^1S$-$2snp\ ^1P^\circ$ have been observed too.

INTRODUCTION

Only few absorption experiments with multiply ionized atoms have been made until now mainly using the technique of ionization of atomic vapours, excited by resonant absorption of tunable laser light, through inelastic collisions with free electrons[1,2].

Recently absorption spectra of multiply ionized species have been obtained by using two laser produced plasmas. This technique, first used by Carroll and Kennedy for studying the Li^+ spectrum[3], has been improved later by Jannitti et al[4]. They used two laser produced plasmas one acting as background continuum radiation source, the other as absorbing medium.

In the soft X-ray (grazing-incidence) region the absorption spectra of Be^{++} (resonance transitions, photoionization and autoionizing lines) and of Be^+ (inner-shell transitions) have been studied[4,5]. Also the photoabsorption spectrum of B^{3+} has been observed.

Here we report on measurements in the normal incidence region of the photoionization cross section of Be^+ $1s^2 2s \rightarrow 1s^2 + \varepsilon p$ and observations of the absorption spectra of B^+ showing transitions of the type $2s2p\ ^3P^\circ-2snd\ ^3D$ and $2s^2\ ^1S-2snp\ ^1P^\circ$.

EXPERIMENT

The sketch of the experiment is outlined in fig. 1.

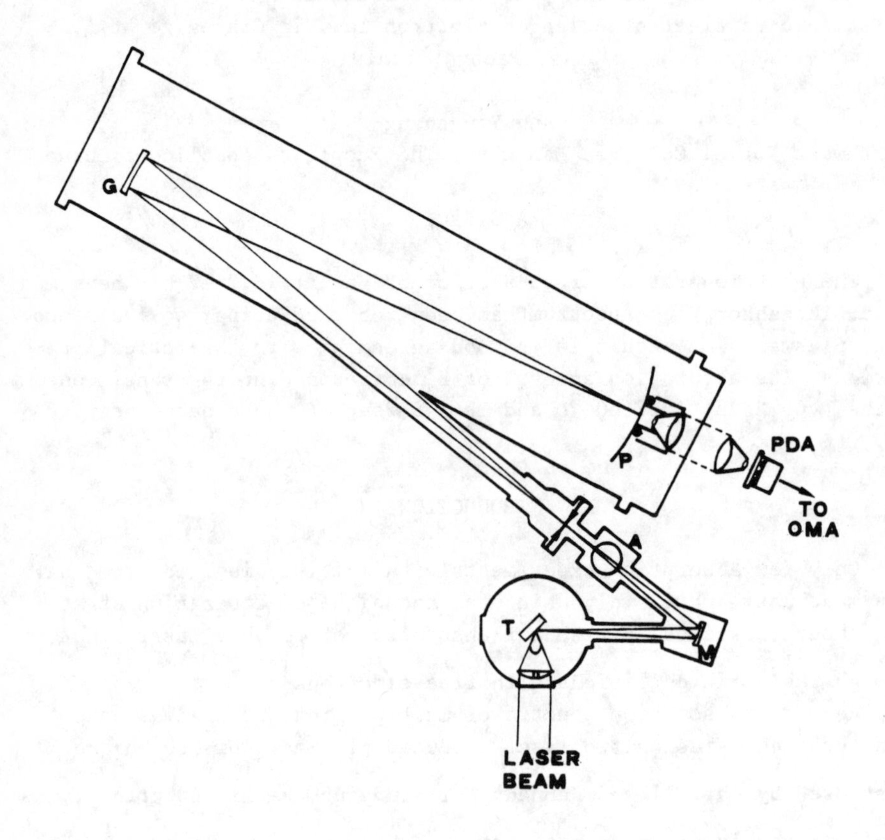

Fig. 1

A ruby laser (10 J, 15 ns) is split into two beams. One, carrying about 70-90% of the total energy, is focused on the plane solid target T, of high atomic number material (Cu, Pb), and generates a plasma, acting as background continuum source. The other, carrying the remaining energy, is focussed on the second target A, with an horizontal surface, generating the absorbing plasma. The time delay between the two beams and the power density on the target for the absorbing plasma can be suitably varied. The spectrum is recorded by a normal incidence spectrograph, fitted with a 600 l/mm, R=2 m grating G and working at variable incidence angle (10°-15°). The target T is oriented with its surface at 45° with respect to both the directions of incidence of the laser beam and that of the observer. In this way the dense and hot plasma region, corresponding to the critical surface, can be observed. The laser beam is focussed on the target A with a spherocylindrical lens linearly, increasing the depth of the absorbing plasma. The radiation emitted by the continuum source is collected with the toroidal mirror M, of radii R=636 mm in the plane of incidence and r=533 mm, whose first astigmatic image (vertical focus) is on the Sirk's focus and the second (horizontal focus) on the entrance slit of the spectrograph. The angle of incidence on the mirror is about 16°. The absorbing plasma is created in the position corresponding to the first astigmatic image of the mirror; consequently just a well defined region of plasma is crossed by the continuum radiation. By varying the distance from the target of this region and the corresponding delay between the two beams, the ionization and the density of the column of absorbing plasma can be suitably changed.

On the focal plane of the spectrograph, a movable trolley carries a scintillator coated plate (TPB). The latter is imaged via two high luminosity (N.A.=0.75) objectives on an intensified photodiode array which is connected to an optical multichannel analyzer. This improvement, with respect to photographic techniques, reduced significantly the duration of the experiment, in addition it provides: a detector of good linearity and high dynamic range; the possibility of visualizing the recorded image in real time and of storing it in a computer for later processing.

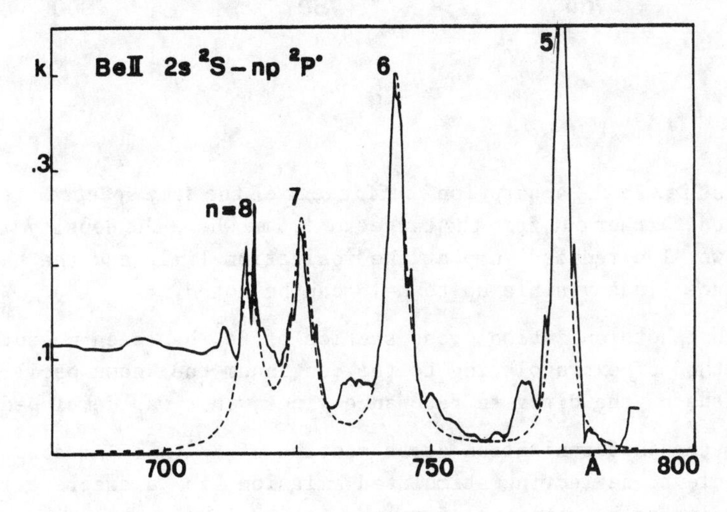

Fig. 2a

In fig. 2a the absorption spectrum of Be^{+}, recorded at $\sim$0.6 mm from the target, is shown. The measured absorption coefficient relative to some transitions of the resonance series ($1s^2 2s\ ^2S - 1s^2 np\ ^2P^\circ$ n=6,7,8) and the ionization limit is reported. It was not possible in these experiments to extend the measurements much beyond the ionization limit because of the quality of the experimental data. The absorption in this spectral region appears influenced by other contributions, that we think due to other photoionization transition like $1s^2 2p \to 1s^2 + \varepsilon d$ and $1s^2 2p \to 1s^2 + \varepsilon s$ of the same ion. However the threshold effect at the ionization limit is clearly visible.

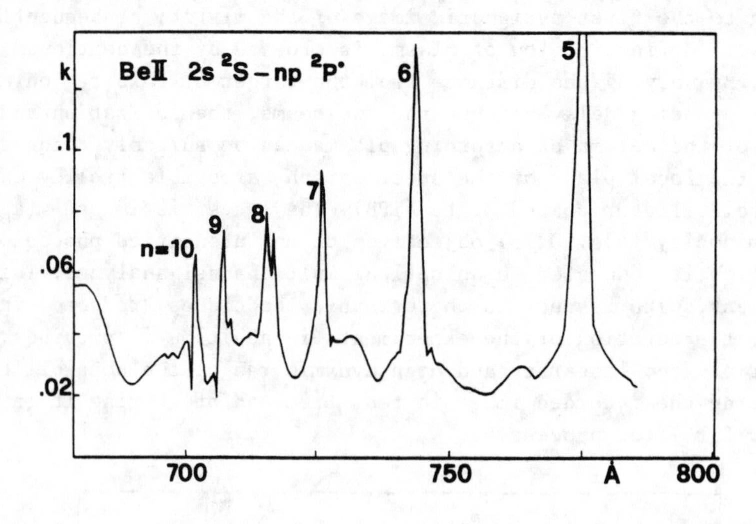

Fig. 2b

In fig. 2b the absorption coefficient of the same spectral region, but recorded farther out from the target at 1.5 mm where the density is lower is shown. The reduced jump at the ionization limit and the narrower resonance lines visible up to n=10 can be noted.

The photoionization cross section of Be^{+} has been measured with the method of extrapolating to the continuum the known oscillator strengths of the discrete resonance lines, that was developed for measuring the photoionization cross section of Be^{++} [5]. The absorption coefficient, neglecting stimulated emission, for a discrete line, corresponding to a transition from the ground level g to an excited level n, is given by[6]

$$k_d^t(\nu) = \frac{\pi e^2}{mc} f_{gn} \phi(\nu) \int_L N_g dl \qquad (1)$$

and that for the photoionization continuum by

$$k_c(\nu) = \sigma(\nu) \int_L N_g dl \qquad (2)$$

where: f_{gn} is the oscillator strength of the line; N_g is the ground state ion density and the integral is calculated along the plasma depth crossed by the continuum radiation; $\phi(\nu)$ is the normalised profile of the line and $\sigma(\nu)$ the photoionization cross section. We can see that by measuring $k_d^t(\nu)$, known f_{gn}, is possible to deduce the value of $\int N_g dl$ from the equation (1). Then $\sigma(\nu)$ can be derived from the equation (2) by measuring $k_c(\nu)$. On the other hand, it is well known that the measured absorption coefficient of a discrete line is affected by the instrumental response, so much that can be strongly distorted[7]. However by measuring carefully the instrumental function, as we did, we can determine the true $k_d^t(\nu)$, i.e. that we'd measure with an instrument of very high resolving power with respect to the line width.

The different steps of the described method are: a) a normalized voigt profile $\phi(\nu)$ is shaped with proper width for each line considered; b) $\phi(\nu)$ is multiplied by a proper coefficient obtaining the true $k_d^t(\nu)$; c) $I(\nu) \propto \exp|-k_d^t(\nu)|$ is convoluted with the previously determined instrumental function; d) a model dependent $k_d^m(\nu)$ is derived for the resulting spectrum and is compared with the experimentally measured one. The procedure stops when a good fitting of the measured $k(\nu)$ is obtained. The profile $\phi(\nu)$, which determines the line shapes, and the line density $\int N_g dl$, on which the peak value of $k(\nu_o)$ depends, are free parameters. Obviously the result is dependent on the quality of the experimental data e.g. if they have low noise and uncertainty.

In fig. 2a the dotted line shows the model dependent absorption coefficient of Be^+ as obtained with the method previously described and in fig. 3 the true absorption coefficient is shown. The agreement with the experimental spectrum is good, especially for the n=6

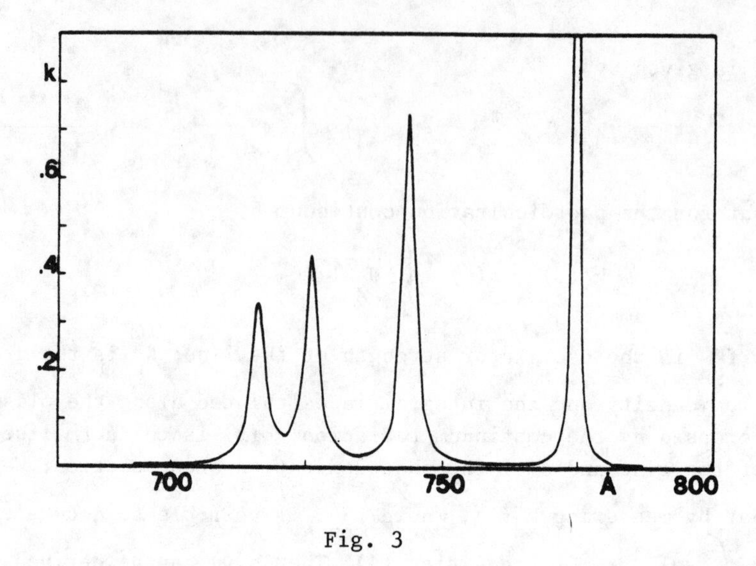

Fig. 3

and 7 lines but is just a little worse for the line $1s^2 2s-1s^2 8p$ which is greatly affected by uncertainties. The line shapes is well approximated with Voigt profiles of increasing width according to the n principal quantum number. The influence of the instrumental function, which has a profile of FWHM ∿1.2 Å, is notable. The resulting line density is 8.2×10^{16} cm^{-2}, consequently $\sigma(thres) = 1.4 \times 10^{-18}$ cm^2 is derived with very good agreement with the theoretical predictions of Reilman & Manson[8], giving $\cong 1.4 \times 10^{-18}$ cm^2 at 20 eV above the ground state. We estimate the accuracy of our measurement within 20%.

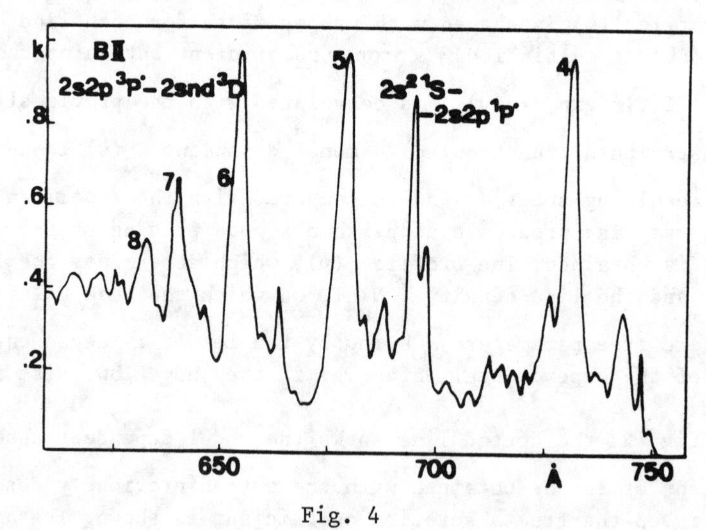

Fig. 4

At the moment only preliminary results of the absorption of B^+ can be reported. In fig. 4 the absorption spectrum between ∿610 Å and 760 Å is shown. We can see the lines arising from the 2s2p $^3P^o$ metastable excited level to the 2snd ^3D levels as well as one of the resonance series. The photoionization jump appears clearly at the limit of the series. A tentative identification of the lines is given.

REFERENCES

1. T.B. Lucatorto and T.J. McIlrath, Appl. Opt. 19, 3948 (1980).
2. T.B. Lucatorto, T.J. McIlrath, J. Sugar and S.M. Younger, Phys. Rev. Lett. 17, 1124 (1981).
3. P.K. Carroll and E.T. Kennedy, Phys. Rev. Lett. 38, 1068 (1977).
4. E. Jannitti, P. Nicolosi and G. Tondello, Physica C. to be publ.
5. E. Jannitti, P. Nicolosi and G. Tondello, Opt. Comm. submitted.
6. J. Cooper, Rep. Prop. Phys. 29, 35 (1966).
7. G.U. Marr, Plasma Spectroscopy, Elsevier (1968).
8. R.F. Reilman and S.T. Manson, Astrophys. J. Suppl. Ser. 40, 815 (1979).
9. K.L. Bell and A.E. Kingston, J. Phys. B4, 1308 (1971).

AMPLIFICATION OF SPONTANEOUS EMISSION IN ALUMINUM AND MAGNESIUM PLASMAS

P. Jaeglé, G. Jamelot, A. Carillon, A. Klisnick, A. Sureau, H. Guennou
Laboratoire de Spectroscopie Atomique et Ionique, Université Paris-Sud, 91405 Orsay and GRECO "Interaction Laser-Matière", Ecole Polytechnique, 91128 Palaiseau, France

ABSTRACT

An experimental method has been carried out in order to perform detailed investigation of electron level populations of multicharged ions in hot dense plasmas produced by lasers. This method provides an accurate diagnostic of the amplification of spontaneous emission (ASE) which may occur in plasmas at X-UV wavelengths as a result of population inversions. Results are presented for lithium-like ions of aluminum and magnesium. A strong departure from Boltzmann's population ratio is observed for several of the levels. Population inversion takes place between the 5f and 3d levels, the upper level being approximately three times more populated than the lower one, while the plasma is recombinating. The measured gain coefficient is about 1 cm^{-1} at 105.7 Å in aluminum for a 20 nanoseconde laser pulse and a 3 GW/cm flux density. Inversions are also observed for 3p - 5d transitions. However, the free bound transitions give rise to a spectrum of continuous absorption lowering the net gain, which can be even cancelled in some cases. A numerical model, using stationary as well as time-dependent solutions of rate equations, has been built. It accounts for the main features of experimental results and is very useful for settling the plasma parameters which are appropriate to a fair yield of ASE.

INTRODUCTION

Experiments performed in recombinating plasmas have shown, for various conditions of plasma production, that population inversions can occur between the levels n=3,4,5 of lithium-like ions [1 - 4]. When the evidence of inversion is supplied by the line intensity ratios along a Rydberg series, like for instance the 2p - nd series, there is no direct information about a possible amplification of radiation through the plasma. However, for the transitions of the ions being in the X-UV range, the research aiming to extend the lasers to very short wavelengths needs utterly reliable methods of measuring absorptions and amplifications of spontaneous emission at the wavelengths of interest. We describe such a method which has been used sucessfully with laser-plasmas of aluminum and magnesium. The time-integrated results which have been obtained regard, not only the transitions able to generate ASE, but also other important transitions of the same ions in such a way that a table of population ratios can be drawn up for comparison with Boltzmann's values. These results are of a great help for understanding the processes which are leading to the inversions. Preliminary time-resolved emission data have also been obtained in using a streak camera whith time-resolution about 100 ps. All the observed features are in good qualitative agreement with

0094-243X/84/1190468-12 $3.00

numerical calculations which include a large number of atomic data and plasma parameter values deduced from hydrodynamical simulation.

METHOD OF MEASUREMENT

As far as small gain is the same thing as a negative absorption, the occurence of ASE will provide a negative peak to the plasma absorption spectrum. However, to investigate X-UV absorption of hot plasmas cannot be done with an ordinary method because the plasma is an inhomogeneous self-emitting medium of very short life-time.

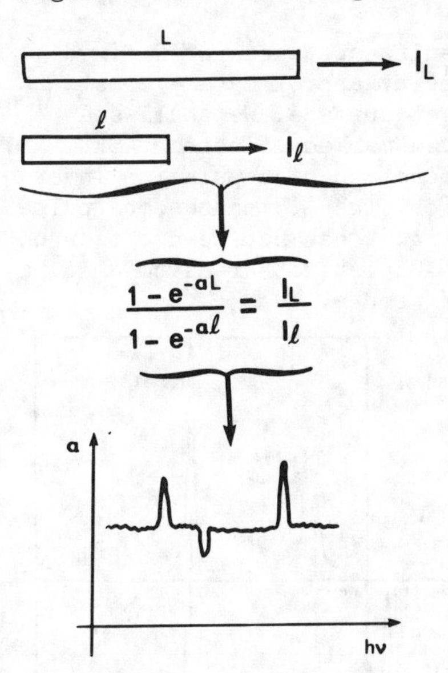

Fig. 1. Sketch of the experimental method of absorption measurements.

For plasmas produced by laser impact on massive target, it is possible to reach a reasonable homogeneity along a fixed direction in making a plasma column by means of a sphero-cylindrical focusing device. A way of working plasma emissivity is to compare the intensities in the axial and transverse directions in order that the change in radiation path length gives the wanted informations about the optical thickness. But the hydrodynamical expansion has different properties in both directions and space-time integrated measurements are not suitable for quantitative issues [3, 5]. Therefore it is better to keep the axial direction of the "column" and to change its length as shown in fig. 1. In practice, this is obtained merely by means of shields set before the focusing lenses; so, varying plasma length does'nt cause either an alteration of surface illumination or a change of target position. Moreover, to remove a possible inhomogeneous absorption in cold expanding shells, the plasma is free end toward the entrance of the detection system.

Let S be the source function of the plasma. From integrating the radiative transfer equation, the intensity emitted in the axis direction of a column of length x is found to be:

$$I(x) = S(1 - e^{-\alpha x})$$

Thus the absorption coefficient α can be deduced from accurate measurement of intensity dependence according to plasma length. The equation which is to be solved has the form:

$$\frac{1 - e^{-\alpha L}}{1 - e^{-\alpha l}} = \frac{I_L}{I_l}$$

where L and l are the two lengths for which the intensities are I_L and I_l. In principle, the larger the ratio L/l is, the better the accuracy of α's measurements is. Nevertheless the plasma length cannot be reduced too much; otherwise the expansion characteristics remain no longer the same and a lack of similarity between both "long" and "short" plasmas does happen. Generally we choose $L/l \simeq 2$.

The experimental device involves a grazing incidence X-UV spectrometer with an Optical Multichannel Analyser which enables to get direct digital recording of the spectra in a tunable wavelength band of width $\simeq 10$ Å.

POSITIVE AND NEGATIVE ABSORPTION LINES OF ALUMINUM LITHIUM-LIKE IONS

First we shall present a set of results obtained for a fixed laser-pulse duration and with a flux density approximately constant from a laser-shot to another on the target surface. We shall see below that the pulse duration has likely a marked effect on ASE, what would be not surprising for the production of population inversion being related to plasma recombination. The 3.5 nanosecond pulse duration used for the results shown in fig. 2 has not been chosen on the ground of its suitability for large ASE but rather as to allow a wide range of flux density with the laser we use.

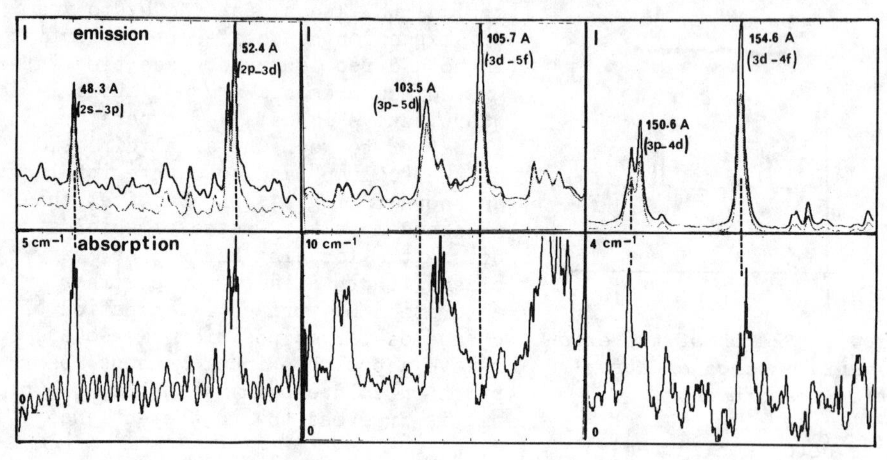

Fig. 2. Above, emission of aluminum plasma "columns" of $L = 8$ mm (solid lines) and $l = 4$ mm (dotted lines) in three wavelength ranges. Below, the absorption spectra deduced from I_L/I_l. Notice the negative peaks on the background absorption at 105.7 Å and, in a less extent, at 103.5 Å.

The figure shows the emission spectra of aluminum plasma columns of 8 mm and 4 mm length; the laser flux density in the focal spot was about 6 GW/cm. Three wavelength ranges, covering respectively the 2 - 3, 3 - 5 and 3 - 4 transitions of Al^{10+} ions, are displayed. Below each emission pattern one can see the plasma absorption spectrum obtained as explained in the previous section. The scale of absorption coefficient is given in cm^{-1}. It must be mentioned that the main experimental errors affect the position of the "zero" absorption, because, i) it comes mainly from weak intensities giving rise to a bad signal-to-noise ratio, ii) most of the errors about column lengths or intensity ratios result in vertical translations of the curves without significant distortion.

The 2s - 3p and 2p - 3d transitions, at 48.3 Å and 52.4 Å, exhibit pronounced absorption lines. The first provides us with valuable informations about the ground level population. In taking into account the transition probability, the line width and in neglecting in this case the upper level population, the absorption at 48.3 Å denotes a 2s population around 4.4×10^{16} cm^{-3}. The second is of interest for the population inversion processes because it is the main escape channel for the electrons cascading from the nf levels. In longer plasma columns, radiation trapping by 2p - 3d transitions could put down inversions with respect to upper levels.

The most interesting feature is seen by comparing both 3d - 5f and 3d - 4f transitions, at 105.7 Å and 154.6 Å: while 3d - 4f, like previous transitions, exhibits a positive absorption line, 3d -5f gives rise to a negative peak in the background absorption. This is the experimental evidence of population inversion between 5f and 3d levels, the observed pattern resulting from competition between ASE at 105.7 Å and larger continuous absorption likely due to free-bound processes. A similar feature is observed at the position of the 3p - 5d line, at 103.5 Å, which is hardly separated of other ion lines in the emission spectrum.

The population ratios deduced from absorption measurements and ground level population are presented in Table I. In addition to our

Table I: Level populations in lithium-like aluminum ions
T_e = 100 - 150 ev, N_e = 10^{19} - 10^{20} cm^{-3}

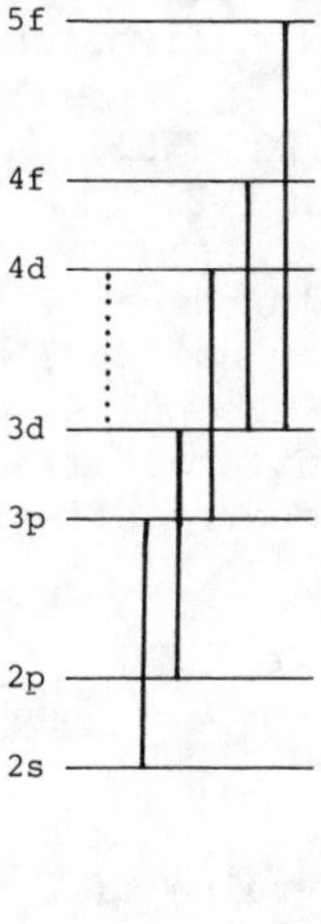

(Reduced Populations)	ETL	EXP
$\frac{N_{2p}}{N_{2s}}$	0.86	0.45
$\frac{N_{3p}}{N_{2s}}$	0.18	0.04 **
$\frac{N_{3d}}{N_{2p}}$	0.21	0.02
$\frac{N_{4d}}{N_{3p}}$	0.58	0.40 *
$\frac{N_{4f}}{N_{3d}}$	0.58	0.78
$\frac{N_{5f}}{N_{3d}}$	0.46	3.3

5f
4f
4d
3d
3p
2p
2s

* N_{4d}/N_{3d} from Kononov and Suckewer

** calculated

own data, we needed an estimation of the N_{4d}/N_{3d} ratio; we used a value of 1.8 in agreement with the inversions reported in ref. 1 and 4 from the 2p - nd emission line intensities. For comparison, the ETL value corresponding to a 150 ev electron temperature is given for each ratio. On the left of the Table, the vertical black lines recall the transitions whose the measured absorption is involved in the calculation.

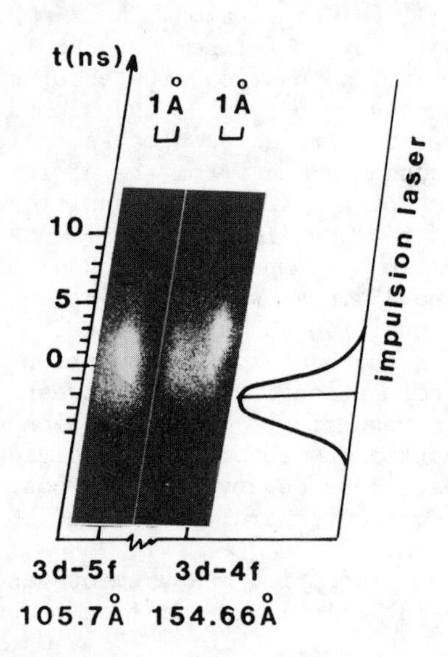

Fig. 3. Time-resolved emission of the 3d - 5f and 3d - 4f lines; the pulse of the laser is plotted on the right.

Considering the prediction of Boltzmann's law, the outstanding fact revealed by Table I is that, whilst both 4f and 5f levels have an extra-population with respect to the 3d level, a population inversion occurs only for the 5f level. The reason for this difference will appear below in discussing the results of a computational model of level populations. The fact is that the dominant processes populating 5f and 4f levels are not the same; in particular, they have not the same free electron density dependence. The difference appears in the time-resolved emission spectra we have recorded with a streak camera. They are shown in fig. 3. The two spots shown by the figure have the same time- origin, the laser pulse being represented on the right; both exhibit a line over a free-bound background. Comparing them, one sees that there is a time interval of 1.4 ns between the tops of 3d - 5f and 3d - 4f emissions. Only the 3d - 5f radiation starts decidedly before the end of the pulse rise, while the 3d - 4f emission lives on later.

VARIATION OF A.S.E WITH FLUX DENSITY AND DURATION OF LASER-PULSE

Now we consider the effect of the flux density and of the laser pulse duration on the negative absorption at 105.7 Å. Let E be the energy of a laser shot, τ, the width of the pulse, x, the length of the focal spot for which a constant transverse size ($\simeq 100\mu$) is assumed in all experiments, γ, the fraction of energy effectively deposited in the plasma. The flux density is defined as:

$$\Phi_{GW/cm} = \gamma \frac{E_{Joules}}{\tau_{ns} \times x_{cm}}$$

Keeping τ constant, we can see that the value of Φ has a marked influence on the behaviour of the 3 - 5 transitions, especially on the negative peak at 105.7 Å. For instance, fig. 4 shows the 0.6 GW/cm and

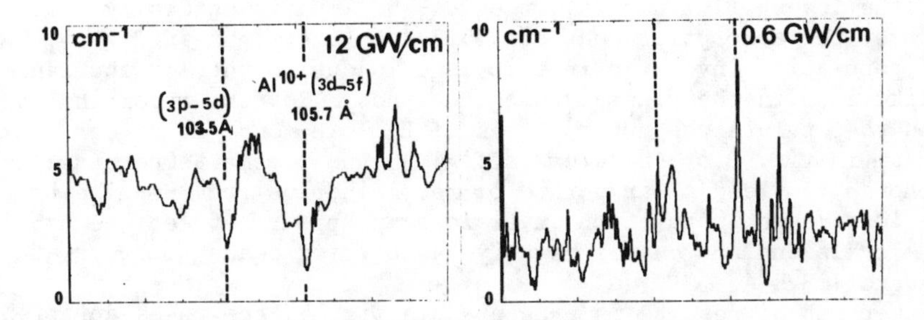

Fig. 4. These results belong to the same series as in fig. 2 but with density flux of 12 GW/cm and 0.6 GW/cm instead of 6 GW/cm.

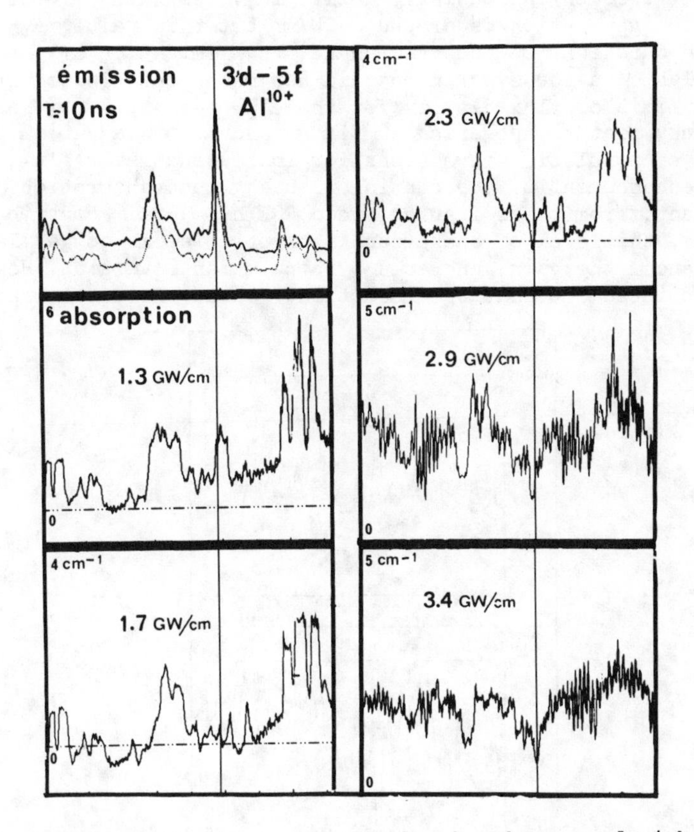

Fig. 5. For these results a 10 ns pulse has been used with an increasing energy. At low density of flux there is no population inversion (positive absorption); ASE contributes clearly to the spectrum at 2.3 GW/cm and more.

12 GW/cm results, the pulse duration being 3.5 ns like in fig. 2. In the case of the lowest flux density, the 3d - 5f line is optically thick, as shown by the absorption peak at 105.7 Å on the right in fig. 4. The lack of population inversion in this case is due to the low abundance of He-like ions in the plasma, which does not allow the recombination cascades to populate the upper levels of Li-like ions. At higher flux, as 6 GW/cm in fig. 2 or 12 GW/cm in fig. 4, the plasma temperature is large enough to yield a large rate of recombination from He-like to Li-like ions; then, population inversions can appear between the n=5 and n=3 levels and ASE causes negative absorption peaks to occur at 105.7 Å and 103.5 Å.

Another set of results is presented in fig. 5, the pulse duration being now 10 ns. It shows a similar dependance of population inversions upon flux density. When the flux increases from 1.3 GW/cm up to 3.4 GW/cm the 3d - 5f transition is firstly optically thick, afterwards optically thin, then it becomes increasingly amplifying. The threshold of population inversion generation is around 2 GW/cm, but this value may depend on the pulse duration. On the other hand, the limitation of the laser energy available in the experiments did not make possible to investigate an upper limit of flux density for these processes.

An unlucky feature appearing in fig. 4 and 5., in addition to the enhancement of population inversions for increasing laser flux, is the concurrent enhancement of the continuous background absorption. As a matter of fact, from these figures, when ASE has a fair contribution to the spectrum, the level of background absorption seems to get large enough to cancel the gain generated by population inversion. However, it must be pointed out that the uncertainty about the determination of

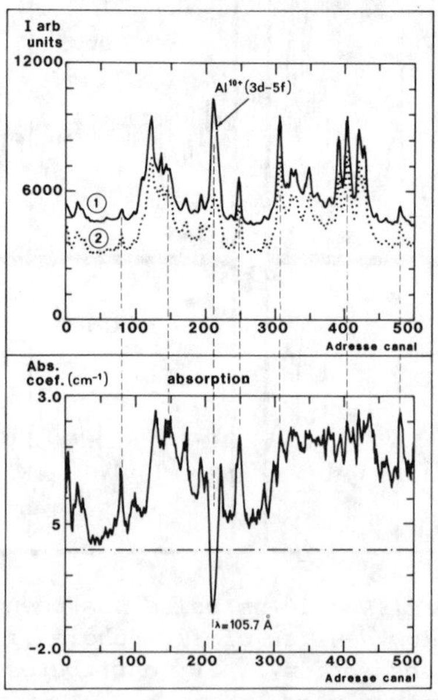

Fig. 6. Result obtained with a 25 ns pulse. The emission spectra for plasma columns of 7.1 mm (solid line) and 4.7 mm (dotted line) are shown in the upper frame. In the lower frame, notice the large peak of negative absorption at λ=105.7 Å. The flux density is about 3 GW/cm.

this level is much larger than about all other features of the spectra. Indeed, in studying the accuracy of the method described in this paper, it can be seen that most experimental errors result in moving the absorption spectra as a whole along the vertical axis with only weak changes of its shape. An improvement of background absorption measurements is needed for the next progress of this work, the main difficulty coming from the small continuous emission signal which is generally available.

Although no attempt could be made to calculate the rate of continuous absorption in our experiments, it can be assumed that, owing to the electron density range under consideration, the free-bound transitions have a large part in the full process. Their rate may depend on the parameters of the laser impulse to the extent that plasma temperature, ionisation and excitation are the functions of these parameters. In the present state of the work, long laser pulses appear to be more propitious to the generation of a net outgoing amplification. Fig. 6 shows a result obtained with a 25 ns pulse and a 3 GW/cm flux density. The absorption spectrum exhibits an outstanding negative peak at 105.7 Å. In this case the measured gain is well above both reabsorption losses and experimental uncertainties. The net gain coefficient is about 1 cm^{-1}.

LITHIUM-LIKE MAGNESIUM

In order to provide a wider support for interpreting the occurence of ASE in recombinating aluminum plasmas, we performed a few experiments in using magnesium instead. This must show if the observed features

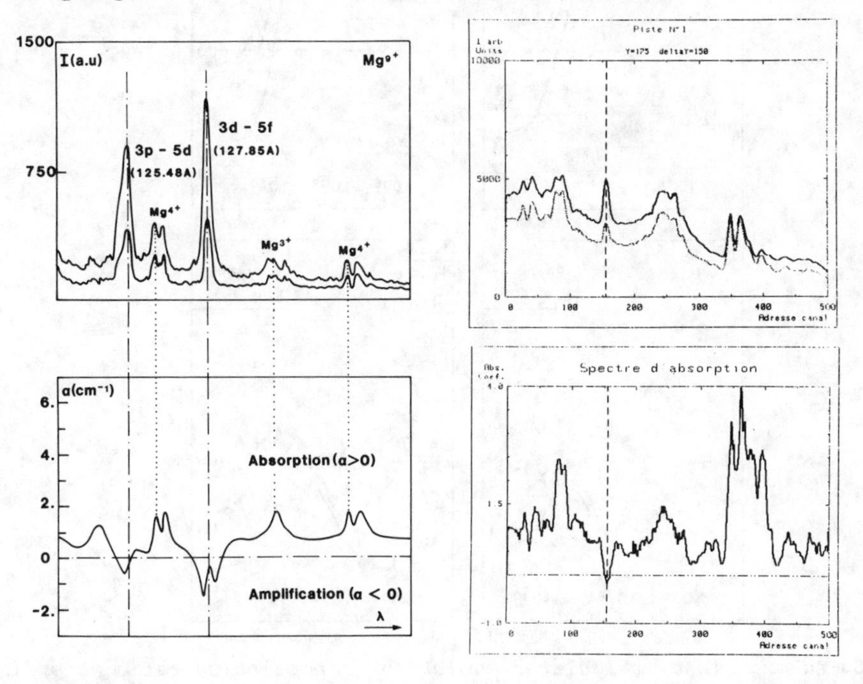

Fig. 7. ASE for 3-5 transitions of lithium-like magnesium. On the left: 6 ns pulse, 10 GW/cm, 250 μ from the target surface; on the right (with a shift of the detector to large wavelengths), 25 ns pulse, 1.4 GW/cm, 750 μ from the target surface.

either are specific of aluminum or can be found for other ions having a lithium-like electronic structure, as for instance Mg^{9+} ion has.Example of results are displayed in Fig. 7. The wavelengths of the 3p-5d and 3d-5f transitions of Mg^{9+} are now respectively 125.5 Å and 127.9 Å. The 3p-5d line is more easily observed than in the case of aluminum because it has less overlap with low Z's spectra. One can see that, with regard to the negative absorption, the behaviour of 3-5 transitions of magnesium is very similar to the one found for aluminum. This suggestes that the explanation of population inversions in these plasmas must involve lithium-like atomic data as an important factor.

RECOMBINATION MODEL ACCOUNTING FOR POPULATION INVERSIONS

For the inverted populations being set forth at levels of main quantum number as high as 5, numerical calculation of population rate in recombinating plasma must include a great deal of levels. We decided to take into account all the levels up to n=7, i.e. 49 levels on the

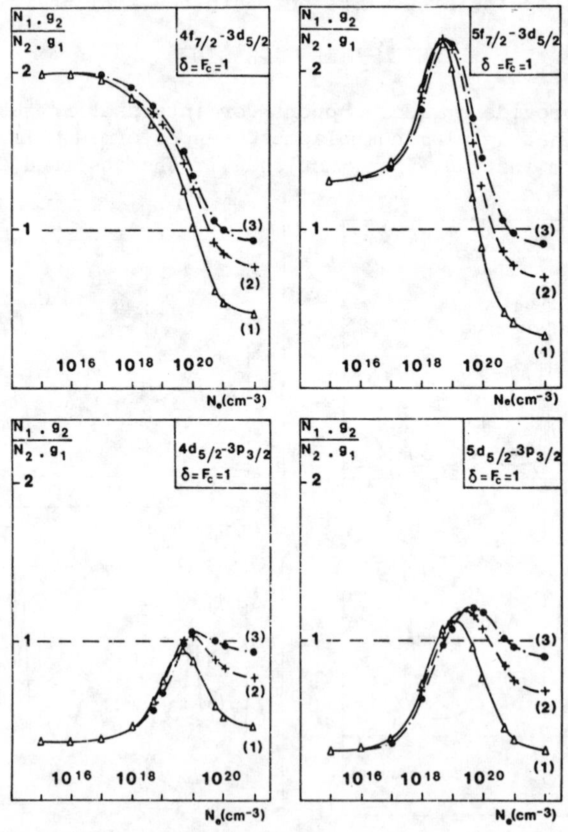

Fig. 8. An example of calculated evolution of population ratio according to plasma electron density in the case of 3-4 and 3-5 transitions of Al^{10+} at three temperatures: 1) 100 ev, 2) 300 ev, 3) 1000 ev; see text for details on atomic calculations and plasma model.

whole. At electron densities of interest, the lowering of the ionization limit makes it unnecessary to consider higher levels.

The first step of calculation has aimed to obtain the energy and the dipole matrix elements for all the levels from n=2 to n=7 of Al^{10+} ion. We used for that a method which gives good results in the case of ions of moderate atomic number ($Z > \simeq 40$) in sparing the long computation time needed in fully relativistic calculations. It consistes in including the term of spin-orbit interaction in the self-consistent processing of multiconfigurational Hartree-Fock equations. This allows to obtain directly the eigenlevels and the eigenenergies of the atomic system described in intermediate coupling. Details on this method are to be found in ref. 6,7.

A main result of calculations is the very short radiative life-time of the $3d_{5/2,3/2}$ levels. Its value is about 1 picosecond while it is at least ten times longer for most of the other levels. In particular, summing nf→3d and np→3d decay probabilities shows that radiative cascades would populate the 3d levels three times slower than they are depopulated by decaying to the $2p_{3/2,1/2}$ levels. Obviously this is a propitious circumstance for the 3d levels getting to be the lower levels in transitions with population inversions. It is to be noticed that the exceptionally fast decay of the 3d levels comes from the nature of the 3d and 2p wave functions which have no nodes along the radial axis. Thus this property is expected to hold true in all the lithium-like isoelectronic sequency.

The second step was to solve exactly the rate equations in the case of aluminum. The processes considered for this calculation are the radiative de-excitations as well as the excitations and de-excitations by electron-ion collisions. Optically forbidden transitions are not included. In order to build a radiativo-collisional equilibrium plasma model, we made the assumption that the higher excited Al^{10+} levels are in thermal equilibrium with the Al^{11+} ground level ("thermal band"), whereas the populations of all other levels but the Al^{10+} ground level result from radiativo-collisional cascades from the highest levels.

Then we have still to fix the ratio of Al^{11+} to Al^{10+} groundlevels. For the sake of convenience, its value has been expressed as a function of a single parameter, δ (see fig. 8), defined as follows: $\delta=1$ if the ratio has the value given by Saha-Boltzmann relation; δ increases linearly according to the departure to this heading value. In pure corona equilibrium, $\delta \gg 1$. However it should not be understood that $\delta=1$ does mean at all that the plasma is in L.T.E.; in fact, the total ion abundance must include the populations of excited levels and the Al^{10+} abundance generally won't obey Saha law for $\delta=1$. Realistic δ-values can be obtained from hydrodynamical computation including ionization rate equations.

The so-called thermal band consisting in the levels of main quantum numbers 6 and 7, and the 2s-ground level of Al^{10+} being fixed by the δ, N_e and T_e values, it remains to solve a system of 23 equations for achieving the detailed populations of the ion. Having in view time-dependent solutions, this is a system of coupled differential equations. From solving it, one sees that the time necessary to get stationary populations varies from 10 ps for $N_e \simeq 10^{20}$ cm^{-3} to 100 ps for $N_e \simeq 10^{17}$ cm^{-3}. These times being much shorter than the laser pulses, one only has to solve the stationary linear system deduced from the previous one.

Fig. 8 shows typical examples of the results obtained from these calculations. The curves represent the reduced-population ratios found for some of the 3-4 and 3-5 transitions at electron densities corresponding to the corona of laser-produced plasmas. The left part of the curves ($N_e \simeq 10^{16}$ cm^{-3}) comes virtually from the radiative cascades alone. As pointed out previously, the fast 3d decay makes it possible inversion to occur between the 3d and the 4f or 5f levels. In shifting to larger densities, one sees the effect of electron-ion collisions upon the population ratios. While the collisions destroy gradually the 3d-4f inversion, they enhance the 3d-5f inversion until a maximum occuring about 10^{19} cm^{-3}. Moreover, though radiative cascades prove to be unable to generate an inversion between the 3p and 5d levels, a small inversion is rising here as the mere consequence of collision enhancement.

Such an unexpected effect of collisions on population ratios, which is the outstanding feature of all the computational modelling, results from the balance of all the processes involved in the rate equations. However, in this model, the reservoir of electrons is the thermal band and a behaviour such as observed for the 3d-5f transition in fig. 8 reveals an especially large collisional coupling between this reservoir and the 5f levels. That the coupling is not so efficient in the case of the 4f levels can be understood from the fact that the energy gap from the thermal band is larger in this case; that is why the collisional population transfer from the thermal band has a much higher rate for n=5 than for n=4. The difference between the 4f- and 5f-population mechanisms is confirmed by the time-dependent calculations which show that the velocity of the 5f-population growing increases much more than the 4f does when the electron density goes from 10^{17} cm^{-3} to 10^{20} cm^{-3}.

We extended this study to a large range of δ-values. The result is that the above effect takes place, at current plasma temperatures, only if $\delta \lesssim 50$. In steady state plasmas this condition appears to be generally fulfilled at densities larger than 10^{20} cm^{-3}, i.e above the limit for population inversions (see fig. 8). However transient conditions in recombinating plasmas can provide the appropriate values of δ. This is shown by time-dependent ionization calculation coupled to the hydrodynamical code FILM(8). Numerical simulations performed for a 2 ns laser-pulse lead to δ=1-10 toward the last part of the pulse, that is to say in the recombinating stage of the plasma life. This result is to be compared to the time-resolved picture of the 3d-5f emission, presented in fig. 3, which shows the emission peak to take place far beyond the top of the pulse.

CONCLUSION

In this work we have given the experimental evidence and the theoretical interpretation of population inversions occuring between the levels of main quantum number 3 and 5 of lithium-like ions in recombinating plasmas. The new experimental method which is described makes possible direct measurements of amplification of spontaneous emission as well as contiguous background absorption. A large extension of this method to various studies of radiation transfer in plasmas is under consideration for next experiments.

The population ratio between the 5f and the 3d levels of lithium-like aluminum has been found to be about 3. The gain coefficient for the discrete transition alone is of the order of 1 cm^{-1} but, in most of the present experiments, the background photoabsorption has had a similar magnitude. Therefore the main issue for the prospect of X-ray laser feasibility is to achieve a positive balance of gain to losses in the plasma itself.

The authors acknowledge the partial financial support given to these researches by the D.R.E.T. under contract N° 79-387

REFERENCES

1. E. Ya. Kononov, K.M. Koshelev, Yu. A. Levykin, Yu. V. Sidel'nikov, S. S. Churilov, Sov. J. Quant. Electron., 6 (1976), 308.
2. P. Jaeglé, G. Jamelot, A. Carillon, Cl. Wehenkel, Japanese Journal of Applied Physics, 17 (1978), Sup. 17-2, 483.
3. G. Jamelot, P. Jaeglé, A. Carillon, A. Bideau, C. Möller, H. Guennou, A. Sureau, Proceedings of the International Conference on LASERS 81, p. 178, Carl B. Collins Editor, STS Press, Melean VA (1981)
4. S. Suckewer, C.H. Skinner, D. Voorhees, H. Milchberg, C. Keane, A. Semet, Princeton Plasma Physics Laboratory Report - 2003, 1983
5. P. Jaeglé, G. Jamelot, A. Carillon, Proceedings of the XVth International Conference on Phenomena in Ionized Gases, Vol. of Invited Papers, p. 25-33, Minsk (USSR), Juillet 1981.
6. A. Sureau, H. Guennou, Com. at Atomic Spectroscopy Symposium, N.B.S. Gaithersburg, 1975
7. H. Guennou, Thesis, May 24th 1983, Orsay (France)
8. J.C. Gauthier, J.P. Geindre, N. Grandjouan, J. Virmont, J. Phys. D 16 (1983) 321

POPULATION INVERSION AND GAIN IN EXPANDING CARBON FIBRE PLASMAS

G.J. Pert, L.D. Shorrock, G.J. Tallents
Department of Applied Physics, University of Hull, Hull, U.K.

R. Corbett, M.J. Lamb, C.L.S. Lewis, E. Mahoney
Department of Physics, Queen's University, Belfast, U.K.

R.B. Eason, C. Hooker and M.H. Key
Central Laser Facility, Rutherford Laboratory, Oxon, U.K.

ABSTRACT

Carbon fibres of a few microns diameter, heated by a Nd:glass laser pulse offer a suitable medium for the generation of gain in the XUV spectral region. Population inversion occurs as recombination is induced by adiabatic cooling in the expansion of the hot fibre. Experimental results and their interpretation using computer modelling have given a good understanding of the underlying physics, and show two regimes of operation: one of high gain places stringent limitations of the uniformity of illumination, the other of lower gain on the length of cylindrical focus. The design constraints on high gain and acceptable experimental tolerance are discussed.

CARBON FIBRE LASER SCHEME

The use of thin carbon fibre irradiated by cylindrically focussed Nd:glass laser radiation, and its harmonic as a working medium for laser action in the XUV has been investigated both experimentally and theoretically. In this system rapid cooling by adiabatic expansion of the fully stripped carbon plasma induces population inversion between the n = 3 and n = 2 levels of the hydrogenic like carbon ion C VI during the recombination cascade. In general the gain in such a system is small, limited by the optical trapping of the lower laser level n = 2. However, if the scheme is to have an application with current technology amplified spontaneous emission of travelling wave mode of operation is necessary, requiring a relatively high gain. This condition can only be achieved if the resonance line can be made optically thin at a reasonably high density.

In the carbon fibre system this handicap is mitigated by two effects. Firstly the carbon is initially fully stripped, and only relatively weakly ($\sim$ 10%) recombined at gain onset to reduce the Lyman α ground state population. Secondly the effective plasma width is kept small. This latter condition is achieved both directly by using fibres of small initial diameter, but more importantly due to the approximately linear dependence of the radial velocity, and the motional Doppler effect, which ensures that only a limited width of the plasma is in resonance at line centre.

0094-243X/84/1190480-09 $3.00

SCALING LAWS AND OPERATING CONDITIONS

It is clear from the above remarks that there is only a restricted region of parameter space in which high gain can be achieved. If the plasma is too cold severe recombination to the ground state gives rise to strong reabsorption of the Lyman α line; on the other hand if too hot recombination population of the upper states is weak. Thus we would expect that for constant plasma dimensions there is an optimum input energy for peak gain. Furthermore since the opacity limitation is weakened, both directly and due to larger velocity gradients, by smaller transverse dimensions, the gain may be increased by using smaller initial fibre mass per unit length, m. This behaviour has been confirmed by computer modelling of uniformly heated plasma cylinders[1], where it was found that the maximum gain scaled roughly as

$$G \sim m^{-17/22} \tag{1}$$

and the energy per unit length necessary to achieve optimum gain scaled as

$$E \sim m^{17/11} \tag{2}$$

for a constant heating pulse profile. At very low masses departures from these scalings occur when the expansion is too rapid to permit complete ionisation to be achieved: typically this limits the mass to that of a fibre of diameter greater than 1.5μm.

Practical realisation of such a scheme requires a heating pulse for the carbon of about 20j/cm in a time of 100ps for a mass of 3μm equivalent diameter. Such energy densities are readily achieved by laser irradiation using a mode-locked Nd:glass system focussed on to a line. The limited gain window in pump energy places limitations on the uniformity of the line focus. For a typical case the half gain width in energy allows a variation of ∿ 25% about the optimum. This is typical of the tolerance allowable in the focus for completely burnt fibre. In practice such uniformity is difficult to achieve with the simple two lens arrangements[2,3] which have been used up to now. In consequence experiments have been limited to fibre lengths of less than 3mm.

During the heating phase absorption of the laser radiation occurs at a density much below that of the solid. The fibre is thus surrounded by a tenuous, hot plasma blanket, which if the initial diameter is sufficiently small completely envelopes the remaining solid. Heat is transferred by thermal conduction radially into the solid thereby forming the hot plasma by ablation: due to the strong non-linear temperature dependence of the thermal conductivity there is a sharp interface between hot and cold components of the fibre[2]. Since the radius of the hot plasma is much larger than that of the initial fibre the hot plasma closely approximates that of a uniformly heated fibre, of the same mass. Following an incomplete burn we may identify two components namely the cold residual core, and the hot plasma; in the latter we may expect an expansion phase having inversion properties similar to those of equivalent uniformly heated fibre of the same mass and energy. Since the heat flow is strongly non-linear the temperature is nearly constant at a value sufficient to ensure complete ionisation in the hot plasma, independent of the fibre diameter, and laser pulse energy, and the ablated hot mass therefore scales roughly proportionally to the absorbed energy

$$m \sim E \qquad (3)$$

which may be compared with the condition for optimum gain (1). If these two conditions are simultaneously satisfied the system may be operated in an optimum mode although burn is incomplete, and indeed give higher gain at a lower energy.

The hot plasma blanketing the fibre plays two further important roles. Firstly the extended dimension of the blanket increases the geometrical overlap of the focus and enhances the absorption. Secondly the azimuthal symmetry of heating is improved to such an extent that for fibres of less than 6 μm diameter one-sided illumination is adequate to obtain a symmetric expansion[2]. To improve the early development of the blanket a two peak laser pulse is used with the first pulse approximately 20% of the second and earlier by between 200 to 400ps.

In an experiment the diameter of fibre used is limited by practical considerations such as visibility of alignment, stiffness, to not less than about 3μm. Such fibres may be relatively thick to heat uniformly, and a limited mass of carbon may be heated (hot plasma) leaving behind a residual core of cold dense material. Since the residual core is cold it remains at approximately the initial fibre radius, and plays no role in the hot plasma expansion. In this way a plasma of smaller effective mass than the fibre mass may be generated: however, in this case the plasma mass and energy are not independent, but the former is determined by the latter in some way depending on the time history of the heating. In this case also angular symmetry cannot be a priori assured, but must be obtained by the structure of the hot plasma itself[2]. Computer calculations allow the construction of a plasma mass/energy plot for given laser pulse characteristics. The general behaviour can be seen from the simple scaling laws (2) and (3). A mass/energy plot of equations (2) and (3) takes the form shown in fig. 1. It is clear that there are two intersections at which the mass and energy of ablated carbon matches the optimum gain condition: one at low mass and high gain, and the other at high mass, but low gain. In practice the mass is limited by the initial fibre mass, the complete burn case, and the high mass intersection

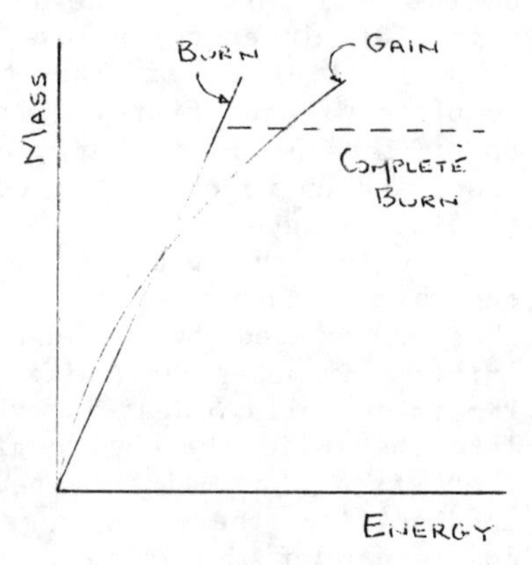

Fig. 1 Schematic representation of the mass/energy relationship for burn and peak gain of a thick fibre.

may not exist. Detailed computer calculations confirm this general behaviour. In a well matched design it is possible to overlap the two conditions over a range of energy within the deviation allowed by the window. In general, however, these are two zones one at low energy, low mass, but high gain and the second at high energy low gain corresponding to a complete fibre burn. Experiments have been performed in both regimes.

LOW ENERGY, LOW MASS EXPERIMENTS

In these experiments a single beam Nd:glass laser of about 10j in a pulse of 180ps duration and with a 20% prepulse 200ps in advance was used to irradiate fibres of diameter 2.5 - 5.0μm in a one-sided illuminated focal spot 40μm wide and 1.5mm long. As these experiments have previously been reported in detail in reference 2 we shall concentrate in this report on their significance in relation to the physical model described above, but including further data confirming the earlier measurements of gain and observations of absorbed energy and effective mass. The time averaged gain was measured by a two spectrograph system as before: absorbed energy by differential plasma calorimeters[3], and the mass from estimates of the mean ion energy given by ion probe signals and the absorbed energy.

The mass/energy diagram for this experimental configuration is shown in fig. 2. It can be seen that the experimentally measured masses are in reasonable agreement with the computationally derived values, particularly in view of a systematic uncertainty as to whether the energy of ionisation should be added to the experimental calorimeter signal. It can be clearly seen that the experimental data corresponds to an absorbed energy of about 1-3j/cm, and an absorption coeffi-

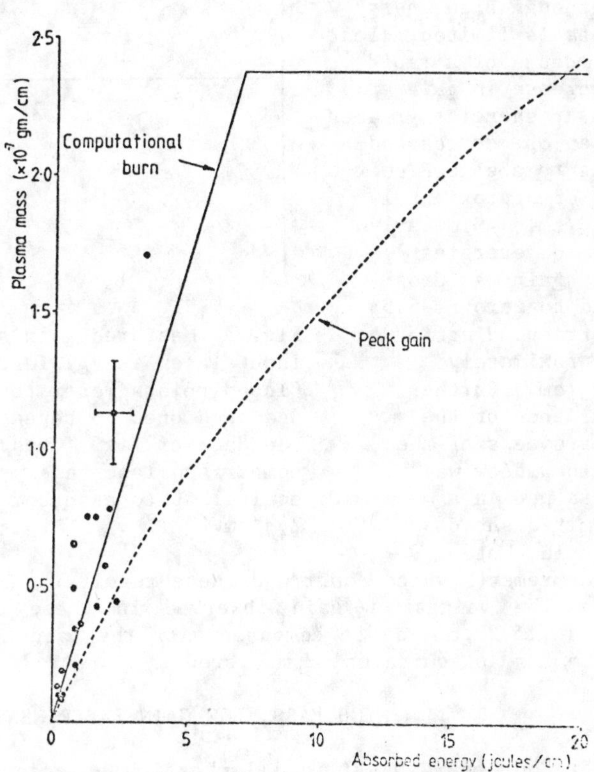

Fig. 2 Mass/energy plot for the low mass experiment. The computational data plotted as lines and the experimental burns the points.

cient from laser energy incident on the field to that absorbed of about 8%. The effective mass in the range (3-10) x 10^{-8} gm/cm corresponds to an equivalent fibre diameter 1.5 -2.5μm.

The optimum gain condition is also plotted in figure 2. it can be seen that the experimental conditions are in a reasonable match near the low mass intersection. However, as the energy is increased the burn and gain curves diverge rapidly and the experiments fall outside the gain window. At the low energy end the gain also falls off rapidly as the equivalent fibre diameter is less than the critical minimum of about 1.5μm.

The experimental measurements of gain shown in fig. 3 confirm this picture. Although high energy data is limited, clear evidence of a rapid decrease in gain at laser energies greater than 8j, corresponding to absorbed energy of approximately 2½j/cm. Similarly at low energies the gain has dropped to zero at 5.5j corresponding to approximately 1½j/cm. Further evidence of the narrowness of the gain window was obtained in a limited set of single shot measurements which confirmed these results.

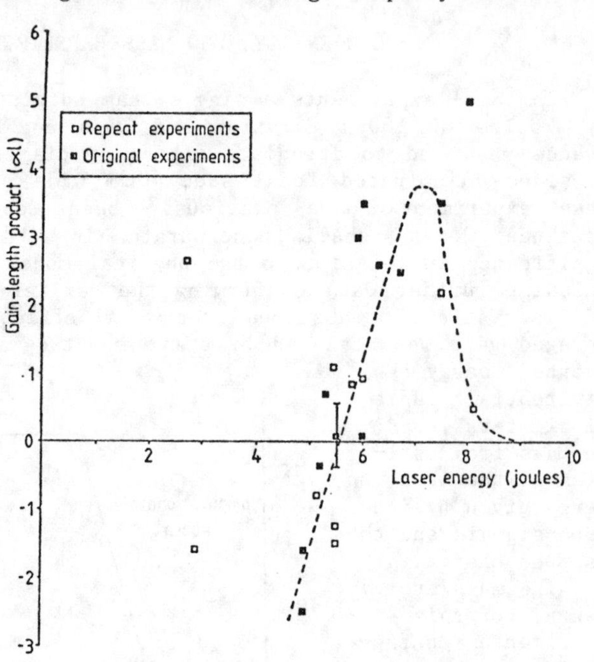

Fig. 3 Measured gain as a function of input laser energy in the low mass regime. Closed points refer to old data of ref. 2 and open ones to repeat experiments in the data of fig. 2 was obtained. For comparison laser energy of 8j is approx. equivalent to absorbed energy density of 2½j/cm.

The values of gain observed in these experiments, namely about 15-20/cm may be compared with the calculated value of about 7/cm for an equivalent 2μm fibre.

HIGH MASS, LOW GAIN EXPERIMENTS

Four beam symmetric illumination by second harmonic Nd:glass radiation of about 30j in a pulse of 200ps duration with a 200% pre-pulse of 200(400)ps spacing was used to illuminate 4(5) μm fibres over a region 1.5(1.8)mm long[3]. Two sets of experiments were carried out, one reports direct measurements of the mass/energy plot, and the other two spectrograph observations of gain.

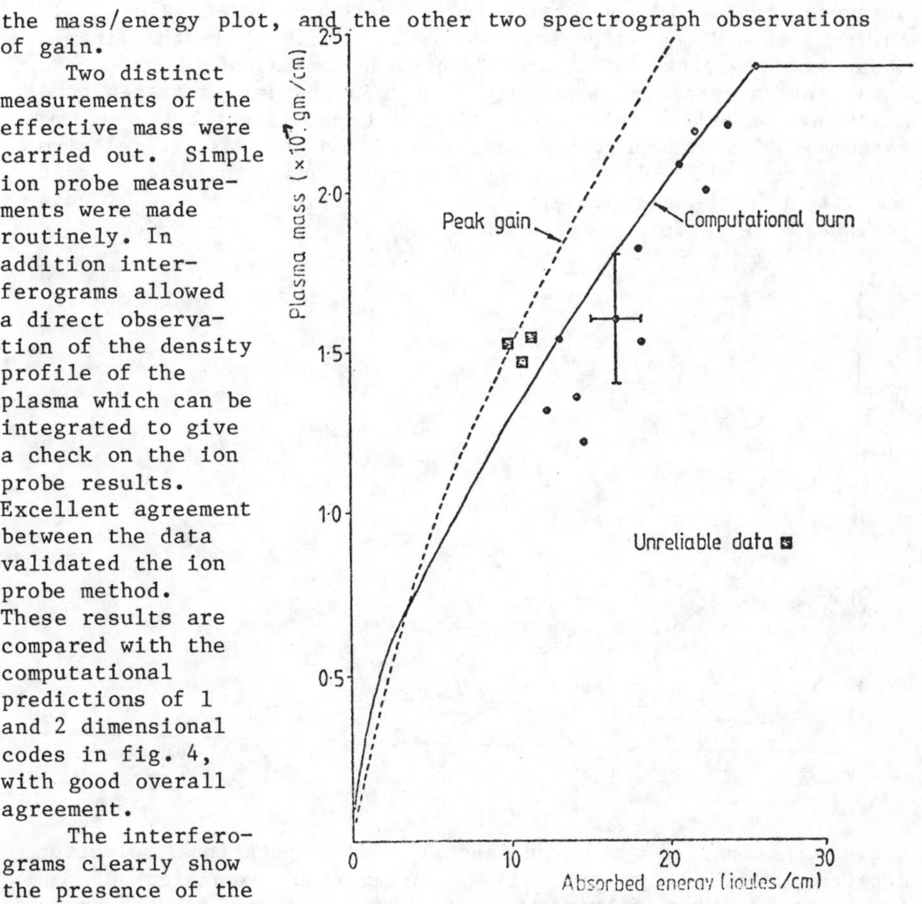

Fig. 4 Mass energy/plot for the high mass interferometry experiment: the data is presented as in fig. 2.

Two distinct measurements of the effective mass were carried out. Simple ion probe measurements were made routinely. In addition interferograms allowed a direct observation of the density profile of the plasma which can be integrated to give a check on the ion probe results. Excellent agreement between the data validated the ion probe method. These results are compared with the computational predictions of 1 and 2 dimensional codes in fig. 4, with good overall agreement.

The interferograms clearly show the presence of the cold dense core, whose radius is much less than that of the expanding plasma. Density profiles measured from two shots with very similar effective mass and absorbed energy at different times are compared with the calculated distribution predicted by the 1 dimensional code in fig. 5. When allowance is made for the uncertainty of the timing zero of the laser pulse, the agreement is most satisfactory.

In addition the overall shape of the density distribution can be reasonably represented by a Gaussian profile, as assumed in the similarity model used in the gain calculations, and the values are consistent between experiment and model. As may be inferred from fig. 4, the gain in these experiments is not optimised. Calculations give a value of $1 cm^{-1}$ under these conditions, yielding a gain-length product ($G\ell$) of only 0.15. This value is too small to be detected by a time averaged two spectrograph experiment. Indeed the measured axial/transverse H_α intensity ratio is approximately 0.9, the optically thin signal being stronger, and is in good agreement with the computed value. In any expansion H_α emission takes place initially while the line is absorbed (negative gain) and transverse emission is stronger. Axial intensity enhancement only occurs in the latter stages, and must have a sufficiently large ($G\ell$) if amplification is to be measured. Although the precise condition depends on relative timing of the H_α emission, calculations indicated that amplification is only observed if $G\ell \geq 0.5$: once beyond this threshold the ratio increases rapidly.

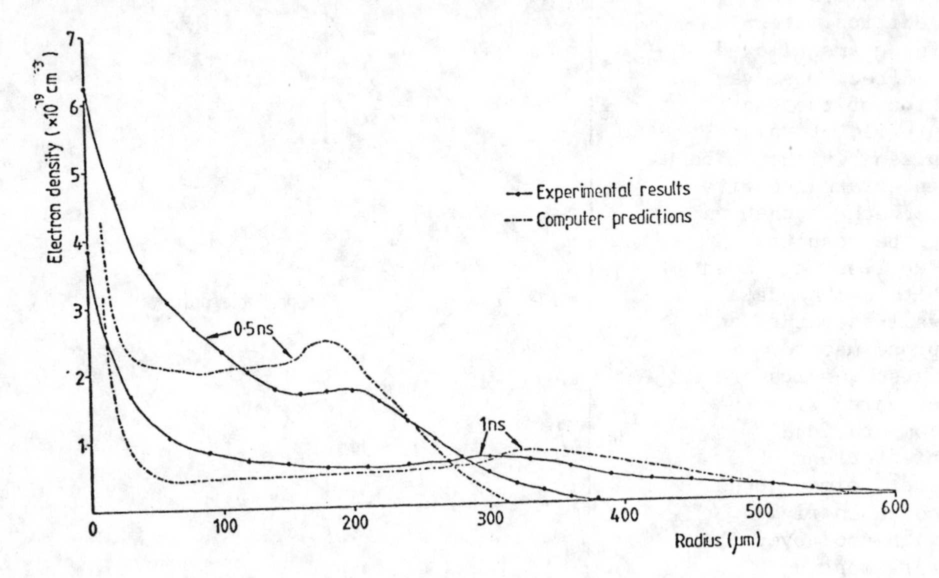

Fig. 5 Comparison of experimental and computational electron density profiles at two times during the expansion of an incompletely burnt fibre with effective mass 1.5×10^{-7} gm/cm and absorbed energy 13j/cm.

Observations of gain in a two spectrograph experiment in the high mass regime have been made with a focal spot length of approximately 3.3mm and prepulse interval of 400ps[3]. In this case a better match of the burn and gain curves is obtained (fig. 6). The predicted value of gain at optimum is about 3.5cm^{-1} giving ($G\ell$) $\sim$ 1 consistent with the general observation of an axial/transverse signal ratio of between 0.9 and 1.4. Unfortunately no measured effective mass data was obtained in these experiments.

CONCLUSIONS

In a thick fibre heating only a fraction of the carbon is heated and contributes to the expansion in which the population inversion takes place. The region of gain is thus a relatively long cylindrical ring whose centre is a cold dense inactive core. The hot plasma mass/energy relationship may be computed by hydrodynamic modelling and checked experimentally. It is found to depend mainly on the laser pulse characteristics. We may superimpose on this the mass/energy plot for uniformly heated plasmas giving optimal gain to identify those regions in which gain can be expected. In general there appear to be two such zones - one at low energy, low mass and one at high, the latter usually corresponding to complete burn.

The highest gains will clearly occur in the low energy region, and have been experimentally observed[2]. However, the window of operation is extremely narrow, as an increase in energy rapidly leads to a region of absorption. This will place stringent limitations on the design of a practical working laser, and recent work[3] has therefore been directed at lower gain, larger energy and mass systems. In these it is difficult to measure gain in a time integrated experiment unless the fibre is about 1cm long.

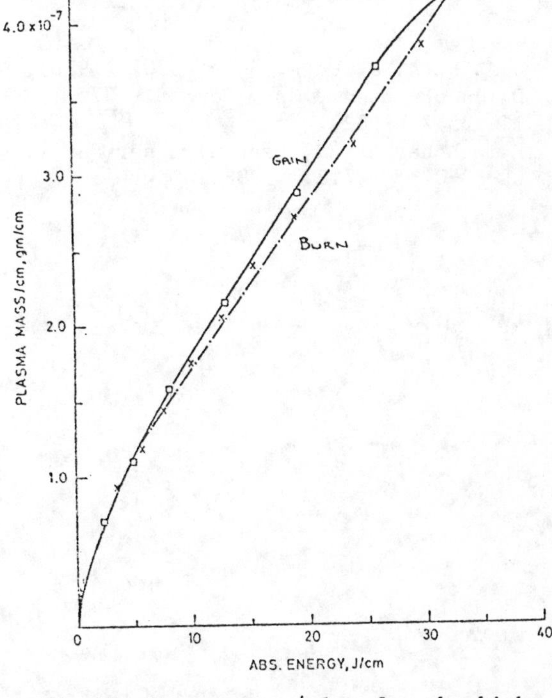

Fig. 6 Mass energy/plot for the high mass gain experiment: the data is presented as in fig. 2.

These constraints can be relaxed by opeating in the incomplete burn regime provided a good match between the burn and gain curves can be found. Computer simulations show that the position of the high mass intersection (fig. 1) is strongly dependent on laser pulse timing, in particular the prepulse interval. Using a longer delay than hitherto two intersections can be found such that over the interval between them the two curves match within the gain window (fig. 6). In this way the stringent restrictions on line uniformity associated with the high gain regime can be relaxed and a practical system of reasonable gain constructed without excessive pump requirements.

ACKNOWLEDGEMENTS

Thanks are due to D. Jacoby, J. Lawrence (low mass) and S. Knight, F. Pinzong and W.T. Toner (high mass) for help in performing these experiments; the assistance of the laser operators at R.A.L. is also gratefully acknowledged. The programme is supported by S.E.R.C.

REFERENCES

1. G.J. Pert, J. Phys. B, 9, 3301 (1976); 12, 2067 (1979).
2. D. Jacoby et al, Opt. Commun., 37, 193 (1981); J. Phys. B, 15, 3557 (1982).
3. E. Mahoney et al, Central Laser Facility Annual Report RL-83-043, p.7.2, (1983).

SPECTROSCOPY OF MATCHED-LINE PUMPING FOR X-RAY LASERS*

R. C. Elton, R. H. Dixon, and T. N. Lee
Naval Research Laboratory, Washington, DC 203375

ABSTRACT

Various conditions necessary for achieving population inversion and gain in the x-ray spectral region by matched-line photon pumping in plasmas are discussed, and some promising test cases are listed. For representative pump lines, measured intensities at the saturation level are reported. For a most promising sodium/neon combination test case two methods of neon plasma production are described, one using frozen layers and the other a low pressure gas. Spectroscopic results for highly-stripped neon ions are reported. Other combinations discussed include oxygen/nitrogen and aluminum/magnesium ions, the former being appropriate for understanding collisional/radiative effects on the pumping of population inversions at high quantum numbers.

I. INTRODUCTION

Ion lasers for the extreme ultraviolet and x-ray spectral regions require a very high density of inverted ions, because of the decreased efficiency of pumping and of reflecting cavities.[1] Such an amplifying medium is found in plasmas formed at high temperatures from moderate-Z atoms which are stripped of outer electrons until fairly simple configurations remain. Optical transitions are favored for achieving quasi-cw operation and for avoiding rapid Auger population-depleting transitions typical of more traditional innershell x-ray sources. To the extent possible, limiting the plasma to the lasant ionic species and maximizing the density of ions with the desired population inversion becomes very important; because "unused" ions tend to become absorbers and also emit resonance radiation that can deplete the population inversions by pumping the lower laser state. Also, in forming each ion multiple electrons are produced which, in abundance, can lead to collisional depopulation and eventually elimination of inverted population densities. Obviously, the generation of excess ions is inefficient; and the pump power density needed to maintain the population inversion against radiation losses demands state-of-the-art technology for lasing in the nanometer wavelength region.

Photon pumping of a single quantum level by a narrow band of x-radiation is therefore a very attractive approach to creating the desired lasing conditions. The narrow-band source can be a filtered continuum; however, intense line emission from a source ion at a wavelength coincident with an absorbing transition ending on the upper laser state in the lasant ion is more efficient, when

*Supported by the Office of Naval Research.
0094-243X/84/1190489-07 $3.00 American Institute of Physics

appropriate matches occur.[2-4] Ideally, the wavelengths of the source and absorbing transitions, as well as the line widths, should match exactly. However, a narrow pump line can pump a homogeneously broadened excited state from a ground state, and a portion of a broad pump line can simulate a narrow band of "continuum". Also, central wavelength mismatches can be adjusted by Doppler shifts if two plasmas with relative motion are used.

In this scheme a good wavelength match is a necessary but not sufficient condition for x-ray laser pumping. Many other considerations, some mentioned above, enter into the selection of a potentially successful combination of pumping and absorbing ions. A high pumping rate requires that the concentration of available energy into the pumping line and the coupling of this to the absorbing ions be very high. Intense resonance lines of one, two and three-electron ions in very high density plasmas are prime candidates as pump sources where matches can be found, and a selection of a few combinations with particularly close wavelength matches ($\Delta\lambda/\lambda_p$) taken from a list[4] of promising combinations is given in Table 1.

Table 1. Some promising line matches for 1-, 2-, and 3-electron pumping*

	Δn_p		λ_p (nm)	$\Delta\lambda/\lambda_p$ x10^4		Δn_a	Δn_L	λ_L (nm)	Ref.
O (H-)	2→1	+	1.9	2.5	→ N(H-)	1s→7p	7→2	8.1	4
K (H-)	2→1	+	0.33	2.7	→ Cl(H-)	1s→4p	4→2	1.9	2,3
Na (He-)	2→1	+	1.1	1.6	→ Ne(He-)	1s→4p	4→2	5.8	2,5
Al (Li-)	3p→2s	+	4.8	0.42	→ Mg(Be-)	2s→4p	4→3	20	6

*The subscripts p, a and L refer to the pumping, absorbing and lasing transitions between the pumped and the lowest levels, respectively.

That resonance line emission from such ionic species can indeed be generated in laser-produced plasmas has been recently demonstrated by the measurement of line-center emissivities approaching unity for various pump candidates.[7]

Generating sufficient pump emission for saturation on such resonance lines requires very high densities and temperatures, such that thermal equilibrium is reached. Such equilibrium is incompatible with inversion conditions in the lasing plasma, which must therefore exist at a lower density and perhaps a different temperature; hence, dual plasmas in close proximity must be produced.

A prime combination that has emerged as a leading test case for matched-line photon pumping in the x-ray region is the helium-like sodium 2p→1s Kα line pumping the helium-like neon 1s→4p Kγ absorbing transition (Table 1), with potential lasing at λ_L = 23, 7.8 and 5.8 nm on 4→3, 3→2 and 4→2 transitions, respectively. All transitions are strong, the wavelength match is good and known to very high precision,[8,9] and the elements are closely matched in nuclear charge, a consideration for adjacent plasmas. For the 4→3 transition at least, the upper laser state is not photoionized by laser emission, which is a decided advantage for initial threshold experiments. Formation of a suitable neon plasma is a problem addressed below.

Another promising shorter λ_L combination listed in Table 1 is hydrogenic potassium and chlorine with λ_L = 6.5 and 1.9 nm for 4→3 and 4-2 lasing transitions, respectively; however the plasma temperature required is just beyond the state of the art at present. A third possibility listed in Table 1 with a close match of transitions involves hydrogenic oxygen and nitrogen; however the n=7 pumped state is subject to collisional depletion, so that this may be most useful as a test case for fluorescent studies directed towards minimization of such collisional/radiative effects. Finally, the lithium-like pump example in Table 1, namely the aluminum/magnesium combination, reportedly[8] has closely matched wavelengths and convenient elements, if 4-electron ground state magnesium ions can be created in usable concentrations.

II. Ne^{8+} PLASMA EXPERIMENTS

A neon plasma suitable[5,14] for a lasant medium should have a temperature of $kT \simeq 100$ eV and an ion density of $N_i \simeq 10^{19} cm^{-3}$. Because measurements are to be made in the vacuum-uv region, the ambient pressure must be typically $\lesssim 1$ Torr ($= 3 \times 10^{16}$ cm^{-3}). The neon may be enclosed at higher gas pressures (as in a pellet) or solidified as described below. Neon ions can also be implanted in host materials, but a low concentration near the surface from which the plasma is generated may be harmful for high gain, as discussed above.

1. Frozen Neon Layers

At temperature below ~ 25 ^oK, neon can be condensed onto a surface in a frozen layer and used as a source for generating a plasma, perhaps by laser irradiation or other methods of pulsed power delivery. We have recently demonstrated[4,10] that such a 2 cm diameter layer can be formed at the tip of a cryogenically-cooled refrigerator (see Fig. 1), and maintained in quasi-equilibrum against sublimation at an ambient pressure of ~ 10^{-5} Torr for at least 10 minutes. A 20 J, 10 ns Nd-glass laser pulse focused to a flux density of 10^{12} W/cm^2 irradiates the frozen neon layer. A portion of a Ne VI-Ne IX typical spectrum obtained is indicated by the microdensitometer tracing shown in Fig. 2.

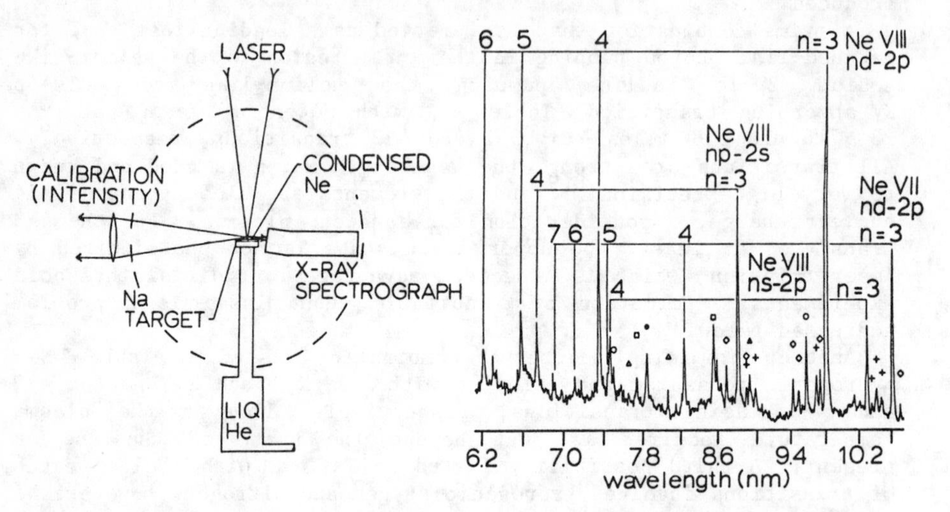

Fig. 1. Schematic of cryogenic neon target and laser assembly. Intensity calibration is by the branching ratio technique (Ref. 14).

Fig. 2. Neon spectrum from a laser-heated frozen layer. The solid dot refers to the 3d→2p Ne IX line. Open symbols and +'s refer to Ne VII and Ne VI lines, respectively.

Indication of the existence of some Ne^{8+} helium-like ions is obtained from the 3d→2p lines at 7.8 nm wavelength. The frozen layer is formed at 1.6 atm pressure and then exposed to the laser under vacuum, a procedure necessary to minimize the surface condensation of impurities such as nitrogen, oxygen and water vapor. We were also able to control the thickness such that a composite plasma of neon and the substrate material could be obtained; however, such a layered target combination would probably not be as versatile for optimizing the combination of conditions needed for laser pumping as would be adjacent materials. Because the neon does not form as readily on thermal insulators as on good conductors, it should be possible to condense neon onto a limited area using a good conducting substrate, along with segments of a poorly conducting sodium compound (such as sodium fluoride). Indeed, such composite targets are now under study for use with a larger dual-beam laser for gain tests.

2. Ambient Gas Ionization

Gaseous neon at a pressure of $\lesssim 1$ Torr is sufficiently transparent for vacuum-uv measurements over a path length of a few centimeters, e.g., to the entrance slit of a good vacuum spectrograph.[11] While the corresponding density is too low for significant gain in neon over a short length,[5,14] pumping can be studied in fluorescence if the Ne^{8+} ions are present in the ground state. We have been able to generate excited Ne^{7+} ions with a 8 J, 10 ns Nd-glass laser focused to a flux density of 4 x 10^{11} W/cm^2 on a graphite target. The spectrum resembles that in Fig. 2, along with the CV and CVI lines from the target. It is observed from space resolved (along the laser axis) spectra that the neon spectral lines both in the extreme-uv and the 200 nm near-uv regions do not exhibit the blast wave characteristics typical of the carbon ion lines. From our experience,[11] such ambient gas ions ionize almost immediately with onset of the laser pulse, so that ionization takes place either at the target region or at greater distances by fast ($v \gtrsim 10^8$ cm/sec) electrons or photons.[11,12] It is non-trivial to estimate ionization times, since innershell ionization followed by autoionization processes may be involved for ions which typically are slow to ionize from outermost shells. Further experiments are underway to understand the ionization mechanisms taking place and to optimize the formation of Ne^{8+} ground state ions for photon pumping from Na^{9+} radiation generated at the target. Should higher ambient pressures be required such that vacuum-uv measurements become difficult, it is also possible that enhanced n=4 population density with photon pumping can be detected with n=4, $\Delta n=0$ lines in the infrared spectral region.

III. $O^{7+} \rightarrow N^{6+}$ COMBINATION

The hydrogenic oxygen-pumping-nitrogen combination shown in Table 1 can be tested at lower laser power, initially at least in fluorescent analysis. Enhancement of the n=7 population density could be detected directly on the 7→6 radiating transition at 252 nm wavelength.[13] This spectral line is similar to the 343 nm line in hydrogenic carbon which is notably intense[14] in recombining plasmas and characterized by a large Stark width,[15] very useful for electron density diagnostics. Both elements can readily be produced at a cold tip as described above for neon or are also conveniently available in compound form for solid targets (e.g., boron nitride and beryllium oxide). Also, as described above for neon, the nitrogen can be provided as an ambient gas at a modest pressure.

IV. $Al^{10+} \rightarrow Mg^{8+}$ COMBINATION

As shown in Table 1, this combination involving lithium-like aluminum pumping with a 3p→2s resonance line the 4-electron beryllium-like magnesium ion has an excellent reported[8] wavelength match. The emission from the pump line at least approaches[7] that of one- and two-electron ion resonance lines, and the materials are convenient and of similar nuclear charge, all advantageous. We have viewed each of these plasmas spectroscopically from laser-heated (25 J, 10 ns, 10^{12} W/cm^2) targets. The aluminum pumping line is reasonably intense, as expected, but the 4p→2s Mg IX absorbing transition line is quite weak, although measurable. It is not unusual in time-integrated exposures of transient plasmas for spectral emission from ions with more than three bound electrons to be low relative to that for the higher stages. More likely, the lithium-like intensities are enhanced by recombination population of excited levels following electron collisional recombination from abundant helium-like ground state ions. We also reduced the laser energy by factors of 2 and 3 times in an attempt to reduce the temperature and increase the Mg^{8+} population; however the line emission did not increase.

V. CONCLUSIONS

Some promising fluorescence and gain experiments are described for matched-line photon pumping of plasma x-ray lasers, and preliminary spectroscopic results are reported. The combinations are chosen both to be within the state of art conditions for reasonable experimental progress and to provide the most promising known matches of transition energy for true test cases. The sodium/neon combination experiments will hopefully yield significant gain in the near future, while the oxygen/nitrogen experiments should shed light on the collisional and radiative processes controlling excited state populations, for which rates such as collisional ionization and recombination are not yet measured in high density plasmas.

REFERENCES

1. R. W. Waynant and R. C. Elton, Proc. IEEE 64, 1059 (1976); R. C. Elton, Optical Engr. 21, 307 (1982); R. C. Elton, "X-Ray Lasers," in Handbook of Laser Science and Technology, Vol. 1, Chapt. 4, M. J. Weber, ed. (CRC Press, Inc., Boca Raton, FL, 1982); R. C. Elton, Comments At. Mol. Phys. 13, 59 (1983).
2. A. V. Vinogradov, I. I. Sobel′man and E. A. Yukov, Sov. J. Quant. Electron. 5, 59 (1975).
3. R. C. Elton, ed., "ARPA/NRL X-ray laser program," Naval Research Laboratory Memorandum Report No. 3482, pp. 92-114 (1977).
4. R. H. Dixon and R. C. Elton, J. Opt. Soc. Am. B, Optical Physics (1984) (in press).
5. J. P. Apruzese, J. Davis and K. G. Whitney, J. Appl. Phys. 53, 4020 (1982).
6. J. F. Seely (private communication, 1983).
7. R. C. Elton and R. H. Dixon, J. Quant. Spectros. and Rad. Trans. (1984) (in press).
8. R. L. Kelly, "Atomic and Ionic Spectrum Lines Below 2000 Å, H through Ar," Oak Ridge National Laboratory Report ONRL-5922 (1982); also R. L. Kelly with L. J. Palumbo, "Atomic and Ionic Emission Lines Below 2000 Å, Hydrogen through Krypton," Naval Research Laboratory Report 7599 (1973).
9. R. L. Kelly, private communication, 1983.

10. R. H. Dixon, J. L. Ford and R. C. Elton (to be published; patent pending).
11. R. H. Dixon and R. C. Elton, Phys. Rev. Lett. 38, 1072 (1977); R. H. Dixon, J. F. Seely and R. C. Elton, Phys. Rev. Lett. 40, 122 (1978).
12. R. C. Elton and R. H. Dixon, Phys. Rev. A, 28, 1886 (1983).
13. J. D. Garcia and J. E. Mack, J. Opt. Soc. Am. 55, 654 (1965); also R.L. Kelly, "Atomic Emission Lines in the Near Ultraviolet, Hydrogen through Krypton," NASA Tech. Memorandum 80268, Goddard Space Flight Center, Greenbelt, MD (1979).
14. R. H. Dixon, J. F. Seely and R. C. Elton, Appl. Optics 22, 1309 (1983); also "Population Density and VUV Gain Measurements in Laser-Produced Plasmas"; and "Calibration and Diagnostic Techniques for VUV Plasma Lasers," in Proc. Topical Conf. on Laser Techniques for Extreme-UV Spectroscopy, AIP Conf. Proceedings No. 90, T.J. McIlrath and R.R. Freeman, eds., pages 277 and 305 (1982).
15. P.C. Kepple and H.R. Griem, Phys. Rev. A 26, 484 (1982).

EMISSION AT 1091 Å IN NEUTRAL CORE-EXCITED Cs

D. E. Holmgren, D. J. Walker, and S. E. Harris
Edward L. Ginzton Laboratory
Stanford University, Stanford, California 94305

ABSTRACT

Certain quartet levels in alkali-like systems retain metastability against autoionization while acquiring large radiative yields. This quasi-metastability occurs through selective coupling to non-autoionizing doublet levels by the spin-orbit interaction. An example of such a level is the $5p^5 5d6s\ ^4P_{5/2}$ level of neutral Cs, which has a calculated branching ratio for radiation at 1091 Å of 43%. Experimentally, we find that this line has an emission intensity equal to 1/6 of that of the strongest ion line of Cs^+, and is a promising candidate for an extreme ultraviolet laser.

INTRODUCTION

It has recently been noted[1] that a sub-class of quartet levels of alkali atoms and ions retain metastability against autoionization and may have large radiative yields. This quasi-metastability occurs through selective coupling to non-autoionizing doublet levels by the spin-orbit interaction. An example of such a level is the $5p^5 5d6s\ ^4P_{5/2}$ level of neutral Cs (Fig. 1). The atomic physics code[2] RCN/RCG predicts a branching ratio for radiation on the $5p^5 5d6s\ ^4P_{5/2} \rightarrow 5p^6 5d\ ^2D_{5/2}$ transition at 1091 Å of about 43%.

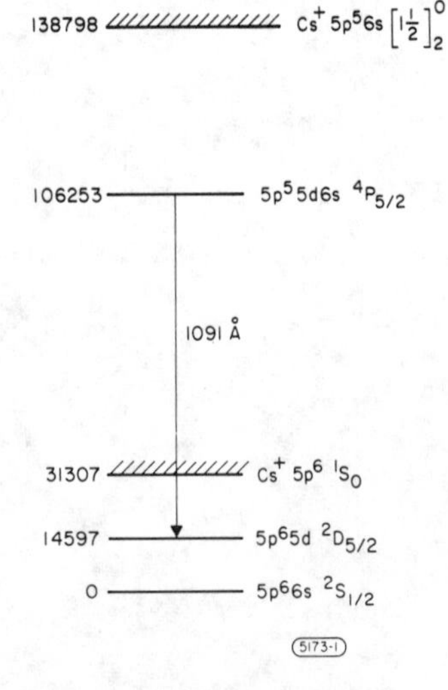

Fig. 1. Energy level diagram of Cs laser.

Experimentally, using a pulsed hollow-cathode discharge we find the ratio of emission on this line to the emission at 927 Å which results from the Cs^+ transition $5p^5 6s\ [1\frac{1}{2}]_1^0 \rightarrow 5p^6\ {}^1S_0$ to be about 16%. Since the cross section for electron excitation of this latter transition is known, we may infer both a strong excitation and a good branching ratio for 1091 Å emission. Since the lower level of this transition, $5p^5 5d\ {}^2D_{5/2}$, may be emptied by an incident laser beam (for example, at 5320 Å), it seems likely that inversion and reasonably high gain should be obtainable at 1091 Å.

QUASI-METASTABILITY

In an alkali-like atom or ion in which L and S are good quantum numbers (pure Russell-Saunders coupled eigenfunctions), several classes of core-excited levels are metastable against autoionization. Among the levels in the doublet manifold, the simultaneous conservation of parity and orbital angular momentum L forbids Coulombic autoionization of pure doublet levels having odd parity and even L, or vice versa, even parity and odd L. Another class of metastable levels consists of all pure quartets which lie below the quartet continuum (below the first triplet level in the next stage of ionization). These levels are metastable against autoionization by the requirement of conservation of spin.

The effect of breakdown of L-S coupling due to the spin-orbit interaction is to cause mixing of doublet and quartet levels. From the properties of angular momenta, the spin-orbit interaction term $L \cdot S$ connects pure L-S basis states satisfying $\Delta L = 0, \pm 1$; $\Delta S = 0, \pm 1$; and $\Delta J = 0$. This mixing to adjacent L levels has the following effect: doublets that were forbidden to autoionize because of parity and angular momentum considerations are mixed with levels which do autoionize rapidly. Consequently, all doublets tend to autoionize. Similarly, quartets are mixed with doublet levels and thereby acquire both autoionizing and radiative character. Tables I and II give the results of single-configuration calculations, using the atomic physics code[2] RCN/RCG for levels of the $5p^5 5d6s$ configuration in neutral Cs. These calculations indicate, even for levels that in a pure L-S scheme would not undergo Coulombic autoionization, that in general the predominant decay mechanism is autoionization.

Table I Autoionizing and radiative rates for odd parity-even angular momentum doublet levels of the Cs $5p^5 5d6s$ configuration

Upper Level	Autoionizing Rates (sec^{-1})	Radiative Rate (sec^{-1})
$({}^1D)\ {}^2D_{3/2}$	1.5×10^{12}	2.6×10^{8}
$({}^1D)\ {}^2D_{5/2}$	2.5×10^{11}	4.3×10^{8}
$({}^3D)\ {}^2D_{3/2}$	6.8×10^{10}	8.4×10^{8}
$({}^3D)\ {}^2D_{5/2}$	3.7×10^{12}	2.5×10^{8}

Table II Autoionizing and radiative rates for quartet levels of the Cs $5p^5 5d6s$ configuration

Upper Level	Autoionizing Rate (sec^{-1})	Radiative Rate (sec^{-1})
$^4P_{1/2}$	1.1×10^{11}	1.8×10^{7}
$^4D_{1/2}$	1.5×10^{13}	8.3×10^{8}
$^4F_{5/2}$	3.7×10^{12}	2.2×10^{8}
$^4P_{5/2}$	5.1×10^{7}	4.3×10^{7}

There exist, however, certain quartet levels in alkali-like systems which in first order couple only to those pure doublet levels which are themselves prohibited from autoionizing; it is these levels that have been termed as quasi-metastable. In second order, they do acquire components of autoionizing doublet levels, and therefore do autoionize; but often sufficiently slowly that the branching ratio for XUV radiation remains large. As an example consider the quasi-metastable $5p^5 5d6s\ ^4P_{5/2}$ level of Cs. In a pure L-S basis, the only doublet levels to which the $^4P_{5/2}$ level has non-zero matrix elements are the singlet and triplet core $^2D_{5/2}$ levels. The coupling to these levels allows $^4P_{5/2}$ to radiate in the XUV, but causes no autoionization. The calculated (RCN/RGN) expansion of the $5p^5 5d6s\ ^4P_{5/2}$ level of Cs is

$$\begin{aligned} ^4P_{5/2} = & - 0.90\ ^4P_{5/2} + 0.35\ ^4D_{5/2} - 0.051\ ^4F_{5/2} \\ & - 0.24\ (^1D)\ ^2D_{5/2} - 0.14\ (^3D)\ ^2D_{5/2} \\ & - 0.0043\ (^1F)\ ^2F_{5/2} + 0.0016\ (^3F)\ ^2F_{5/2} \end{aligned} \quad (1)$$

The much smaller $(0.0043)^2$ and $(0.0016)^2$ components of the autoionizing $^2F_{5/2}$ levels which appear in second order through the diagonalization, result in the relatively slow $^4P_{5/2}$ autoionizing rate. As listed in Table II, the level Cs $5p^5 5d6s\ ^4P_{5/2}$ autoionizes at a calculated rate of $5.1 \times 10^7\ s^{-1}$ and radiates on the $5p^5 5d6s\ ^4P_{5/2} \rightarrow 5p^6 5d\ ^2D_{5/2}$ transition at a rate of $4.3 \times 10^7\ s^{-1}$ at a calculated wavelength of 1075 Å.

XUV EMISSION SPECTRUM

Using a pulsed hollow cathode discharge and a McPherson monochromator we have observed the XUV emission spectrum of Cs. As described elsewhere,[3,4] the hollow cathode operated at a voltage of 3 kV, a current of about 300 amps, and a Cs pressure of about 2 torr. The observed emission spectra are shown in Figs. 2 and 3. The strongest features are those of Xe-like Cs^+ at 901.2 and 926.7 Å. The emission at 1091 Å had an intensity typically equal to about 1/6 of the 927 Å line and is identified as the $5p^5 5d6s\ ^4P_{5/2} \rightarrow 5p^6 5d\ ^2D_{5/2}$ transition. This identification is based upon the observed fine-structure splitting

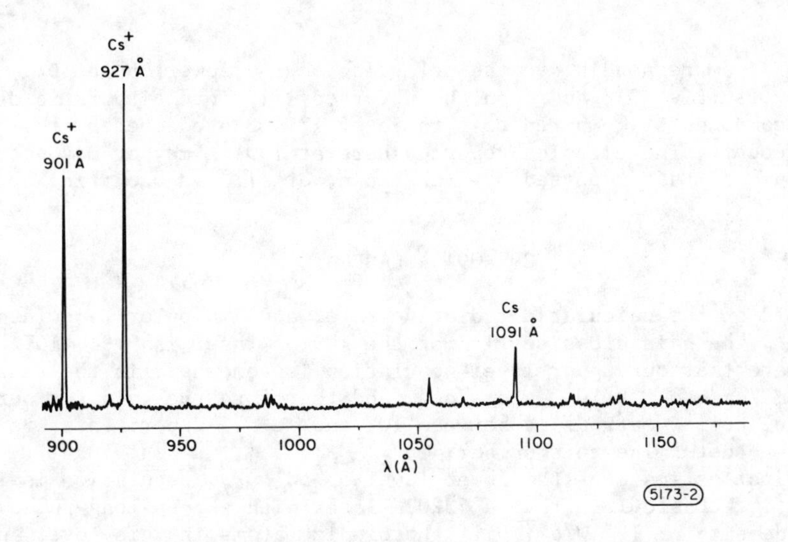

Fig. 2. Emission scan of Cs from pulsed hollow-cathode discharge.

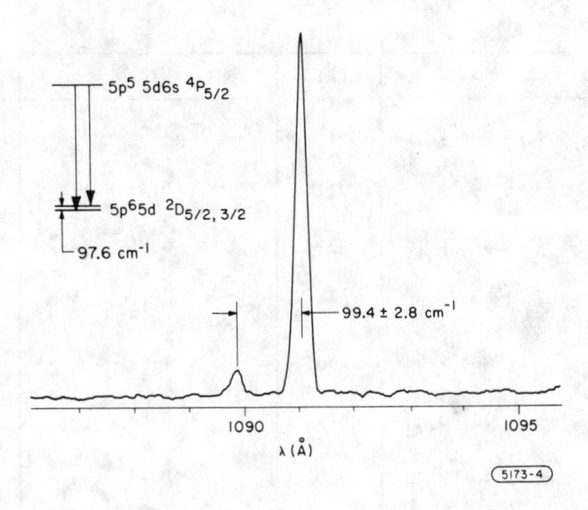

Fig. 3. High resolution scan of Cs near 1091 Å.

of 99 cm^{-1}, corresponding to the splitting of the lower $5p^6 5d\ ^2D_{5/2}$ and $^2D_{3/2}$ levels (Fig. 3); and upon the observed 1/12 intensity ratio of the two components, compared to a ratio of 1/14.3 predicted by the RCN/RGN code.[2] The emission reported here at 1091 Å may or may not be the same as that reported by Aleksakhin, et al.[5] as occurring at 1085 Å.

Cs 1091 Å LASER

Based on the calculated radiative rate, and a Doppler linewidth of 0.15 cm^{-1}, the gain cross section of the 1091 Å laser is 3.9×10^{-14} cm^2. We estimate that our upper level population is somewhere in the range of 10^{11} atoms/cm^3 to 5×10^{12} atoms/cm^3, and therefore the gain for our 30 cm long hollow cathode is between 10% and e^5. Experiments are now underway to determine this number.

A first estimate of the lower level $5p^6 5d\ ^2D_{5/2}$ population is 10^{13} atoms/cm^3. We calculate that a 5320 Å laser with a 5 ns long pulse and a power density of 10^8 W/cm^2 will photoionize atoms in this level and reduce the population by a factor of 5000.

TRANSITIONS FROM QUASI-METASTABLE LEVELS IN OTHER ELEMENTS

Using the code RCN/RGN, we have taken a first look at transitions in other alkali atom and alkali-like ions which originate from quasi-metastable levels.[1]

Table III gives the transition wavelength, Doppler width, and gain cross section σ for XUV laser transitions which originate from the lowest quasi-metastable level of each element. The Doppler width is calculated at a temperature corresponding to a vapor pressure of 10 torr.

Table III Gain cross sections of transitions from quasi-metastable levels

Transition	λ(Å)	Doppler Width (cm^{-1})	$\sigma(cm^2)$
Na $2p^5 3s3p\ ^4S_{3/2} \rightarrow 2p^6 3p\ ^2P_{5/2}$	415	1.04	1.6×10^{-16}
K $3p^5 3d4s\ ^4P_{5/2} \rightarrow 3p^6 3d\ ^2P_{5/2}$	711	0.43	4.4×10^{-16}
Rb $4p^5 5s5p\ ^4S_{3/2} \rightarrow 4p^6 5p\ ^2P_{3/2}$	821	0.24	9.1×10^{-15}
Cs $5p^5 5d6s\ ^4P_{5/2} \rightarrow 5p^6 5d\ ^2D_{5/2}$	1075	0.15	3.9×10^{-14}
Mg^+ $2p^5 3s3p\ ^4S_{3/2} \rightarrow 2p^6 3p\ ^2P_{3/2}$	256	1.78	8.71×10^{-17}
Ca^+ $3p^5 3d4s\ ^4P_{5/2} \rightarrow 3p^6 3d\ ^2D_{5/2}$	529	0.83	1.72×10^{-16}
Sr^+ $4p^5 4d5s\ ^4P_{5/2} \rightarrow 4p^6 4d\ ^2D_{5/2}$	620	0.45	5.6×10^{-15}
Ba^+ $5p^5 5d6s\ ^4P_{5/2} \rightarrow 5p^6 5d\ ^2D_{5/2}$	771	0.27	2.4×10^{-14}

ACKNOWLEDGEMENTS

The authors gratefully acknowledge helpful discussions with J. Reader, T. Lucatorto, A. Mendelsohn, K. Pedrotti, J. Spong, and J. F. Young. The work described here was supported by the U.S. Air Force Office of Scientific Research and the U.S. Army Research Office.

REFERENCES

1. S. E. Harris, D. J. Walker, R. G. Caro, A. J. Mendelsohn, and R. D. Cowan, "Quasi-Metastable Quartet Levels in Alkali-Like Atoms and Ions," Opt. Lett. (to be published).
2. Robert D. Cowan, The Theory of Atomic Structure and Spectra (University of California Press, Berkeley, 1981), Secs. 8-1, 16-1, and 18-7.
3. R. W. Falcone and K. D. Pedrotti, Opt. Lett. 7, 74 (1982).
4. R. W. Falcone, D. E. Holmgren, and K. D. Pedrotti, in Laser Techniques for Extreme Ultraviolet Spectroscopy, R. R. Freeman and T. J. McIlrath, eds. (AIP, New York, 1982).
5. I. S. Aleksakhin, G. G. Bogachev, I. P. Zapesochnyl, and S. Yu. Ugrin, Sov. Phys. JETP 53, 1140 (1981).

HIGH RESOLUTION LASER SPECTROSCOPY OF $Xe(5p^6 \rightarrow 5p^5\ 11s')$ At 94.2 nm

Keith D. Bonin, Mark B. Morris and T. J. McIlrath
Institute for Physical Science and Technology
University of Maryland, College Park, MD. 20742

ABSTRACT

A high resolution Fizeau wavemeter and tunable, coherently-generated 94.2 nm radiation were used to measure, for the first time, the width of the 11s' autoionizing resonance in xenon. An absolute wavelength determination of the resonance of 2 parts in 10^6 has been achieved.

INTRODUCTION

Much progress has been achieved in recent years in the production of tunable, coherent radiation in the Vacuum Ultraviolet (VUV) by non-linear mixing in gases. The advantage of these sources include their high brightness, narrow spectral bandwidth and the ability to establish the VUV wavelengths from measurements in the visible spectral region. Although much high resolution work has now been done, especially on molecular systems, in the region between 100 nm and 200 nm, work below 100 nm in the Extreme Ultraviolet (XUV) has been much more limited. This is related to the fact that the driving wavelengths must be shorter, the fact that many non-linear media are self-absorbing below 100 nm (e.g. Xe, NO, CO etc.) and the difficulties in working in a region where there are no transparent bulk materials for windows or lenses. In this paper we report the use of tunable coherent radiation near 94.2 nm for high resolution spectroscopic studies of autoionizing transitions in Xe. The generation of the XUV radiation is discussed in another paper in this volume[1] and the properties of the radiation will only be summarized here with the emphasis of the paper on the high resolution studies of the $5p^6\ ^1S_o \rightarrow 5p^5(^2P_{1/2})11s'$ transition in Xe.

The relevant Xe levels are shown in Fig. 1. The first ionization limit to $Xe^+\ 5p^5\ ^2P_{3/2}$ occurs at 97833.7 cm^{-1} (102.2 nm) and the second limit to $Xe^+\ 5p^5\ ^2P_{1/2}$ is at 108370.8 cm^{-1} (92.28 nm). There are two absorption channels from the $5p^6$ initial state corresponding to the $5p^5(^2P_{1/2})$ ns' and $5p^5(^2P_{1/2})$ nd' series converging on the second limit and those lines which lie above the first $^2P_{3/2})$ limit autoionize to the underlying continua. The result is a series of very broad features, the nd' channel, and comparatively sharp features, the ns' channel (See Fig. 2). The d' channel is easily resolved using conventional spectroscopic techniques but the profiles of the s' transitions cannot be obtained with the best of the classical instruments, although values have been obtained for

0094-243X/84/1190502-07 $3.00

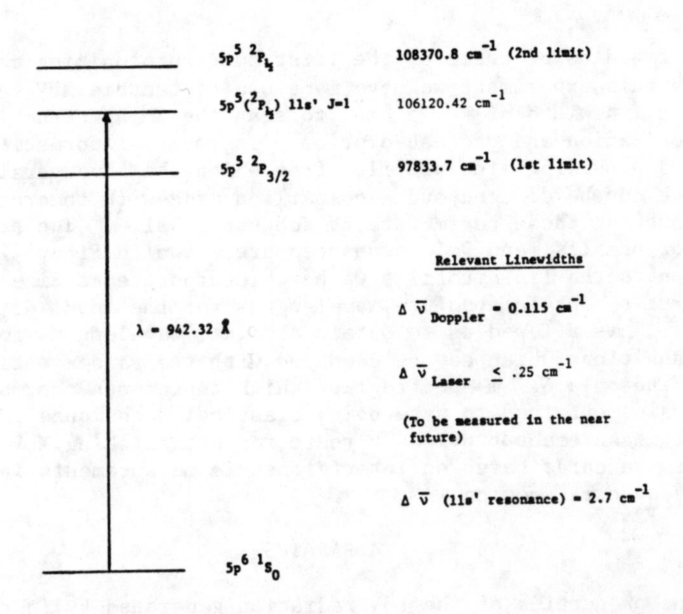

Fig. 1. Xe 11s' autoionizing resonance energy level diagram with relevant linewidths from present experiments

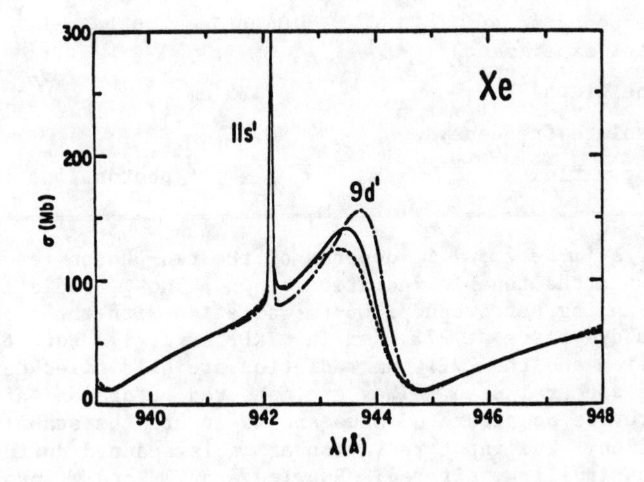

Fig. 2. Xenon photoionization cross sections.
Solid - theoretical (includes 5p, 5s and 4d correlations)
Dashed - theoretical (5p shell correlation only)
Dash-dot-experimental data obtained by J.H.D. Eland with a photon resolution of 8 cm^{-1}. (Fig. is from Ref. 3).

the half widths of three of the first four autoionizing members[2].

In this experiment we have made use of tunable XUV radiation having a linewidth of ≈0.25 cm^{-1} to scan the Xe spectrum in both photoionization and photoabsorption. Intensity information is obtained providing line profiles from which the Fano parameters of the line can be deduced and a comparison made with theoretical calculations for these parameters by Johnson et al.[3]. The predicted shape of the 11s' and 9d' resonances are shown in Figure 2. In addition to the line profiles we have incorporated a Fizeau interferometer to obtain absolute wavelengths for the input driving waves[4]. This allowed us to obtain absolute wavelengths for the XUV transitions which can be compared with the values obtained by K. Yoshino on a 6.7 m spectrograph which represents the most accurate value obtained to date using classical techniques[5]. This type of measurement provides a route for establishing XUV wavelengths standards based on interferometric measurements in the visible spectral region.

APPARATUS

The properties of the XUV radiation generated by four-wave mixing in krypton are shown in Table 1. The frequency of the

Table 1. Properties of XUV radiation source

XUV frequency scan (current experiment)	106,115 - 106,145 cm^{-1}
XUV line width	0.25 cm^{-1}
XUV absolute frequency	0.25 cm^{-1}
XUV photon flux	5 x 10^9 photons/5ns pulse

generated XUV at $\omega_4 = 2\omega_1 + \omega_2$ depends on the two-photon resonance radiation $_1$ and the tunable radiation at ω_2. The ω_1 radiation is produced by summing narrowband 1.06 μm radiation with the second harmonic of a dye laser at 272.0 nm in a KDP crystal. Both the 1.064μ radiation and the 272.0 nm radiation are held fixed during the 10 minute spectral scan. They are measured before and after the run to provide an accurate value and to ensure the stability of the ω_1 radiation. The input radiation at ω_2 is scanned during the run and is continually monitored. Knowledge of ω_1 and ω_2 provide a knowledge of ω_4 in the XUV throughout the scan. Measurements in practice are made and recorded approximately once every 10 shots during a run having a scan rate of ∿4 x 10^{-3} cm^{-1} per shot. The wavelength measurements were made with a Fizeau wedge interferometer which has been adapted for operation with pulsed input radiation[4] and found to have a resolution with pulsed sources of better than 1 part in 10^6.

The XUV radiation is generated by four-wave mixing in krypton at a pressure of ≈5 torr. The generating cell is separated from the experimental chamber by a rapidly rotating disk (see Ref. 1). A schematic of the experimental chamber is shown in Fig. 3. The

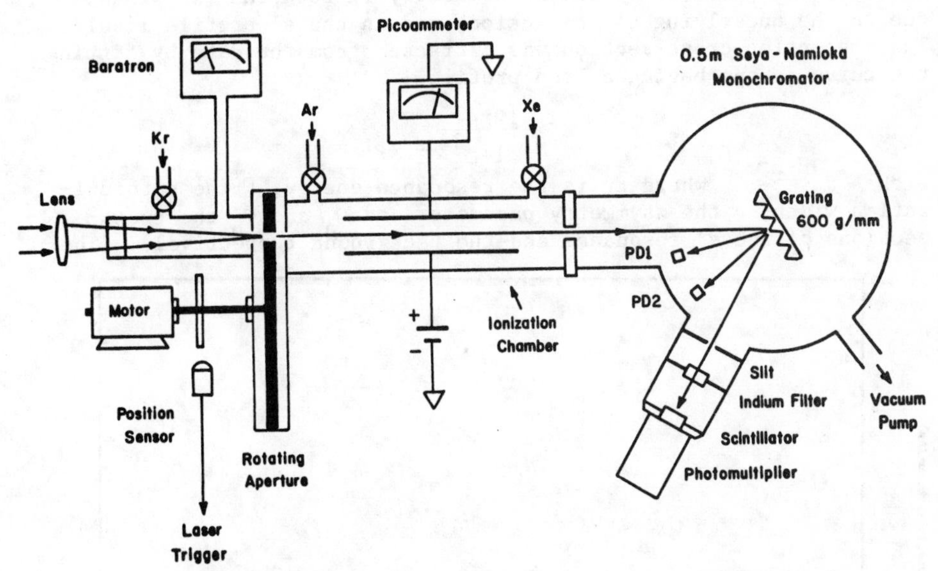

Fig. 3. VUV generating cell, xenon ionization chamber, and detection apparatus.

absorption of radiation above the first ionization limit is accompanied by ionization and a pair of plates with a 10V potential were placed just inside the experimental chamber. The ion signal was recorded on a shot by shot basis and provides a measure of the photoionization signal. The radiation was then dispersed using a 0.5 m Seya-Namioka monochromator using an indium filter to remove scattered light. The monochromator was used essentially as a narrowband filter having a pass band of ≈20 Å in the XUV (2000 cm^{-1}). The XUV was detected by fluorescence of a sodium salicylate scintillator followed by a photomultiplier to provide a photoabsorption signal. Although the photoionization and photoabsorption signals were separately recorded and compared to verify that the resonance maxima occured at the same wavelength for both processes, it was generally assumed that the quantum yield for ionization was unity and a ratio of the XUV transmission to the ionization signal was monitored. This ratio eliminated fluctuations in the XUV intensity from the data.

RESULTS

Fig. 4 shows the ratio of the ionization signal to the transmitted XUV for a typical scan. Although the line is nearly symmetric it is possible to see an asymmetry in both the far wings, due to the underlying d' absorption, and in the s' profile itself. The absorption cross-section was extracted from the data by fitting the curve to one having a Fano profile

$$\sigma = \sigma_a \frac{(q+\varepsilon)^2}{1+\varepsilon^2}$$

with $\varepsilon = \frac{E-Er}{\Gamma/2}$ where Er is the resonance energy, Γ the autoionization width, q the asymmetry parameter and σ_a and σ_b the cross-sections of the s' resonance and the background respectively. The

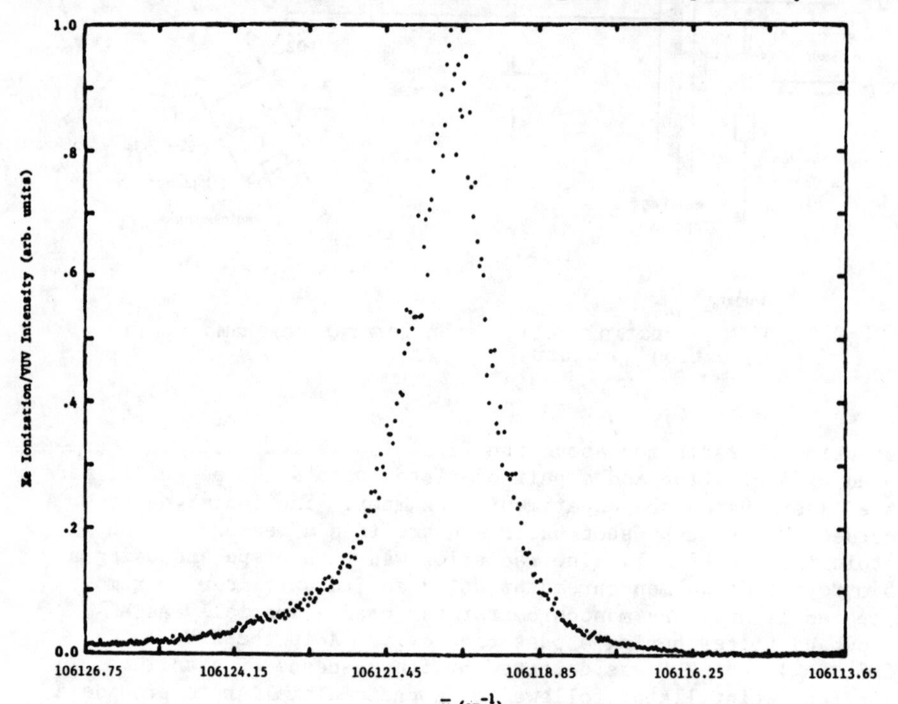

Fig. 4. Xenon ionization/VUV intensity vs. wavenumber: raw data

terms Er, Γ, q, and σ_a were treated as variables in the curve fitting routine with $_b$ assumed to have a linear dependence on ω_4 with a slope and magnitude determined by the d' resonance. The results of the curve fitting are shown in Fig. 5 where the fitted curve is superposed onto the experimental data and in Table 2 where the parameters are given for the fitted curve. It is clear from Fig. 4 that the asymmetry parameter q is positive in sign here.

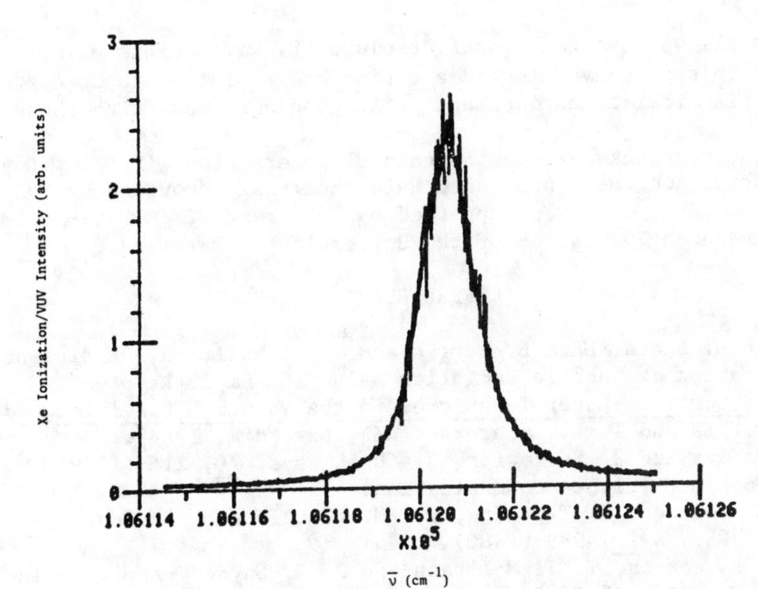

Fig. 5. Xenon ionization/VUV intensity vs. wavenumber: data and fitted curve.

Table 2. Fano parameters for X_e $5p^6\ {}^1S_o$ $5p^5({}^2P_{1/2})$ 11s' transition

	Bonin et al.	K. Yoshino[5]
$\overline{\nu}_R$	106120.42 ± 0.25 cm^{-1}	106120.8 ± 0.6 cm^{-1}
Γ	2.7 ± 0.3 cm^{-1}	3.38 ± 0.5 cm^{-1}
q	12 ± 4	

Table 2 also compares the values derived from this experiment with absolute wavelengths obtained by Yoshino and it is seen that the agreement is within experimental uncertainty. Current work involves measurements of higher s' transitions in Xe as well as d' transitions near the ionization limits where d' widths are no longer obtainable by classical techniques.

CONCLUSIONS

We have shown that narrowband, tunable, XUV radiation below 100 nm can be used to obtain profiles of absorption and ionization features in atoms with a resolution near the Doppler limit. Furthermore, the use of high accuracy wavemeters with the visible driving radiation allows measurements of absolute XUV wavelengths which, even with this prototype instrumentation has a similar accuracy to the best classical techniques and is in good agreement with those values.

The authors acknowledge the helpful interaction with J.J. Snyder and K.Yoshino and the unpublished data generously provided by K. Yoshino. This work was supported by NSF Grant NSF CPE 81-19250 and the Research Foundation of the University of Maryland.

REFERENCES

1. Keith D. Bonin, Mark B. Morris and T. J. McIlrath, "Efficient Generation of Tunable Radiation below 100 nm in Krypton", Proc. Conf. on Laser Techniques in the Vacuum Ultraviolet, ed. S. Harris and T. B. Lucatorto (AIP, New York, 1984).
2. K. Radler and J. Berkowitz, J. Chem. Phys. 70, 216 (1979); K. Yoshino, Private communication.
3. W. R. Johnson, K. T. Cheng, K. -N. Huang and M. Le Dourneuf, Phys. Rev. A 22, 989 (1980).
4. Mark B. Morris, T. J. McIlrath and J. J. Snyder, Appl. Optics, to be published; J. J. Snyder, Laser Focus, May, 1982, p. 55.
5. K. Yoshino, Private communication.

POSSIBILITIES FOR XUV OPERATION AT THE 1.0 GeV STANFORD FEL RING

S. Benson, D. A. G. Deacon, and J. M. J. Madey
High Energy Physics Laboratory
Stanford University, Stanford, CA 94305

ABSTRACT

The capabilities of present high energy electron storage rings should make possible operation of storage ring FEL oscillators at wavelengths to 500 Å and below in the XUV. We describe the characteristics of a possible XUV storage ring FEL using the FEL accelerator facilities presently under development at Stanford.

Synchrotron radiation, the radiation emitted by relativistic electrons when deflected by a strong magnetic field, is now heavily utilized as a source of radiation between 1 and 1000 Å for laboratory research.[1] The popularity of synchrotron radiation seems largely to be based on the broad range of wavelengths which can be covered, and on the available intensity, which is substantially higher than attainable using existing thermal UV and XUV sources, or conventional X-ray sources. But it is also clear that laser sources, where available, could offer experimentors yet higher intensities and the further advantage of coherence. The manifest appeal and value of laser UV and XUV sources has inspired considerable theoretical and experimental development activity, as evident in the contributions at this conference.

Under appropriate circumstances, a synchrotron radiation source can be adapted to generate coherent UV and XUV radiation via the free electron laser (FEL) mechanism. As described by Madey, Colson, Hopf and Renieri,[2] electrons passing through an undulator or wiggler magnet are capable of amplifying light on the long-wavelength side of the synchrotron radiation lineshape. If properly designed, and if an appropriate level of feedback can be provided, gains sufficient for oscillator operation are available, in principal, at wavelengths extending deep into the UV and XUV. In experiments to date, conducted at wavelengths from millimeter waves to the visible[3], the mechanism has been exploited to increase the intensity of the synchrotron radiation emitted by the electrons by factors in excess of 10^8 while simultaneously reducing the synchrotron radiation linewidth and suppressing all but the lowest order radial modes of the radiation field.

The basic requirements of XUV FEL operation are generally similar to those for operation of high brightness incoherent XUV undulator and wiggler sources, specifically, low electron beam energy spread and emittence to minimize the electrons' longitudinal velocity spread in the undulator, and high peak electron current.[4] The similarity of requirements clearly has important practical implications for those interested in developing XUV FEL's: much of the accelerator design and development work required to support XUV FEL operation has already been carried out to satisfy the requirements of existing or proposed incoherent undulator sources.[5] Probably the only unique requirements for laser operation relate to the length of the interaction region which will likely have to be of the order 10-20 meters, a factor of

2-4 longer than contemplated in most existing undulator-based synchrotron radiation source designs.

Based on the interests in and requirements for XUV FEL operation, and related requirements for cost-effective accelerator facilties for general synchrotron radiation and longer-wavelength FEL research, we have begun the development of dedicated storage ring and linear accelerator facilities in the High Energy Physics Laboratory to serve these requirements. The storage ring is based on a design developed by Helmut Wiedemann in 1980. The principal characteristics of the design are listed below:[6]

$$i_{peak} = 270 \text{ amperes}$$

$$\varepsilon_x = 1.7 \times 10^{-6} \text{ cm/radians}$$

$$\varepsilon_y = 1.7 \times 10^{-7} \text{ cm/radians}$$

$$E = 1.0 \text{ GeV}$$

$$\Delta E/E = 6 \times 10^{-4}$$

A similar ring, adapted to reduce the emittance and energy spread of the electron and positron beams for the linear collider program, has been in operation at SLAC since January, 1983.

The gain available in the XUV assuming these parameters and an optimized undulator is plotted in Figure 1 as a function of wavelength and the length of the interaction region. Given normal incidence reflectances of the order of 50%, sufficient gain should be available to support oscillator operation to at least 500 Å, and possibly to 100 Å, assuming a 20 meter interaction length.[7] As evident in Figure 1, the strong dependence of gain on interaction length places a premium on the length; 20 meters is probably the minimum length required to operate with confidence in the region below 1000 Å.

The practical realization of an XUV FEL will, of course, be contingent on both the attainment of the e^- beam paramerers listed above and the realization of critical elements of the supporting technology. So far as the FEL interaction is concerned, the most sensitive parameter is the electron beam emittance. As shown in Figure 2, a factor of two increase in the emittance can reduce the short wavelength gain by a factor of $2^{3/2}$. Although comparable emittances have been demonstrated in the SLAC damping ring, it will be necessary to take all available steps, probably including the use of positrons, to insure the attainment of the lowest possible emittance in an XUV storage ring FEL. The optical elements of the resonator cavity will clearly also require close attention; though adequate reflectances have probably been demonstrated at least to 500 Å (using silicon carbide), the figure requirements for a 20 meter resonator and the mechanical and chemical effects of prolonged exposure to intense synchrotron radiation and XUV laser radiation remain to be studied. Finally, the fabrication and trimming of the very long undulators required for these systems will require substantial further development of engineering, construction, and calibration techniques to insure the attainment of the theoretically available gain.

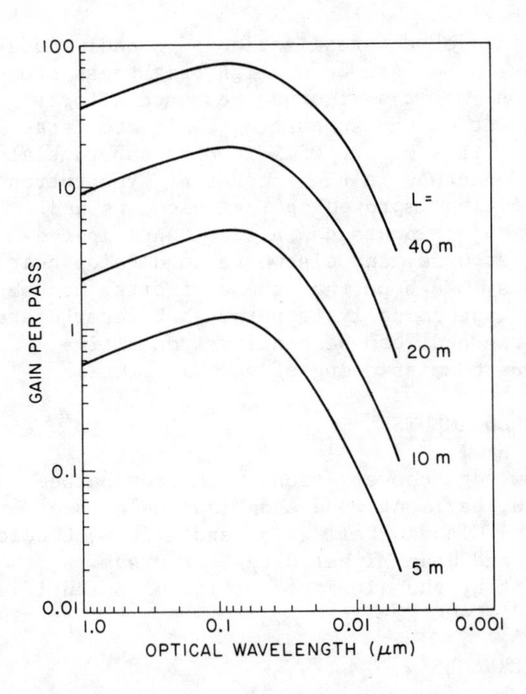

Figure 1: Small signal FEL gain assuming an optimized undulator and the e⁻ beam parameters of the Stanford FEL storage ring for undulator lengths of 5, 10, 20 and 40 meters.

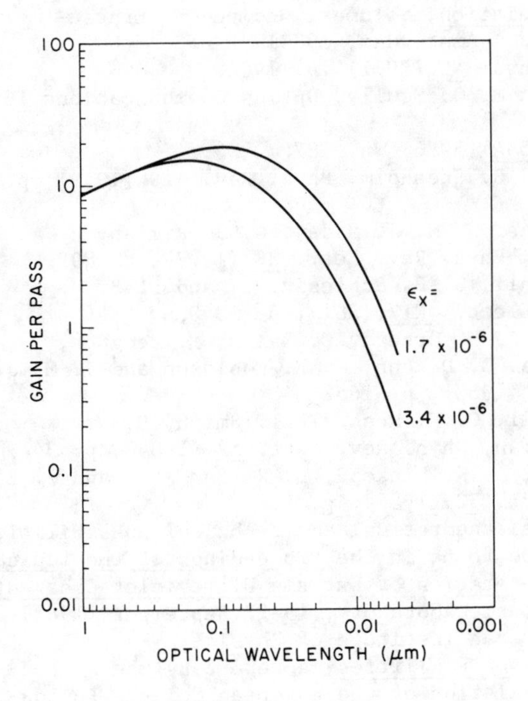

Figure 2: Effects of a factor of two increase in the emittance on the small signal gain for the 20 meter undulator in Figure 1.

We believe the possibilities for the realization of broadly tuneable XUV laser sources raised by the existence of high brightness storage rings and the FEL interaction merit a vigorous research effort. As was the case for the development of the supporting basic accelerator technology, it is likely that this effort will involve substantial inputs from the efforts already underway in the incoherent synchrotron radiation research program to develop improved optical elements and undulator and wiggler designs. Major inputs can also be anticipated from the programs now beginning at Orsay and elsewhere to develop near UV FELs on existing storage rings. We hope that these efforts, and the specialized new efforts required specifically to pursue XUV laser operation in our new laboratory at Stanford, can be concluded on a time scale to match the scheduled 1988 commissioning of our facilities.

ACKNOWLEDGEMENTS

It is a pleasure to acknowledge conversations with Troy Barbee and Helmut Wiedemann of Stanford, David Attwood and Klaus Halbach of the Lawrence Berkeley Laboratory, Claudio Pellegrini and Alfredo Luccio of Brookhaven, and Yves Petroff and Jean Michel Ortega of Orsay.

This work supported in part by the Air Force Office of Scientific Research under Contract F49620-84-C-0012.

REFERENCES

1. Handbook on Synchrotron Radiation, Volume 1, edited by Ernest-Eckhard Koch (North Holland: Amsterdam, 1983).
2. J. M. J. Madey, J. Appl. Phys. 42 (1971), p. 1906;
F. A. Hopf, P. Meystre, and M. O. Scully, Optics Communications 16 (1976), p. 413;
W. B. Colson, Phys. Lett. 59A (1976), p. 187;
A. Bambini, A. Renieri, and S. Stenholm, Phys. Rev. A19 (1979), p. 2013.
3. D. A. G. Deacon, L. R. Elias, J. M. J. Madey, G. J. Ramian, H. A. Schwettman and T. I. Smith, Phys. Rev. Lett. 38 (1977), p. 892;
D. S. Birkett, T. C. Marshall, S. P. Schlesinger, and D. B. McDermott, IEEE J. Quant. Elect. QE17 (A81), p. 1348;
M. Billardon, P. Elleaume, J. M. Ortega, C. Bazin, M. Bergher, M. Velghe, Y. Petroff, D. A. G. Deacon, K. E. Robinson and J. M. J. Madey, Phys. Rev. Lett. 51 (1983), p. 1652;
J. A. Edighoffer, G. R. Neil, C. E. Hess, T. I. Smith, S. W. Fornaca, and H. A. Schwettman, Phys. Rev. Lett. 52 (1984), p. 344;
S. H. Gold, D. L. Hardesty, A. K. Kinkead, L. R. Barnett, and V. L. Granatstein, Phys, Rev. Lett. 52 (1984), p. 1218.
4. A description of the general theoretical and technical possibilities for XUV FEL operation can be found in the Proceedings of the Topical Meeting on Free Electron Generation of Extreme Ultraviolet Coherent Radiation (Brookhaven National Laboratory, 19-22 September, 1983), to be published by the American Institute of Physics.
5. European Synchrotron Radiation Facilities, a report by the ad hoc committee on synchrotron radiation of the European Science Foundation (Chairmen: B. Buras, Y. Farge, and D. J. Thompson); European Science Foundation, Strasbourg , June, 1979)

5. (continued)
W. H. Backer, "Some Aspects of the Orbits in an Electron Synchrotron Used as a Synchrotron Radiation Source," (Eindhoven University of Technology, 1979);
National Center for Advanced Materials Conceptual Design Report (Lawrence Berkely Laboratory, 1983) report number PUB5084.
6. H. Wiedemann, J. de Physique Colloq. 44 (1983) p. 201.
7. J. M. J. Madey, "Conceptual System Design of XUV FEL's" to be published in the proceedings of the Topical Meeting on Free Electron Generation of Extreme Ultraviolet Coherent Radiation (Brookhaven National Laboratories, 19-22 September, 1983).

DESIGN CONSIDERATIONS FOR OPTICALLY PUMPED, QUASI-CW, UV AND XUV LASERS IN THE Be ISOELECTRONIC SEQUENCE

Mahadevan Krishnan and James Trebes
Yale University, New Haven, Connecticut 06520

ABSTRACT

Intense line radiation from plasmas of MnVI, PIX, AℓV, AℓVIII, AℓIX, and AℓXI may be used to selectively pump population inversions in plasmas of Be-like CIII, NIV, FVI, NeVII, NaVIII, and MgIX. Quasi-cw lasing is possible on 4p-3d and 4f-3d transitions at wavelengths from 2177 Å to 230 Å. At the XUV wavelengths, 1 J, 10 ns laser output pulses at 10^8 W power levels are shown possible with existing discharge technology. Since all six laser ions are in the Be isoelectronic sequence, detailed studies of the optical pumping process at UV wavelengths in CIII would provide scaling parameters for the less accessible XUV wavelengths.

INTRODUCTION

The concept of optical pumping with line radiation predates the invention of the laser by three decades.[1,2] Optical pumping with line radiation uses intense line radiation from a source medium to selectively pump a nearly coincident transition from the ground state to an excited state in an adjacent medium. The pumped level may then be inverted with respect to lower levels. In 1930, Boeckner[2] described fluorescence in CsI which was selectively pumped by 3889 Å line radiation from a Helium discharge lamp. Jacobs, et al.[3] measured optical amplification of about 4% at 3.2 μ in CsI using such direct optical pumping, and subsequently Rabinowitz, et al.[4] constructed a CsI laser oscillator at 7.19 μ. Following these pioneering achievements at infrared wavelengths, the concept of selective optical pumping of population inversions using line radiation has not been extended to visible wavelengths because of the lack of suitable line coincidences and the relative ease and flexibility of other pumping mechanisms such as direct collisional excitation or excitation-transfer. In the course of the development of short wavelength laser media, the old concept of selective optical pumping using line radiation reemerged in 1975, when Vinogradov, Sobelman, and Yukov[5] and Norton and Peacock[6] proposed the use of x-ray line radiation in one ion species to pump inversions at soft x-ray wavelengths in another ion species. Matthews, et al.[7] have identified several other optically pumped soft x-ray laser schemes and are exploring some of these schemes experimentally. Trebes and Krishnan[8,9] demonstrated UV fluorescence due to optical pumping and excitation in CII ions in a vacuum arc discharge, using AℓIII line radiation from a laser produced plasma. In a companion paper in these proceedings,[10] the collisional-radiative kinetics in the pumped CII ions are described in detail and the feasibility of lasing at UV wavelengths is discussed.

This paper examines design criteria for a proposed new class of optically pumped, quasi-cw, UV and XUV lasers in six ions of the Be isoelectronic sequence. The possible wavelengths range from 2177 Å to 230 Å in CIII, NIV, FVI, NeVII, NaVIII, and MgIX. In 1964, McFarlane[11] reported laser oscillation on the 3p-3s transitions in CIII at 4647.45 and 4650.16 Å, and in NIV at 3478.67 and 4097.32 Å. Elton[12] discussed the feasibility of extending these 3p-3s ion lasers into vacuum ultraviolet wavelengths. The 3p-3s transitions were chosen both because of the favorable lifetime ratio of these levels and because strong collisional excitation of the 3p upper level is possible from the 2s ground state of the Be-like ions. An important feature of ions of the Be isoelectronic sequence is that in addition to the 3p-3s transitions, the 4p-3d and 4f-3d transitions also have favorable lifetime ratios[13] for sustaining cw laser oscillation. Selective population of the 4p and 4f states could thus lead to a new class of XUV and soft x-ray lasers not considered earlier. This paper describes how intense, line radiation from an ion species in one plasma may be used to resonantly pump Be-like ions in an adjacent plasma from the $2s^2\ {}^1S$ ground state to the $2s4p\ {}^1P^0$ upper state. Lasing is then possible on the $4p\ {}^1P^0$ -- $3d\ {}^1D$ transition, with the 3d lower level decaying rapidly to the 3p and 2p levels. Figure 1 shows the relevant energy levels in CIII. At appropriate electron densities and temperatures in the Be-like plasmas, collisions rapidly transfer the $4p\ {}^1P^0$ excitation to the $4p\ {}^3P^0$ and the 4d and 4f levels. Lasing is therefore also possible on the $4p\ {}^3P^0$ -- $3d\ {}^3D$ and 4f-3d (singlet and triplet) transitions, as shown in Fig. 1.

PUMP CANDIDATES FOR Be-LIKE IONS

The first requirement of an optically pumped laser is the availability of a line in some ion species which is nearly coincident in wavelength with the absorption transition in the pumped species. Such a pump line should be intense so that the stimulated absorption excites a large fraction of the ground state population to the upper laser level. To achieve this, it is desirable that the pump plasma medium be optically thick at the pump line wavelength. When the upper level of the pump transition is in LTE with the ground state, the pump plasma radiates like a blackbody at this wavelength. In essence, one can create a population ratio of excited state to ground state in the pumped plasma which is characterized by the temperature of the pumping plasma rather than the temperature of the pumped plasma. If the temperature of the pumped plasma is much lower than that of the pumping plasma, strong inversions may be produced. This feature combined with selectivity are advantages of optical pumping over three-body recombination pumping or collisional pumping in which many levels

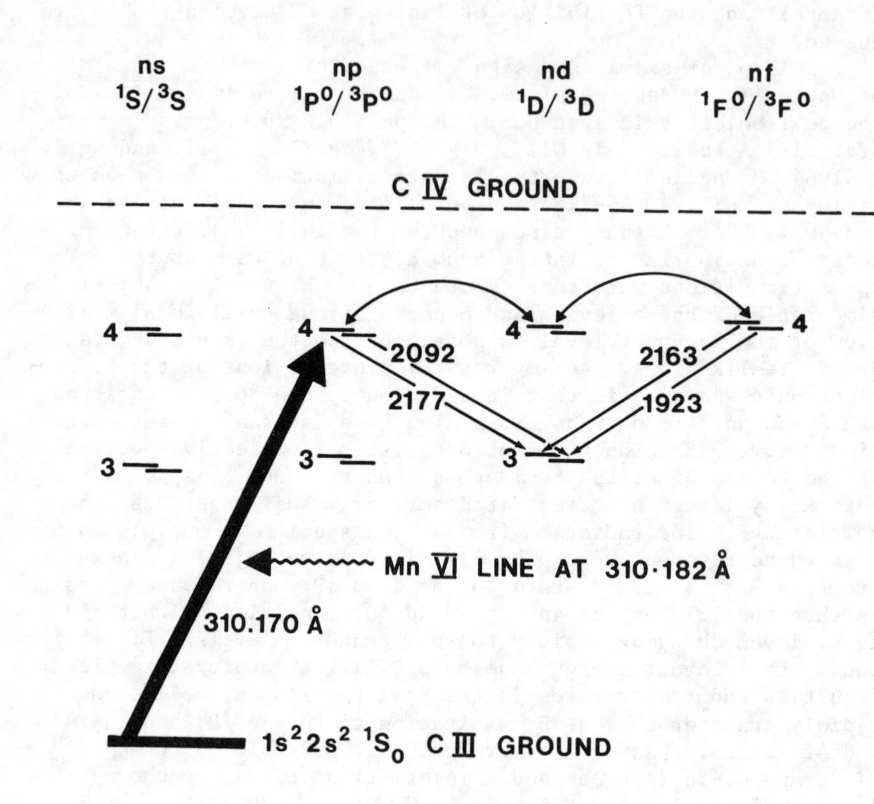

Fig. 1. Energy levels in CIII. Optical pumping of the $2s^2\ ^1S$ -- $2s4p\ ^1P^0$ transition is accompanied by collisional transfer to the $4p\ ^3P^0$, 4d and 4f levels. Quasi-cw lasing is possible on the 4p-3d and 4f-3d transitions.

are pumped. In He-like and Be-like ions in particular, electron collisional pumping with a thermal distribution will tend to populate the 3p levels as well as the 4p levels. Collisional transfer from 3p to 3d can then destroy the inversions.

Table I lists selected optical pumps for ions of the Be-like isoelectronic sequence. OV is omitted because a suitable pump was not found. In each pump ion the pump transition is optically allowed and terminates on one of the ground state configurations.[14] This allows for the creation of an optically thick and intense pump line. Table I also lists selected laser wavelengths from 4p-3d and 4f-3d transitions. Laser wavelengths from the UV to the XUV region are possible with these schemes. He-like ions[7] extend the concept to soft x-ray wavelengths.

Optical pumping with line radiation is best achieved by using two distinct plasmas, with only line radiation at a selected fre-

TABLE I: Optical pumps for the $2s^2\ ^1S$ -- $2s4p\ ^1P^0$ transition in Be-like ions, and typical wavelengths of possible laser transitions, 4p-3d and 4f-3d.

LASER SPECIES	$2s^2$ -- 2s4p WAVELENGTH (Å)	PUMP ION	4p-3d LASER WAVELENGTH (Å)	4f-3d LASER WAVELENGTH (Å)
CIII	310.17	MnVI	2177	2163
NIV	197.23	PIX	1284	1079
FVI	99.203	AℓV	554	513
NeVII	75.765	AℓVIII	404	360
NaVIII	59.759	AℓIX	308	285*
MgIX	48.34	AℓXI	250*	230*

*scaled hydrogenically

quency propagating from the pump plasma into the pumped, lasing medium. In practice, the proximity of the two plasmas results in collisional and other interactions as well, which can spoil the selectivity of the optical pumping and destroy the inversion. To minimize these extraneous interactions, an experimental arrangement is proposed which consists of two coaxial discharges imbedded in a strong axial magnetic field, as shown in Fig. 2. The lasing medium is produced by an arc discharge between hollow electrodes, with N, F, and Ne introduced as gases through one electrode. C, Na, and Mg plasmas may be produced by vacuum arc discharges from electrodes composed of these species. The outer, pump plasma may be generated by an exploding wire array or by a vacuum arc between electrodes composed of the pump element. As shown in Fig. 2, an insulating barrier is interposed between the plasmas to prevent radial breakdown between the electrodes. Furthermore, the strong axial magnetic field acts to confine the plasmas and minimizes radial interactions between them. The use of two separate electrical networks to generate the two plasmas allows independent control over the plasma parameters. In general, the two plasmas require distinctly different densities and temperatures.

The pump ion density and temperature must be high enough to render the pump plasma optically thick to the pump line, in order to maximize its intensity. The high temperature is also necessary

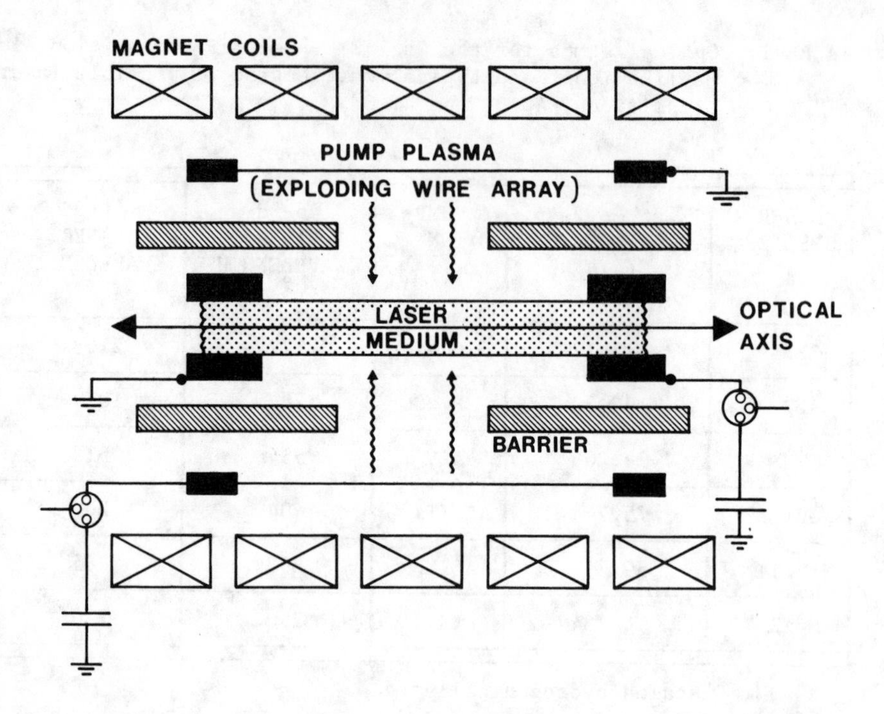

Fig. 2. Schematic diagram of proposed coaxial discharge arrangement.

to overcome the wavelength mismatch between the pump and absorption wavelengths by Doppler broadening. For the pumped plasma, since the ground state of the pumped ion is the primary source of the population inversion, the ion (and hence electron) density should be as high as possible to maximize gain but low enough to avoid optical trapping of the lower laser level and collisional depletion of the upper laser levels. Such depletion can occur by excitation, ionization, and by super-elastic de-excitation. The temperature of the pumped plasma must be low enough to minimize collisional depletion of the inversion, while maintaining the desired ground state density of the laser ions. In general, these disparate requirements are best met by creating two separate plasmas. Laser produced plasmas are also suited to these optical pumping schemes, since they enable the production of plasmas with a wide range of densities and temperatures, corresponding to different laser wavelengths and pulse durations. Two separate lasers are ideal. If a single laser beam is used to produce both plasmas, careful target design is required to satisfy the disparate requirements. Figure 3 shows one suggested experimental configuration for laser produced plasmas. The incident laser beams are focused to two separate line foci by two cylindrical lenses

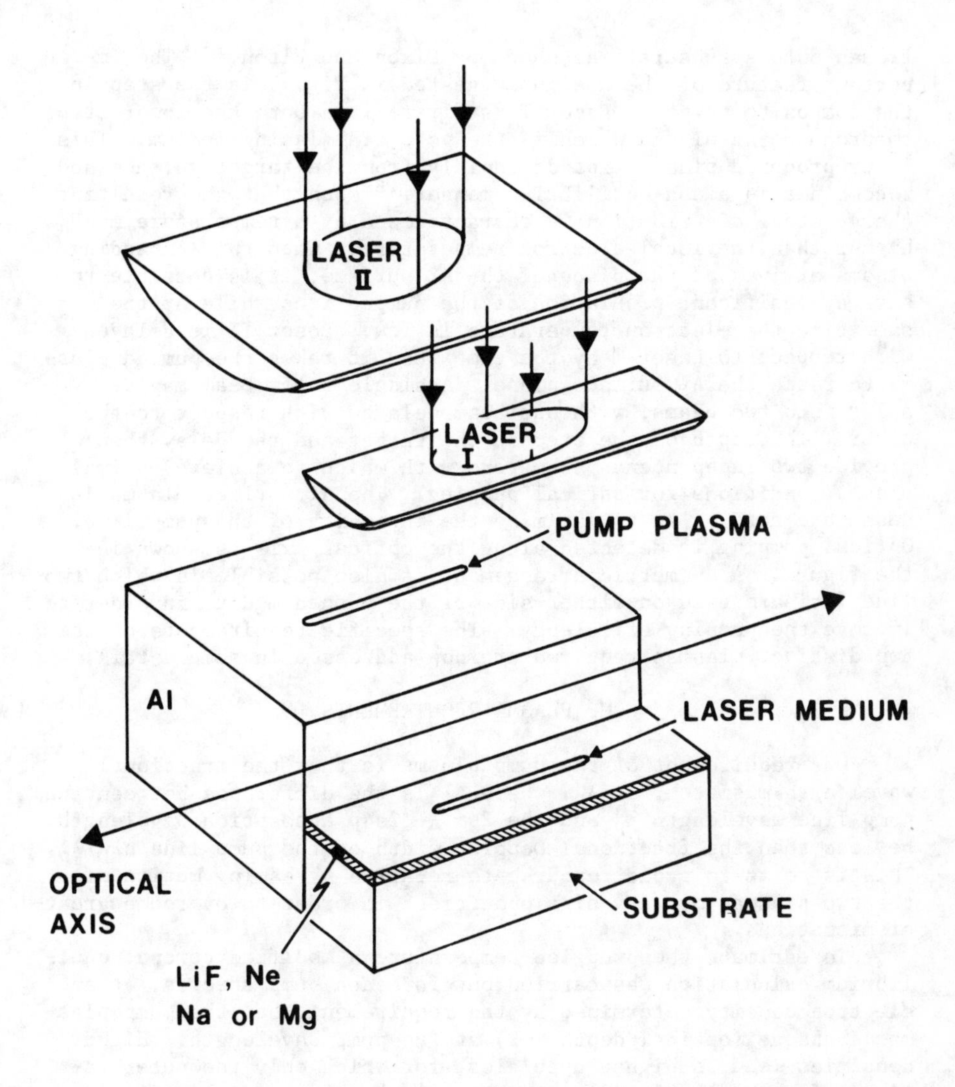

Fig. 3. Proposed experimental configuration for laser produced plasmas. Two different laser may be used, with wavelengths and focal geometries chosen to produce optimal conditions in the two distinct plasmas. Or, a single laser may be split and the two beams delayed with respect to each other to produce the required plasmas.

as shown. One line focus is on Aℓ, to produce surface plasmas of AℓV, AℓVIII, AℓIX, or AℓXI. The other line focus produces the laser medium, FVI, NeVII, NaVIII, or MgIX. For FVI, a lithium fluoride target is convenient. For NeVII, a film of Ne may be frozen onto a substrate as shown by Dixon and Elton.[15] The important feature of the design suggested in Fig. 3 is the step in the composite target. Laser I is first fired onto the lower step to produce the plasma which is the potential lasing medium. This laser produced plasma expands rapidly from the target surface and recombines in a non-equilibrium manner,[16] such that the resultant charge state distribution is characterized by a temperature much higher than the local electron temperature. When this expanding plasma arrives at the plane of the Aℓ surface, it is possible to have a significant population of the pumped ions while at the same time the electron temperature is low. Laser II is delayed with respect to Laser I by the time that it takes the pumped plasma to reach the Aℓ surface plane. A single laser beam may be split into two beams, with one beam delayed with respect to the other. The depth of the step in the target and the delay then provide two independent parameters with which to achieve optimal plasma conditions for optical pumping. The Aℓ surface plasma is dense and hot, so as to maximize the intensity of the pump line. Optical pumping is detected along the optical axis as shown in the figure. A symmetric arrangement is also possible in which two line foci are used on either side of the pumped medium in order to improve the pumping efficiency. The specific requirements of the two distinct plasmas required are now addressed in some detail.

PUMP PLASMA REQUIREMENTS

One requirement of the pump plasma is that the fractional wavelength mismatch, $\Delta\lambda/\lambda_p$, where $\Delta\lambda$ is the difference between the pump line wavelength λ_p and the $2s^2$ -- 2s4p absorption wavelength, be less than the fractional Doppler width of the pump line $\Delta\lambda_D/\lambda_p$. This is so as to avoid recourse to relative streaming motion of the two plasmas or very high opacities[6] in order to overcome greater mismatches.

To estimate the pump ion temperature a modified coronal equilibrium calculation was carried out for each pump species, at an electron density determined by the requirement that the pump plasma be opaque (optical depth ≃ 2) at the pump wavelength. Higher densities lead to higher opacities, for which only the outer regions of the pump plasma contribute to the pump radiation field. The coronal model includes for each ion species: collisional ionization, three-body recombination and radiative recombination. For carbon, di-electronic recombination was also included and was found to influence the results. For the other ions, the electron densities are sufficiently high that collisional modification of the di-electronic rates is significant, and accurate rates are not available. Comparison of results without inclusion of di-electronic recombination with similar calculations which include this process[17] revealed little difference over the range of densities and temperatures considered. A typical charge state distribution in an Aℓ plasma is shown in Fig. 4, for an electron density of

1×10^{18} cm^{-3}. Consider, for example, the AℓIX charge state, to be used as a pump for NaVIII. Figure 4 shows that this charge state is dominant at temperatures between 65 eV and 85 eV. Because the distribution is double valued, there are two temperatures at which the opacity is the same. The higher of the two temperatures is preferable because it maximizes the pump line intensity. Thus for this case the incident laser pulse width and focal power density should be adjusted to produce a surface plasma consisting mostly of AℓIX at a temperature of 85 eV. Temperatures were similarly determined for the other pump ions and are listed in Table II. Also shown in the Table are the quantities $\Delta\lambda/\lambda_p$ and $\Delta\lambda_D/\lambda_p$, defined above. For each candidate pair except NeVII - AℓVIII, the Doppler width of the pump line is greater than the mismatch, allowing for good optical coupling. In AℓVIII, the mismatch is about 1.5 Doppler widths. Increasing the opacity of the Aℓ plasma would broaden the line further and overcome the mismatch. A second requirement of the pump plasma is that the pump line intensity must be high enough so that the characteristic time for optical pumping by stimulated absorption be shorter than the radiative lifetime of the 4p level. Then a large fraction of the ground state population is optically pumped to the upper laser levels.

TABLE II: The fractional wavelength mismatch between the pump line and the $2s^2$ -- 2s4p line in the lasing ion is compared with the fractional Doppler half width of the pump line. The pump ion temperature is determined by a coronal equilibrium calculation.

LASER SPECIES	PUMP ION	FRACTIONAL WAVELENGTH MISMATCH, $\Delta\lambda/\lambda_p$ ($\times 10^{-5}$)	PUMP ION TEMPERATURE (eV)	FRACTIONAL DOPPLER WIDTH $\Delta\lambda_D/\lambda_p$ ($\times 10^{-5}$)
CIII	MnVI	4	25	4.6
NIV	PIX	10	90	13
FVI	AℓV	3	22	7
NeVII	AℓVIII	17	70	12
NaVIII	AℓIX	3	85	14
MgIX	AℓXI	5	130	16

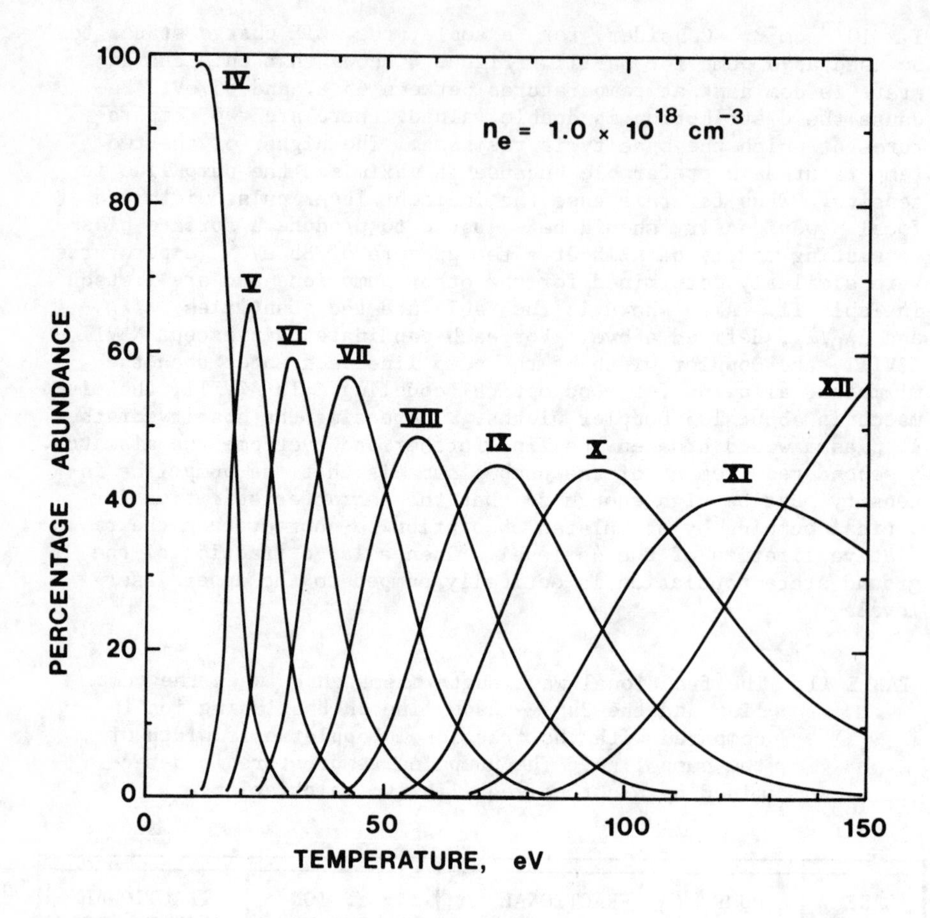

Fig. 4. Charge state distribution in an Aluminum plasma, obtained from a modified coronal equilibrium model. The electron density is 10^{18} cm^{-3}.

To show this, consider the rate equation for the optically pumped upper level, viz.:

$$\frac{dn_u}{dt} = \left(n_\ell - \frac{g_\ell}{g_u} n_u\right) B_{\ell u} I - n_u A_{u\ell} \tag{1}$$

where n_u, n_ℓ, g_u, and g_ℓ are the densities and statistical weights of the upper and lower levels, respectively, I is the pump line intensity integrated over the line profile and over solid angle, and B and A are the Einstein coefficients. Equation (1) is written assuming that the upper level lifetime is determined predominantly by radiative decay to ground. In steady-state:

$$\frac{n_u}{n_\ell} = \frac{B_{\ell u} \quad I}{A_{u\ell} + \frac{g_\ell}{g_u} B_{\ell u} \quad I} \tag{2}$$

For blackbody line intensity, $I = \frac{8\pi h\nu^3}{c^2} \left(e^{\frac{h\nu}{kT}} - 1 \right)^{-1}$,

and Eq. (2) reduces as shown by Apruzese, et al.[17] to:

$$\frac{n_u}{n_\ell} = \frac{g_u}{g_\ell} e^{-\frac{h\nu}{kT}} \tag{3}$$

where ν and T are the pump line frequency and blackbody temperature, respectively. Equation (3) shows that when $h\nu/kT << 1$, the populations approach the ratio of their respective statistical weights. Such a strong pump condition is not satisfied by all of the pump lines chosen, but significant optical pumping is still possible. For example, in MnVI, the coronal estimate of T = 25 eV and $h\nu$ = 40 eV. For these conditions, if the pump line is absorbed over 4π steradians, $n_u/n_\ell = 0.6$ in CIII. Similarly, in the AℓXI plasma, T = 130 eV and $h\nu$ = 256 eV. Now $n_u/n_\ell = 0.4$ in MgIX. While these ratios are less than the maximum possible ratio of 3 for the 2s-4p absorption transition, they are still quite sufficient to produce high gain lasers, as discussed further below. In summary, Table II shows that at temperatures typical of the pump plasmas, the wavelength mismatch is readily overcome and the pump line intensity is sufficient to pump a large fraction of the ground state population to the 2s4p upper state.

REQUIREMENTS OF THE PUMPED MEDIUM

Turning now to the pumped medium, for each of the laser ions, the temperature T_e was first determined from the coronal equilibrium. Figure 5 shows a charge state distribution in a sodium plasma at a density of 1×10^{17} cm^{-3}. The NaVIII ions are found to be abundant between 45 and 65 eV. In this case, the lower of these two temperatures is chosen so as to minimize deleterious electron collisional effects on the optical pumping. The temperatures appropriate to the other laser ions were similarly determined. These values of T_e are listed in column two of Table III. An upper bound on the electron density was determined by considering four processes which could destroy the inversion: optical trapping of the 3d-2p line, collisional ionization from the pumped levels, collisional de-excitation from the upper laser level, and collisional excitation of the 3p level from the ground state followed by transfer to the 3d level. The most deleterious process was optical trapping, for which the electron density n_e was determined by requiring that optical trapping not increase the 3d lifetime to more than the 4f lifetime. This preserves the possi-

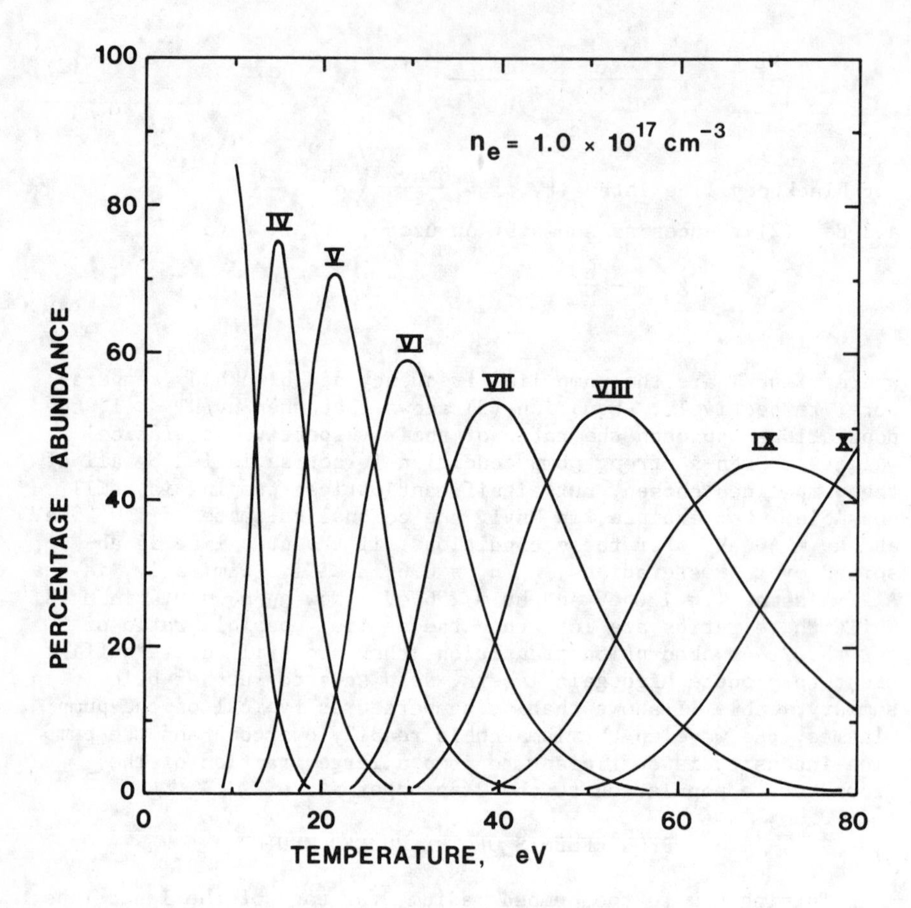

Fig. 5. Charge state distribution in a Sodium plasma, obtained from a modified coronal equilibrium model. The electron density is 10^{17} cm^{-3}.

bility of quasi-cw lasing on the 4f-3d transitions. As a worst case of optical trapping, the 2p $^3P^0$ level was assumed to be in LTE with the ground state, which maximizes its population. For a given density of the 2p $^3P^0$ level, the optical depth of the laser medium at the 3d ^3D -- 2p $^3P^0$ wavelength was computed. Using a Holstein escape factor,[19] the modified lifetime of the 3d ^3D level was then computed. The maximum allowed density of the 2p $^3P^0$ level was that which increased the 3d lifetime to just equal the 4f lifetime. Quasi-neutrality then yielded the upper bound on n_e. The values of n_e so determined are listed in column three of Table III. Also listed are the radiative lifetimes (in ns) of the lasing transitions. The lifetimes, energies, and collisional rate coefficients in CIII were obtained from Morgan.[20]

TABLE III: Parameters of the optically pumped laser media. T_e is determined by a coronal equilibrium; n_e is given by requiring that optical trapping not increase the 3d lifetime to larger than the 4f lifetime, so that quasi-cw lasing is possible; the time, τ, for collisional transfer from 4p to 4f, and the required inversion density, n_{4f}, for specified gain, α, are also computed.

LASER SPECIES	T_e (eV)	n_e (cm^{-3}) (x 10^{15})	4p-3d LIFETIME (ns)	4f-3d LIFETIME (ns)	τ (ns)	α (cm^{-1})	n_{4f} (cm^{-3}) (x 10^{13})
CIII	4	1	11	1	1	0.01	0.004
NIV	7	4	4	0.25	0.5	0.01	0.01
FVI	20	70	0.7	0.06	0.07	1	4
NeVII	30	100	0.4	0.03	0.07	10	70
NaVIII	45	200	0.2	0.02	0.04	10	100
MgIX	55	200	0.15	0.01	0.03	10	150

For the rest of the ions, the oscillator strengths were assumed to be equal to those in CIII and the lifetimes were scaled hydrogenically.

For the chosen values of n_e and T_e, the times τ for collisional transfer from the 4p to 4f levels are listed in column six of Table III. Since these collisional transfer times are up to a factor of ten shorter than the 4p-3d radiative lifetimes and are comparable with the 4f-3d radiative lifetimes, most of the optically pumped 4p population will be transferred to 4f. The dominant lasing transitions will be the 4f-3d transitions. Under these conditions of n_e and T_e, the time for ionization from the 4f levels was estimated to be about a hundred times longer than the 4f-3d radiative lifetime. Finally, the inversion density n_{4f} required to produce a specified gain α on the 4f-3d lines was computed and is also shown in Table III. For CIII and NIV, α of 0.01 cm^{-1} was chosen because reflecting optics can be used. FVI also allows a hole coupled reflecting cavity, but the lower reflectivities available at 500 Å demand α of 1 cm^{-1}. For NeVII, NaVIII, and MgIX, an ASE source is required, with $\alpha > 10$ cm^{-1}. Comparison of the values of n_e and n_{4f} shown in Table III reveals that all these ion species are capable of lasing even with weak optical pumping. For example, the ratio of 4f to $2s^2$ populations in CIII required is 1×10^{-4}. Strong optical pumping would result in a ratio of 0.6. For MgIX, the required ratio is 0.06. Again, the saturated ratio would be 0.4. For the high gain lasers, gain saturation of the lasing transitions would increase the required inversion densities and strong optical pumping to saturation would probably be required.

CONCLUSION

The above analysis has shown that Be-like, optically pumped plasmas offer the potential for high gain, quasi-cw lasers at wavelengths from 2177 Å down to 230 Å. The new class of lasers described here may be tested using laser produced plasmas or by creating coaxial arc discharges. The longer wavelength candidates may be studied at relatively low input powers with long pulse durations. Detailed spectroscopic studies of the optical pumping and subsequent kinetics would reveal key scaling laws that may be directly applicable to the design of the isoelectronically scaled, less accessible shorter wavelength candidates. The design criteria summarized in Table III suggest that in MgIX, an inversion density of 1.5×10^{15} cm^{-3} may be maintained by strong pumping from a ground state population of 3×10^{16} cm^{-3}. If the laser medium is 0.5 cm in diameter and 4 cm long, the quasi-cw output power at 230 Å is $\sim 10^8$ W. The power required for optical pumping is then $\sim 10^9$ W. If only 10% of the total power radiated by the AℓXI pump line is absorbed by MgIX, the total power required in the AℓXI pump line is then $\sim 10^{10}$ W, which for a 10 ns laser pulse, requires 100 J of total energy in the pump line. The 10 ns output duration is 1000 times longer than the radiative lifetimes of the 4f levels. Recently,[21] single soft x-ray lines with energies >100 J have been measured in many elements in terawatt, imploding plasma discharges. Existing pulsed power technology or high power lasers may thus be used to test the XUV laser schemes proposed in this paper.

ACKNOWLEDGMENTS

This research was supported by the Air Force Office of Scientific Research (Grant # 81-0077). Helpful discussions with W.R. Bennett, Jr. and R. Jensen are gratefully acknowledged. We are indebted to W.L. Morgan of Lawrence Livermore Labs. for the atomic data on CIII. After preparation of this manuscript, we were informed by R.C. Elton that one of the schemes proposed here, the MgIX - AℓXI combination, was suggested earlier[15] by J.F. Seely.

REFERENCES

1. W.R. Bennett, Jr., personal communication
2. C. Boeckner, J. Res. Natl. Bur. St. 5, 13 (1930).

3. S. Jacobs, G. Gould, and P. Rabinowitz, Phys. Rev. Lett. 7, 451 (1961).
4. P. Rabinowitz, S. Jacobs, and G. Gould, Appl. Opt. 1, 513 (1962).
5. A.V. Vinogradov, I.I. Sobelman, and E.A. Yukov, Sov. J. Quant. Electron. 5, 59 (1975).
6. B.A. Norton and N.J. Peacock, J. Phys. B 8, 6 (1975).
7. D.L. Matthews, P. Hagelstein, E.M. Campbell, A. Toor, R.L. Kauffman, L. Koppel, W. Halsey, D. Phillion, and R. Price, IEEE J. Quant. Elect. QE-19, 1786 (1983).
8. J. Trebes and M. Krishnan, Phys. Rev. Lett. 50, 679 (1983).
9. J. Trebes and M. Krishnan, IEEE J. Quant. Elect. QE-19, 1870 (1983).
10. J. Trebes and M. Krishnan, these proceedings.
11. R.A. McFarlane, Appl. Phys. Lett. 5, 91 (1964).
12. R.C. Elton, Appl. Opt. 14, 97 (1975).
13. M. Duguay, Laser Focus 9, 45 (1983).
14. R.L. Kelly and L.J. Palumbo, NRL Rept. 7599 (1973).
15. R.H. Dixon and R.C. Elton, J. Opt. Soc. Am.-B 1, April 1984.
16. Y.B. Zeldovich and Y.B. Raizer, Physics of Shock Waves and High Temperature Hydrodynamic Phenomena, Vol. 2 (Academic Press, New York, 1967), p. 571.
17. R. Hulse, personal communication.
18. J.P. Apruzese, J. Davis, and K.G. Whitney, J. Appl. Phys. 53, 4020 (1982).
19. T. Holstein, Phys. Rev. 72, 1212 (1947).
20. W.L. Morgan, personal communication.
21. R.J. Dukart, 14th Winter Colloquium on Quantum Electronics, Snowbird, Utah, January 10-13, 1984.

AUTHOR INDEX

AIP Conference Proceedings

No.	Title		
No. 97	The Interaction Between Medium Energy Nucleons in Nuclei-1982 (Indiana University)	83-70649	0-88318-196-7
No. 98	Particles and Fields - 1982 (APS/DPF University of Maryland)	83-70807	0-88318-197-5
No. 99	Neutrino Mass and Gauge Structure of Weak Interactions (Telemark, 1982)	83-71072	0-88318-198-3
No. 100	Excimer Lasers - 1983 (OSA, Lake Tahoe, Nevada)	83-71437	0-88318-199-1
No. 101	Positron-Electron Pairs in Astrophysics (Goddard Space Flight Center, 1983)	83-71926	0-88318-200-9
No. 102	Intense Medium Energy Sources of Strangeness (UC-Santa Cruz, 1983)	83-72261	0-88318-201-7
No. 103	Quantum Fluids and Solids - 1983 (Sanibel Island, Florida)	83-72440	0-88318-202-5
No. 104	Physics, Technology and the Nuclear Arms Race (APS Baltimore-1983)	83-72533	0-88318-203-3
No. 105	Physics of High Energy Particle Accelerators (SLAC Summer School, 1982)	83-72986	0-88318-304-8
No. 106	Predictability of Fluid Motions (La Jolla Institute, 1983)	83-73641	0-88318-305-6
No. 107	Physics and Chemistry of Porous Media (Schlumberger-Doll Research, 1983)	83-73640	0-88318-306-4
No. 108	The Time Projection Chamber (TRIUMF, Vancouver, 1983)	83-83445	0-88318-307-2
No. 109	Random Walks and Their Applications in the Physical and Biological Sciences (NBS/La Jolla Institute, 1982)	84-70208	0-88318-308-0
No. 110	Hadron Substructure in Nuclear Physics (Indiana University, 1983)	84-70165	0-88318-309-9
No. 111	Production and Neutralization of Negative Ions and Beams (3rd Int'l Symposium, Brookhaven, 1983)	84-70379	0-88318-310-2
No. 112	Particles and Fields-1983 (APS/DPF, Blacksburg, VA)	84-70378	0-88318-311-0
No. 113	Experimental Meson Spectroscopy - 1983 (Seventh International Conference, Brookhaven)	84-70910	0-88318-312-9
No. 114	Low Energy Tests of Conservation Laws in Particle Physics (Blacksburg, VA, 1983)	84-71157	0-88318-313-7
No. 115	High Energy Transients in Astrophysics (Santa Cruz, CA, 1983)	84-71205	0-88318-314-5
No. 116	Problems in Unification and Supergravity (La Jolla Institute, 1983)	84-71246	0-88318-315-3
No. 117	Polarized Proton Ion Sources (TRIUMF, Vancouver, 1983)	84-71235	0-88318-316-1

No. 118	Free Electron Generation of Extreme Ultraviolet Coherent Radiation (Brookhaven/OSA, 1983)	84-71539	0-88318-317-X
No. 119	Laser Techniques in the Extreme Ultraviolet (OSA, Boulder, Colorado, 1984)	84-72128	0-88318-318-8